Irrigation Systems Management

Dean E. Eisenhauer
Derrel L. Martin
Derek M. Heeren, General Editor
Glenn J. Hoffman

Manufactured in the United States of America
Print date: 6 December 2021

DOI: 10.13031/ISM.2021
International Standard Book Number (ISBN) 978-1-940956-42-8
ASABE Publication 801M0221OA

For more information and material for instructors,
see https://www.asabe.org/ISM.

Copy editing and layout by Peg McCann
Cover design by Melissa Miller
Cover photos:
Drip lateral, photo courtesy of Toro
Lake McConaughy, photo courtesy of Steve Melvin, Nebraska Extension
Weather station, canal, furrow irrigation in corn, and center pivot photos by the authors

ASABE is an educational and scientific organization dedicated to the advancement of engineering applicable to agricultural, food, and biological systems.
For more information visit www.asabe.org.

ASABE, 2950 Niles Rd., St. Joseph, MI 49085-9659 USA
phone: 269-429-0300 fax: 269-429-3852 e-mail: hq@asabe.org

This book is dedicated to our wives and children for their love and support:

Maria, Emily, and April

DEE

JoAnn, Jennifer, and Kimberly

DLM

Amber, David, Elizabeth, Nathan, and Joshua

DMH

Maria, Kimberly, Karen, and Sheryl

GJH

Contents

Foreword

Agriculture is the largest consumer of global freshwater resources, currently estimated to account for 75 percent of all water diverted for human use. The global population is anticipated to peak at approximately 10 billion in 2050, and our food systems will need to evolve to respond to the increasing population, improving diets, climate change and other political and social changes. The demand for water for food, fiber and fuel is projected to increase by another 50 percent in that time frame. Today, one third of the world's food is produced on 21 percent of its cultivated land as a result of effective irrigation systems. This clearly highlights the import role irrigation will continue to play in intensifying agricultural production and feeding the world.

Irrigation has a long history. Around the world, water has been diverted for irrigation for thousands of years. In the United States reminants of irrigation infrastructure dating back over 3200 years can be found in the Southwest. For the last 250 years, the total area under irrigation in the U.S. has continuously increased. Despite this expansion and increasing crop productivity, the amount of water used for agriculture has remained relatively stable since the 1980s, showing an improvement in how agricultural water is managed.

With all the gains achieved through irrigation, there remain downsides to consider— and there are many if close attention is not paid to the system, its management, and the sustainability of vital soil and water resources. Over-extraction—or taking more water than can be replaced through precipitation—of both groundwater and surface water is a widespread issue around the world. This is especially true in the absence of robust water accounting to determine the resources available and ensuring overall consumptive use does not exceed system limits. Mismanagement of the water, fertilizer and other agricultural chemicals can lead to degradation of water quality and contamination of ground and surface water resources. That said, well managed irrigation presents the opportunity to minimize the level of contamination.

Farmers' access to irrigation is a crucial component of a highly productive agricultural system—one that reduces risk and increases resilience. For it to be effective, the system must integrate into the broader farm system, provide the farmer with a solid return on their investment and sustain the vital natural resources upon which the enterprise depends.

Drawing on decades of collective experience in research, teaching, outreach and practice, the authors present the knowledge and technical insights into the development and management of the common irrigation systems adopted across the U.S. and many other parts of the world. The understanding of these systems—combined with the relevant knowledge in other contexts—are critical to addressing water and food security for the future.

Peter G. McCornick, Executive Director
Daugherty Water for Food Global Institute (DWFI)
at the University of Nebraska

Preface

Like most textbooks, this book grew out of our desire to have written material that matches the educational needs of both the students and the instructor of a college course, in this case a course entitled *Irrigation Systems Management*. The book is the culmination of course notes which have been in development and use for nearly 30 years.

The emphasis of this book is on the management of irrigation systems that are used for agricultural crop production. There are two distinct components of the book, starting with the soil-water-plant-atmosphere system and how soil water should be managed to achieve the desired crop production outcomes. This includes in-depth presentations on soil water storage and movement, plant water use, managing the soil water reservoir through irrigation scheduling, and salinity management. The book then shifts to the second component, which is the description and management of the various forms of agricultural irrigation systems along with their water supply. Whether it be a surface, sprinkler, or microirrigation system, the irrigation manager must not only know how much water to apply but also how to manage the system itself to achieve efficient application. High application efficiency can only be realized by minimizing runoff, deep percolation, evaporation, and drift onto non-target areas. Since energy costs are an integral part of the management equation, one chapter in the book deals with the hydraulics and energy requirements of pumping and distributing water. One of the key themes spread throughout the book is providing guidance to irrigation managers on how to improve irrigation water productivity (production per unit of irrigation water) and minimize water resource contamination.

Our goal is for the reader to understand the complexities of irrigation systems and how they are to be managed to meet the water needs of the crop production system. This is not an irrigation engineering design book; we have purposely minimized the presentation of design steps and the supporting equations. The intended audience of the book is upper-level undergraduate students and graduate students who are pursuing degrees in Agricultural or Natural Resource Sciences. Example majors include Agricultural Systems Technology, Agronomy, Crop Science, Mechanized Systems Management (or equivalent), Natural Resources Management, Soil Science, and Water Science. We expect the reader to have a basic understanding of soils, crops, physics, and the application of algebraic equations. We have also tried to add enough advanced material to challenge graduate students when the book is used in courses that are taught simultaneously at the undergraduate and graduate level. We hope the book will match the needs of students who plan to work in irrigation and related industries, university extension and outreach, private consulting, government service, or production agriculture and that it will continue to serve as a useful reference to them following completion of their formal education.

The book is being published by the American Society of Agricultural and Biological Engineers (ASABE) as an open-access book and will be available online at no charge so that it can be used globally for a wide array of applications. Specific chapters, for example, may be useful for international workshops, industry training sessions, employee onboarding, non-governmental organizations involved in irrigation development, continuing education, etc. The book uses a mix of the U.S. Customary System of Units (USCS) and the International System of Units (SI), although USCS is used most frequently, reflecting the context in which the book was developed. However, we have added helpful unit conversions to assist readers in countries where SI units are common.

The notes from which this book was developed have been tried and tested for many years, not just at the University of Nebraska-Lincoln (UNL), but also at other land-grant universities in the U.S. including Kansas State University, Oklahoma State University, South Dakota State University, Texas A & M University, University of Missouri, and Washington State University. For four years the book was used for continuing education of statewide field staff of the Natural Resources Conservation Service in Nebraska. In addition, selected chapters have been used regularly in our *Irrigation Laboratory and Field Course,* a course that we originally developed specifically for international students who were studying at the IHE Delft Institute for Water Education in Delft, The Netherlands. We appreciate the feedback for improvement from all students and instructors who have used the draft of the book.

The authors of the book have a combined and balanced experience of over 150 years in teaching, research, extension, and consulting. A very high proportion of their experience is in the Midwest, Great Plains, and Western region of the U.S. Thus, they are accustomed to larger-scale farm irrigation systems. However, the authors also have significant international experience through various assignments and projects in countries throughout the world, giving them a wider view of farm irrigation systems, including smallholder farms, which influenced the approaches taken in the book.

One final note is on the arrangement of names for the author order of the book. This can sometimes be an awkward dilemma: who contributed the most? This book was truly a team effort with all authors making significant original writing and editorial contributions. Eisenhauer, Martin, and Hoffman, now Emeriti faculty members of the Department of Biological Systems Engineering at UNL, initiated the development of the book. That said, the truth is the completion of the book would never have occurred without the final push and motivation by Derek Heeren, Associate Professor in Biological Systems Engineering and the current instructor of *Irrigation Systems Management.* Appropriately he is listed as General Editor because of his extra efforts in working through the publication process with our team and the publisher, ASABE.

Dean E. Eisenhauer
Derrel L. Martin
Derek M. Heeren, General Editor
Glenn J. Hoffman
May, 2021

Acknowledgements

The development of a textbook of this magnitude is only possible with the help of many people, including the support staff who prepared and formatted the manuscript and provided editorial suggestions. Biological Systems Engineering Office Associates who provided immeasurable assistance in this process include Lorna Pleasant, Teresa Ryans, and Julie Thompson. Sheila Smith, Illustrator in Biological Systems Engineering, developed most of the original artwork and drawings. Some of the concepts and questions in the book were developed during class laboratory activities; technical support for these activities was provided primarily by Alan Boldt, Research Engineer and Lab Manager in Biological Systems Engineering. We are forever grateful for the assistance of these staff members.

Over the years many student hourly employees have provided assistance with proof reading, providing editorial comment, and helping prepare graphics and equation formatting. At the risk of inadvertently leaving out the names of some, we gratefully acknowledge the help of the following former students: Sonya Cooper, Clayton Blagburn, Troy Nelson, and Eric Wilkening. Eric was especially instrumental in the final push towards publication with his help in updating and refining the figures and formatting the text. The input of these students was extremely useful.

We are indebted to our publisher, ASABE. This includes staff members Donna Hull, Joseph Walker, and Peg McCann. Their guidance in navigating through the process and assistance with the book's development was so very helpful. We appreciate the patience of this group, especially Donna who we're sure gave up on us years ago. Also, the book's quality was greatly enhanced by incorporating comments made by reviewers who had been selected by the publisher: Dr. Thomas Scherer, Dr. Michael Kizer, Dr. Donald Slack, Dr. Allen Thompson, and Dr. Trisha Moore.

We are especially grateful for the financial support needed to get the book into a publishable form. Support was provided by the ASABE Foundation's Harold Pinches and Glenn Schwab Teaching Materials Fund and the Daugherty Water for Food Global Institute at the University of Nebraska.

Finally, we would like to thank the UNL administration in Biological Systems Engineering and the Institute of Agriculture and Natural Resources for allowing us to pursue and successfully complete this project.

Dean E. Eisenhauer
Derrel L. Martin
Derek M. Heeren, General Editor
Glenn J. Hoffman
May, 2021

About the Authors

Dr. Dean E. Eisenhauer is an Emeritus Professor of Biological Systems Engineering at the University of Nebraska-Lincoln where he specialized in hydrologic and irrigation engineering with appointments in teaching, research, and extension. He received B.S. and M.S. degrees in Agricultural Engineering from Kansas State University and a Ph.D. degree in Agricultural Engineering from Colorado State University. He was on the University of Nebraska-Lincoln faculty from 1975 until he retired in 2015. Dean taught courses in irrigation management, watershed management, engineering hydrology, hydrologic modeling, and vadose zone hydrology. His research interests include hydrologic impacts of land use and water management practices in agricultural regions, infiltration and overland flow, water flow in the vadose zone, engineering of vegetative buffers for riparian and upland ecosystems, chemigation, management of furrow irrigation, and water measurement technology for irrigation.

Dean advised or co-advised 33 graduate students and served on 73 graduate student committees during his career. He has co-authored 57 journal articles and 4 book chapters. He is a registered Professional Engineer in Nebraska. Dean is a Faculty Fellow in the Robert B. Daugherty Water for Food Global Institute (DWFI) at the University of Nebraska. At DWFI he was a co-developer of a successful partnership with IHE Delft Institute for Water Education in Delft, Netherlands (IHE). Dean's pastimes and hobbies include spending time with family, vegetable gardening, bicycling, traveling, and reading.

Dr. Derrel L. Martin is Professor Emeritus of Irrigation and Water Resources Engineering in the Department of Biological Systems Engineering at the University Nebraska-Lincoln. He earned B.S. and M.S. Degrees in Agricultural Engineering from the University of Nebraska in 1975 and 1979. He received his PhD from Colorado State University in 1984. Derrel has also worked for a state agency and a consulting company before attending graduate school. Dr. Martin is a Professional Engineer, a Fellow of the ASABE, and served as Chair of the Soil and Water Division of the ASABE.

Dr. Martin worked in the Department of Biological Systems Engineering at the University of Nebraska-Lincoln for forty-one years. He taught courses in irrigation engineering and management, and vadose zone hydrology. His research focused on irrigation engineering and management, protection of groundwater quality by minimizing pollution from agricultural chemicals, deficit irrigation due to water limitations, evapotranspiration and crop water use, and modeling the soil-water-plant system to assess productivity. He spent over fifteen years delivering Extension programs on irrigation, water resources management and energy use in irrigated agriculture. He developed software used by state and local water management agencies. He served as an expert witness in regional litigation and United States Supreme Court actions. He grew up on an irrigated farm and ranch in western Nebraska and continues to work with family partners. Derrel's interest include spending time with family, attending collegiate sporting events, and volunteering.

Dr. Derek M. Heeren is an Associate Professor and Irrigation Engineer in the Department of Biological Systems Engineering at the University of Nebraska-Lincoln (UNL), and a Faculty Fellow of the Daugherty Water for Food Global Institute (DWFI). He graduated from South Dakota State University in 2004 with a B.S. in Agricultural and Biosystems Engineering and an M.S. in 2008. Before graduate school, Derek spent two years working at a geotechnical engineering firm near St. Louis, Missouri. He obtained his Ph.D. in Biosystems Engineering from Oklahoma State University in 2012, before coming to UNL.

Derek has taught or co-taught eight courses, including Irrigation Systems Management, Advanced Irrigation Management, Irrigation Laboratory Field Course, and Modeling Vadose Zone Hydrology. Derek's research has addressed irrigation management, sprinkler irrigation systems, irrigation technology, vadose zone hydrology, water quality, and surface water-groundwater interaction, with projects in the United States, India, Malawi, Zambia, and Rwanda. Derek has published 43 peer-reviewed journal articles. He has served as advisor or coadvisor for 12 graduate students from six countries and has served on graduate committees for an additional 15 graduate students. Derek is the Partnership Coordinator for the partnership between DWFI and the IHE Delft Institute for Water Education, Delft, Netherlands. Outside of work, Derek enjoys time with family, outdoor hobbies, and church activities.

Dr. Glenn J. Hoffman received combined BS and MS degrees in Agricultural Engineering in 1963 from Ohio State University followed by a PhD from North Carolina State University. He started in 1966 as a research engineer at the USDA/ARS (Agricultural Research Service) Salinity Laboratory in Riverside, California. His research led to recognition as an international authority on the management of irrigated salt-affected soils and crop salt tolerance. In 1984, Glenn led scientists at the USDA/ARS Water Management Laboratory in Fresno, California on studies of subsurface drip irrigation and the water requirement and salt tolerance of tree crops.

Glenn was selected head of the Agricultural Engineering Department at the University of Nebraska-Lincoln in 1989. He led faculty members to change the department to Biological Systems Engineering. Undergraduate enrollment increased 5-fold to 250 and graduate student numbers doubled. The department's annual budget, including grants, increased to more than $6 million. He retired in 2003.

Glenn is the author of 170 scientific publications, including 17 book chapters and lead editor of 2 irrigation monographs. He consulted on salinity management in Israel, Australia, Spain, Pakistan, Egypt, Iran, India, Brazil, China, and Argentina. Personal interests include family activities, foreign travel, mineral collecting, and bridge.

Common Unit Conversions for Irrigation

International System of Units (SI) and U.S. Customary System of Units (USCS)

Length Units

1 km (kilometer) = 0.6214 mi (miles)
1 mi = 1.609 km
1 km = 1,000 m (meters)
1 mi = 5,280 ft (feet)
1 m = 3.281 ft
1 ft = 0.3048 m
1 m = 100 cm (centimeters)
1 ft = 12 in (inches)
1 in = 2.54 cm

Area Units

1 ha (hectare) = 10,000 m^2
1 ha = 2.471 ac (acres)
1 ac = 0.4047 ha
1 ac = 43,560 ft^2

Volume Units

1 L (liter) = 0.2642 gal (gallons)
1 gal = 3.785 L
1 m^3 = 1,000 L
1 m^3 = 264.2 gal
1 ft^3 = 7.481 gal
1 m^3 = 35.31 ft^3
1 ft^3 = 0.02832 m^3
1 ac-ft (covers 1 acre with 1 foot of water) = 1,233 m^3
1 ac-ft = 325,851 gal
1 ac-in (acre-inch) = 27,154 gal
1 ha-cm (hectare-centimeter) = 100 m^3
1 ac-in = 1.028 ha-cm (1 for practical purposes)
1 bu (bushel) = 35.24 L
1 bu = 9.309 gal

Flow Units

1 L/s (liter/second) = 15.85 gpm (gallons/minute)
1 gpm = 0.06309 L/s
1 cms (m^3/s) = 1,000 L/s
1 cms = 35.31 cfs (ft^3/s)
1 cfs = 448.8 gpm (450 for practical purposes)
1 cfs = 0.02832 cms
1 ac-in/hr = 1.008 cfs (1 for practical purposes)
1 ac-in/hr = 452.6 gpm (450 for practical purposes)

Weight and Mass Units

1 t (metric ton) = 1,000 kg (kilograms)
1 T (U.S. ton) = 2,000 lb (pounds)
1 t = 1.102 T
1T = 0.9072 t
1 kg = 2.205 lb
1 lb = 0.4536 kg

Pressure Units

1 atm (atmosphere) = 101.3 kPa (kilopascal)
1 bar = 100 kPa
1 bar = 0.9869 atm (1 for practical purposes)
1 kPa = 0.1450 psi (pounds per square inch)
1 psi = 6.895 kPa
1 m (meter of water head) = 9.804 kPa
1 ft (foot of water head) = 0.4335 psi
1 kPa = 0.3346 ft = 0.1020 m (0.1 for practical purposes)
1 psi = 2.307 ft = 0.7032 m (0.7 for practical purposes)

Crop Yield Units

Rice, cotton, alfalfa, and similar crops

1 lb/ac (lb dry weight/ac) = 1.121 kg/ha (kg dry mass/ha)
1 kg/ha = 0.8922 lb/ac
1 T/ac (T dry weight/ac) = 2.242 t/ha (t dry mass/ha)
1 t/ha = 0.4461 T/ac

Tree fruit, grapes, vegetables, and similar crops

1 lb/ac (lb wet weight/ac) = 1.121 kg/ha (kg wet mass/ha)
1 T/ac (T wet weight/ac) = 2.242 t/ha (t wet mass/ha)

Maize/corn[a]

1 bu/ac (bushel/ac) = 0.0530 t/ha (t dry mass/ha)
1 t/ha = 18.9 bu/ac

Soybean[a]

1 bu/ac = 0.0585 t/ha (t dry mass/ha)
1 t/ha = 17.1 bu/ac

[a] The unit of bu/ac is volume per area. The volume (bu) is calculated using the wet weight of the grain, a wet bulk density of 56 lb/bu (maize) or 60 lb/bu (soybean), and an assumed grain moisture content (wet basis) of 15.5% (maize) or 13% (soybean).

Temperature Conversions

°C	°F	°F	°C
0	32	30	−1
5	41	40	4
10	50	50	10
15	59	60	16
20	68	70	21
25	77	80	27
30	86	90	32
35	95	100	38
40	104	110	43

U.S. Public Land Survey System

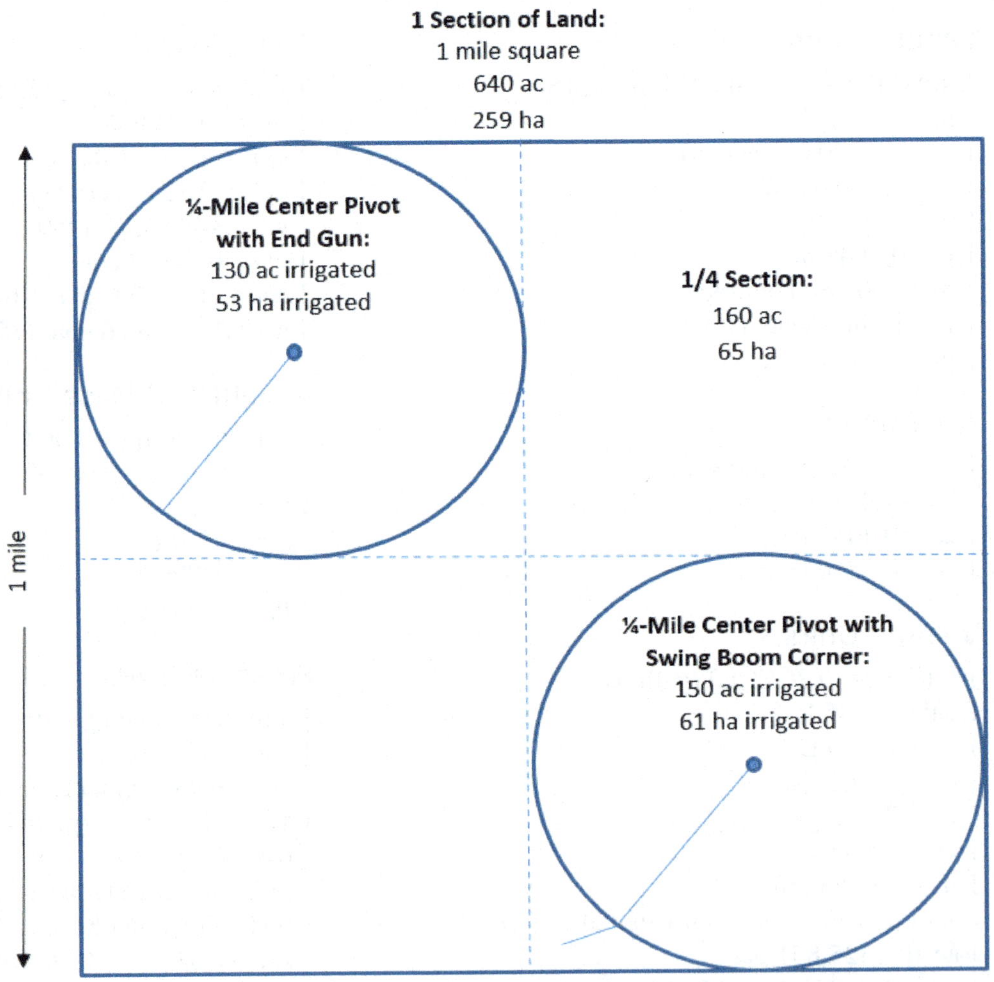

Chapter 1
Introduction to Irrigation

1.1 Introduction

Irrigation is the supply of water to crops by artificial means. It is designed to permit the desired plant growth in arid regions and to offset drought in semiarid regions or subhumid regions. Even in areas where average seasonal precipitation may seem ample, rains are frequently unevenly distributed, or soils have low water holding capacities so that traditional rainfed agriculture is a high-risk enterprise. Irrigation provides a means for stable food production. In some areas, irrigation prolongs the effective growing season. With the security provided by irrigation, additional inputs like higher producing varieties, additional fertilizer, better pest control, and improved tillage, become economically feasible. Irrigation reduces the risk of these expensive inputs being wasted by drought.

On a global scale, irrigation has a profound impact on fresh water supplies, world food production, and the aesthetics and value of landscapes. One-third of the world's food comes from the 21% of the world's cultivated area that is irrigated (Table 1.1). In the U.S., irrigated agriculture accounted for about half of the total value of crop sales on 28% of harvested crop land in 2012 (USDA, 2019).

Table 1.1. Worldwide distribution of irrigated areas in 2017 (adapted from FAO, 2021).

	Irrigated Area (millions of acres)	Percent of Cropped Lands	Percent of World Total
Asia	574	39	71
America	128	14	16
Europe	56	8	7
Africa	39	6	5
Oceania	8	10	1
World	806	21	100

Irrigation has turned many of the earth's driest and most fertile lands into important crop producing regions. For example, Egypt could grow virtually no food without water drawn from the Nile or from underground aquifers. California's Central Valley and the Aral Sea basin—the fruit and vegetable baskets of the United States and the former Soviet Union—would produce little without irrigation. The world's major grain producing areas of northern China, northwest India, and the U.S. Great Plains would drop by one-third to one-half without irrigation to supplement rainfall. Irrigation fills a key role in feeding an expanding world population and seems destined to play an even greater role in the future.

As practiced in many places, however, irrigation is still based largely on traditional methods which fail to measure and optimize the supply of water to satisfy plant water demands. Unmeasured irrigation tends to waste water, nutrients, and energy, and may cause soil degradation by waterlogging, erosion, and salination. The vital task of assuring adequate global food production must include a concerted effort to modernize irrigation systems and improve water management. These improved techniques will help achieve sustainable and efficient production while protecting the environment. New systems must be based on sound principles and designs to optimize irrigation in relation to essential inputs and operations while guaranteeing sustainability of irrigated agriculture. Water and soil must be recognized as vital, precious, and vulnerable resources and managed accordingly.

In recent years, revolutionary developments have taken place in the design and management of irrigation. Understanding of the interactive relationships among soil, plant, and climate regarding the ideal disposition and utilization of water continues to evolve. These scientific developments have been paralleled by a series of technical innovations in water control which make it possible to establish and maintain nearly optimal soil moisture conditions.

1.2 Role of Irrigation

The irrigation process consists of introducing water to the soil profile where plants can extract it to meet their needs, mainly evapotranspiration. An important goal of irrigators is to design and manage their irrigation system to optimize placement and timing of applications to promote growth and yield while protecting against soil erosion, salination, water quality degradation, or other detrimental environmental impacts. Since physical circumstances and socioeconomic conditions are site specific, there is no single answer to designing, developing, and managing an irrigation system. In all circumstances, however, the factors and principles involved are universal.

The practice of irrigation has evolved gradually toward improved control over plant, soil, and even weather variables. The degree of control possible today is still only partial because of unpredictable extremes in the weather. Modern irrigation is a sophisticated operation, involving the monitoring and manipulation of numerous factors impacting crop production. With the continuing loss of suitable land and water and the rising demand for agricultural products, the search for new knowledge on how to improve irrigation and the need to apply this new knowledge have become increasingly urgent.

Any attempt to irrigate must be based on a thorough understanding of soil-water-plant relationships. The movement of water, once applied, consists of a sequence of dynamic processes beginning with the entry of water into the soil, called infiltration. The rate of infiltration is governed by the rate at which water is applied to the soil surface, as long as the application rate does not exceed the capacity of the soil to absorb it. An important criterion for a sprinkler or microirrigation system is to deliver water at a rate that will prevent ponding, runoff, and erosion.

After infiltration, water normally continues to move because of gravity and hydraulic gradients in the soil. Water moves downward and, with some irrigation systems, laterally in a process called redistribution. In this process the relatively dry deeper zone of the soil profile absorbs water draining from wetter zones above. Within a few days (depending on the irrigation system and management) the rate of flow becomes so low as to be negligible. The water content of the wetted zone as flow becomes negligible is termed the field capacity and represents the upper limit of the soil's capacity to store water. Field capacity is normally higher in clay than in sandy soils.

Any water draining below the root zone is generally considered to be a loss from the standpoint of immediate plant water use. It is not necessarily a final loss, however. If the area is underlain by an exploitable aquifer, the water percolating below the root zone may eventually recharge the aquifer and be recovered by pumping. Some deep percolation may later return to streams or drainage systems. This quantity of water plus surface runoff from irrigated agriculture is called return flow. Where the water table is close to the soil surface, some water may enter the root zone by capillary rise up from the saturated zone below the water table and supply a portion of the crop's water requirement. This process of subirrigation, however, may infuse the root zone with salts. Water flowing down through the root zone may leach soluble salts or crop nutrients and degrade the quality of groundwater.

Properly designed and managed, modern irrigation methods can increase crop yields while avoiding waste, reducing drainage, and promoting integration of irrigation with essential concurrent crop management operations. The use of degraded water has become more feasible, and coarse-textured soils, steeply sloping lands, and stony soils, previously considered not irrigable, are now productive. Such advances and their consequences were unforeseen only a few decades ago.

1.3 Irrigation Development

For thousands of years, irrigation has contributed substantially to world food production. Historians note that irrigation was one of the first modifications of the natural environment undertaken by early civilizations. Several millennia ago, irrigation permitted nomadic tribes to settle in more stable communities with assurance of annual crop productivity. Initial attempts at irrigation were rudimentary, consisting of ponding water in basins enclosed by low earthen dikes.

The earliest societies to rely successfully on irrigation were located in four major river basins: the Nile in Egypt around 6,000 B.C.E., the Tigris and Euphrates in Mesopotamia about 4,000 B.C.E., the Yellow River in China around 3,000 B.C.E, and the Indus in India approximately 2,500 B.C.E. In Mexico and South America, irrigation was practiced by the Maya and Inca civilizations more than 2,000 years ago. In Iran, ganats, 3,000 year-old tunnels to bring water from the mountains to the valley, are used to this day (Kuros, 1984). Earthen dams to store surface water were first constructed in the second and third centuries in Japan to irrigate rice and were constructed as early as the third century B.C.E. in Sri Lanka (i.e., the Abhaya Wewa reservoir). In Central Europe, irrigation was documented as early as the third century C.E. (Csekö and Hayde, 2004).

In North America, irrigation is known to have existed among Native Americans of the southwest as early as 1200 B.C.E. Early Spanish explorers found evidence of irrigation canals and diversion points along rivers. The Spaniards introduced new irrigation methods and irrigated crops such as grapes, fruits, vegetables, olives, wheat, and barley. As in other areas of the world, irrigation made it possible for Native Americans to develop settlements and enjoy a more secure food source.

At the beginning of the 1800s, the total irrigated area in the world was estimated at about 20 million acres (Gulhati, 1973). Up to that time most irrigation works were small systems. Irrigation began to expand in many countries in the nineteenth century and took on new dimensions in terms of the amounts and methods of water diversion and management. The first barrages, short diversion dams, were built in the Nile Delta in about 1850. About the same time in India, several irrigation canal systems were constructed. The Lower Chanab Canal in Pakistan was the first canal system intended strictly for arid land not previously cultivated. In 1847 Mormon colonies began irrigating in Utah. Their efforts expanded into California, Nevada, Idaho, Wyoming, Arizona, New Mexico, and Canada. German immigrants started an irrigation colony in Anaheim, California, in 1857, and an irrigation colony was started in 1870 at Greeley, Colorado. At the end of the nineteenth century, irrigation in the world was estimated at 100 million acres, a fivefold increase during the century (Gulhati, 1973).

Historians sizing up the twentieth century will almost certainly include irrigation as one of the century's characteristics. During the first half of the century, irrigated area worldwide rose to more than 230 million acres. The surge continued in the second half of the century with over 800 million acres in 2017 (Table 1.2).

Many countries—such as China, Egypt, India, Indonesia, Israel, Japan, Korea, Pakistan, and Peru—rely on irrigation for more than half of their domestic food production. Countries with 10 million irrigated acres or more are tabulated in Table 1.3. Large areas of irrigated lands in southeast Asia lie

Table 1.2. Growth in irrigated land and world population since 1900 (adapted from FAOSTAT, 1999; FAO, 1998, 2021).

Year	Irrigated Area (millions of acres)	Population (billions)
1900	100	1.5
1950	235	2.5
1970	422	3.7
1990	598	5.3
1997	669	5.9
2017	806	7.5

in the humid equatorial belt. These areas have monsoon climates with very large totals of annual rainfall, but portions of the year are dry. In these countries, paddy or flooded rice is the dominate irrigated crop. Countries like China, Korea, Japan, Indonesia, and the Philippines have long been noted for this type of irrigated agriculture. Irrigated area in each country (as a percentage of cultivated area) is shown in Figure 1.1.

Table 1.3. Top 10 irrigated countries in the world in 2017 (adapted from FAO, 2021).

Country	Irrigated Area (millions of acres)	Population (millions)
India	174	1,339
China	173	1,453
United States	66	325
Pakistan	49	208
Iran	22	81
Indonesia	17	265
Thailand	16	69
Mexico	16	125
Turkey	13	81
Brazil	11	208

At the beginning of the twentieth century, irrigation in the western United States amounted to about 3 million acres. Early Caucasian settlers in the western United States were no different than people of ancient civilizations. They developed cooperative irrigation practices and formed communities, especially in southern California and Utah. Irrigation development in the west in the twentieth century was tied closely to the 1902 Reclamation Act which provided capital and the expertise to construct major water supply facilities. During the first three decades of the twentieth century, large multipurpose federal water projects were designed and built for irrigation, flood control, power generation, wildlife and fish habitat, and water-based recreation. Examples include the Colorado River, the Columbia Basin, Central Utah, the Missouri Basin, the Minakoka Project of Idaho, and the Salt River Project of Arizona. Following these projects, private development of pump irrigation from extensive natural underground reservoirs (aquifers) in the plains states, ranging from the Dakotas south to the high plains of Texas, permitted a major increase in irrigation from 1950 to 1980. In the last decades of the twentieth century, irrigation in southeastern states like Florida, Georgia, and South Carolina, where crops grown extensively on sandy soils are at risk during periods of drought, increased rapidly.

The distribution of irrigation in 2017 in the United States from the USDA Farm and Ranch Irrigation Survey is shown in Figure 1.2. The irrigated areas of 20 leading states are presented in Table 1.4, as well as the percentage change in irrigated area for these states over a 15-year period (2002 to 2017). The data for several western states, like California, Arizona, Wyoming,

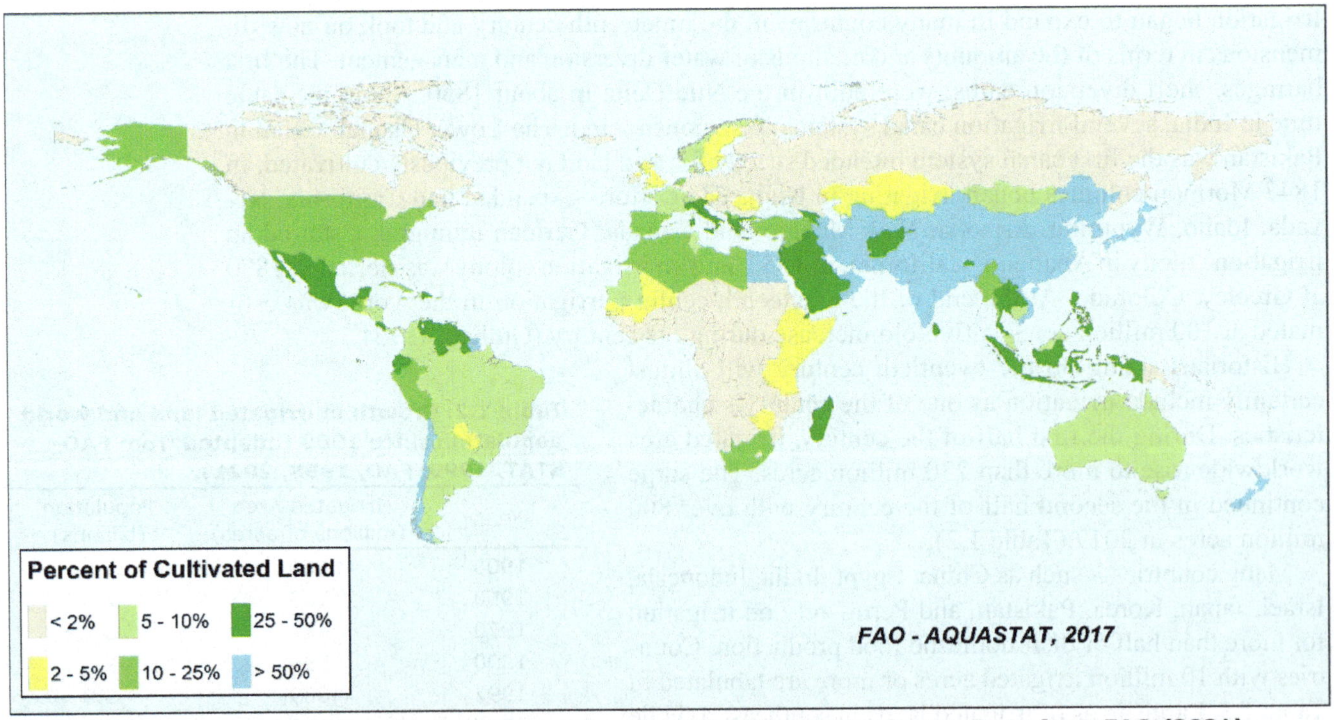

Figure 1.1. Global distribution of irrigation as a fraction of cultivated land area. Data from FAO (2021).

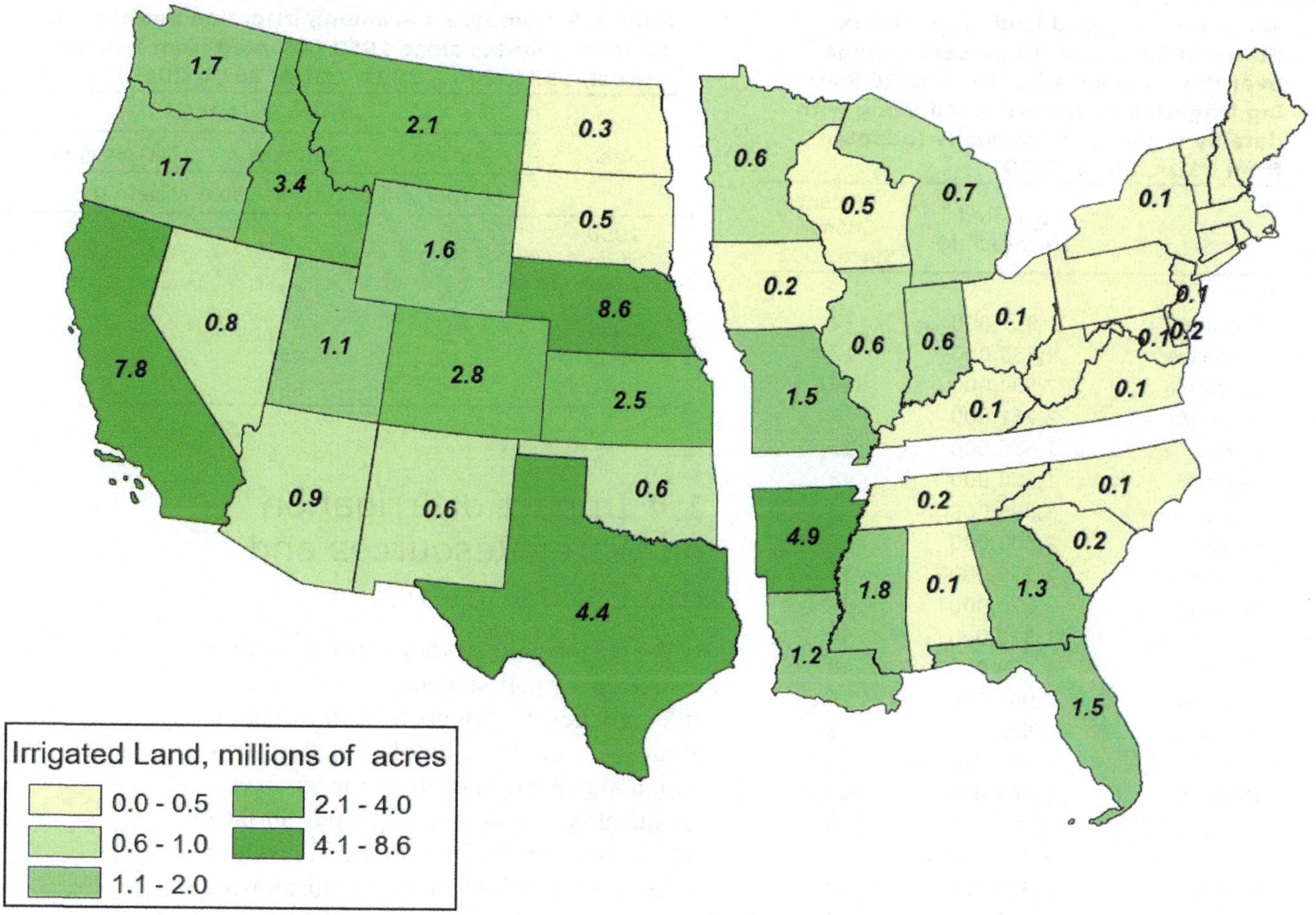

Figure 1.2. Irrigated farmland by state in the United States in 2017 (data from USDA, 2019).

Colorado, Montana, Idaho, Texas, and Utah, indicate that the size of the irrigated area either increased slowly or decreased. This indicates that land and water resources were developed near the maximum possible area under the socioeconomic conditions of the time. Areas with large increases in irrigation were near or just east of the hundredth meridian, the line on the globe that roughly divides the semiarid West from the subhumid Midwest in the United States. The states of Arkansas, Georgia, Louisiana, Michigan, Mississippi, Missouri, and Nebraska had large percentage increases in irrigation during the 15-year period. This increase, in large part, was a consequence of groundwater being tapped by irrigation wells. As you review the 20 leading irrigation states, you will also notice the amount of irrigated lands in southeastern states, like Georgia, Mississippi, and Louisiana, increased dramatically during the 15-year period. By 2017 there were over 58 million irrigated acres in the United States (Table 1.4).

The type of irrigation system (Chapter 5) in the United States has also changed with time. Table 1.5 summarizes the percent of irrigated land using surface, sprinkler, and microirrigation since 1950. After a maximum of 98% surface irrigation (Chapter 10), this type of irrigation system has declined to 35% of the irrigated area in 2018. Meanwhile, the amount of sprinkled land has increased from 2%of the irrigated land in 1950 to 55% 2018. The amount of sprinkle irrigation (Chapters 11-13) now surpasses that of surface irrigation in the U.S. Microirrigation (Chapter 14), which includes drip/trickle, microspray, and similar systems, has increased from its infancy in the 1960s to 3 million acres at the turn of the century. Microirrigation accounted for 10% of the irrigated area in the U.S. in 2018.

Table 1.4. Irrigated land in the United States in 2017 and the percent change over the previous 15 years. The 20 leading irrigation states are listed along with data by region and nationally (adapted from USDA, 2014, 2019).

	Irrigated Land (acres)	Percent Change Since 2002
State:		
Arizona	911,000	-2
Arkansas	4,855,000	17
California	7,834,000	-10
Colorado	2,761,000	7
Florida	1,519,000	-16
Georgia	1,288,000	48
Idaho	3,398,000	3
Kansas	2,503,000	-7
Louisiana	1,236,000	32
Michigan	670,000	47
Mississippi	1,815,000	54
Missouri	1,529,000	48
Montana	2,061,000	4
Nebraska	8,588,000	13
Nevada	790,000	6
Oregon	1,665,000	-13
Texas	4,363,000	-14
Utah	1,097,000	1
Washington	1,689,000	-7
Wyoming	1,568,000	2
Region:		
19 Western states	41,234,000	-2
9 Southeastern states	11,393,000	20
22 Northeastern states	5,387,000	41
Nation:		
U.S. total	58,014,000	5

Table 1.5. Comparisons among irrigation methods in the United States since 1950 (adapted from Irrigation Journal, 1971, 2000, 2001; USDA, 2014, 2019).

Year	Irrigation Method		
	Surface	Sprinkler	Microirrigation
	(percent of total irrigated area in U.S.)		
1950	98	2	-
1970	81	19	-
1990	56	42	2
2000	45	50	5
2008	39	54	7
2018	35	55	10

1.4 Impact of Irrigation on Water Resources and the Environment

As responsible stewards of our natural resources, irrigation managers should consider any negative impacts from irrigation along with the benefits of irrigation. Irrigation may have a negative impact on water quantity and/or water quality. Surface water diversion for irrigation will result in reduced streamflow downstream and reduced water volume in downstream water bodies. The Aral Sea, between Uzbekistan and Kazakhstan, is an extreme example, now being less than 10% of its original size. Irrigation from groundwater pumping will result in declining groundwater levels and stream flow depletion if annual pumping exceeds the annual groundwater recharge (Figure 1.3). Reduced groundwater levels may result in reduced baseflow in nearby streams (Chapter 9). Crop water use (Chapter 4) is often the largest use of water in agricultural watersheds. It is important for water resources managers to understand that reductions in water diversions for irrigation do not always result in a reduction in consumptive use of water resources (Chapter 5).

Water quality concerns include both groundwater and surface water. Irrigation often results in deep percolation, resulting in the leaching of soluble fertilizers or other chemicals (Chapter 5). In some areas, nitrate leaching has resulted in groundwater nitrate concentrations above the maximum concentration allowed for human consumption. Deep percolation can be minimized with good irrigation scheduling (Chapter 6). Runoff from irrigation can contain nutrients, pesticides, and sediments. This can particularly be a problem in surface irrigation systems if the runoff is not collected and reused on additional fields (Chapter 10). Chemigation needs to be managed well to prevent chemicals from entering surface water or contamination of the water source through backflow from the irrigation system (Chapter 15).

Finally, soil quality is also a concern. Irrigation systems that result in runoff may also trigger soil erosion. In arid regions, salt accumulation in the soil can be a significant concern

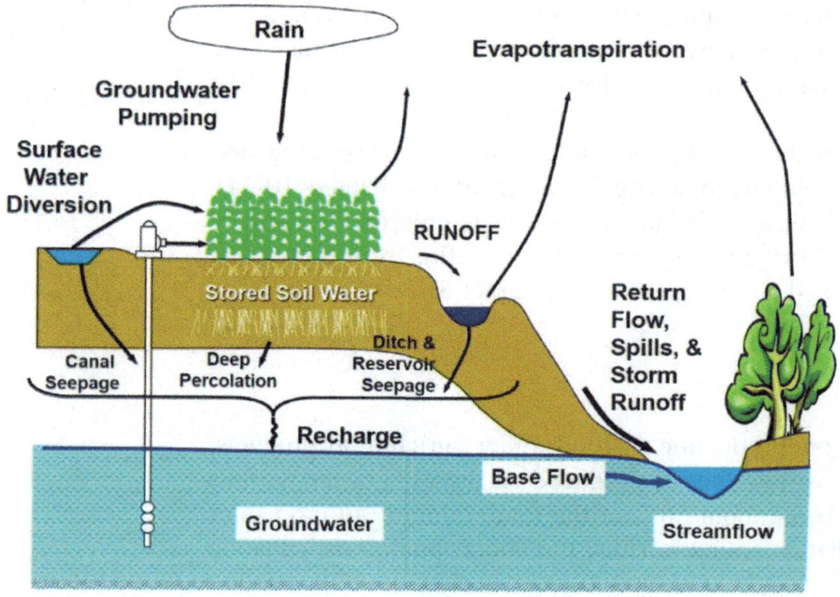

Figure 1.3. Large-scale water balance, showing the water cycle and interactions between irrigation and surface and groundwater resources.

depending on the salinity of the irrigation water (Chapter 7). In these situations, it is often necessary to include subsurface drainage along with the irrigation system.

Good irrigation management should seek to increase food production (and farm profits) while minimizing negative impacts on water resources and the environment. In many situations, best management practices provide methods to achieve both of these goals simultaneously.

1.5 Irrigation Management Concepts

Modern irrigation management is based on the concept of soil-plant-water relations. This concept is a unification system in which all processes are interdependent. In this unified system, called the soil-plant-atmosphere relations, the availability of soil water is not a property of the soil alone, but a function of the plant, soil, and environment. The rate of water uptake by the plant depends on the root's ability to absorb water from the soil, the soil's ability to transmit water toward the roots, and the evaporative demand of the atmosphere. These, in turn, depend on: (1) characteristics of the plant such as rooting density, rooting depth, rate of root growth, and the plant's ability to maintain its vital functions under water stress; (2) properties of the soil like hydraulic conductivity, soil bulk density, soil texture, soil layers, water retention, and available water capacity; and (3) weather conditions which dictate the rate of transpiration from the crop and soil evaporation. These components of the continuum will be presented and discussed for soil water (Chapter 2) and plant water use (Chapter 3).

Irrigation scheduling is the term that describes the procedure by which an irrigator determines the timing and quantity of water application. It is possible to schedule irrigations based on monitoring the soil, the plant, and/or the microclimate. By monitoring soil moisture, the idea is to measure the reserve of water within the crop root zone as it is diminished following each irrigation to ascertain when the soil water has been depleted to a prescribed minimum level. Sensing the water status of the plant is a second method to detect the beginning of plant water stress. There are many plant sensing techniques available today to measure or infer plant water status ranging from specialized equipment to visual observation. As important as the earliest detection of plant water stress is, it does not give information on how much water to apply. A third technique is to monitor the meteorological conditions that impose the evapotranspirational demand on the crop. Accumulating the amount of water lost to the atmosphere by the crop will estimate the amount of water to apply. The timing of the irrigation is established by knowing the capacity of the soil to store water or monitoring the water status of the plant. These various measuring techniques and strategies for scheduling irrigations will be presented in detail (Chapter 6).

The later chapters of the text are devoted to descriptions of the various types of irrigation systems, emphasizing the major methods employed in irrigated agriculture (Chapters 10-14). The application of agricultural chemicals through irrigation systems is presented separately (Chapter 15). Effective management of irrigation systems also benefits from a working knowledge of the hydraulics of pipeline and pumping systems (Chapter 8).

1.6 Summary

Irrigation is extremely important in the production of food, other agricultural products, ornamentals, and turf. One-third of the world's food is produced on the 21% of the world's cultivated area that is irrigated. In the U.S., about 50% of the total value of all crop sales comes from the 28% of the cropland that is irrigated. Thus, the understanding of irrigation and its management are critical to all of us.

Here, the basic concepts to understand are that water is applied, distributed in the soil, and stored for plant use. Various irrigation systems and their operation and management are then presented in the context of each system's advantages and disadvantages. Procedures for determining when and how much irrigation water to apply are discussed in detail throughout this text to assist the reader in being as efficient as possible when utilizing this precious resource, water.

Questions

1. Name the three states west of the Mississippi River with the largest irrigated areas.
2. Name the three leading states east of the Mississippi River with the largest irrigated areas.
3. Which state lost the largest amount of irrigated land from 1969 to 1999?
4. Which state gained the most acres of irrigated land?
5. Name three states where microirrigation is a popular irrigation method. Where is sprinkler irrigation practiced and why?
6. List three benefits and three negative consequences from irrigation development.
7. Explain why the irrigation industry is changing from a development era to a management era.

References

Cseko, G., & Hayde, L. (2004). *Danube Valley: History of irrigation, drainage and flood control.* New Delhi, India: International Commission on Irrigation and Drainage.

FAO. (1998). *Production yearbook* (Vol. 52). Rome, Italy: Food and Agricultural Organization of the United Nations.

FAO. (2021). AQUASTAT database. Rome, Italy: Food and Agricultural Organization of the United Nations.

FAOSTAT. (1999). FAOSTAT statistical database. Rome, Italy: Food and Agricultural Organization of the United Nations. Retrieved from http://apps.fao.org

FAOSTAT. (2005). FAOSTAT statistical database. Rome, Italy: Food and Agricultural Organization of the United Nations. Retrieved from www.faostat.fao.org

Gulhati, N. D. (1973). Introduction. In V. A. Kovda, C. Van den Berg, & R. M. Hagan (Eds.), *Irrigation/drainage, and salinity* (pp. 1-14). London: Hutchinson.

Irrigation Journal. (1971). Irrigation survey. *Irrig. J., 21*(5), 10-17.

Irrigation Journal. (2000). 1999 Annual irrigation survey. *Irrig. J., 50*(1), 16-31.

Irrigation Journal. (2001). 2000 Annual irrigation survey. *Irrig. J., 51*(1), 12-30, 40-41.

Kuros, G. R. (1984). Qanats—A 3000 year-old invention for development of groundwater supplies. *Proc. Special Session, Int. Committee on Irrigation and Drainage.*

USDA. (2014). 2013 Irrigation and water management survey. Vol. 3. Special Studies, Part 1. AC-12-SS-1. Washington, DC: USDA.

USDA. (2019). 2018 Irrigation and water management survey. Vol. 3. Special Studies, Part 1. AC-17-SS-1. Washington, DC: USDA.

Chapter 2
Soil Water

2.1 Introduction

Like humans, plants need water to survive. But, from where do plants get their water? Well, the soil beneath your feet is the answer. The soil within the plant root zone serves as a reservoir for storing precipitation and irrigation water for future use by plants. Effective management of an irrigation system requires the understanding and use of the basic concepts of soil water. Without an adequate understanding of these concepts, the irrigator will not know how much water to apply or when to irrigate. These are fundamental concerns in irrigation management. After a review of these concepts, water entry into the soil will be discussed followed by a presentation of field methods for measuring soil water.

The goal of irrigation management is to maintain the amount of water in the soil between wet and dry limits to satisfy the plant's water requirements. The wet soil limit occurs when plants suffer because of decreased aeration, and the dry soil limit occurs when plants have difficulty obtaining the water they need. Thus, it is necessary to determine the amount of soil water available to plants and the proper amount of irrigation water to be applied.

Two measures of soil water are important for managing irrigation systems. The first is the amount of water in the soil, which is commonly referred to as the ***soil water content***. The second property is the ***soil water potential***, which is a measure of how hard plants have to work to remove water from the soil.

Before considering these two measures of soil water, the impact of the components of the soil and their impact on the ability of soil to store water must be understood. The basic components of the soil are the solid mineral particles, organic matter, the voids among the particles, and water and air occupying the voids. The capacity of the soil to store water depends upon the volume of the voids present.

2.2 Soil Composition

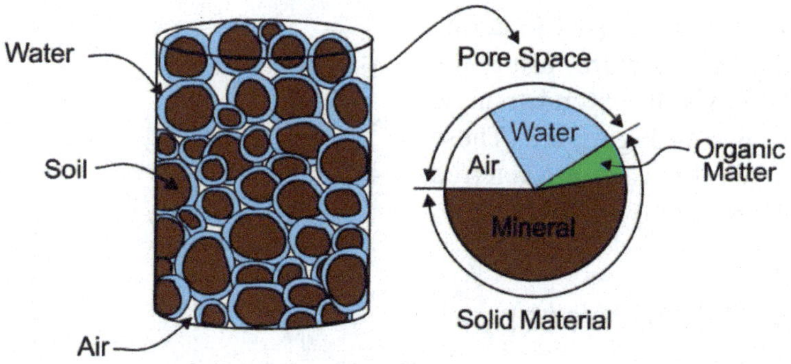

Figure 2.1. Composition of a soil volume.

As Figure 2.1 illustrates, soil is composed of three major components: soil particles, air, and water. The fractions of water and air are contained in the voids between soil particles. The ratio of the volume of pores (voids) to the total (bulk) volume of a soil is the ***porosity*** (φ). One way to determine porosity is to measure the volume of a soil that is composed of soil particles and the fraction made up of the pores. The porosity may also be determined using the soil ***bulk density*** (ρ_b) Bulk density is the density of the undisturbed (bulk) soil sample described by:

$$\rho_b = \frac{M_s}{V_b} \tag{2.1}$$

where: ρ_b = soil bulk density (g/cm^3),
M_s = mass of dry soil (g), and
V_b = volume of bulk soil sample (cm^3).

Given the bulk density of a soil, the porosity (in percent) can be calculated as:

$$\varphi = \left(1 - \frac{\rho_b}{\rho_p}\right) 100\% \tag{2.2}$$

where: φ = soil porosity
ρ_p = soil particle density (a common value for mineral soils is 2.65 g/cm^3).

Example 2.1

An undisturbed soil sample with a volume of 80 cm^3 is taken from an irrigated field. The mass of the soil sample after drying is 100 grams. What is the soil bulk density? What is the porosity?

Given: M_s = 100 g
V_b = 80 cm^3

Find: ρ_b and φ

Solution:

$$\rho_b = \frac{M_s}{V_b} = \frac{100\text{ g}}{80\text{ cm}^3} = 1.25\text{ g/cm}^3$$

$$\varphi = \left(1 - \frac{\rho_b}{\rho_p}\right) 100\% = \left(1 - \frac{1.25}{2.65}\right) 100\% = 53\%$$

The soil sample in Example 2.1 consists of 47% soil particles and 53% pore space (air and/or water).

The mineral fraction of the soil volume is composed of ***sand***, ***silt***, and ***clay*** separates. With the USDA classification system, the equivalent diameter size limits are: Clay < 0.002 mm, silt 0.002 to 0.05 mm, and sands 0.05 to 2.0 mm. The relative proportions of the various soil separates are used to define the ***soil texture*** using the USDA soil textural triangle shown in Figure 2.2. These textural classes are referred to frequently in this book.

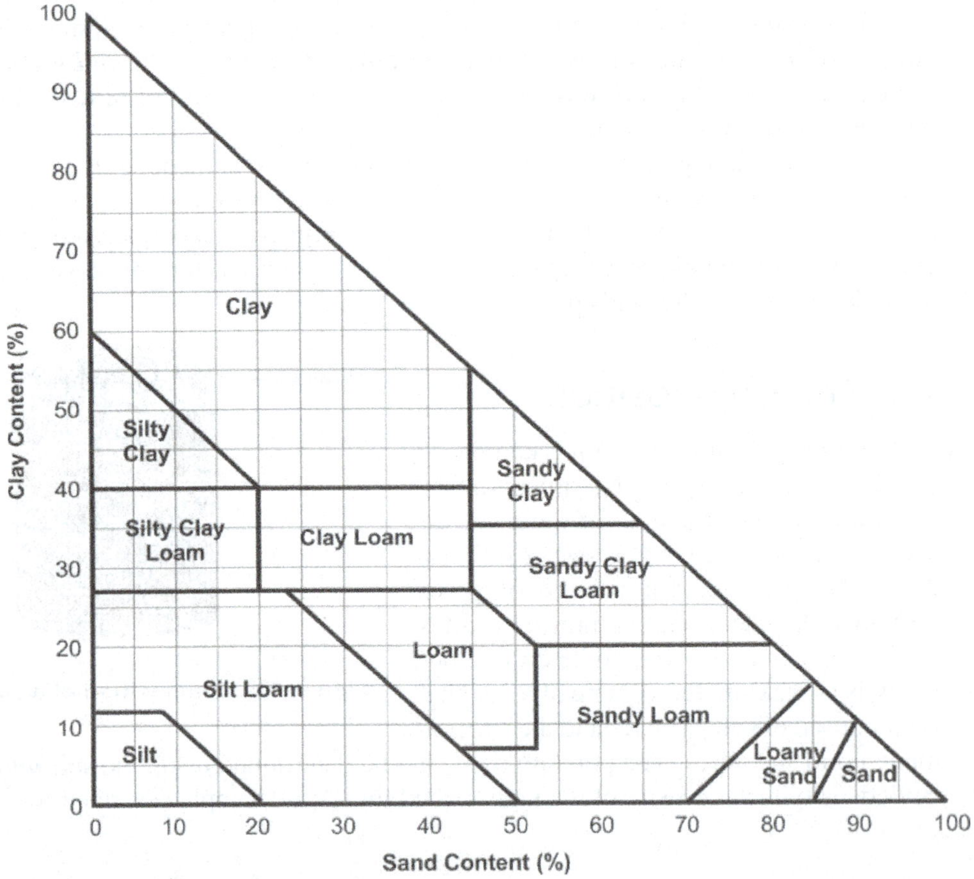

Figure 2.2. USDA soil textural triangle.

2.3 Soil Water Content

The amount of water in a soil can be expressed in many ways, including a dry soil basis (mass water content), a volumetric basis (volumetric water content), fraction of the available water remaining, and fraction of the available water depleted. With so many different terms, confusion is bound to arise. Irrigation managers must understand all of these terms to interpret soil water status correctly.

The ***mass water content*** or ***gravimetric water content*** (θ_m) is the ratio of the mass of water in a sample to the dry soil mass, expressed as either a decimal fraction or as a percentage. Mass water content is determined by weighing a field soil sample, drying the sample for at least 24 hours at 105°C, and then weighing the dry soil. The decrease in mass of the sample due to drying represents the mass of water in the soil sample. The mass of the sample after drying represents the mass of dry soil. The mass water content is found by:

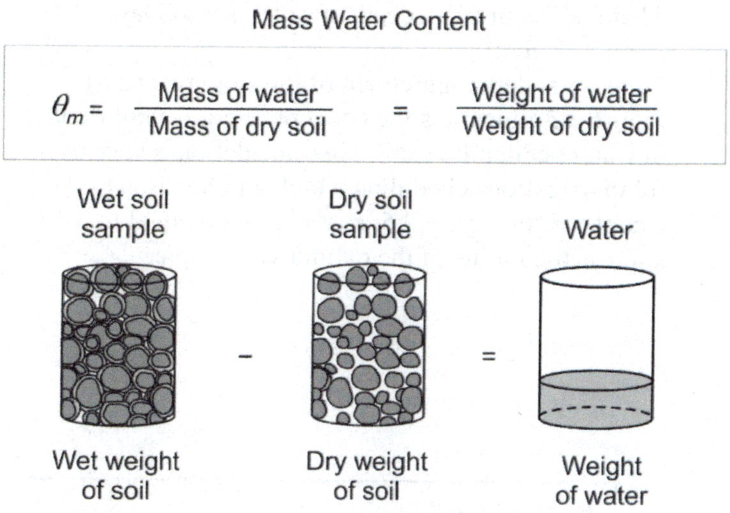

Figure 2.3. Concept of mass water content.

$$\theta_m = \frac{M_w}{M_s} \tag{2.3}$$

where: θ_m = mass water content,
M_w = mass of water lost during drying (g), and
M_s = mass of dry soil (g).

Figure 2.3 illustrates the relationship between the weight of the water in the soil and the dry weight of the soil when determining mass water content.

The ***volumetric water content*** (θ_v) represents the volume of water contained in a volume of undisturbed soil. The volumetric water content is defined as:

$$\theta_v = \frac{V_w}{V_b} \tag{2.4}$$

where: θ_v = volumetric water content,
V_w = volume of water (cm^3), and
V_b = bulk volume of soil sample (cm^3).

Figure 2.4 illustrates the volume of the components needed to calculate θ_v.

To find the volumetric water content, the volume of the undisturbed soil sample must be determined, which is sometimes difficult to measure. The mass water content is more easily determined but is often not as useful as the volumetric water content. Therefore, the following equation, which connects mass water content to volumetric water content, is convenient.

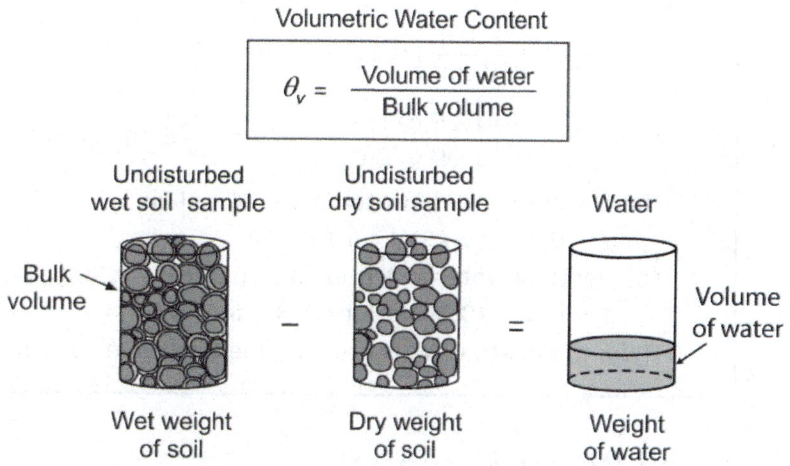

Figure 2.4. Concept of volumetric water content.

$$\theta_v = \frac{\rho_b}{\rho_w}\theta_m \tag{2.5}$$

where: ρ_w = density of water, which is 1 g/cm^3.

When comparing water amounts per unit of land area, it is frequently more convenient to speak in equivalent depths of water rather than water content. The relationship between volumetric water content and the equivalent depth of water in a soil layer is:

$$d = \theta_v L \tag{2.6}$$

where: d = equivalent depth of water in a soil layer (cm)

L = depth increment of the soil layer (cm).

Figure 2.5 illustrates the concept of equivalent depth of water per depth of soil. This calculation is very useful in irrigation scheduling which will be discussed in Chapter 6. In Figure 2.5, π is a constant equal to 3.14 and r is the radius of the cylindrical sample.

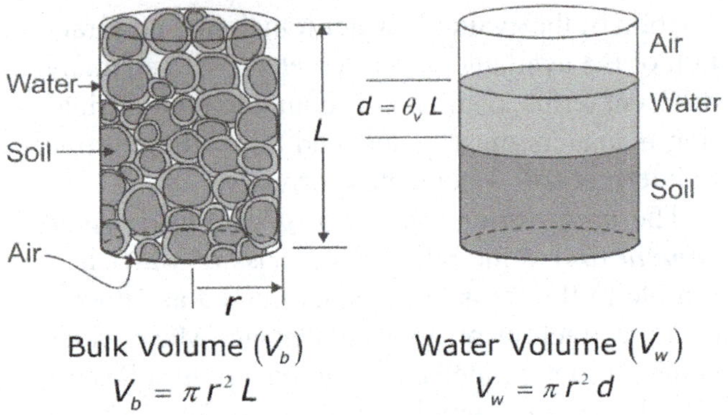

Figure 2.5. Concept of depth of water contained in a soil layer.

Example 2.2

A field soil sample prior to being disturbed has a volume of 80 cm³. The sample weighed 120 grams. After drying at 105°C for 24 hr, the dry soil sample weighed 100 grams. What is the mass water content? What is the volumetric water content? What depth of water must be applied to increase the volumetric water content of the top 1 ft of soil to 0.30?

Given: $M_s = 100$ g

$M_w = 120\text{ g} - 100\text{ g} = 20$ g

$V_b = 80\text{ cm}^3$

Find: θ_m

θ_v

d

Solution:

$$\theta_m = \frac{M_w}{M_s} = \left(\frac{20\text{ g}}{100\text{ g}}\right) = 0.20\text{ g of water / g of soil}$$

$$\rho_b = \frac{M_s}{V_b} = \left(\frac{100\text{ g}}{80\text{ cm}^3}\right) = 1.25\text{ g/cm}^3$$

$$\theta_v = \frac{\rho_b}{\rho_w}\theta_m = \left(\frac{1.25\text{ g/cm}^3}{1.00\text{ g/cm}^3}\right) \times 0.20 = 0.25\text{ cm}^3\text{ of water / cm}^3\text{ of soil}$$

The current depth of water in 1 ft of soil is:

$d = \theta_v L = (0.25)(12\text{ in}) = 3$ in

The depth of water in 1 ft of soil when $\theta_v = 0.30$ will be:

$d = \theta_v L = (0.30)(12\text{ in}) = 3.6$ in

Thus, the depth of water to be added is 0.6 in (3.6 in − 3.0 in).

2.4 Soil Water Potential

The amount of water in the soil is not the only concern in irrigation management. Plants must be able to extract water from the soil. ***Soil water potential*** (Ψ_t) is an indicator or measure of the energy status of soil water relative to that of water at a standard reference (Hillel, 1980). This energy is due to the position of the water relative to the reference and the internal state of the water and is often expressed as energy per unit of volume (***pressure***) or energy per unit of weight (***head***). Common units of pressure and head, and their equivalents are shown in Table 2.1. The standard reference is often denoted at a high energy level and assigned a value of zero. Thus, soil water potential and its components can all have negative values. A high energy level of water potential will have a smaller negative value (lower magnitude) than a water potential at a lower energy level. For example, in a wet soil, matric potential (discussed below), Ψ_m, will have a small negative value, say $\Psi_m = -0.3$ bar, while in a dry soil Ψ_m may be –15 bars.

Table 2.1. Common units of pressure and head and their equivalents.

Unit	Pressure Equivalent	Water Head Equivalent
1 atmosphere (atm)	101.3 kPa (kilopascals)	1034 cm H_2O
	1.013 bar	34 ft H_2O
	101.3 cb (centibar)	76 cm Hg
	14.7 psi (lb/in^2)	29.9 in Hg
1 psi (lb/in^2)	6.89 kPa	2.31 ft H_2O

The three major components of soil water potential are gravitational potential (ψ_g), ***matric potential*** (ψ_m), and osmotic potential (ψ_o). The ***soil water potential*** (ψ_t) then is:

$$\psi_t = \psi_g + \psi_m + \psi_o \tag{2.7}$$

Equation 2.7 ignores the impact of overburden pressure on soil water potential. The gravitational potential is due to the force of gravity pulling downward on the water in the soil. Matric potential is a result of the forces the soil particles place on the water by adhesion and surface tension at the soil-air interface. These combined forces cause capillarity, which is sometimes referred to as soil water tension. Soil water tension is expressed as a positive value. Osmotic potential is caused by dissolved solids (salts) in the soil water. The osmotic potential affects the availability and movement of water in soils when a semipermeable membrane (like plant roots) is present. This topic is discussed in more detail in Chapter 7.

Where rainfall is significant and irrigation water is nearly free of salts, the concentration of salts in the soil is generally low, so the osmotic potential is near zero. The osmotic potential does not influence the flow of water through the soil profile. It does, however, have an effect on water uptake by plants and on evaporation. During evaporation, water changes from liquid to vapor at the soil-air interface near the soil surface but salts are left behind in the soil. The higher the salt content of soil water (lower osmotic potential) the lower the rate of evaporation. Water uptake through plant roots is also influenced by the osmotic potential; the higher the salt concentration in the soil solution the more work a plant has to do to absorb water from the soil. Thus, where soil salinity is appreciable, osmotic potential must be considered for evaluating plant water uptake or where water vapor flow is important.

The component of soil water potential that dominates the release of water from soil to plants when salts are minimal is the matric potential. Several forces are involved in the retention of water by the soil matrix. The most strongly held water is adsorbed around soil particles by electrical forces. This water is typically held too tightly for plants to extract. Water is also held in the pores between soil particles by a combination of attractive (surface tension) and adhesive forces. The strength of the attractive force depends on the sizes of the soil pores. Large pores will freely give up pore water to plants due to the much higher matric potential in the soil or to drainage due to the gravitational potential (Martin et al., 2017). For a given amount of water in a particular soil, there will be a corresponding matric water potential. Here we will express the magnitude of the matric potential as soil water tension thus in the positive realm. The curve representing the relationship between the water tension within the soil and its volumetric water content is referred to as the soil water release or soil water retention curve. The soil water release curves in Figure 2.6 show that water is released (volumetric water content is reduced) by the soil as the tension increases.

Soil water release curves are often used to define the amount of water available to plants. Two terms are used to define the upper and lower limits of water availability. The upper limit, ***field capacity*** (θ_{fc}), is defined as the soil water content where the drainage rate, caused by gravity, becomes negligible. Thus, the soil is holding all of the water it can without any significant loss due to drainage. The ***permanent wilting point*** (θ_{wp}), the lower limit, is the water content below which plants can no longer extract water from the soil. At this point (WP) and at higher tension values, plants will wilt permanently and will not recover if the water stress is relieved. Neither of these two limits are exact. The WP has traditionally been defined as the water content corresponding to 15 bars of soil water tension or 1,500 cb. This is a reasonable working definition because the water content varies only slightly over a wide range of soil water tension near 15 bars. For example, if the plants permanently wilt at 20 bars of tension, the water content is not much different than at 15 bars and the error in the estimate of water available to plants is small. Example values of θ_{fc} and θ_{wp} are given in Figure 2.6 for several soil types.

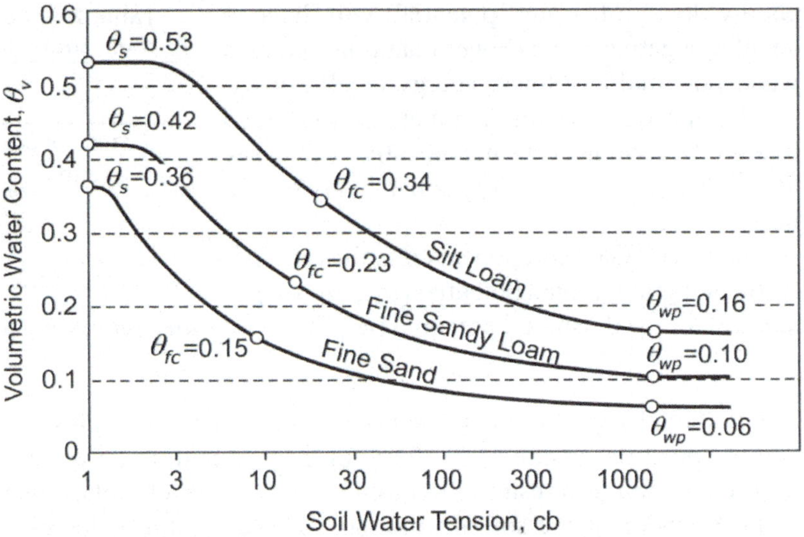

Figure 2.6. Example soil water release curves for three soil textures showing the values of θ_s, θ_{fc}, and θ_{wp} for each soil.

Field capacity is often considered to be the water content at a matric potential of minus one-third bar or a tension of 33 cb. This is *not* a good definition for all soils. This tension value for FC is fairly good for some fine-textured soils but is too large a tension for medium- and coarse-textured soils. The field capacity values shown in Figure 2.6 are more representative than a strict one-third bar definition. Methods will be presented later where the actual field conditions will be used to estimate field capacity for irrigation management.

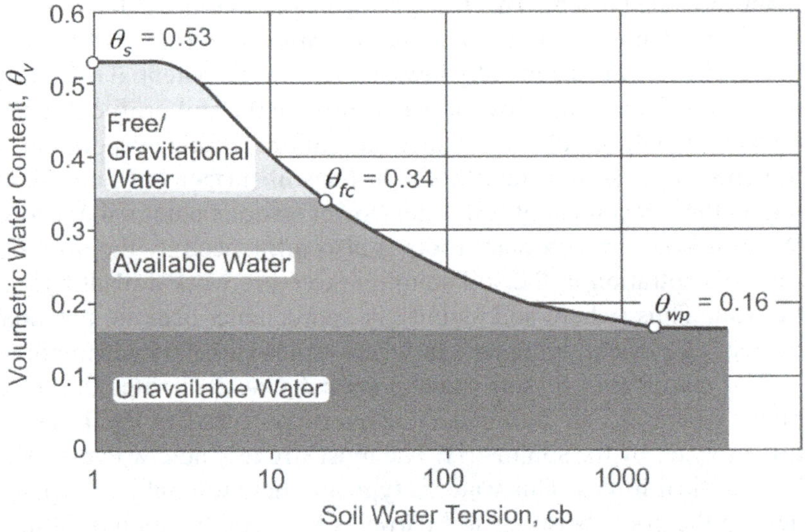

Figure 2.7. Graphical representation of free-draining water, water available for plant uptake, and unavailable water on a soil water release curve.

Users of soil water measurements must keep in mind that there is a difference between volumetric water content at field capacity (θ_{fc}) and volumetric water content at saturation (θ_s). If the voids are completely filled with water and air is absent, the soil is said to be saturated. The volumetric water content equals porosity at saturation, i.e., $\theta_s = \varphi$. As gravity causes drainage to occur, air enters the soil and soil water content reaches θ_{fc} as drainage from gravity ceases. Thus, θ_{fc} is less than θ_s.

The relationships among the soil water that is free to drain due to gravity, the soil water available for plant water use, and the soil water that is not available to be extracted by plant roots is illustrated in Figure 2.7 on a soil water release curve. The water that is free to drain by gravity is between θ_s and θ_{fc}. Available water is that water between θ_{fc} and θ_{wp}, and unavailable water is that water between θ_{wp} and 0.

2.5 Available Water and the Soil Water Reservoir

Irrigation managers can view the soil as a reservoir for holding water. Figure 2.8 illustrates the analogy between a reservoir and a soil. Soil without any water would be like an empty reservoir (Figure 2.8a). Pores in the soil, measured as porosity, provide space for the storage of water. When saturated, the entire void space (reservoir) is filled with water as in Figure 2.8b. After 1 to 3 days of drainage, the water content reaches field capacity (Figure 2.8c). Water leaves through the drain tube on the side by gravity in the analogy. A plant can easily extract water between field capacity and a specific water content represented by the bottom of the large tube extending into the reservoir (Figure 2.8d). This specific water content is referred to as minimum balance and the difference between field capacity and minimum balance is ***allowable depletion*** (AD). As the water content decreases below the minimum balance (Figure 2.8e), the plant must work harder to extract the water it requires. The stress a plant experiences below the minimum balance causes a reduction in yield potential. If the reservoir is not replenished, the water content will continue to decrease and eventually reach permanent wilting, which is represented by the bottom of the small tube in Figure 2.8f. Once this point is reached, a plant can no longer recover even if water is added.

The water held between field capacity and the permanent wilting point is called the ***available water capacity*** (AWC), i.e., available for plant use. The AWC of the soil is expressed in units of depth of available water per unit depth of soil, for example in/in or cm/cm. The AWC is calculated by:

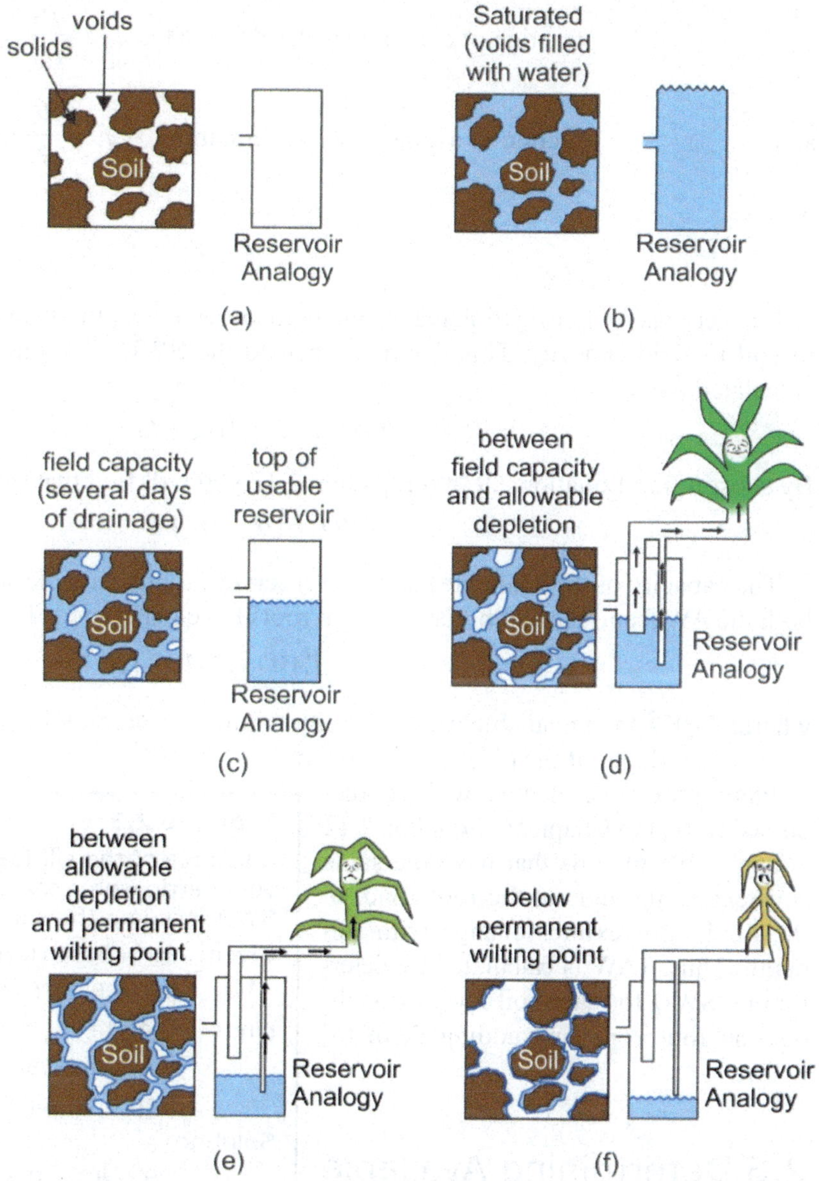

Figure 2.8. Reservoir analogy of soil water.

$$AWC = \theta_{fc} - \theta_{wp} \tag{2.8}$$

For the fine sandy loam soil shown in Figure 2.6, the volumetric water content at field capacity (θ_{fc}) is 0.23 and the volumetric water content at WP (θ_{wp}) is 0.10. Thus, the available water capacity for that soil is 0.13 in/in or cm/cm (0.23 – 0.10). You should read this as 0.13 inches of water per inch of soil depth.Field soils are generally at a water content between the FC and the WP. Commonly used terminology in irrigation management is ***soil water depletion*** (SWD) or ***soil water deficit*** (SWD). SWD refers to how much of the available water has been removed, i.e., the difference between θ_{fc} and θ_v, the actual soil water content. The difference between θ_v and θ_{wp} is the amount of available water remaining.

Often the depleted and remaining water values are expressed as a fraction or percentage. The equations for determining the fraction of available water depleted and the fraction of available water remaining are as follows:

$$\text{fraction of available water depleted} = f_d = \frac{\theta_{fc} - \theta_v}{\theta_{fc} - \theta_{wp}} \tag{2.9}$$

and

$$\text{fraction of available water remaining} = f_r = \frac{\theta_v - \theta_{wp}}{\theta_{fc} - \theta_{wp}} \tag{2.10}$$

Also,

$$f_r = 1 - f_d \tag{2.11a}$$

or

$$f_d = 1 - f_r \tag{2.11b}$$

It is very useful in irrigation management to know the depth of water required to fill a layer of soil to field capacity. This depth is equal to the SWD. Do you see why? SWD can be calculated by:

$$SWD = f_d\,(AWC)\,L \tag{2.12}$$

By substituting Equation 2.9 into Equation 2.12 you will find that this is equivalent to:

$$SWD = (\theta_{fc} - \theta_v)\,L \tag{2.13}$$

The capacity of the available soil water reservoir, total available water (TAW), depends on both the AWC and the depth that the plant roots have penetrated. The relationship is:

$$TAW = AWC\left(R_d\right) \tag{2.14}$$

where: TAW = total available water capacity within the plant root zone (cm), and
R_d = depth of the plant root zone (cm).

Plant root zone depths will be discussed further in Chapter 6. Equation 2.14 is applicable to soils that have the same soil texture throughout the root zone. In the field, soil textures change with soil depth. Thus, TAW is calculated by determining SWD for each soil layer throughout the root zone and adding them together.

Example 2.3

A sample of the silt loam soil characterized in Figure 2.6 has a volumetric water content of 0.26. Calculated f_d, f_r, AWC, and SWD. Assume the soil is 36 inches deep.

Given: θ_{fc} = 0.34 (from Figure 2.6)
θ_{wp} = 0.16 (from Figure 2.6)

Find: f_d and f_r
Available water capacity (AWC)
Depth of soil water depletion (SWD)

Solution:

$$f_d = \frac{\theta_{fc} - \theta_v}{\theta_{fc} - \theta_{wp}} = \frac{0.34 - 0.26}{0.34 - 0.16} = 0.44$$

$$f_r = 1 - f_d = 1 - 0.44 = 0.56$$

$$AWC = \theta_{fc} - \theta_{wp} = 0.34 - 0.16 = 0.18 \text{ in/in}$$

$$SWD = f_d\,(AWC)\,L = 0.44\,(0.18 \text{ in/in})\,36 \text{ in} = 2.85 \text{ in}$$

2.6 Determining Available Water Capacity

The values of θ_{fc} and θ_{wp} of a soil used to calculate AWC, can be determined by field and laboratory methods. Discussion of the various techniques to measure these variables is beyond the scope of this book. The reader should refer to Bruce and Luxmore (1986) and Klute (1986) and references therein for detailed information. Relatively simple experiments for approximating these variables are explained below.

Field capacity may be determined by flooding a small area of land, covering it to suppress evaporation, waiting several days for drainage to become negligible, and then sampling to determine the water content throughout the soil profile. When flooding ceases, the water content falls rapidly as the largest soil pores are quickly drained by gravity. After the rate of drainage slows in 1 to 3 days, the water content remains nearly constant. This is field capacity. At this time, the soil should be sampled for water content. As a rule of thumb, 1 day of drainage will generally be adequate for sandy soils, 2 days for silt loam soils, and 3 days for silty clay loam soils. A simpler field method of determining field capacity is to take soil samples at intervals following a thorough irrigation or rain in a fallow field. When θ_v remains nearly constant the value is θ_{fc}.

The water content at WP can be determined by measurements in areas where the available soil water has been exhausted. In this case, an area that experiences severe water stress would be a good location to take a soil sample. The sample could be analyzed for θ_v at that time to determine the θ_{wp} throughout the soil profile.

If field capacity is known, θ_{wp} can be estimated by subtracting AWC from θ_{fc}. Suppose the θ_{fc} is 0.30 in/in and the AWC is 0.18 in/in. Wilting point, θ_{wp} is then 0.30 minus 0.18 or 0.12 in/in. Often, AWC is tabulated in soil survey reports and textbooks.

Generally in irrigation management, the same value of θ_{wp} is used throughout the root zone for calculating water requirements. At the same time, we use root zones shallower than what is explored by plant roots. This creates a margin of safety, and to some extent, accounts for the fact that the permanent wilting point in the upper portion of the root zone is often higher than in the lower portion. This simplification makes water balance calculations much easier and has worked well in scheduling and designing irrigation systems.

2.7 Tabulated Values of Typical Soil Properties

Data for soil properties are available from various sources. For example, in the U.S., county-level Soil Survey Reports normally list many of the soil properties described in this chapter. These reports are available electronically for application with geographic information systems and on the internet such as the Web Soil Survey (http://websoilsurvey.nrcs.usda.gov/app/). An example listing is shown in Table 2.2.

Generalized values of AWC, θ_{fc}, and θ_{wp} for a range of soil textures are given in Table 2.3.

Table 2.2. Examples of soil properties for Platte County, Nebraska (USDA, 1988).

Soil Name	Soil Depth (in)	USDA Texture	Bulk Density (g/cm³)	Permeability (in/hr)	Available Water Capacity (cm³/cm³, in/in, or m/m)
Geary	0–11	Silty clay loam	1.30–1.40	0.6–2.0	0.18–0.23
	11–34	Silty clay loam/clay loam	1.35–1.50	0.2–2.0	0.17–0.20
	34–60	Silty clay loam/clay loam/silt loam	1.30–1.40	0.6–2.0	0.15–0.19
Hobbs	0–8	Silt loam	1.20–1.40	0.6–2.0	0.21–0.24
	8–60	Silt loam/silty clay loam/very fine sandy loam	1.20–1.40	0.6–2.0	0.18–0.22
Gothenburg	0–4	Sandy loam	1.40–1.50	2.0–6.0	0.13–0.22
	4–60	Sand and gravel	1.70–1.90	> 0	0.02–0.04
Boel	0–12	Fine sandy loam	1.50–1.70	2.0–6.0	0.16–0.18
	12–60	Fine sand, loamy fine sand, coarse sand	1.50–1.60	6–20	0.05–0.10
Inavale	0–6	Loamy fine sand	1.50–1.60	6–20	0.10–0.12
	6–18	Fine sand, loamy sand/loam fine sand	1.50–1.60	6–20	0.06–0.11
	18–60	Fine sand, loamy sand/loam fine sand	1.50–1.60	6–20	0.05–0.10
Valentine	0–11	Fine sand	1.70–1.90	6–20	0.07–0.09
	11–60	Fine sand, loamy fine sand, loamy sand	1.70–1.90	6–20	0.05–0.11

Table 2.3. Example values of soil water characteristics for various soil textures.[a]

Soil Texture	θ_{fc}	θ_{wp}	AWC
	cm³/cm³, in/in, or m/m		
Coarse sand	0.10	0.05	0.05
Sand	0.15	0.07	0.08
Loamy sand	0.18	0.07	0.11
Sandy loam	0.20	0.08	0.12
Loam	0.25	0.10	0.15
Silt loam	0.30	0.12	0.18
Silty clay loam	0.38	0.22	0.16
Clay loam	0.40	0.25	0.15
Silty clay	0.40	0.27	0.13
Clay	0.40	0.28	0.12

[a] Example values are given. You can expect considerable variation from these values within each soil texture.

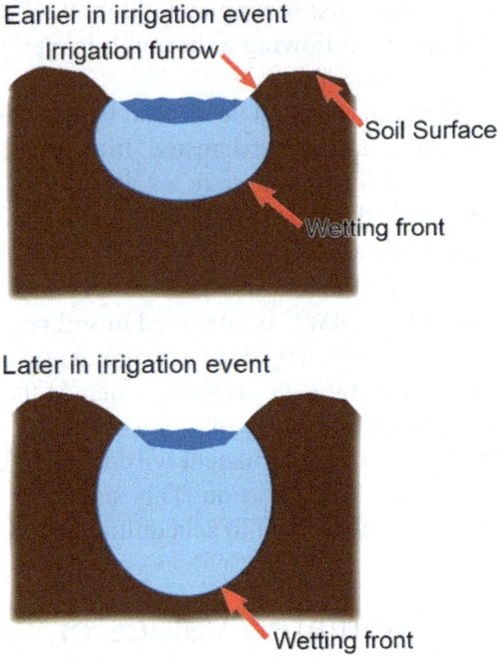

Figure 2.9. Wetting patterns early and late during a furrow irrigation water application.

2.8 Infiltration

The reservoir of water in the soil is generally replenished by the process called ***infiltration***, the entry of water through the soil surface. Infiltration is very important in irrigation since the goal is to supply water to the root zone to meet plant needs. In most cases, the goal is that all of the applied irrigation and rain enters the soil, thereby minimizing the amount of water that runs off the soil surface.

What causes water to enter the soil? Two things drive infiltration: capillarity and gravity. During the initial stages of a water application, capillary forces dominate water movement into the soil. Capillary forces work equally in all directions. Thus, capillary forces pulling water into the soil are the same in the horizontal and vertical directions. As time progresses, the capillary forces diminish, and gravity becomes the dominant force. This change in the dominant force is illustrated in Figure 2.9a where a wetted pattern under an irrigated furrow is almost semicircular in the early stages of an irrigation, but as infiltration progresses, the wetted pattern elongates in the vertical direction (Figure 2.9b). The elongation is due to the dominance of the gravitational force over capillary forces with time.

Infiltration can be described in terms of either the rate of infiltration, which is the depth of water that infiltrates per unit of time, or the cumulative amount of water infiltrating over time. Cumulative infiltration is the total depth that has infiltrated after a specific time has elapsed. The curves shown in Figure 2.10 illustrate infiltration rates with time for several soil types. This figure applies where the soil surface is ponded instantaneously as would be the case for surface irrigation (i.e., furrows, borders, and basins). The curves show that initially the infiltration rate is very high and as time progresses, or more correctly, as the amount of water that has infiltrated increases, the rate of infiltration decreases. Therefore, a decay curve

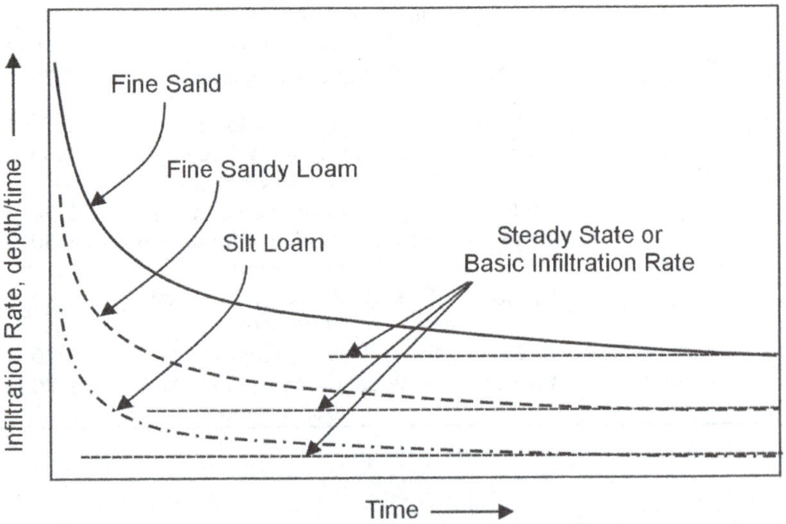

Figure 2.10. The rate of infiltration as an irrigation event proceeds and the steady rate of infiltration for three soil textures.

results with a decreasing rate of infiltration. As time continues, the infiltration rate will approach a nearly steady rate, sometimes called steady-state rate or ***basic infiltration rate*** or ***basic intake rate***. Does the infiltration rate go to zero after a long period of application? No. It can only be zero if the soil is completely impermeable or if there is no gravity (outer space).

Cumulative infiltration, or the total depth of water infiltrated over time, is shown in Figure 2.11. The curves in Figure 2.11 show that cumulative depth increases with time, but it is not a straight line. Infiltration accumulates at a fast rate early and then slows later in the irrigation or rainfall event. The slope of the curve approaches the steady-state infiltration rate shown in Figure 2.10. Be careful not to confuse the soil cumulative infiltration or depth of infiltration with the depth to which water has penetrated in the soil. View it as you would water in a rain gauge. The depth of infiltration is analogous to the depth of water in the rain gauge. It is the volume of water that is infiltrated per unit of land area.

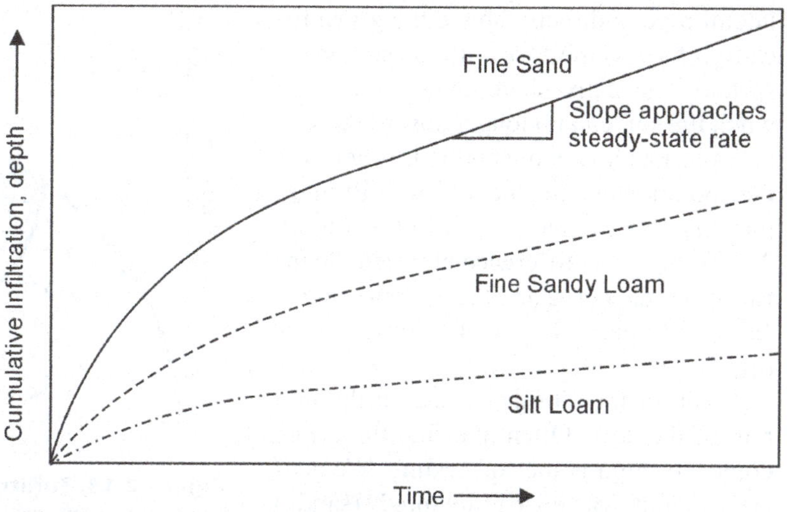

Figure 2.11. Examples of cumulative infiltration for three soil textures.

What if ponding does not occur instantaneously such as with a gentle rain or with a stationary sprinkler system that has a constant rate of application? Initially, all of the water that falls from the rain or from the sprinklers will infiltrate the soil. However, if the application period is long, the intensity of the rain or the application rate of the irrigation system may exceed the infiltration capacity of the soil. When this occurs, water will pond on the surface (surface saturation). Once the surface layer is saturated, the infiltration rate begins to follow a curve similar in shape to the ponded water case. Figure 2.12 shows a situation where a stationary sprinkler system applies water at a rate which infiltrates initially, but then, at some point, surface ponding occurs. The irrigation system now is applying water at a rate faster than can be absorbed by the soil. From that time forward, the water not infiltrated is referred to as potential runoff. There is water on the soil surface, and after all the surface depressions are filled, ***runoff*** will begin. An ideal stationary sprinkler irrigation system would be designed so that the ***application rate*** does not exceed the steady-state infiltration rate of the soil. Thus, no runoff would ever occur. The ideal irrigation system, however, is rarely achieved.

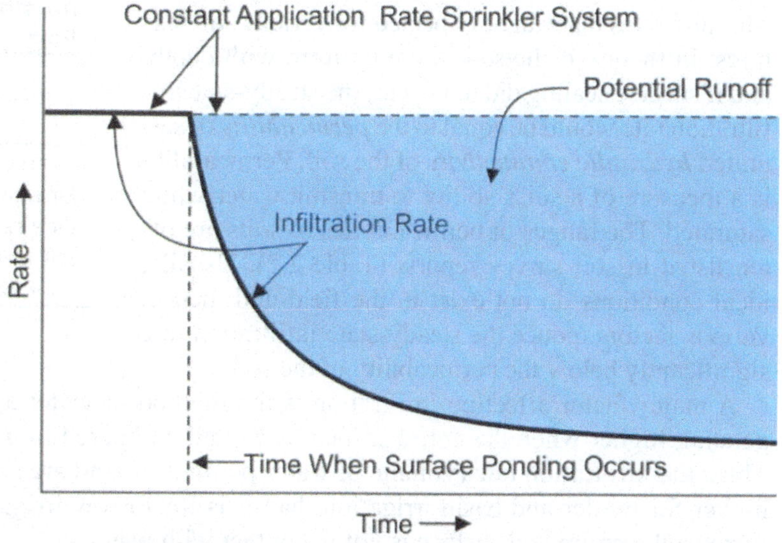

Figure 2.12. Infiltration rate over time for a constant rate of water application that eventually exceeds the infiltration rate of the soil surface.

For a moving sprinkler system, such as a center pivot or a traveling gun, the application pattern would appear similar to that shown in Figure 2.13. The application rate of the system increases with time as the irrigation system approaches a given location until it reaches a peak or maximum, after which it begins to decrease. It creates a symmetrical application rate versus time relationship in the absence of wind. For center pivots, the maximum occurs when the lateral pipe is directly above the given location. As was the case with the stationary sprinkler system, there may be a time when the soil can no longer absorb the water as fast as it is being applied. When surface ponding occurs, the rate of infiltration into the soil decreases as it did in Figure 2.11. Again, the difference between the infiltration rate curve and the system's application rate curve is potential surface runoff.

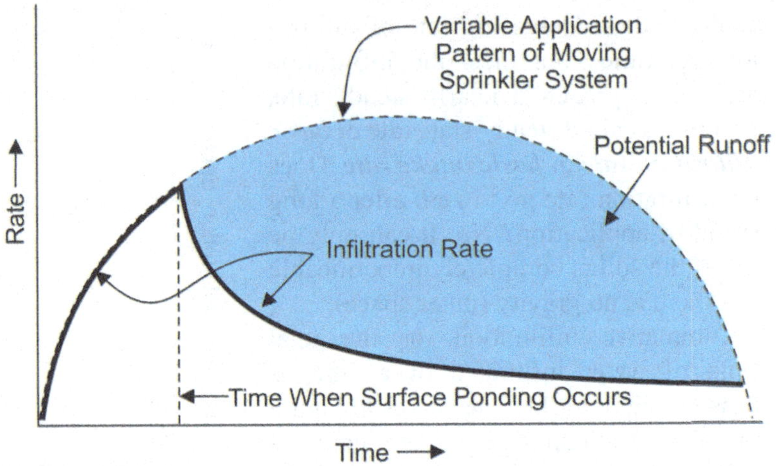

Figure 2.13. Infiltration rate as a function of time for a moving sprinkler system.

What factors influence the infiltration rate of the soil? Often, the first thing that comes to mind is the soil texture. We generally think of coarser-textured (sandy) soils having higher infiltration rates than fine- (clay) and medium-textured (loam) soils (Figures 2.10 and 2.11). Table 2.4 shows typical steady-state infiltration rates that can be expected for various soil textures. In theory, if the soils were uniform with depth, and if surface sealing did not occur, the steady-state infiltration rate would be equal to the ***permeability*** or saturated ***hydraulic conductivity*** of the soil. Permeability is a measure of a soil's ability to transmit water while saturated. The ranges of permeabilities of soils are often listed in soil survey reports (Table 2.2). Usually, ideal conditions do not exist in the field and, hence, various factors reduce the steady-state infiltration rate significantly below the permeability of the soil.

Table 2.4. Basic or steady-state infiltration rates for stationary sprinkler systems (adapted from Pair, 1983).

Soil Texture	Minimal Surface Sealing (in/hr)	Some Surface Sealing (in/hr)
Coarse sand	0.75–1.00	0.40–0.65
Fine sand	0.50–0.75	0.25–0.50
Fine sandy loam	0.35–0.50	0.15–0.30
Silt loam	0.25–0.40	0.13–0.28
Clay loam	0.10–0.30	0.05–0.25

A major factor affecting infiltration is the method of water application. Infiltration is, in general, higher when the entire surface is wetted compared to only a portion of the surface. Thus, the infiltration rate (volume of water per unit of land area per unit of time) is generally higher for border and basin irrigation than it is for furrow irrigation, because with irrigated furrows the entire soil surface is not in contact with water.

Surface sealing is another factor influencing infiltration rate. Surface sealing occurs in both surface and sprinkler irrigation. With surface irrigation the shearing effect of the flowing water causes the soil aggregates on the surface to decompose into smaller aggregates and individual particles which tend to form a thin layer with low permeability on the soil surface. It is common to find large differences between infiltration during the first irrigation event and infiltration during later irrigation events due to surface sealing.

With sprinkler irrigation and rainfall, surfacing sealing is caused by the impact of the falling water drops on any exposed soil aggregates. Again, the aggregates are broken into smaller aggregates and individual particles, thus, forming a surface seal.

Another factor that has a large influence on infiltration is soil cracking. Soils that contain fine soil particles (clays) shrink when drying and swell during wetting. The shrinking soil cracks as it dries. These cracks cause the initial infiltration rate to be high as water flows freely into them. As the soil wets, the clay particles swell and the cracks close, which causes a rapid decrease in the infiltration rate.

Tillage also has a large impact on infiltration rate and, in fact, is often performed to enhance the infiltration rate. Conservation tillage practices that leave crop residues on the soil surface can also enhance infiltration. Crop residue on the surface protects the soil from the impact of water drops from sprinkler irrigation or rainfall, thus, reducing the formation of a surface seal. Likewise, deep tillage (chiseling) is sometimes used to enhance infiltration.

Soil water content is another factor that influences infiltration. The wetter the soil, the lower the infiltration rate. The initial infiltration rate of a moist soil is, in general, lower than the initial infiltration rate of an identical dry soil. As time progresses, the infiltration rate of these two conditions will converge to the same steady-state value.

Water temperature is also known to influence infiltration rates because temperature changes the viscosity of water. As temperature increases, the viscosity decreases, hence, the infiltration rate increases (Duke, 1992). The influence of this factor is often noticed where relatively cool groundwater is applied at the head of a surface irrigated field and begins to advance across the field. On a hot day, as the water moves across the field, it is warmed. As the water warms, the infiltration rate can go up and, thus, create a differential between the infiltration rate at the inlet end and the downstream end of the field. Sometimes this phenomenon is noticed when water in furrows that has already advanced to nearly the end of the field during the early, cool part of a day may actually recede back up the field as the day progresses. This observation is often incorrectly associated with a higher rate of evaporation from the water surface. In reality, the increased water temperature has increased the infiltration rate.

2.9 Storage of Infiltrated Water

Where does the water go once it has infiltrated into the soil? How deep will it penetrate into the plant root zone? Will it penetrate beyond the root zone?

Although an oversimplification, water applied to a soil can be viewed as filling the soil profile in layers as illustrated in Figure 2.14. Even if a soil layer is wetted to saturation, it is assumed that it quickly (in a few days) drains to field capacity. The excess water (excess of FC) from a soil layer drains to the layer immediately beneath it. This sequence continues until all of the water has been stored or reaches the groundwater. Within a few days, water in each layer drains to field capacity or a lower water content.

Water that penetrates deeper than the root zone is referred to as ***deep percolation***. One goal in irrigation management is to minimize the amount of deep percolation. Deep percolation means that more water has been applied than necessary. Deep percolation transports chemicals below the root zone, a process called leaching. Leaching

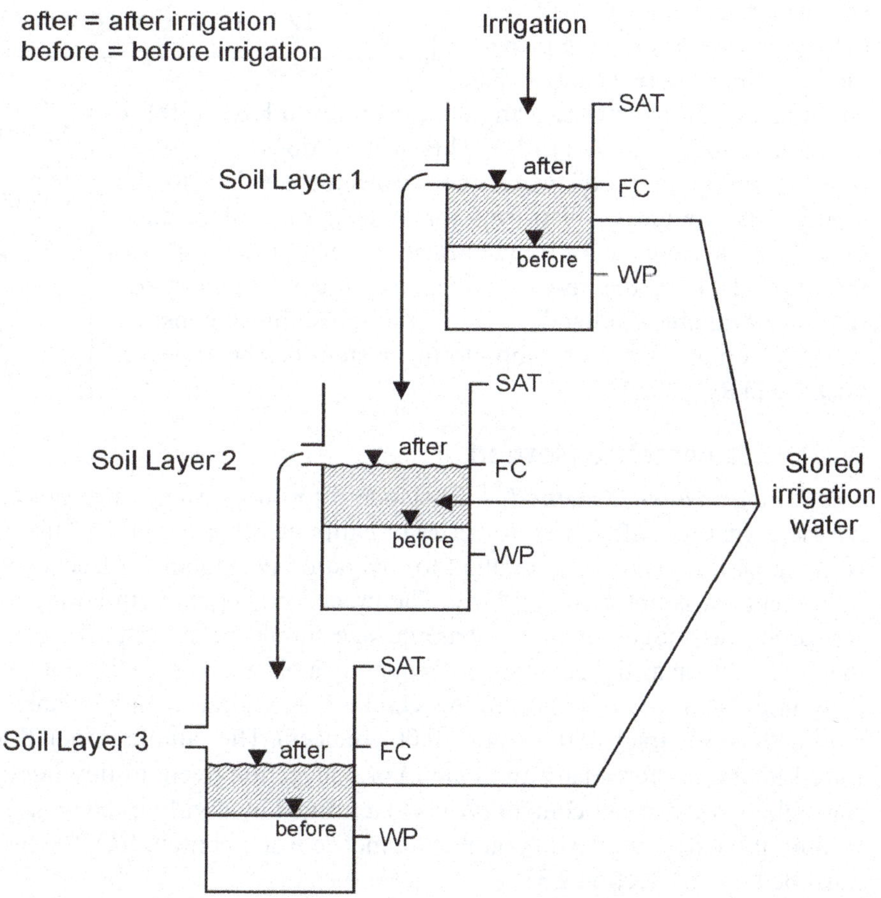

Figure 2.14. "Simplified" storage of water infiltrating into a soil profile with three layers.

is generally unwanted but is sometimes required to remove excess salts from the root zone (as will be discussed in Chapter 7).

Example 2.4

Given a soil with the following characteristics, calculate the depth to which 4 inches of infiltrated water would penetrate.

Layer	Depth (in)	θ_{fc}	θ_v
1	0-12	0.34	0.20
2	12-30	0.40	0.33
3	30+	0.30	0.24

Solution:

Using Equation 2.13:

$SWD_1 = (0.34-0.20)\,12 \text{ in} = 1.7 \text{ in}$

$SWD_2 = (0.40-0.33)\,18 \text{ in} = 1.3 \text{ in}$

1.7 in + 1.3 in = 3.0 in is required to fill the first two layers

The remaining water is:

4.0 in – 3.0 in = 1.0 in

To find the depth penetrated in the third layer (L_3) use the same equation, but solve for L_3 when SWD_3 = 1.0 in:

$$L_3 = \frac{1.0}{(0.30-0.24)} = 16.7 \text{ in}$$

The depth from the surface penetrated by a 4-inch application is then 46.7 in:

12 in + 18 in + 16.7 in = 46.7 in

2.10 Measuring Soil Water Content and Matric Potential

Measuring soil water content and matric potential is important in irrigation management. Measuring soil water content is useful for determining whether soil water content is being kept within allowable bounds (to be discussed in Chapter 6), when the next irrigation should occur, and how much water the soil can hold without deep percolation. Many methods are available for measuring soil water content. We will only discuss a few of the more proven methods. For more detailed discussions of soil water measuring devices, refer to Evett (2007), Gardner (1986), or Ley (1994). This section does not discuss systems for logging soil water data, transmitting data to the cloud, data storage, or platforms for viewing and interpreting data. These systems make it much easier to incorporate soil water sensor data into day-to-day farm management, and many options are available from industry. A list of questions to consider when selecting a soil water monitoring system has been developed by ITRC (2019).

Example 2.5

From the information in Example 2.4, calculate the depth of water that was lost by deep percolation if the depth of the crop root zone is 36 inches.

Solution:

46.7 in – 36 in = 10.7 in

d = (0.30 – 0.24)10.7 in = 0.6 in

2.10.1 Gravimetric Method

The ***gravimetric method*** is the standard for measuring soil water content. By standard, we mean that it is often used to verify or calibrate other methods. This does not mean that it is the most frequently used method for irrigation management. Because of its high labor requirements, it is not used regularly. The procedure begins with taking a soil sample using a soil probe, soil auger, or shovel. Sample size should be at least 100 g (¼ pound). The soil is then sealed in an airtight container (frequently a plastic bag) so that moisture is not lost before weighing. Next, the wet mass of the sample is measured with a balance or scale that can be read with an accuracy of 0.5 grams (0.004 ounces). The sample is then dried at 105°C (220°F) for 24 hours in a forced air (preferable) or convection oven. Following drying, the sample is reweighed. Mass water content (θ_m) is determined by dividing the weight of the water by the weight of the dry soil. To determine volumetric water content (θ_v), the bulk density of the soil must be known (Section 2.3).

Gardner (1986) describes a method using a microwave oven for drying, which is helpful when results are needed quickly. The drying time is dependent on the initial water content and sample size. A typical drying time ranges from 10 to 30 minutes. A precaution is that a rapid rise in temperature occurs in the sample once the moisture has been driven out. If the temperature gets too high, some organic matter may burn which, in the calculations, might be erroneously mistaken for water loss.

Since sampling locations are not fixed or permanent, the gravimetric approach has the advantage that soil samples can be taken at any desired location within the field or irrigated area each time sampling occurs. Also, the method can give an accurate measure of volumetric water content, within 1%, if bulk density is known and a reliable balance is used for weighing. A disadvantage is the labor required to take the soil samples, especially at deeper depths. Another disadvantage is that the results are not immediately available.

2.10.2 Feel and Appearance

The ***feel and appearance method*** also requires the collection of soil samples at the desired depths. The soil sample is crumbled into small pieces and then squeezed by hand to form a ball. The cohesiveness of the ball is an indication of the soil's wetness. Also, whether it leaves an imprint in the palm of the hand after squeezing should be noted. The soil is then ribboned out between the thumb and the forefinger. Table 2.5 provides a detailed explanation of how to interpret the soil water content by the feel and appearance method.

This method requires a great deal of judgement and experience for good estimates of soil water. Nevertheless, it is widely used. Experienced users probably achieve an accuracy of f_r plus or minus 0.10. Thus, if estimated $f_r = 0.55$, the true value probably ranges from 0.45 to 0.65. This method is low in cost and allows moisture measurements to be taken quickly at multiple locations in the field. Considering the spatial variability of soil water in a field, the method can be adequate for irrigation management, especially if measurements are checked against a more accurate method periodically. A major disadvantage of this method is the need for experience before confidence is gained and accuracy is achieved.

Table 2.5. Guide for judging how much water is available for crops (taken from USDA, 1972).

Fraction of Available Soil Water Remaining	Feel or Appearance of Soil			
	Loamy Sand or Sand	Sandy Loam	Loam and Silt Loam	Clay Loam or Silty Clay Loam
0 Wilting point	Dry, loose, single grained, flows through fingers.	Dry, loose, flows through fingers.	Powdery dry, sometimes slightly crusted but easily broken down into powdery condition.	Hard, baked, cracked, sometimes has loose crumbs on surface.
0.25	Appears to be dry, will not form a ball with pressure.	Appears to be dry, will not form a ball.	Somewhat crumbly but holds together from pressure.	Somewhat pliable, will ball under pressure.
0.50	Appears to be dry, will not form a ball with pressure.	Tends to ball under pressure but seldom holds together.	Forms a ball somewhat plastic, will sometimes slick slightly with pressure.	Forms a ball, ribbons out between thumb and forefinger.
0.75	Tends to stick together slightly, sometimes forms a very weak ball under pressure.	Forms weak ball, breaks easily, will not slick.	Forms a ball, is very pliable, slicks readily.	Easily ribbons out between fingers, has slick feeling.
1 Field capacity	Upon squeezing, no free water appears on soil but wet outline of ball is left on hand.	Upon squeezing, no free water appears on soil but wet outline of ball is left on hand.	Upon squeezing, no free water appears on soil but wet outline of ball is left on hand.	Upon squeezing, no free water appears on soil but wet outline of ball is left on hand.

Note: Ball is formed by squeezing a handful of soil very firmly.

2.10.3 Neutron Scattering

An accurate method for measuring soil water is the neutron scattering or attenuation technique, which uses an instrument called a ***neutron probe***. With this method, a radioactive source is lowered into an access tube installed vertically into the soil (Figure 2.15). The source is lowered to the desired depth of measurement and emits neutrons traveling at high speed. The speed of the neutrons is attenuated or slowed by hydrogen ions present in soil water. The rate of attenuation is dependent on the amount of water present. A detector, located near the source, counts the number of slow-moving neutrons over a short count period, 30 seconds to 2 minutes. There is a good correlation between the count of slow-moving neutrons and θ_v.

An advantage of this method is the size of the soil volume sensed by the instrument. In effect, the probe samples a sphere with a diameter of 6 to 10 in, depending on soil water content. Neutron probes are also more accurate (within 1%) than most other soil water sensors. Disadvantages of the method include: (1) high initial cost, (2) a license is required to operate an instrument that is radioactive, (3) a calibration curve (Figure 2.16) must be developed for a given access tube material (usually aluminum, steel or polyvinyl chloride plastic) and for the soil of interest, (4) measurements within the top 6 to 8 in of soil are not reliable and require a separate calibration, and (5) measurements can only be made where the access tubes have been installed. The last item can be an advantage if repeated measurements at the same location in the field are desired. Neutron probes are often used in irrigation research.

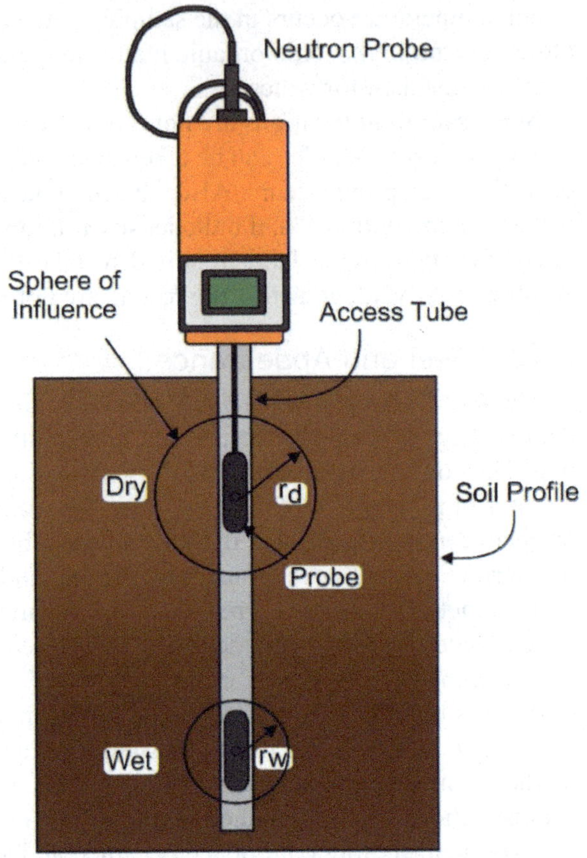

Figure 2.15. Neutron attenuation method for measuring volumetric water content in a soil profile.

2.10.4 Time Domain Reflectometry

One soil water measurement technique that takes advantage of the fact that a soil's apparent dielectric permittivity (ε_a) is dependent on θ_v is ***time domain reflectometry*** (TDR). TDR requires the placement of two parallel rods (wave guides) into the soil. An electromagnetic wave is pulsed along the wave guides. The reflected signal from the tip of the wave guide is captured with a fast oscilloscope, recording voltage as a function of time. The travel time of the recorded wave must be calculated with a graphical interpretation of the waveform (with software) as part of the TDR method (Evett, 2007). The travel time provides a direct measurement of ε_a. The wave will travel faster in a dry soil than in a wet soil, with a lower travel time and a lower ε_a. The ε_a is comprised of the permittivity of the water, the permittivity of the soil, and the permittivity of air, and the water has a much larger influence on ε_a than the soil or air. Therefore, ε_a is directly proportional to θ_v. Because of the strong correlation between ε_a and θ_v, TDR is an accurate method for sensing θ_v (within 2%). Its use was initially limited to research due to high costs, but ongoing technology development is reducing the price of TDR sensors. It

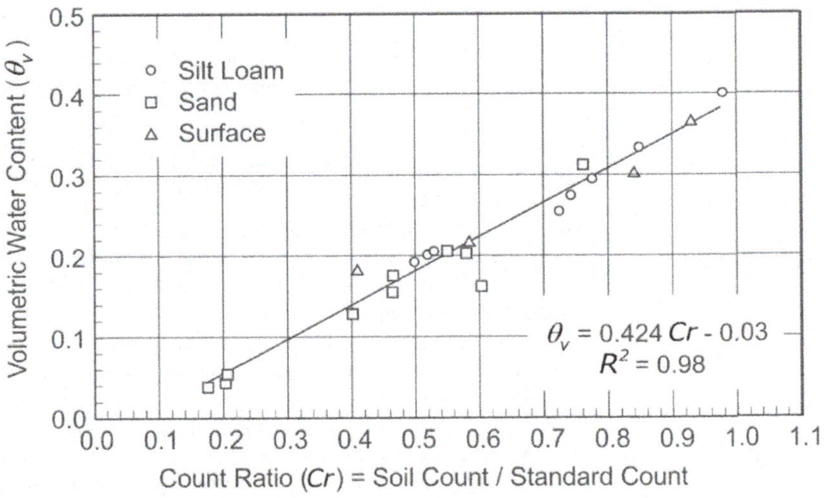

Figure 2.16. Calibration curve for a neutron probe.

has the advantage of not using a radioactive source, so licensing is not required. The measurement volume is approximately cylindrical and is dependent on the length of the rods and the spacing between rods. The diameter of the cylinder is about 1.5 times the spacing between rods.

2.10.5 Capacitance Probes

Similar to TDR, capacitance probes also take advantage of the correlation between ε_a and θ_v. However, instead of measuring ε_a directly with travel time, it is estimated indirectly by quantifying capacitance and frequency, which is why these sensors are known as ***capacitance probes*** (or ***frequency domain reflectometry***). This method uses the soil as a dielectric and measures the capacitance of the soil (Evett, 2007). The capacitance circuit is pulsed with high-frequency radio waves. A natural resonant frequency is established which is dependent on the capacitance. The measured frequency is used to calculate the capacitance, which is used to determine the ε_a, which is correlated to θ_v. Capacitance probes can be an easy-to-use option for monitoring trends in θ_v; however, for accurate determination of the magnitude of θ_v, capacitance probes are highly dependent on a calibration for the specific soil in which it is installed.

There are two forms of capacitance probes. One form has two or three electrodes which are inserted directly into the soil. The probe can be permanently installed at the desired depth in the soil profile, or it can be a portable device with the electrodes inserted at the soil surface. The measurement volume is dependent on the length and spacing of the electrodes. The second form requires an access tube (Rudnick et al., 2016), similar to neutron scattering. This allows soil water to be determined at multiple depths in the soil profile; however, the sensing volume is much smaller since the sensor is not in direct contact with the soil. Several types of capacitance probes are produced by industry for irrigation management.

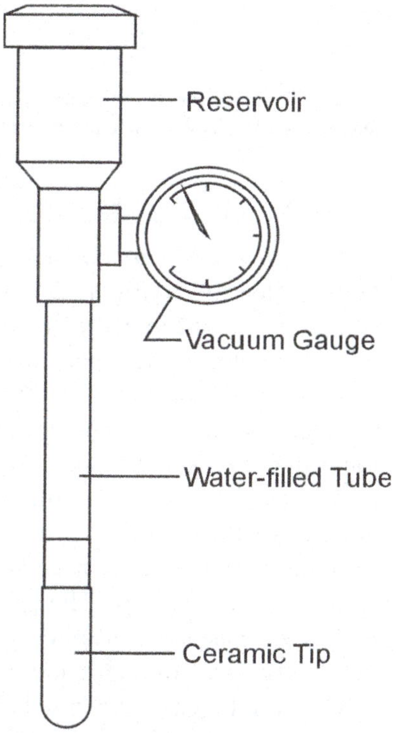

Figure 2.17. Components of a typical tensiometer.

2.10.6 Tensiometers

Soil water tension (matric potential, ψ_m) can be measured by several methods. The oldest tool, and one that measures tension directly, is the ***tensiometer***. Tensiometers (Figure 2.17) have three components: a water filled tube (usually transparent); a porous cup (usually ceramic) at one end of the tube; and a vacuum gauge (or manometer) at the other end. The tube is sealed at the gauge end. The tensiometer is installed in the field so that the porous cup is at the desired soil depth. The cup must have direct contact with the surrounding soil so that the water in the cup is hydraulically connected to the water in the soil. As the soil dries, water is "pulled" out of the tensiometer. Since the tube is sealed at the gauge end, vacuum increases in the tube as water is being pulled out. Flow continues until there is equilibrium between the water in the tensiometer and the soil water. The vacuum gauge is a direct indicator of soil water tension. Usually, the vacuum is registered in centibars (cb) and the scale reads from 0 to100 cb. As the tension or vacuum approaches 1 bar, dissolved air in the water is released. The air accumulates in the top of the tube. When this happens, the readings are no longer reliable. Thus, the practical operating range for this instrument is 0 to 75 cb. A zero reading corresponds to a saturated soil, while a reading of 8 cb corresponds to FC for fine sand soils and a reading of about 20 cb is FC for silt loam soils, as shown in Figure 2.6. By using Figure 2.6, you should be able to demonstrate that, for fine sand, about 70% of the AWC has been depleted at 75 cb (the upper limit of the instrument), but only about 45% of the AWC has been depleted for silt loam at 75 cb. As will be discussed in Chapter 6, a common criterion for irrigation is to allow up to 50% depletion of the AWC before irrigation. This criterion indicates why the tensiometer has some limitations for irrigation management on finer-textured soils.

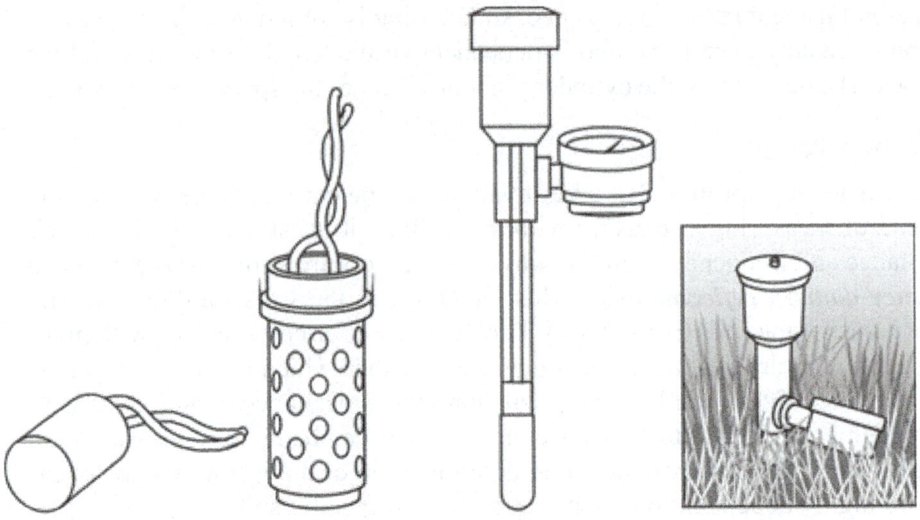

Figure 2.18. Various sensors for measuring soil water tension (from left to right): a gypsum electrical resistance block, a granular matrix sensor, a tensiometer, and a tensiometer installed in the soil.

2.10.7 Electrical Resistance Blocks and Granular Matrix Sensors

Electrical resistance blocks consist of a porous material, usually gypsum, with two embedded electrodes (Figure 2.18). The blocks are buried at the desired soil depth. As with tensiometers, good contact with the surrounding soil is essential. When the soil water equilibrates with the water in the block, an ohmmeter with an AC current source can be used to measure electrical resistance between the electrodes. There is a relationship between the measured resistance and the water content of the gypsum, and the water tension in the gypsum is equal to the water tension in the soil. Therefore, the soil water tension (Ψ_m) and the measured electrical resistance are related. You might ask, why not just embed the electrodes directly into the soil and bypass the use of the gypsum? The problem with this approach is the effect of electrolytes in the soil on the resistance. Thus, electrical resistance in the soil is dependent on both soil water and soil salinity. The gypsum somewhat buffers the effect of the salts in the soil on observed resistance. In saline soils the effect of salts on the measured resistance cause inaccurate estimates of matric potential.

Gypsum blocks have largely been replaced by ***granular matrix sensors***. One limitation of resistance blocks is that the gypsum matrix is a very fine material. Thus, the usable range is limited to high soil water tensions, usually greater than 50 cb. To overcome the limitation of gypsum blocks to the wet range, blocks composed of a coarser media, such as sand, have been developed. These coarser blocks, referred to as granular matrix sensors, have a usable range of 5 to 200 cb (Evett, 2007). Granular matrix sensors have a longer usable life than resistance blocks. Another advantage is that granular matrix sensors are low cost compared to most other soil water sensors. The low cost makes it possible to install a large number of sensors in a field, in order to better account for spatial variability in soils. Also, on a small scale (cm to m), the spatial variability in Ψ_m is somewhat lower compared to θ_v, so a measurement of soil water tension may represent a larger volume of soil than a θ_v sensor with a relatively small measurement volume. A disadvantage is that, if θ_v is desired, a soil water release curve is needed to convert Ψ_m to θ_v, which introduces more uncertainty along with the normal uncertainty from soil water sensor data. For this reason, irrigation scheduling based on granular matrix sensors often uses Ψ_m directly, comparing it to a threshold Ψ_m where crop stress would be expected to occur. University extension guides have been developed with specific guidance on using granular matrix sensors (Irmak et al., 2016).

2.10.8 Thermal Dissipation Blocks

Another, less common, approach that uses porous blocks has a heater and temperature sensor embedded within the block. The porous block installation must result in good contact with the soil, allowing water tension in the block to come into equilibrium with the water tension in the surrounding soil. The block is heated by passing current through the heater. The rate of heat dissipation in the block is then measured. The rate of heat dissipation is directly related to the water content in the block, and, since the porous block has a known release curve, the water content of the block is directly related to water potential in the block and soil. An advantage of heat dissipation blocks is that they are sensitive to soil water over a wide range. Unfortunately, the heat dissipation blocks must be individually calibrated and they are considerably more expensive than granular matrix sensors. A potential application of this concept is using heated cables to determine water content at many locations along the cable (Sayde, 2010). Since the heat dissipation occurs in the soil (instead of a block), heat dissipation would be correlated to θ_v instead of Ψ_m. This method is still under development.

2.10.9 Placement of Soil Water Sensors

The above methods of soil water measurement require that representative sites be selected for sampling. This means that sampling must consider the variability in soils, the variability of water applications, and the variability of plant populations within the irrigated area. The microclimate around the area to be measured should also be considered. This is especially important in landscape applications where buildings and streets can greatly affect the environment surrounding the irrigated plants.

Soil water measurements must be taken at depths that represent the plant root zone. Estimates of soil water content secured with the heel of your boot usually are inadequate to describe what is really happening within the plant's root system. In Chapter 6, root zone depths for various crops will be presented. Installation of soil water sensors should be done with great care, since a good installation is required in order to obtain high-quality data.

One of the frustrations of measuring soil water is the large number of samples that are required before you feel comfortable with how well the measurements represent the soil water conditions in the irrigated area. Because of natural variability of soil properties and the variability in depth of rainfall and irrigation applications within the irrigated area, considerable variability in measured soil water can be expected. Another problem is the number of locations that must be monitored to truly represent the plant root zone. A minimum of two soil depths should be measured, and often three or four are required to properly represent root zone soil water conditions. One approach to reduce the uncertainty in soil water data is to focus more on the trends over time rather than the magnitude of the measurement. If possible, determine the FC from the sensor data after it is installed (looking at the trend after a large wetting event which saturates the soil profile), and track the amount of depletion below FC (i.e., calculate SWD) over time.

2.10.10 Remote Sensing

Ongoing research is investigating remote sensing as a possible method to determine θ_v for irrigation management. Generally speaking, this could include satellite remote sensing, airborne remote sensing (manned or unmanned aircraft), and proximal remote sensing (sensor placed near the crop). For above-ground, moving irrigation systems, proximal remote sensing can include sensors mounted on the irrigation system itself (e.g., mounted on the lateral of a center pivot). The primary advantage provided by remote sensing is the spatial dataset, allowing the user to quantify spatial patterns in soil water. The primary limitation is that measurement of θ_v with remote sensing is typically restricted to a shallow layer of soil at the soil surface (i.e., top 10 to 25 cm), which can be problematic if using irrigation to manage the entire root zone for deep-rooted crops (e.g., 100 cm). One proximal technology is the cosmic-ray neutron probe (Franz et al., 2015); ongoing research is estimating root zone θ_v based on shallow θ_v from the cosmic-ray neutron probe (Franz et al., 2020). Microwave remote sensing can also be used to estimate θ_v, with sensors mounted on a center pivot (Qiao et al., 2016). The Soil Moisture Active Passive (SMAP) satellite (U.S. National Aeronautics and Space Administration) and the Soil Moisture and Ocean Salinity (SMOS) satellite (European Space Agency) are examples of satellites which are used to produce soil water data products (Al-Yaari et al., 2019).

2.11 Summary

One of the most important functions of a soil is to serve as a reservoir for storing precipitation and irrigation water for use by plants. Water is stored in the void spaces between soil particles. When the voids are filled with water, the soil is said to be saturated. A saturated soil rapidly drains to a, more or less, constant moisture level called field capacity (FC). Plants can extract water from the soil until the soil water content reaches the permanent wilting point (WP). At permanent wilt, plants will not recover even if the soil is rewet. The difference between the water content at field capacity and wilting point is called available water capacity (AWC). Finer-textured soils have a higher AWC than coarser-textured soils. AWC ranges from 0.05 to 0.25 inches of water per inch of soil.

The equivalent depth of water stored in a soil layer of known thickness can be determined if the volumetric water content (θ_v) is known. Total available water capacity (TAW) of the root zone is the product of the root zone depth and the AWC. Soil water deficit (SWD) is the amount of water that has been depleted from the TAW. Allowable depletion (AD) is the maximum soil water deficit that should occur before water is applied. Plant water stress will occur if SWD exceeds AD.

Water enters the soil by infiltration which is due to capillary and gravitational forces. The process is affected by the method of irrigation. When cumulative infiltration exceeds SWD, water percolates below the reach of plant roots and, simultaneously, leaches dissolved salts and chemicals from the root zone.

Soil water content can be determined by a variety of methods. The simplest and least accurate is by feel. The standard method is collecting soil samples and weighing them before and after oven drying. There are a number of sensors or devices that are buried in the soil from which readings are made to infer soil water content or soil water potential.

Questions

1. Would a fine-textured soil have a higher or lower available water capacity than a coarse-textured soil? How would the bulk densities compare? Explain why the differences, if any, occur.
2. Describe total soil water potential and its components. Why is matric potential important in irrigation management?
3. Will water infiltrate into the soil even if the root zone is at field capacity? If so, where will the water be stored? Explain.
4. Repeat Question 3 for a saturated soil.
5. Show mathematically how Equation 2.13 is derived from Equation 2.12.
6. The use of wetting agents has often been suggested to enhance infiltration. Wetting agents act by reducing the surface tension of the liquid. What effect does this have on capillary forces and infiltration? Explain.

7. An irrigation of 2.5 in of infiltration is followed by 1 in of rainfall infiltration. If a clay loam soil had a 50% depletion of the available water ($f_d = 0.5$) prior to the water application and the root zone depth is 30 in, how much water would deep percolate?

8. If the average count ratio for neutron scattering measurements was 1.0, how much water needs to be infiltrated to bring a silt loam soil to field capacity?

9. A soil to be irrigated has two layers: the top layer is a silt loam 12 in deep and the other is a silty clay with a thickness of 3 ft. If both layers were at the permanent wilting point, how much water could be applied without water draining below the first layer? Second layer?

10. If you install a granular matrix sensor into the silt loam soil depicted in Figure 2.6 and the reading was 35 cb, what would be the volumetric water content of this soil? If the bulk density of the soil was 1.35 g/cm^3, what would be the mass water content?

11. If the average count ratio from a series of measurements with the neutron probe whose calibration is shown in Figure 2.16 on a golf course was 0.5 before irrigating and 0.9 after irrigating, how much water was added to a soil profile 1 ft deep?

12. A soil sample was taken just prior to irrigation and weighed wet then dried and re-weighed.

    The following data were obtained:

    Wet mass = 240 g
    Dry mass = 200 g

    The soil has the following characteristics:

    $\theta_{fc} = 0.30$ cm^3/cm^3
    $\theta_{wp} = 0.10$ cm^3/cm^3
    $\rho_b = 1.25$ g/cm^3

    Determine the following:

    θ_m = mass water content
    θ_v = volumetric water content
    f_r = fraction of available water remaining
    f_d = fraction of available water depleted

13. Tensiometers are placed in a fine sandy loam soil at depths of 6, 18, and 30 in. The following readings were taken:

Depth (ft)	Tensiometer Reading (cb)
6	30
18	70
30	50

Use Figure 2.6 to help answer the following questions. Assume each tensiometer reading represents 1 ft of soil.

a. Determine the available soil water remaining at each depth (in/ft).

b. Determine the total available soil water remaining, in inches, in the 3-ft profile.

c. What is the fraction depleted in each layer?

d. How much water would have to be applied to bring the soil water level to field capacity to a depth of 3 ft?

14. Answer the following:

a. Using the soil water release curves shown in Figure 2.6, determine the expected soil moisture tensions at $f_d = 0.3$ and $f_d = 0.6$ for the silt loam soil.

b. Repeat (a) for the fine sand.

c. Would a tensiometer work properly for the four cases in (a) and (b)? Explain your answer.

15. The feel and appearance method was used to estimate the soil water in the following layers:

Depth (ft)	f_r
0–1	0.55
1–2	0.65
2–3	0.60
3–4	0.80
4–5	0.80
5–6	0.80
6–7	0.80

a. If the volumetric water content at field capacity and wilting point is 0.35 and 0.13, respectively, how deep would 4.2 in of infiltrated water penetrate into the soil profile? Give your answer in inches.

b. If the root zone depth is 30 in, how many inches of water would percolate below the root zone?

References

Al-Yaari, A., Wigneron, J. P., Dorigo, W., Colliander, A., Pellarin, T., Hahn, S., Mialon, A., Richaume, P., Fernandez-Moran, R., Fan, L., Kerr, Y. H., & De Lannoy, G. (2019). Assessment and inter-comparison of recently developed/reprocessed microwave satellite soil moisture products using ISMN ground-based measurements. *Remote Sens. Environ., 224*, 289-303.

Bruce, R. R., & Luxmoore, R. J. (1986). Water retention: Field methods. In A. Klute (Ed.), *Methods of soil analysis. Part I. Physical and mineralogical methods.* Agron. Monogr., 9 (2nd ed., pp. 663-686). Madison, WI: ASA, CSSA, SSSA.

Duke, H. R. (1992). Water temperature fluctuations and effect on irrigation infiltration. *Trans. ASAE, 35*(1), 193-199.

Evett, S. R. (2007). Soil water and monitoring technology. In R. J. Lascanood, & R. E. Sojka (Eds.), *Irrigation of agricultural crops.* Agronomy No. 30. Madison, WI: ASA.

Franz, T. E., Wahbi, A., Zhang, J., Vreugdenhil, M., Heng, L., Dercon, G., & Wagner, W. (2020). Practical data products from cosmic-ray neutron sensing for hydrological applications. *Frontiers Water, 2*, 9.

Franz, T. E., Wang, T., Avery, W., Finkenbiner, C., & Brocca, L. (2015). Combined analysis of soil moisture measurements from roving and fixed cosmic ray neutron probes for multiscale real-time monitoring. *Geophys. Res. Lett., 42*(9), 3389-3396.

Gardner, W. H. (1986). Water content. In A. Klute (Ed.), *Methods of soil analysis. Part I. Physical and mineralogical methods.* Agron. Monogr. 9 (2nd ed., pp. 493-544). Madison, WI: ASA, CSSA, SSSA.

Hillel, D. (1980). *Fundamentals of soil physics.* New York, NY: Academic Press.

Irmak, S., Payero, J. O., VanDeWalle, B., Reese, J., Zoubek, G., Martin, D. L., Kranz, W. L., Eisenhauer, D. E., & Leininger, D. (2016). Principles and operational characteristics of Watermark granular matrix sensor to measure soil water status and its practical applications for irrigation management in various soil textures. Ext. Circular EC783. Lincoln: Nebraska Extension.

ITRC. (2019). Irrigation Consumer Bill of Rights™: Soil and plant moisture monitoring systems. ITRC Report No. R 19-004. San Luis Obispo, CA: Irrigation Training and Research Center, California Polytechnic State University.

Klute, A. (1986). Water retention: Laboratory methods. In A. Klute (Ed.), *Methods of soil analysis. Part I. Physical and mineralogical methods.* Agron. Monogr. 9 (2nd ed., pp. 635-662). Madison, WI: ASA, CSSA, SSSA.

Ley, T. W. (1994). An in-depth look at soil water monitoring and measurement tools. *Irrig. J., 44*(8), 8-20.

Martin, D. L., Kranz, W. L., Smith, T., Irmak, S., Burr, C., & Yoder, R. (2017). Center pivot irrigation handbook. Ext. Circular EC3017. Lincoln: Nebraska Extension.

Pair, C. H. (ed.). (1983). *Irrigation.* Silver Spring, MD: Irrigation Association.

Qiao, X., Khalilian, A., Payero, J. O., Maja. J., M., Privette, C. V., & Han, Y. J. (2016). Evaluating reflected GPS signal as a potential tool for cotton irrigation scheduling. *Adv. Remote Sensing, 5*, 157-167.

Rudnick, D. R., Irmak, S., Haghverdi, A., Burr, C. A., Shaver, T. M., & Dorr, T. J. (2016). Soil water sensors for irrigation management. Ext. Circular EC3002. Lincoln: Nebraska Extension.

Sayde, C., C. Gregory, M. Gil-Rodriguez, N. Tufillaro, S. Tyler, N. van de Giesen, M. English, R. Cuenca, and J. S. Selker. (2010). Feasibility of soil moisture monitoring with heated fiber optics. *Water Resour. Res.* 46, W06201. doi:10.1029/2009WR007846.

USDA. (1972). Irrigation Guide for Nebraska. Washington, DC: USDA Soil Conservation Service.

USDA. (1988). Soil Survey of Platte County, Nebraska. Washington, DC: USDA Soil Conservation Service.

Chapter 3
Measuring Water Applications

3.1 Introduction

3.1.1 Need for Water Measurement

In Chapter 2, we discussed soil water storage and related this storage to an "equivalent depth" of water on the soil surface. You can envision this "equivalent depth" as the water measured in a rain gauge. In Chapter 4 we will present how this depth relates to plant water needs. In this chapter we discuss how we can determine the depth of water applied with an irrigation system.

Have you ever wondered what it would be like to drive an automobile without a speedometer and an odometer? You might feel somewhat lost. You would not know how fast you were going, nor how far you've traveled. Irrigating without water measurement is much the same way. Without knowing the water flow rate, you do not know how fast you are applying water. And, without measured volumes, you cannot determine the depth of application. Good water management begins with accurate water measurement. Unfortunately, because of the regulatory implications, some water users have an unfavorable attitude towards water measurement. Good water managers use water measurement to evaluate how efficiently they are using the water that they apply.

Energy management is another reason to measure water. To evaluate the energy efficiency of pumping systems, you need to know both the energy input and the output from the pumping system. The output includes the water flow rate.

3.1.2 Depth Volume Relationships

Irrigators commonly measure and discuss rainfall depth. Since irrigation is artificial rainfall, it is also useful to express irrigation water application as a depth. Equations 3.1 and 3.2 relate the depth applied and applied volume to the land area irrigated:

$$d = \frac{V}{A} \quad (3.1)$$

$$V = A \times d \quad (3.2)$$

where: d = depth of water applied,

V = volume of water applied, and

A = area irrigated.

The concepts of Equations 3.1 and 3.2 are shown in Figure 3.1.

Since the volume of water applied is the product of system flow rate and the time of application, Equation 3.2 is often expressed as:

$$Q \times t = A \times d \quad (3.3)$$

where: Q = system flow rate and

t = time of water application.

This equation assumes that the flow rate is constant over the

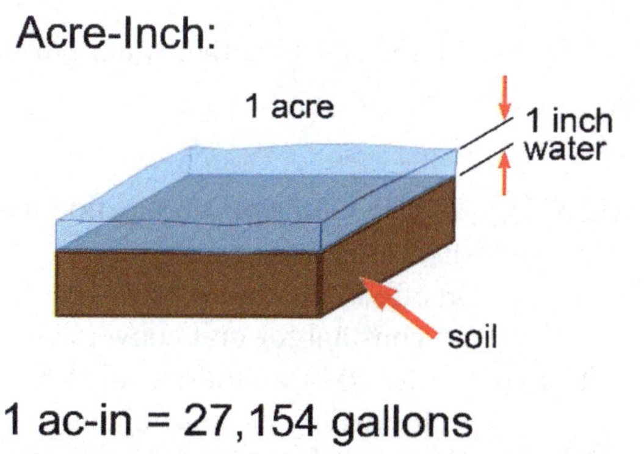

Figure 3.1. Relationship between volume of water applied, land area, and depth applied.

Table 3.1. Conversion factors used in water measurement.

U.S. Customary System of Units (USCS)	Conversion between USCS and SI Systems	International System of Units (SI)
	Volume Units	
1 gal = 8.33 lb 1 ft^3 = 7.48 gal 1 ac-in = 3,630 ft^3 1 ac-in = 27,154 gal 1 ac-ft = 43,560 ft^3 1 ac-ft = 325,851 gal	1 ft^3 = 0.02832 m^3 1 L = 0.264 gal 1 gal = 3.79 L 1 m^3 = 264.2 gal 1 ac-in = 1.028 ha-cm	1 L = 1,000 cm^3 1 cm^3 = 0.001 L 1 m^3 = 1,000 L 1 L = 0.001 m^3 1 ha-cm = 100 m^3
	Flow Units	
1 cfs = 449 gpm (450 for practical purposes) 1 cfs = 1 ac-in/hr 452 gpm (450 for practical purposes) = 1 ac-in/hr 1 gpm = 0.00223 cfs 1 gpm = 0.00221 ac-in/hr	1 cfs = 0.02832 cms 1 cms = 35.31 cfs 1 gpm = 0.06309 L/s 1 L/s = 15.85 gpm 1 gal/h = 63.1 mL/s	1 L/s = 1,000 mL/s 1 cms = 16.7 L/min
	Length Units	
1 mile = 5280 ft 1 rod = 16.5 ft	1 ft = 0.3048 meters 1 meter = 3.281 ft	1 cm = 0.01 meter 1 meter = 100 cm 1 km = 1,000 meter
	Area Units	
1 ac = 43,560 ft^2	1 ac = 0.4047 ha 1 ha = 2.471 ac	1 ha = 10,000 m^2

cfs = cubic feet per second; gpm = gallons per minute
cms = cubic meters per second; L/s = liters per second

application time. Use of Equation 3.3 is basic to efficient irrigation management. Although straightforward, Equation 3.3 requires that the user apply the appropriate conversion factors to make the units consistent. When using Equation 3.1 and 3.2, it is most convenient to convert volume units to acre-inch (ac-in) when working in agriculture. Table 3.1 lists common unit conversion factors for volume and flow rate. When using U.S. units, approximately 450 gallons per minute (gpm) equals 1 ac-in/hr. An ac-in is the volume of water that covers 1 acre 1 inch deep. Before using Equation 3.3 for an agricultural application, the system flow rate (Q) should be converted to ac-in/hr. One ac-in/hr is also equal to 1 cubic foot per second (cfs), another common flow unit used in agricultural irrigation.

Equation 3.3 can be rearranged to calculate both depth per unit of time (d/t) and average application intensity.

$$\frac{d}{t} = \frac{Q}{A} \tag{3.4}$$

For sprinkler heads and other water emitters, Equation 3.4 often takes the form:

$$A_r = \frac{96.3\, q}{A} \tag{3.5}$$

where: A_r = ***application rate***, application intensity, or precipitation rate (in/hr),
q = discharge rate (gpm),
A = in effect, the area irrigated by the device (ft^2), and
96.3 = constant for unit conversion.

The area irrigated by an individual sprinkler head equals the spacing between heads on the lateral (ft) multiplied by the spacing between laterals (ft).

The flow rate (Q) is a volume per unit of time and can be metered with flow measuring devices. For a known flow rate, Equation 3.3 can then be used to determine depth applied.

Using Equation 3.3 to determine depth requires that the flow rate remain constant over the entire application time. Records must be kept of both time of application and flow rate. If the flow measuring device includes a volume totalizer, record keeping is much simpler. Volume totalizers register the total volume that has passed through the device much like an odometer measures total miles traveled in an automobile.

Equation 3.1 would be used to calculate depth as shown in the next example.

Example 3.1

An irrigation system delivers 900 gpm. If 30 acres is irrigated every 24 hr, determine the total depth applied in inches.

Given: t = 24 hr

A = 30 ac

Find: Depth applied in inches

Solution:

Using Equation 3.3: $Q \times t = A \times d$ or $d = \frac{Qt}{A}$

First, convert the flow rate from gpm to acre-inch/hour:

$$900 \text{ gpm} \times \frac{1 \text{ ac-in/hr}}{450 \text{ gpm}} = 2 \text{ ac-in/hr}$$

$$d = \frac{(2 \text{ ac-in/hr})(24 \text{ hr})}{30 \text{ ac}} = 1.6 \text{ in}$$

Example 3.2

A flow meter has a volume totalizer. If 90 acres were irrigated and the totalizer registered 12,590,900 gallons after an irrigation and 8,925,100 gallons before the irrigation, what was the depth of application in inches?

Given: Volume after = 12,590,900 gal

Volume before = 8,925,100 gal

Find: Depth in inches

Solution:

$$d = \frac{V}{A}$$

Volume applied = 12,590,900 gal − 8,925,100 gal = 3,665,800 gal

$$V = 3{,}665{,}800 \text{ gal} \frac{(1 \text{ ac-in})}{(27{,}154 \text{ gal})} = 135 \text{ ac-in}$$

$$d = \frac{135 \text{ ac-in}}{90 \text{ ac}} = 1.5 \text{ in}$$

3.2 Basic Principles of Flow Measurement

3.2.1 Velocity-Flow-Area Relationship

The flow rate in an irrigation water conduit can be expressed as:

$$Q = V_m A_f \tag{3.6}$$

where: V_m = mean velocity of flow in the channel or pipeline and

A_f = cross-sectional area of flow.

The concept of this equation is shown in Figure 3.2. This equation is called the continuity equation and is fundamental to water measurement. Velocity (V_m) is the average or mean velocity within the pipeline or channel. The use of this equation is illustrated in Example 3.3.

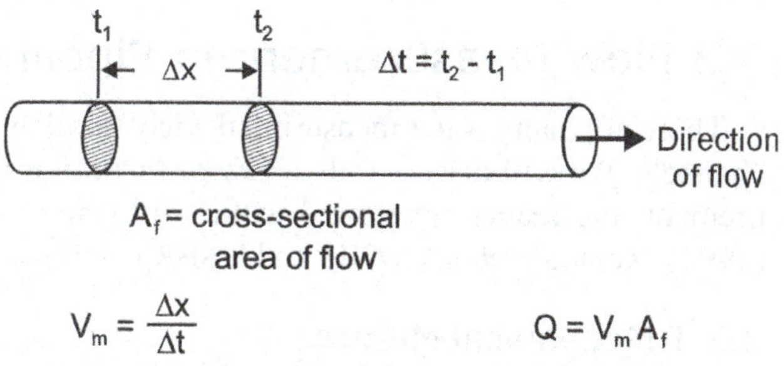

Figure 3.2. The continuity principle for flow.

3.2.2 Measurement of Mean Velocity

With most water measuring devices, the fundamental measurement is the velocity of the flowing water. Using the continuity principle (Equation 3.6), flow velocity is converted to flow rate. There are many methods used to estimate flow velocity. These include mechanical devices such as impellers, paddle wheels, bucket wheels, vanes, floats, and the measurement of pressure differences within hydraulic structures to infer the flow velocity. Newer devices, e.g., ultrasonic meters, use either the Doppler principle or the time of travel of an ultrasonic wave to estimate the velocity. These devices will be discussed in more detail in Sections 3.3 and 3.4.

Example 3.3

Determine the flow rate (gpm) in a circular pipeline that has an inside diameter (ID) of 8 in and a mean velocity of flow of 5 ft/s.

Given: ID = 8 in

V_m = 5 ft/s

Find: The flow rate (Q) in gallons per minute (gpm)

Solution:

$$A_f = \frac{\left[\pi\left(ID^2\right)\right]}{4}$$

ID = 8 in

$$A_f = \frac{\pi(8)^2\ \text{in}^2}{4} = \frac{64\pi\ \text{in}^2}{4} = 50.27\ \text{in}^2$$

$$A_f = 50.27\ \text{in}^2 \frac{1\ \text{ft}^2}{144\ \text{in}^2} = 0.349\ \text{ft}^2$$

$$Q = (5\ \text{ft/s})\left(0.349\ \text{ft}^2\right) = 1.75\ \text{cfs}$$

$$= (1.75\ \text{cfs})\frac{450\ \text{gpm}}{1\ \text{cfs}} = 785\ \text{gpm}$$

3.2.3 Distribution of Velocity

The water velocity in a pipeline or in an open channel is not constant throughout its cross section. Typically, the velocity in a closed circular pipeline is highest in the middle of the pipeline and then gradually goes to zero at the wall of the pipeline. This is illustrated in Figure 3.3. Likewise, open channels also have nonuniform velocities within the flow area. Again, the velocity is zero at the wall of the channel and then gradually increases towards the center. In Figure 3.4 you see the illustration of the nonuniform distribution of velocity in an open channel. The variations of velocity within the flow conduit affect where velocity should be sensed or how to correct the measured velocity to obtain the mean velocity as required to use the continuity equation.

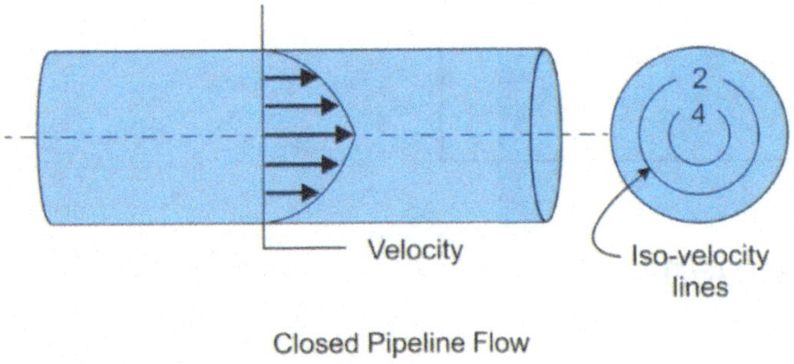

Figure 3.3. Velocity distribution in a closed circular pipeline (pipe is full).

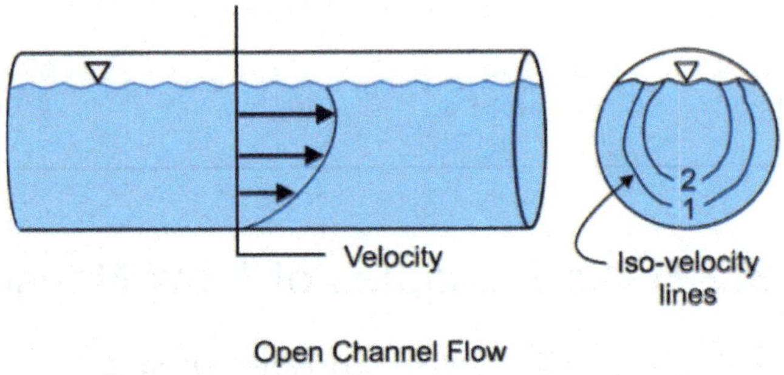

Figure 3.4. Velocity distribution in a circular pipe with open channel flow (pipe is not full).

3.3 Flow Measurement in Pipelines

There are many water measuring devices available for both pressurized pipelines and open channels. We will discuss only a few of them. For more detailed discussions of water measurement, the reader is referred to the following: ASME (1971), Bos et al. (1984), Miller (1996), Replogle et al. (1990), and USBR (1997).

3.3.1 Mechanical Meters

Propeller meters (impeller meters) and turbine meters are common methods for measuring pipeline flow in agricultural irrigation and in municipal water distribution systems. The force of the flowing water turns the propeller. The propeller is sensing the velocity in the pipeline. A propeller meter is illustrated in Figure 3.5. The rotations of the propeller are converted to flow rate by proper gear ratios in the meter head. The diameter of the propeller is usually slightly smaller than the inside diameter of the pipeline. This gives a good estimate of the mean velocity in the pipeline and allows the meter to operate over a wide range of flows.

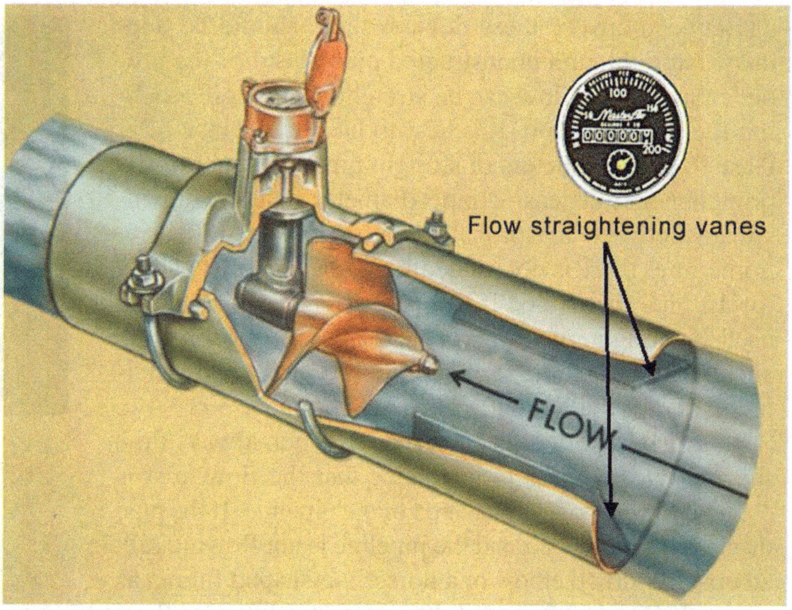

Figure 3.5. Typical irrigation propeller meter (image supplied courtesy of Sparling Instruments LLC, El Monte, CA).

The register in the meter head of these devices comes in various configurations. Four common ones are shown in Figure 3.6. The one on the left has the volumetric totalizer combined with a sweep hand, which can be used for timing the rate of flow. Each revolution of the sweep hand corresponds to a known volume of water that has passed through the meter. The second register from the left contains two components: the totalizer and the flow rate indicator. The third register from the left has three components on one meter head: the totalizer, the flow rate indicator, and an index hand or sweep hand for timing. The register on the right represents a digital display which contains both a flow rate indicator display and a volume totalizer. One key advantage of electronic register heads is that they lend themselves to remote monitoring through cellular or satellite communication.

Usually the accuracy of the meter is based upon what is registered on the totalizer. Since the totalizer and the sweep hands are directly connected, the true flow rate is best obtained by either timing the sweep hand or timing the rate that the numbers are changing on the odometer. The least accurate, or the poorest representation of the meter's accuracy, is the flow rate indicator. The flow rate indicator is helpful to observe changes in flow rate and as an indicator of excessive spiraling or disturbed flow. The latter condition is noticed by significant needle movement or bounce.

Proper selection and installation of flow measuring devices are very important. Propeller meters should be located away from pipeline fittings that cause spiraling or disturbance of water flowing in the pipe such as pumps, elbows, valves, etc. A flow disturbance refers to the disruption or distortion of the parabolic velocity distribution an example of flow caused by a pipe elbow is shown in Figure 3.7.

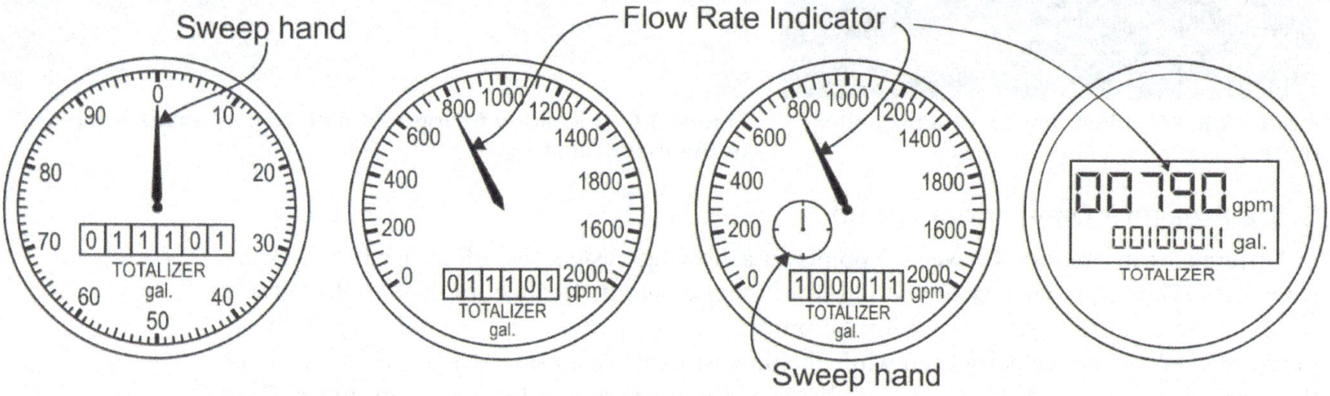

Figure 3.6. Options available for registers on a propeller meter.

Between any of these devices there should be adequate, straight, and unobstructed pipe ahead of the propeller so that the flow can be straightened before reaching the meter section. It is best to have a distance of at least 10 pipe diameters of straight pipe upstream of the propeller and at least 1 pipe diameter distance downstream between the propeller and a flow disturbance. Sometimes there is not adequate room available to allow for 10 pipe diameters. If not, a shorter distance can be used if straightening vanes are placed in the pipeline ahead of the propeller (Figure 3.5). A typical field installation of a propeller meter is shown in Figure 3.8.

It is also important that the meter section always flow full of water. This is to guarantee that the flow area is equal to the cross-sectional area of the pipeline. If the pipe discharges into the air and the pipeline is not flowing full, an upward turned elbow or a horseshoe-shaped fitting, as shown in Figure 3.9, is useful to guarantee full pipe flow.

Another approach to measuring flow in pipelines is paddlewheel meters. These mechanical meters usually have a magnetic pickup to measure the number of revolutions of the paddle wheel. The paddlewheel movement then is converted to flow rate by the velocity area relationships. Like the propeller meters, the paddlewheel should be installed with adequate piping ahead of the meter so that the velocity profile can be established before the water reaches the meter.

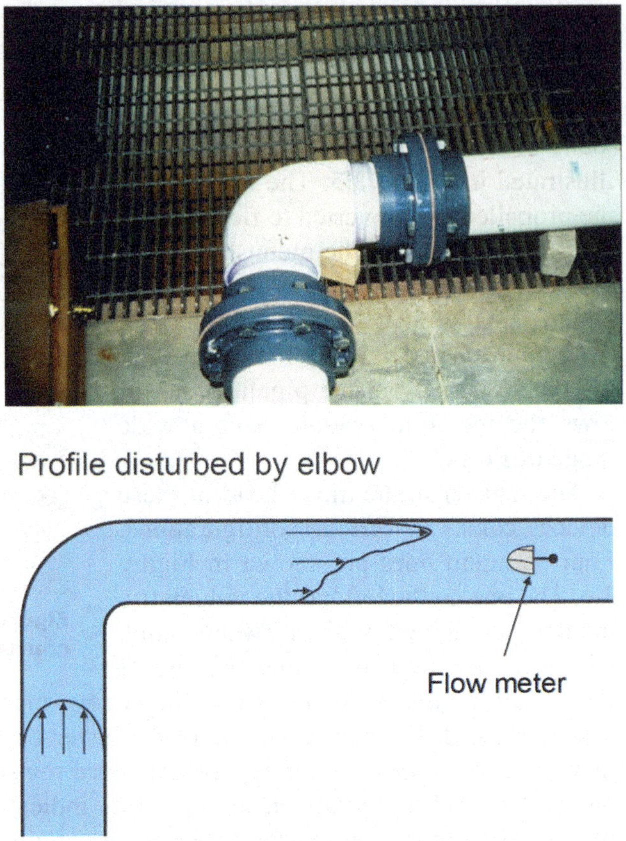

Figure 3.7. Flow disturbance of velocity profile caused by a pipe elbow.

Figure 3.8. Field installation of a propeller meter.

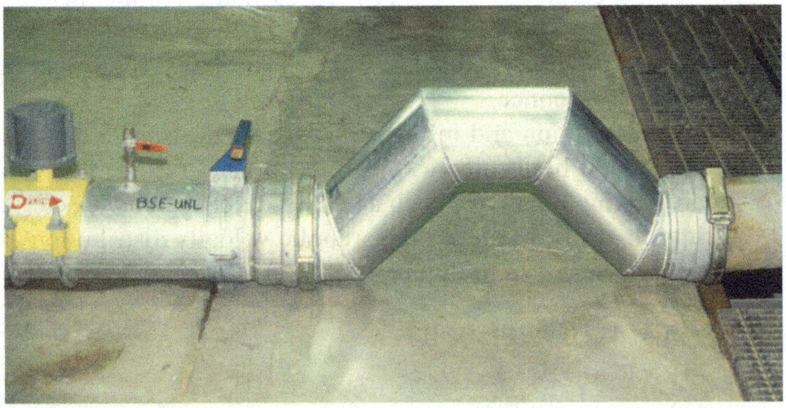

Figure 3.9. U-shaped fitting installed to guarantee full pipe flow in the meter section.

3.3.2 Pressure Differential Methods

Differences in pressure between 2 points in a flowing system are often used to measure flow rates. The mean velocity is inferred from the pressure difference. In the simplest case a pitot tube is used (Figure 3.10). With the pitot tube the upstream sensor picks up both the pressure head and the velocity head while the downstream sensor only senses pressure. Thus, there is a difference in head or pressure between the upstream and downstream tubes. Pitot tubes come in various configurations. It is best to sense velocity at several positions in the pipeline to obtain a good estimate of the average velocity.

In Chapter 8, we will discuss the relationships between the various forms of energy in flowing water: the pressure energy, relative energy due to elevation, and velocity energy. The change of forms of energy from pressure to velocity will be illustrated by using Bernoulli's' Energy Equation. As the velocity increases in a pipeline, the pressure is usually reduced. This principle is used quite often in flow measurement.

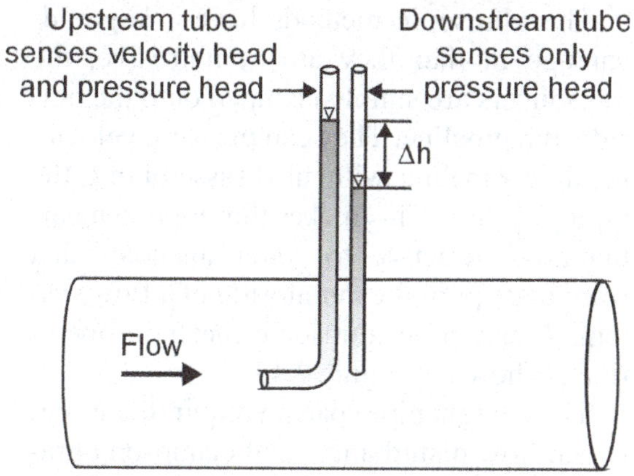

Figure 3.10. Pitot tube for measuring flow velocity.

A Venturi, such as the one shown in Figure 3.11, can be used to measure flow. The upstream pressure is higher than the pressure in the Venturi throat because of the high velocity in the throat of the Venturi. There is a correlation between the difference in pressure, or head, between the upstream sensing position and the throat of the Venturi. The pressure differential is directly related to the velocity of the fluid in the pipeline.

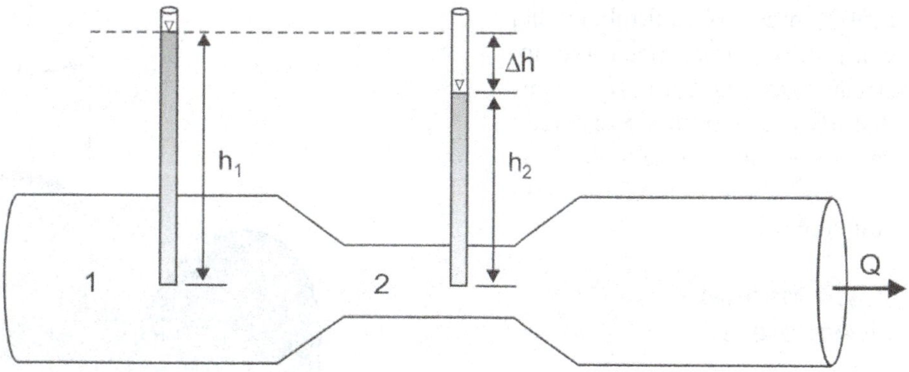

Figure 3.11. Venturi for flow measurements.

Another pressure differential device is the orifice meter. The orifice shown in Figure 3.12 discharges to the air. In this case the head is measured upstream. The downstream head is zero (atmospheric pressure). The flow of an orifice follows the following relationship:

$$Q = KA_0\sqrt{2g\Delta h} \tag{3.7}$$

where: K = flow coefficient,

A_0 = cross-section area of the orifice,

g = gravitational constant, 32.2 ft/s^2, and

Δh = head of water upstream of the orifice.

An orifice meter does not have to discharge to the air, but rather it can be imbedded within a pipeline and the difference in pressure head upstream and immediately downstream of the orifice can be measured to determine differential head.

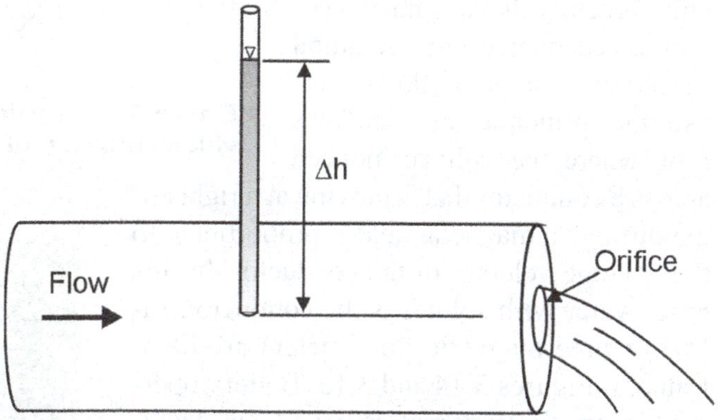

Figure 3.12. Pipe end-cap orifice meter.

3.3.3 Ultrasonic Measurement

Another approach for measuring flow in a pipeline is to use ultrasonic energy. With this method ultrasonic waves are transmitted through the pipe wall and into the flow. One burst of energy is transmitted upstream while another burst is sent downstream. The travel time of the two waves are measured and compared. The difference in wave velocity is directly related to the velocity of the water. Another ultrasonic approach takes advantage of the Doppler principle. A high-frequency signal is transmitted into the liquid. Suspended particles or gas bubbles reflect the wave. The frequency of the reflected wave is measured. The difference in transmitted and reflected frequencies is directly proportional to the liquid's flow velocity.

The ultrasonic methods have a large advantage in that they are nonintrusive; the transducers are simply clamped onto the outside of a pipeline. They can measure velocity inside a pipeline without disassembling the piping system. This makes this approach particularly attractive to water agencies that want to do periodic monitoring of a flow system. A clamp-on ultrasonic meter in operation is shown in Figure 3.13.

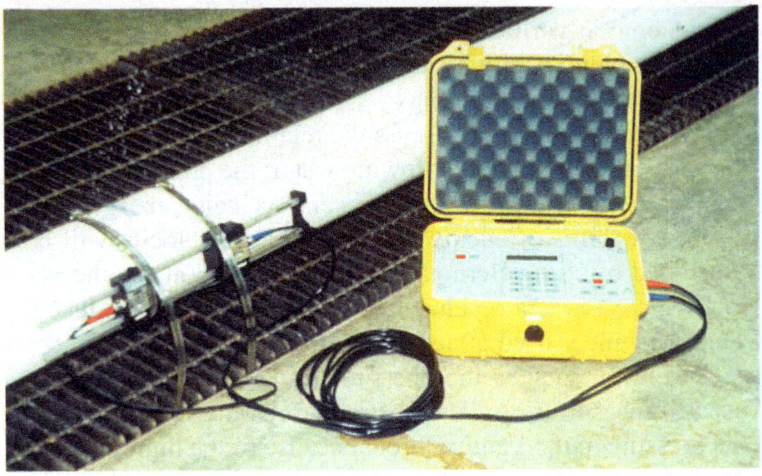

Figure 3.13. Clamp-on ultrasonic meter installed on a pipeline.

The straight pipe spacing requirements between flow disturbances and clamp-on ultrasonic meters are the same as propeller meters, 10 pipe diameters upstream of the meter and 1 pipe diameter downstream. Eisenhauer (2008) presents multipliers that can remove meter bias if the upstream spacing between a flow disturbance and clamp-on meter cannot be met. These multipliers were based on research by Johnson et al. (2001).

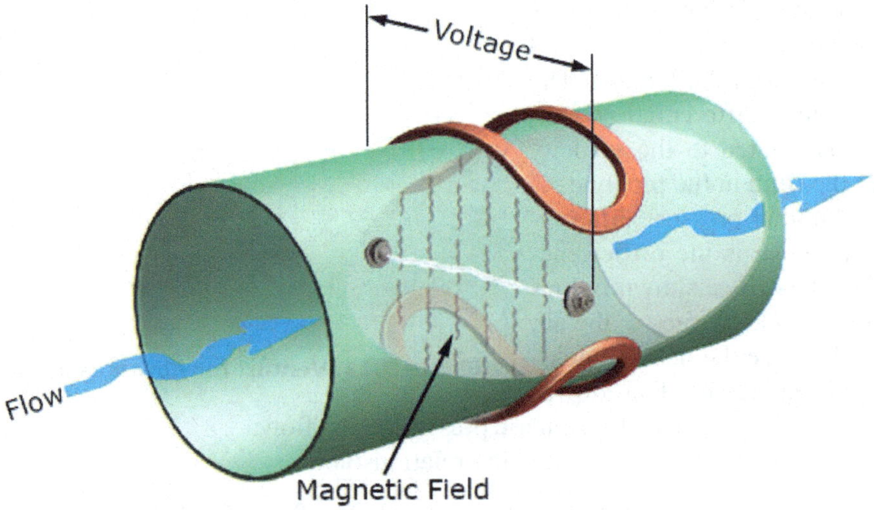

Figure 3.14. In-line magnetic flowmeter. (Modification of diagram provided courtesy of McCrometer Corporation.)

3.3.4 Magnetic Flowmeters

Magnetic flowmeters have been used for the measurement of pipeline flows for many years but only recently have gained economic acceptance in agricultural irrigation. Magnetic flowmeters use the principle of Faraday's Law where the voltage induced across a conductor that is moving at a right angle through a magnetic field is proportional to the average velocity of the conductor. In this case, water with solutes is the conductor. Inline or tube magnetic flow meters are illustrated in Figures 3.14 and 3.15. Battery operated magnetic flow meters have been developed to increase their applicability in irrigation. Research has shown that the in-line magnetic flow meters require less upstream distance from flow disturbances as compared to other meter types. For example, rather than needing 10 pipe diameters upstream and 1 downstream of straight unobstructed pipe between a flow disturbance and the metering section (as is the case for propeller meters), only 2 are needed upstream and 1 downstream for magnetic flow meters. Depending

Figure 3.15. Electromagnetic flowmeter installed in the field.

upon the situation, these shorter required distances can be a significant savings in the costs of retrofitting a piping system to accommodate metering.

3.4 Flow Measurement in Open Channels

Open channel flow is distinguished from pipeline flow by the fact that the water surface is at atmospheric pressure. With closed conduit or pipeline flow, the surface of the water is contained by the conduit's wall which causes the water pressure to exceed atmospheric pressure.

3.4.1 Velocity Methods

Probably the most common method for measuring open channel (stream) flow is to use a current meter (Figure 3.16). The current meter that measures the water velocity at predetermined positions in the channel can either be a mechanical device, such as the cup-type meter illustrated in Figure 3.16, or ultrasonic devices that utilize the Doppler principle. In the latter case, the water velocity is assumed equal to the velocity of suspended particles. Water velocity and depth of the measurement are determined at various points across the stream. The flow rate in subsections of the channel is computed using the velocity and flow depth data.

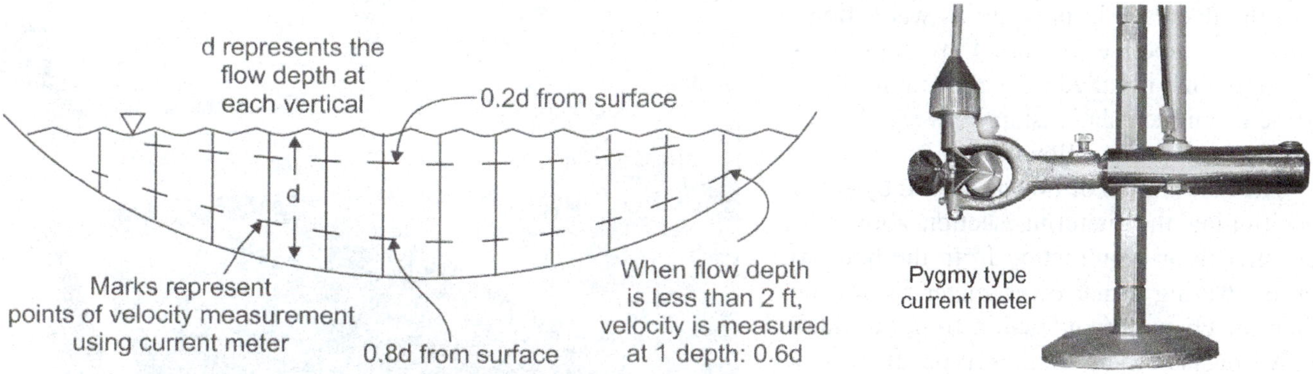

Figure 3.16. Current meter method for measuring flow rate in an open channel.

A simpler approach to velocity measurement of an open channel is to use a float on the water surface. The float speed is timed between two positions in the flow, such as shown in Figure 3.17. The float speed is the water velocity at the surface. Since the surface velocity (V_s) is not the average velocity (V_m), the float velocity has to be corrected. The average velocity can be calculated using Equation 3.8 where K_f is the velocity correction factor.

$$V_m = K_f V_s \qquad (3.8)$$

K_f typically ranges from 0.65 to 0.8. The 0.65 correction factor applies to depths of 1 foot or less and 0.8 for water depths of 20 feet or more. The float method can be used as a quick estimate of flow, but it is normally not sufficiently accurate for good water management.

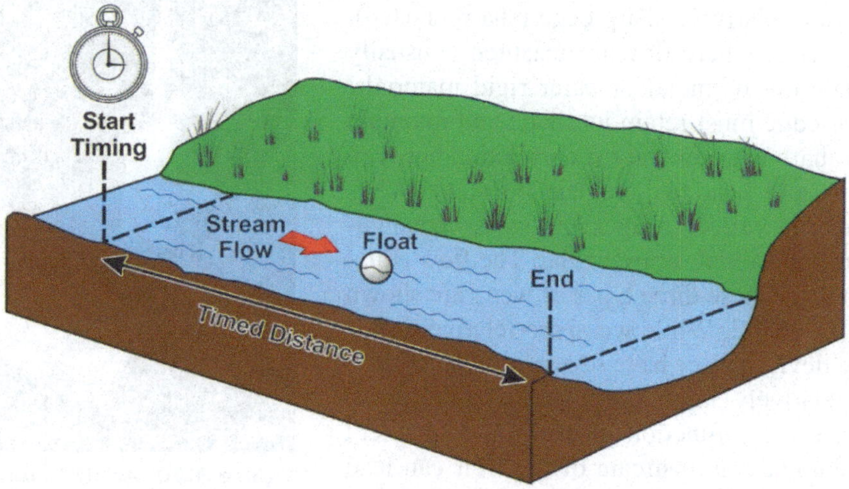

Figure 3.17. Float method for determining surface velocity in a channel.

3.4.2 Pressure Differential Methods

Like flow in pipelines, the pressure differential concept can be used to measure with open channel flow. With open channel devices, velocity is usually not computed but is imbedded in the equations of flow. The equations of flow then account for both the shape of the metering section and the implied velocity. There are two general classes of pressure differential devices used in open channels: weirs and flumes. An example of a weir is shown in Figure 3.18. Figures 3.19 and 3.20 are pictures of flumes. With both classes of devices, a head or depth of water is measured upstream of the metering section. Since the metering section causes a contraction of flow, there is a lowering of the water surface elevation through the metering section, much like the decrease in pressure as water flows through a pipeline Venturi. Flow must pass through what is called critical depth for there to be a unique relationship between the upstream head and the flow rate.

The contraction of flow is caused by either positioning the metering section above the channel floor (contraction from the bottom) or by having a narrower metering section than the channel (contraction from the side). Flow measurement flumes typically use a side contraction. Weirs always have a bottom contraction. Often, weirs use both a side and bottom contraction (Figure 3.18) while flumes sometimes have both side and bottom contractions.

Table 3.2 presents various shapes of weirs that are used to measure flow. These weirs have a relatively sharp edge (sharp crested). The edge where flow is measured is usually made out of metal or other rigid materials. The edge must retain its shape and maintain its sharp edge so that the correlation between flow and head will remain constant. Weirs come in various shapes and sizes: rectangular, trapezoidal, or triangular. The flow equations for these three types of weirs are shown in Table 3.2. While weirs are relatively simple devices, they have several disadvantages; a relatively large head loss is required to make them function properly and sediment accumulation upstream of the weir can lead

Figure 3.18. A weir for measuring water flow in an open channel.

Figure 3.19. A Parshall flume to measure open channel flow.

Figure 3.20. An RBC flume to measure open channel flow.

Table 3.2. Weir shapes and discharge formulas.

Measuring Device (all sharp crested)	Views	Formula
Rectangular Weir (without side contraction)	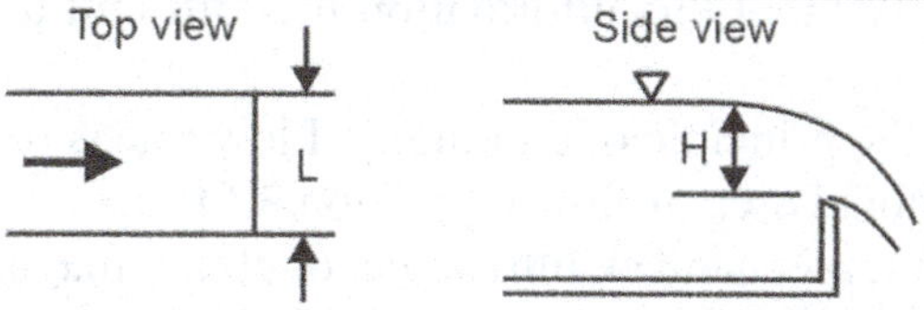	$Q = 3.33LH^{3/2}$
Trapezoidal Weir	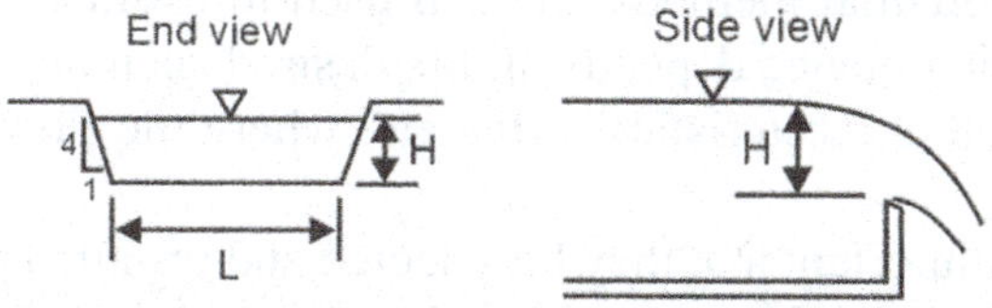	$Q = 3.37LH^{3/2}$
90° Triangular Weir	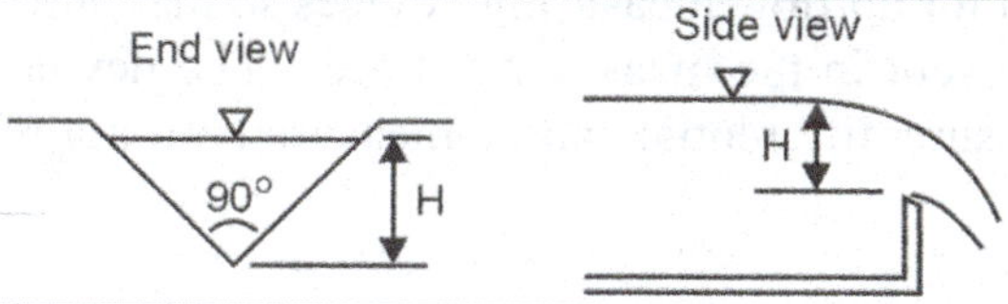	$Q = 2.49H^{3/2}$

to a change in the weir's head-flow relationship. The nappe of water leaving the crest of the weir must spring free of the weir for the unique head discharge relationship. If downstream water submerges a weir, the calculated flow may be incorrect. Flumes usually have a much higher tolerance to downstream submergence than weirs.

Like weirs, there are many shapes and designs of flow measuring flumes available for flow measurement. The Parshall flume is common in irrigation; it is illustrated in Figure 3.19. Parshall flumes come in various sizes with throat widths from 1 inch to 50 feet. A big advantage is that sediment flows freely through Parshall flumes.

Another approach to flow measurement is the RBC flume. The RBC flume was designed to utilize a small ramp or bottom contraction within a prismatic channel or flume. This is illustrated in Figure 3.20. One advantage of this flume is that if an irrigator has a trapezoidal irrigation channel with stable sides, such as a concrete-lined ditch, the flow measuring device can be created by installing a ramp and a staff gauge upstream of the ramp section. An important feature of this type of flume is that the calibration is very predictable once the dimensions and materials of the metering section are known. Calibration equations and tables are available for Parshall and RBC flumes of numerous sizes. Example calibrations are given in Tables 3.3 and 3.4.

Table 3.3. Flow rate of 1-foot Parshall flume.

Upstream Head (ft)	Flow Rate[a] (cfs)	Flow Rate[a] (gpm)
0.25	0.46	207
0.50	1.35	605
0.75	2.53	1140
1.00	3.95	1770
1.25	5.58	2500
1.50	7.41	3320
1.75	9.40	4220
2.00	11.60	5190

$Q = 3.95H^{1.55}$

Q = flow rate in cfs and H = head in ft [a] Assumes that free flowing criteria is met.

Table 3.4. Flow rate of 8-inch fiberglass RBC flume.

Upstream Head (ft)	Flow Rate[a] (cfs)	Flow Rate[a] (gpm)
0.10	0.06	29
0.20	0.21	93
0.30	0.42	188
0.40	0.70	314
0.50	1.04	469
0.60	1.45	651
0.70	1.92	860

$Q = 3.575\ (H + 0.01259)^{1.8419}$

Q = flow rate in cfs and H = head in ft

[a] Assumes that free flowing criteria is met.

3.5 Summary

Flow measurement is important in irrigation so that both the rate that water is being applied and the depth of application are known. Without this information it is difficult to be a good water manager.

Flow measurement typically relies on the principle of continuity. Flow rate is related to the velocity of water flow and the cross-sectional area of flow (Equation 3.6).

For both pipeline and open channel flow, mechanical, ultrasonic, or electromagnetic meters are used to sense velocity. Pressure differential methods are also used to estimate velocity. The selection of the proper flow measuring device depends on the desired accuracy, the cost of the measuring device, and the physical characteristics of the site where the flow is to be measured.

With all flow measuring devices, it is important that they be selected and installed properly. Upstream conditions must be considered for all flow measuring devices so that unreasonable flow disturbance and spiraling is not present in the measurement area. The devices should also be selected so that an adequate pressure differential can be measured but not result in a large energy or head loss in the conduit.

Questions

1. List three reasons for measuring water.
2. What would you consider to be an acceptable accuracy for water measurement in irrigation? Explain your answer.
3. What is the fundamental physical law used by most flow measuring devices?
4. Which is more useful for determining depth applied in irrigation, a volume totalizer, or a flow rate indicator? Why?
5. What assumption is made about flow rate when using Equation 3.3 to calculate depth applied?
6. a. Show how Equation 3.3 can be rearranged to determine the depth applied.
   b. Show how Equation 3.3 can be rearranged to determine the time required to apply a desired depth.
   c. Show how Equation 3.3 can be rearranged to determine the flow rate required to apply a desired depth in a given time period.
7. Why are long sections of straight pipe and straight channel, free of obstructions, required of upstream flow measuring devices?
8. Why must the metering section of a propeller meter flow full?
9. How many gallons per minute are required to apply 1 million gallons in a day?
10. Why is it better to time the totalizer or a timing hand (index hand) of a propeller meter than to read the flow rate indicator directly to determine flow rate?
11. A totalizer on a flow meter is timed to determine flow rate. The last digit represents 100 gallons. Ten numbers are allowed to pass during timing. The time was 1 minute, 30 seconds. Determine the flow rate in:
    a. gpm (gallons per minute)
    b. cfs (cubic feet per second)
    c. m^3/s (cubic meters per second)

d. L/s (liters per second)
e. ac-in/hr (acre-inch per hour)
f. ha-cm/hr (hectare-cm/hour)

12. A 130-ac field was irrigated. The totalizer on the system's flow meter read:
    After irrigation: 60,325,100 gallons
    Before irrigation: 57,324,600 gallons
    Calculate the gross depth applied in inches.
13. A 200-ac field was irrigated. The totalizer on the system's flow meter read:
    After irrigation: 2,425 ac-in
    Before irrigation: 2,121 ac-in
    Calculate the gross depth applied in inches.
14. A 7,000 square foot lawn was watered. The household meter registered 500,300 ft^3 before watering and 500,708 ft^3 after watering. Calculate the gross depth applied in inches (assume that other uses of water in the house were insignificant during the water application).
15. A golf course irrigation system irrigates 60 ac and the flow rate is 1200 gpm.
    a. How many hours of irrigation will be required to apply 1 inch of water?
    b. If you can only irrigate 8 hours per day, how many days will it take to apply 1 inch of water?
    c. Suppose ET is 0.25 in/d and you want to apply this amount each day (assume you can only irrigate 8 hours per day). How many gpm would be needed?

References

American Society of Mechanical Engineers. (1971). *Fluid meters, their theory and application* (6th ed.). New York, NY: ASME.

Bos, M. G., Replogle, J. A., & Clemmens, A. J. (1984). *Flow measuring flumes for open channels*. New York, NY: John Wiley & Sons.

Eisenhauer, D. E. (2008). Using ultrasonic flow meters in irrigation applications. NebGuide G1426. University of Nebraska-Lincoln Extension.

Johnson, A. L., Benham, B. L., Eisenhauer, D. E., & Hotchkiss, R. H. (2001). Ultrasonic water measurement in irrigation pipelines with disturbed flow. *Trans. ASAE, 44*(4), 899-910.

Miller, R. W. (1996). *Flow measurement engineering handbook* (3rd ed.). Boston, MA: McGraw-Hill.

Replogle, J. A., Clemmens, A. J., & Bos, M. G. (1990). Measuring irrigation water. In G. J. Hoffman, T. A. Howell, & K. H. Solomon (Eds.), *Management of farm irrigation systems*. St. Joseph, MI: ASAE.

U. S. Bureau of Reclamation. (1997). *Water measurement manual* (3rd ed.). Washington, DC: USBR, U.S. Department of the Interior.

Chapter 4
Plant Water Use

4.1 Introduction

How much irrigation water is required for a 100-acre field next week? Is one inch enough or do you need two? Do you need a pump capable of delivering 900 gallons per minute or will 750 be adequate? How different is water use for small plants compared to a fully developed crop? How long does water need to be applied to the fairway on the eighteenth fairway? How large of canal is needed for a reservoir to supply an irrigation district? These questions can be answered if we know how plants use water. ***Evapotranspiration*** is the term used to describe plant water use.

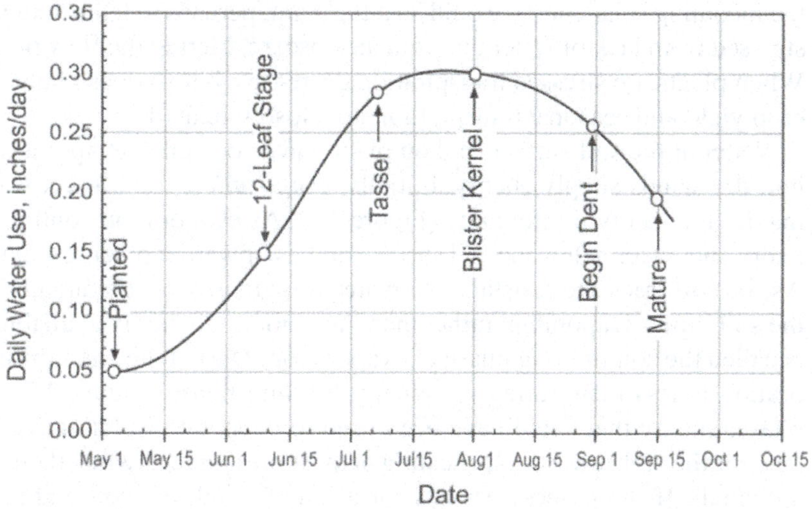

Figure 4.1. Example of water use and time of growth stages for corn in the Northern High Plains of the U. S.

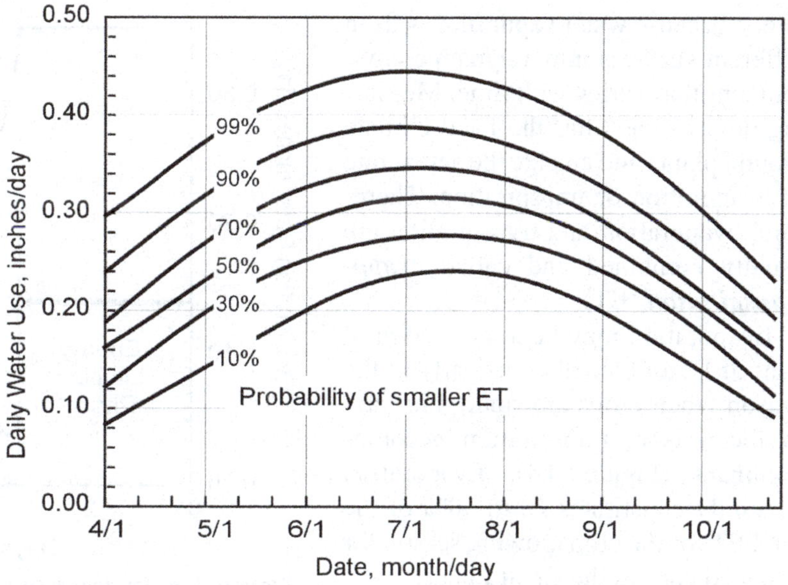

Figure 4.2. Frequency distribution of water use for well-watered alfalfa with full cover in southern Idaho (adapted from Wright and Jensen, 1972).

The seasonal water use pattern is critical for irrigation management. If the rate of water use is known, managers can determine when to irrigate and the depth of water to apply. Curves that show the rate of water use during the season can be used to estimate future water needs (Figure 4.1). It is important to know the seasonal amount of water use to plan for irrigation requirements, cropping area, and other management decisions. The seasonal total is especially important where water supplies are limited or regulated as often occurs for irrigation projects supplied from reservoirs.

The rate of water use varies annually; therefore, the average water use curve is frequently inadequate. A distribution of the water use rate for well-watered alfalfa is shown in Figure 4.2. The 50% line represents the average water use rate. The 90% line represents a rate that will only be exceeded once in 10 years. Curves such as Figure 4.2 are useful in deciding the risk involved with a management strategy. Water and money can be saved by reducing the amount applied, i.e., using an average water use rate. However, the manager would be more confident that the crop would not be stressed if a higher probability were used.

4.2 Water Use Processes

Understanding how plants use water and evaluating the effect of weather on water use require consideration of fundamental processes. Plants extract water from the soil and transport water to the leaves. The stomata, very small openings, located on the upper and lower surfaces of the leaves, allow for the intake of carbon dioxide required for photosynthesis and plant growth. Water vapor is lost from the plant leaves by evaporation in the stomatal cavity and the flow of water vapor from the stomata into the atmosphere. This is transpiration.

Transpiration is necessary to cool the plant and maintain productivity. Converting liquid water to vapor (i.e., evaporation) requires a large amount of energy. If plants did not transpire, the incoming solar energy would heat the plant, perhaps to lethal temperatures. When plants are stressed from lack of water the stomata close, restricting the flow of water and carbon dioxide. When plants are stressed transpiration decreases, but so does photosynthesis. For this reason, crop yield and seasonal transpiration are closely related.

Water at the soil surface and on plant leaves or mulch evaporates when ***solar radiation*** or hot, dry winds supply energy. Initially, ***evaporation*** from a wet soil surface progresses at a maximum energy limiting rate (Figure 4.3). As evaporation continues the soil surface begins to dry and water below the soil surface moves upward replacing soil water lost by evaporation. As the soil dries the resistance to water flow increases. Eventually the rate of water flow in the soil limits evaporation rather than the amount of energy available to evaporate water. This is called the soil limiting phase of evaporation. During the soil limiting phase the rate of evaporation is less than during the energy limiting phase (Figure 4.3). During the soil limiting phase, energy that could have been used to evaporate water is available to heat the soil and air near the soil surface. The heating is most pronounced when there is no crop or when plants are small. If the process persists for a long period, the soil and air become quite hot as in desert climates.

Evaporation and transpiration are difficult to measure or predict separately, because water vapor moves from different surfaces into a dynamic environment that varies with time. Measuring devices can alter the local climate around plants and change the actual rate of evaporation or transpiration. Therefore, evaporation and transpiration are usually combined and called ***evapotranspiration*** (ET).

Evaporation may be a large component of ET of annual crops early in the season when crops are small, but later in the season, transpiration becomes dominant (Figure 4.4). Evaporation generally constitutes 20 to 30% of the total ET for the crop growing season for irrigated corn in the Great Plains.

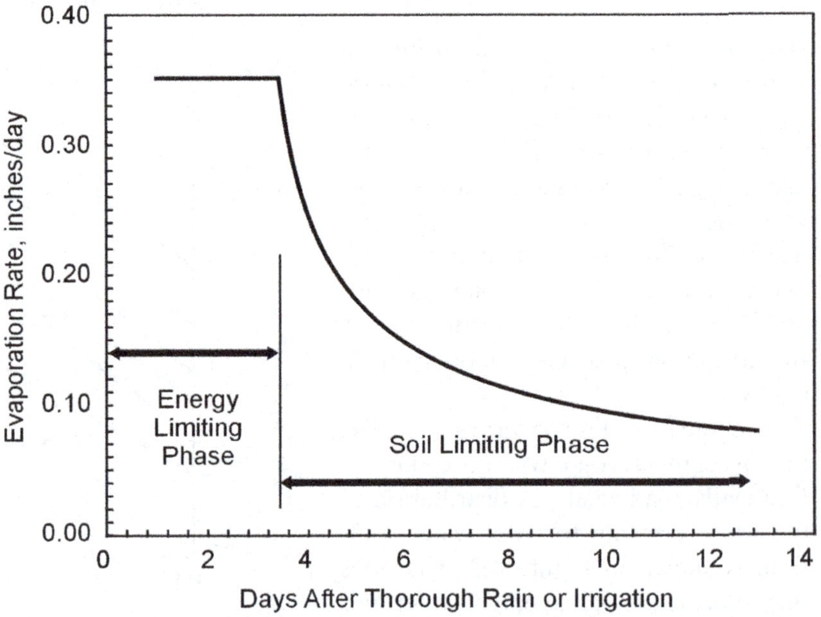

Figure 4.3. Example of the stages of evaporation from a bare soil.

Transpiration and evaporation from soil, plant leaves, and mulch are evaporative processes. Considerable energy is required to evaporate water. The energy absorbed by plants on a sunny and windy summer day would evaporate enough water to cover the soil surface to a depth of approximately 0.4 inches. For an area of one acre, this would equal about 11,000 gal/d.

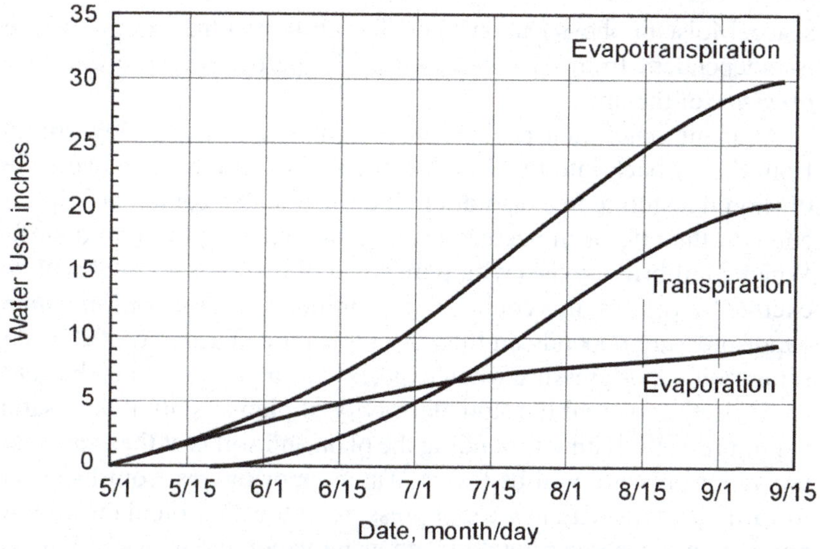

Figure 4.4. Example of seasonal patterns of evaporation, transpiration, and ET for irrigated corn in the Great Plains.

The energy available for ET comes from several sources (Figure 4.5). Much of the energy comes from extraterrestrial radiation emitted by the sun. Some extraterrestrial radiation is absorbed or reflected in the atmosphere. The radiant energy that ultimately reaches the crop canopy is called ***solar radiation.*** Plant and soil surfaces reflect some solar radiation back into the atmosphere. The portion of the solar radiation absorbed varies depending on the color of the surface and other soil and plant properties. The fraction of the solar radiation reflected to the atmosphere is called the albedo. The albedo for plant and soil surfaces ranges from 35% for snow covered soils to 10% for dark soils that are wet. A commonly used value for the albedo of actively growing crops is 23%. In this case, 77% of the solar radiation is absorbed and used for ET and photosynthesis.

Long-wave radiation is the second component of the radiation balance. Energy is transferred due to the temperature difference between objects. In both cropped and uncropped landscapes, the exchange is between the plant and soil surfaces to the atmosphere. Because the atmosphere is cold relative to the surface of the earth, long-wave energy is lost from the plant-soil system.

Radiant energy available for ET is called ***net radiation*** equal to the absorbed solar radiation minus the emitted long-wave radiation.

Advection is the lateral or horizontal transfer of mass, heat, or other property. Hot, dry winds supply energy for ET due to advection. The amount of energy transferred depends on the wind speed and the vapor pressure of the air. According to Dalton's law of partial pressure, the pressure exerted by a mixture of gases is equal to the sum of the pressures exerted by each gas if it alone occupied the

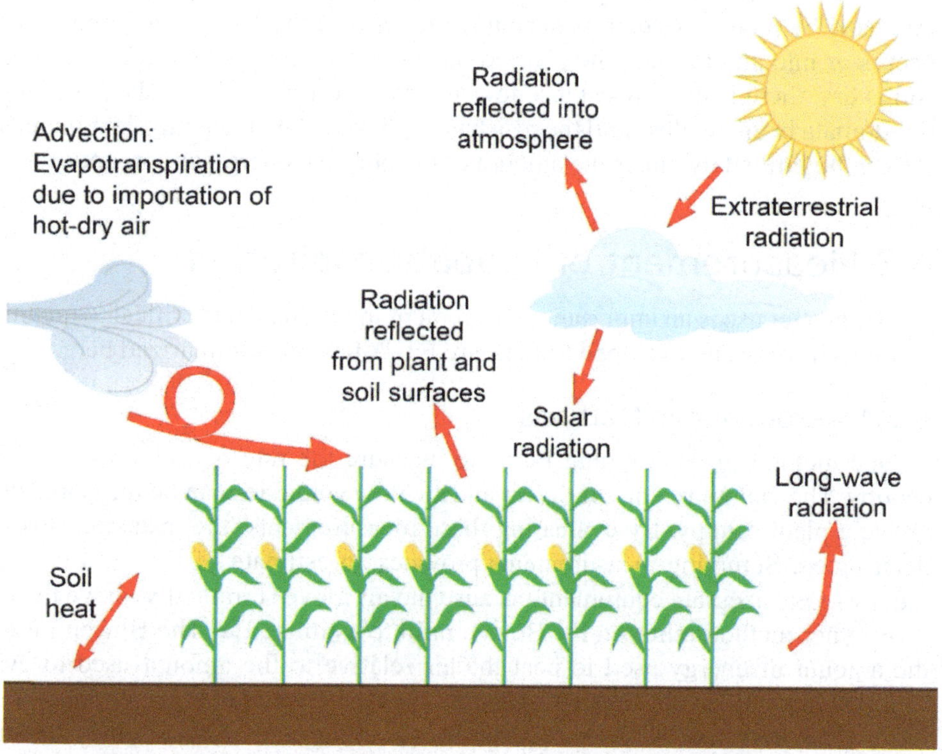

Figure 4.5. Diagram of energy sources for evapotranspiration.

space. Moist air obeys Dalton's law. The portion of the barometric pressure due to water vapor is independent from the other gases. The partial pressure due to water vapor is the ***vapor pressure*** of the air.

At an air water interface, water molecules continually flow from the water into the air and from the air back into the liquid. If the air is dry, more molecules leave the liquid than enter the liquid resulting in evaporation. If air in a sealed container is left in contact with water long enough, the rate of molecules leaving and entering the liquid surface reach an equilibrium. When equilibrium exists with pure water, the air is saturated with water vapor. The pressure exerted by vapor at this equilibrium condition is the ***saturation vapor pressure***. The saturation vapor pressure depends on the air temperature. The ratio of the actual vapor pressure to the saturation vapor pressure when expressed as a percentage is the ***relative humidity***.

Air in the soil and the stomatal cavity of plants is often near saturation; thus, it has a high vapor pressure. If air surrounding the plant and soil is at the same temperature, but much drier, the vapor pressure will be lower. Water vapor moves from locations of high vapor pressure toward locations with low vapor pressure. If the air around the crop were contained in a chamber, it would become saturated with water vapor and ET would then be negligible because the air could not hold any additional water. If the saturated air were replaced with dry air, ET would resume. The more rapidly the air is exchanged and the drier the air, the higher the ET rate. In windy-arid locations, advection may contribute as much to ET as radiation. However, in humid locations or in areas with little wind, advection may be quite low.

Two other energy sources for ET are the exchange of heat between plants and the soil (called soil heat flux), or between plants and the surrounding air. For example, if the soil is warmer than plants, energy is transferred from the soil to the plants. This energy may increase transpiration. Conversely, if the canopy is warmer than the soil, energy flows toward the soil and transpiration may decrease. The same type of energy transfer occurs between plants and air. Plants that are not stressed for water are generally cooler than the ambient air during the middle of the day. However, if stressed for water, plants will often be warmer than the ambient air (USDA-SCS, 1993).

Two additional factors impact ET. First, there must be a source of water in the soil to supply that used by plants. Second, water must move from the soil to the point where evaporation occurs or into and through the plant to the stomatal cavity where transpiration occurs. If the soil is dry, there is more resistance to water transport in the soil. Also, as plants are stressed, the stomata begin to close and the resistance to water flow from the plant increases. Therefore, ET can be limited by either the amount of evaporative energy or amount of water in the soil.

4.3 Measurement of Evapotranspiration

Plant water use is an important management input; thus, it is critical to quantify ET. Several methods have been developed to measure ET. A few are summarized here.

4.3.1 Aerodynamic Methods

One method of determining ET is to measure the rate of water vapor leaving the plant canopy. The vapor pressure of the air and air flow velocities can be measured at several levels above a plant canopy. By evaluating these measurements, the instantaneous ET rate can be determined. Summing measurements provides an estimate of ET for a day. This technique requires very accurate equipment because the air moves erratically above the canopy.

Another method relies on the Bowen ratio to estimate ET. The Bowen ratio is the ratio of the amount of energy used to heat the air relative to the amount used to evaporate water. Equipment has been developed to measure the Bowen ratio and to compute ET. A major problem is that advection is ignored in the Bowen ratio method which may be an unacceptable assumption for some locations.

4.3.2 Soil Water Methods

Soil water is the source for ET, and several methods have been used to relate changes in soil water to plant water use. The primary components of the soil water balance are illustrated in Figure 4.6. The soil water balance can be expressed as:

$$ET = AW_b - AW_e + P + d_g + U_f + R_i - R_o - d_p \tag{4.1}$$

where: ET = amount of ET during the period,
AW_b = amount of available soil water in the root zone at the beginning of a period,
AW_e = amount of available soil water in the root zone at the end of a period,
P = total precipitation during the period,
d_g = gross irrigation during the period,
U_f = groundwater contribution to water use during the period,
R_i = surface water that runs onto the area during the period,
R_o = surface runoff that leaves the area during the period, and
d_p = deep percolation from the root zone during the period.

The ET can be estimated from Equation 4.1 if all other terms are known or can be approximated. If the groundwater table is more than 6 ft below the soil surface, the contribution from groundwater can be ignored. Rain and irrigation from sprinklers are usually measured with rain gages or similar devices. Measuring devices have been developed for surface irrigation applications. Soil water content can be measured using neutron scattering or other techniques described in Chapter 2. Deep percolation is difficult to measure and is often assumed to be insignificant unless large rains occur, or large irrigations are applied. A significant problem with the soil water balance technique is that repetitive measurements must be made throughout the season. One week is usually the shortest period for using the soil water balance method to estimate ET. If deep percolation or runoff is significant, the soil water balance method is further limited because of the lack of measuring capabilities.

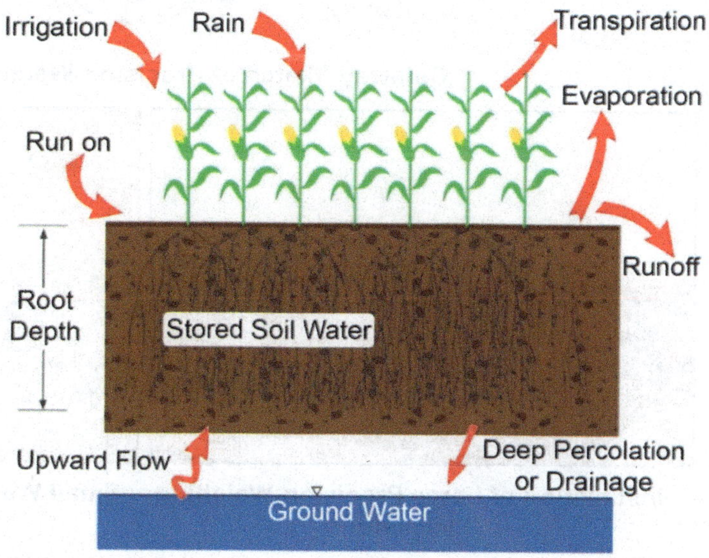

Figure 4.6. Diagram illustrating the components of the soil water balance.

4.3.3 Lysimetry

Lysimeters are specially designed open-top tanks that are filled with soil, preferably undisturbed soil, and planted to the same crop as the surrounding area. The tanks are buried in the field. Water used for ET by plants grown in the lysimeter must come from the soil water within the tank. ET can be measured by monitoring soil water contents and water applications from irrigation or rain. The soil tank is used to isolate soil water from the surrounding area and to prevent run on, runoff, upward groundwater flow, and drainage. For some applications drainage is allowed and the volume of deep percolation is measured. The soil water within the tank can be measured with traditional methods such as neutron probes. The amount of water in the tank can also be determined by weighing the tank, soil, plants, and soil water. Since soil water is the only item that changes significantly over short time periods, the change in weight equals the amount of water used for ET.

Various types of lysimeters have been utilized to measure ET. The most elaborate and accurate lysimeters are called weighing lysimeters (Figure 4.7). These lysimeters use weighing

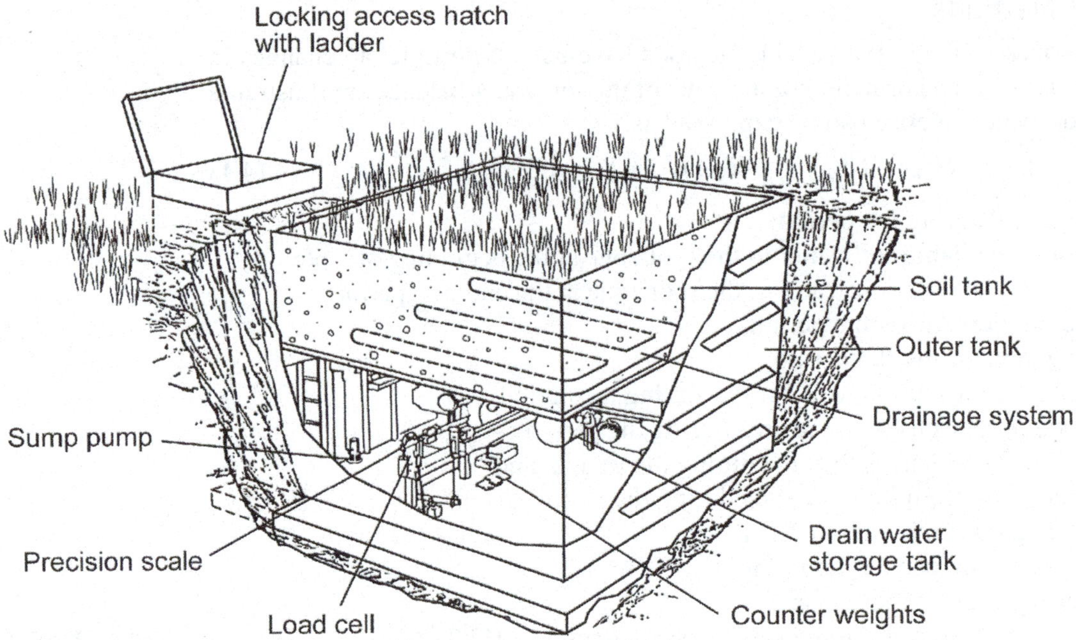

Cutaway Sketch of Precision-Weighing Lysimeter

Installation of Large Precision-Weighing Lysimeter

Small Weighing Lysimeter for Turf Grass Measurements

Figure 4.7. Examples of weighing lysimeters (picture of large lysimeter is courtesy of USDA-ARS, Bushland, Texas).

devices to measure water lost from the soil tank. Large plants with deep root zones usually require large lysimeters. Short plants, with shallow root systems, can be measured by lifting small lysimeters and weighing with a scale. The most sophisticated weighing devices are high precision and can be used to measure small changes of weight. A good description of precision lysimeters is given by Marek, et al. (1988). Such systems have counter balanced weighing systems resulting in a measurement accuracy approaching 0.001 inches of ET. The high accuracy is required for daily measurements.

Other types of lysimeters do not weigh the soil-plant-water system. Non-weighing lysimeters function the same as the field water balance method except upward flow of groundwater and runoff or run on are prevented by the sides and the bottom of the lysimeter. A drainage system is usually installed in the bottom of the lysimeter to measure deep percolation and to prevent water from ponding at the bottom of the lysimeter. Water table lysimeters, common in humid regions, are a second type. With this design, deep percolation is prevented, and a water table is maintained in the lysimeter. Changes in soil water and the elevation of the water table are measured along with other soil water balance terms. Non-weighing and water table lysimeters are usually only accurate enough for estimating the amount of ET over a period of approximately 1 week. More elaborate methods are needed to measure daily or hourly ET rates.

Besides being expensive to install and operate, lysimeters pose several problems. Using lysimeters to measure ET was summarized by Allen, et al. (1991). The best lysimeters are those filled with an undisturbed soil column. These are termed monolithic lysimeters, and if they are large, their filling can be difficult and expensive (Figure 4.7). Regular and careful maintenance of the lysimeter and the surrounding area is required to maintain accuracy. Spatial variability can be significant when measuring ET and several lysimeters may be required. Lysimeters are usually research tools and are too complex and labor intensive for water management.

4.3.4 Plant Monitoring Methods

Plant transpiration can be measured using several techniques. One of these is the autoporometer. With this instrument, a small chamber is clamped onto a growing plant leaf and changes in the humidity and temperature of the air within the chamber are used to compute transpiration during that period. The transpiration rate and other plant responses change very rapidly due to external factors. Therefore, the porometer can only remain on the leaf for a few minutes. Another limitation of the porometer is that only a small part of one leaf is used for measurement. Characterizing the transpiration for an entire crop canopy requires numerous measurements. Further, these measurements only provide instantaneous transpiration rates. Generally, irrigation management requires plant water use for daily and longer time periods. Thus, porometers are primarily used in experiments to investigate plant response to stress and for very short-term water use estimates.

A second method uses infrared thermometers to predict transpiration based upon the difference between the plant temperature and the air temperature. The infrared thermometer has been used successfully to detect plant stress and to predict irrigation timing. If the incoming solar radiation and other energy terms are known, the ET rate can be estimated using the techniques of Hatfield (1983) and Jackson (1982). These techniques are complex and require extensive calculation as well as continuous monitoring of plant temperature. The infrared plant monitoring method can be used to help schedule and manage irrigation but needs further development to estimate ET.

4.4 Calculating ET

Knowledge of plant water use rates is essential to manage irrigation systems accurately. Because measurement of ET is difficult and time consuming, equations have been developed to predict water use rates. These equations are based on weather factors, plant species, and stage of plant development and soil water status. The equations can estimate past water use and forecast future water use which are both essential in planning, designing, and scheduling irrigations.

The simplest equation to predict plant water use is based on two factors:

$$ET = K_c\, ET_o \tag{4.2}$$

where: ET = actual crop or plant ET rate,

K_c = crop coefficient, and

ET_o = reference crop ET rate.

Crop coefficients are used to describe the behavior of agricultural crops. The reference ET represents the amount of energy available for ET. This is expressed as the water use rate of a reference crop.

4.5 Reference Crop ET

Reference crop ET is defined as the ET rate from a large expanse of a uniform canopy of dense, actively growing, vegetation provided with an ample supply of soil water. The reference is a hypothetical crop (vegetation) (Allen et al., 1998). Two references are commonly used: (1) a short crop (grass clipped to maintain a height of 5 inches) and (2) a tall crop (alfalfa that is about 20 inches tall). For this text, the short reference crop is used.

Other terms have been used to represent the amount of energy in the environment that is available to evaporate water. Potential ET was widely used historically to represent this energy. Currently some authors are beginning to use reference surface ET. Both terms are synonymous with reference crop ET.

Reference crop ET can be predicted using a standardized equation that utilizes appropriate coefficients and standardized procedures. Numerous methods have been developed to estimate reference crop ET (ET_o). The simplest methods generally use average air temperature. The most complex methods require hourly data for solar radiation, air temperature, wind speed, and vapor pressure. There are many approaches between these extremes. The Penman-Monteith equation (Jensen and Allen, 2016) has proven to be reliable for computing reference crop ET for most locations.

The Penman-Monteith equation and associated relationships for calculating coefficients were presented by Allen, et al. (1998). They utilized a short reference crop and presented procedures for either daily or hourly computations. Only the daily version of the procedure is presented here. The short reference crop and daily time steps can be used for many situations. If computations are necessary for mountainous or coastal regions, readers should refer to Allen, et al. (1998) or Jensen and Allen (2016) for appropriate methods.

The Penman-Monteith equation to predict water use of the reference crop was given by Allen, et al. (1998) as:

$$ET_o = \left[\frac{\Delta(R_n - G) + K_t\,\rho_a\,c_p\left(\dfrac{e_s - e_a}{r_a}\right)}{\Delta + \gamma\left(1 + r_s / r_a\right)} \right] / \lambda \tag{4.3}$$

where: ET_o = ET for a short reference crop,

R_n = calculated net radiation at the crop surface,

G = soil heat flux density at the soil surface,

e_s = saturation vapor pressure of air,

e_a = actual vapor pressure of the air,

Δ = slope of the saturation vapor pressure versus temperature curve,

γ = psychrometric constant,

K_t = unit conversion constant,

ρ_a = density of air,

c_p = specific heat of air,

r_a = aerodynamic resistance of water vapor movement,

r_s = bulk resistance of crop and soil surfaces, and

λ = heat of vaporization of water.

The amount of energy available to evaporate water determines ET rates. However, the rate of water movement from the soil and plants into the atmosphere also depends upon the resistance to the movement of water vapor within and out of the plant canopy. Wind in the atmosphere above the crop canopy causes air movement to be turbulent which results in mixing of air in the atmosphere. Thus, any water vapor that enters the atmosphere above the canopy readily mixes with air in the atmosphere and there is little resistance to water vapor movement. The flow of air within the upper portion of the crop canopy and in the air layer immediately above the crop is much less turbulent. Since there is little mixing of air from the lower portion of the plant with the air at the top of the canopy, the only way that water vapor can leave the soil plant system is due to vapor pressure gradients. Water vapor flows from areas of high vapor pressure to locations with low vapor pressure. The rate of transfer can be estimated if the difference in vapor pressure is known along with the resistance to the flow of water vapor. The aerodynamic resistance (Figure 4.8) represents the resistance to water vapor movement in the boundary layer just above the crop.

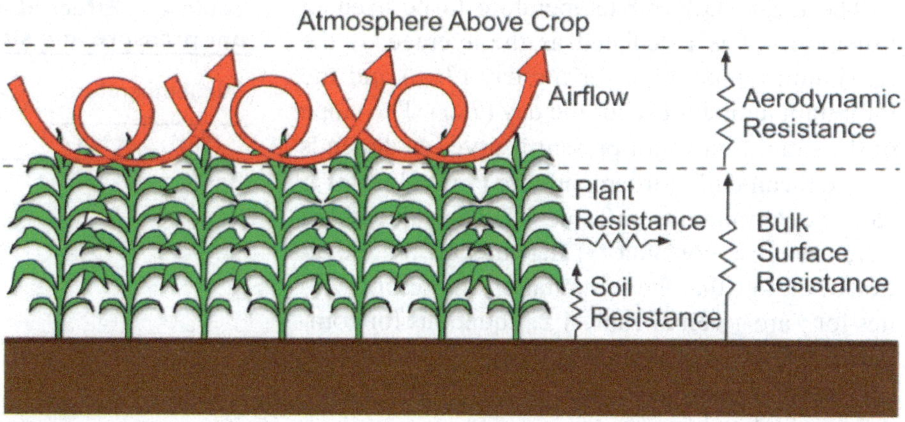

Figure 4.8. Diagram of resistances to water vapor flow for evapotranspiration.

Below the boundary layer the resistance to water vapor flow is controlled by soil and plant properties. For evaporation to occur water must flow through the soil pores to reach the soil/air interface. Resistance to water vapor flow also occurs within the stomata and cuticle of plant leaves. Stomatal resistance varies with the degree of water stress that plants experience, while soil resistance varies with water content. However, for reference conditions, the combined effect of these resistances can be combined into the bulk surface resistance as in Figure 4.8.

While the form of the Penman-Monteith equation in Equation 4.3 is the most accurate, it requires extensive calculations. Equations associated with calculation of variables and the derivations of constants required in Equation 4.3 are very complicated and are explained in more detail by Allen et al. (1998). An American Society of Civil Engineers task force reviewed the use of the Penman-Monteith equation for 82 site-year combinations across the United States (ASCE-EWRI, 2004). They found that a simplified form of the equation provided acceptable results while simplifying calculation procedures. The reduced form of the Penman-Monteith equation for computing daily ET for a short reference crop (clipped grass approximately 5 inches tall) is given by:

$$ET_o = \underbrace{\left[\frac{\Delta\ R_n / 62.23}{\Delta + \gamma\ (1 + 0.00634\ U_2)}\right]}_{\text{Radiation Term}} + \underbrace{\left[\frac{\gamma\ U_2 \left(\frac{30.18}{T + 460}\right)(e_s - e_a)/25.4}{\Delta + \gamma\ (1 + 0.00634 U_2)}\right]}_{\text{Aerodynamic Term}} \qquad (4.4)$$

where: ET_o = daily reference crop ET (in/d),
R_n = net radiation (MJ/m^2/d),
U_2 = daily wind run measured 2 m above the ground (mi/d),
T = mean daily air temperature measured at height of 1.5 to 2.5 m (°F),
e_s = saturation vapor pressure of air (kPa),
e_a = actual vapor pressure of the air (kPa),
Δ = slope of the saturation vapor pressure versus temperature curve (kPa/°C), and
γ = psychrometric constant (kPa/°C).

The left bracketed term in Equation 4.4 corresponds to the radiation component of the reference ET while the right bracketed term corresponds to the aerodynamic component.

This reduced form of the Penman-Monteith equation does not include the soil heat flux. The soil heat flux for daily data can generally be neglected; however, if shorter or longer time steps (i.e., hourly, or monthly) are considered, the soil heat flux should be included into the computation of ET_o as described by Jensen and Allen (2016).

The mean daily air temperature to be used in Equation 4.4 is calculated as the average of the maximum temperature for the day (T_{max}) and the minimum temperature for the day (T_{min}). The slope of the saturation vapor pressure curve (Δ) depends on the mean daily air temperature (T). Values of Δ can be determined from Table 4.1. The value of the psychrometric constant (γ) depends on the elevation of the site that the computations represent. Values for γ are given in Table 4.2. Equations for computing Δ and γ are provided by Allen, et al. (1998).

The saturation vapor pressure (e_s) for Equation 4.4 is calculated from:

$$e_s = \frac{e^o(T_{max}) + e^o(T_{min})}{2} \qquad (4.5)$$

where $e^o(T_{max})$ and $e^o(T_{min})$ are the saturation vapor pressure at the maximum (T_{max}) and minimum (T_{min}) air temperatures for the day, respectively. The actual vapor pressure (e_a) is computed as the saturation vapor pressure at the dew point temperature of the air (T_{dew}). The dew point temperature is usually a direct input from the weather data or is derived from the weather data, then Table 4.1 can be used with T_{dew}.

Computation of the net radiation (R_n) for estimating ET_o involves several steps. First, the amount of solar radiation that would be received on a clear day is determined as a function of the day of the year and the elevation of the site above mean sea level. Figure 4.9 can be used to determine the clear-sky radiation (R_{so}). Then, the net outgoing long-wave radiation (R_{nl}) can be determined from Figure 4.10 using the maximum, minimum, and dew point temperatures along with the ratio of the measured solar radiation for the day (R_s) compared to the clear-sky radiation for that date. The net radiation is then computed from:

$$R_n = (1 - \alpha)\, R_n - R_{nl} \qquad (4.6)$$

where α is the albedo equal to 0.23 for the short reference crop. Use of these figures and equation 4.6 will be illustrated in Example 4.1.

Determination of the reference ET using Equation 4.4 involves numerous computations. A graphical procedure has been developed to accomplish the calculations. Equation 4.4 contains two bracketed sections. The left portion, within the first set of brackets, represents the reference ET that results from solar radiation. The right portion of the equation, in the second set of brackets, represents ET due to the humidification of the air. The second portion is referred to as the aerodynamic component, i.e., the ET due to advection. Figure 4.11 can be used to determine the amount of reference ET from radiation and Figure 4.12 can be used to determine the amount of reference ET from humidifying the air. Example 4.1 illustrates the procedure.

Table 4.1. Effect of air temperature on saturation vapor pressure and slope of saturation vapor pressure.

Air Temperature (°F)	Air Temperature (°C)	Saturation Vapor Pressure, e_s (kPa)[a]	Slope of Saturation Vapor Pressure, Δ (kPa /°C)
20	-7	0.37	0.029
25	-4	0.46	0.034
30	-1	0.56	0.041
35	2	0.69	0.049
40	4	0.84	0.059
45	7	1.02	0.070
50	10	1.23	0.082
55	13	1.48	0.097
60	16	1.77	0.113
65	18	2.11	0.132
70	21	2.50	0.154
75	24	2.96	0.178
80	27	3.50	0.206
85	29	4.11	0.237
90	32	4.81	0.272
95	35	5.62	0.311
100	38	6.55	0.354
105	41	7.60	0.403
110	43	8.79	0.457
115	46	10.14	0.518
120	49	11.67	0.584

[a] Note a pressure of 1 kPa = 0.145 lb/in^2. The atmospheric pressure at sea level averages about 101 kPa.

Table 4.2. Value of the psychrometric constant (γ) as a function of elevation. at the location of consideration.

Elevation Above Sea Level (ft)	Psychrometric Constant, γ (kPa/°C)
0	0.067
500	0.066
1000	0.065
1500	0.064
2000	0.063
2500	0.062
3000	0.060
3500	0.059
4000	0.058
4500	0.057
5000	0.056
5500	0.055
6000	0.054
6500	0.053
7000	0.052

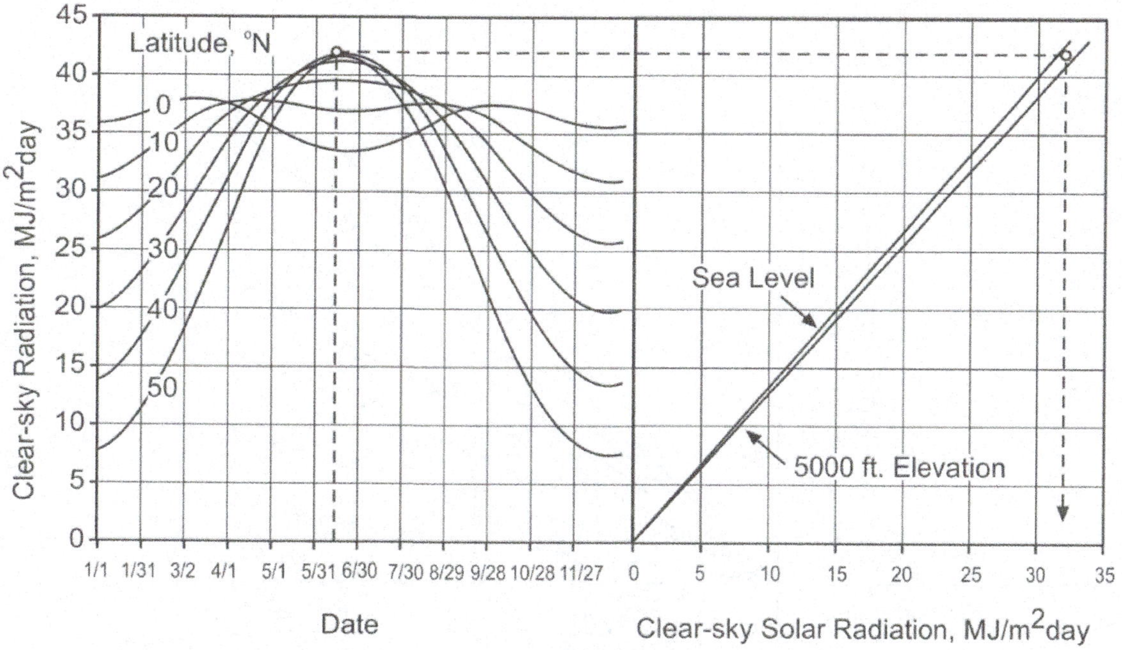

Figure 4.9. Diagram to determine clear-sky radiation (R_{so}) for the northern hemisphere.

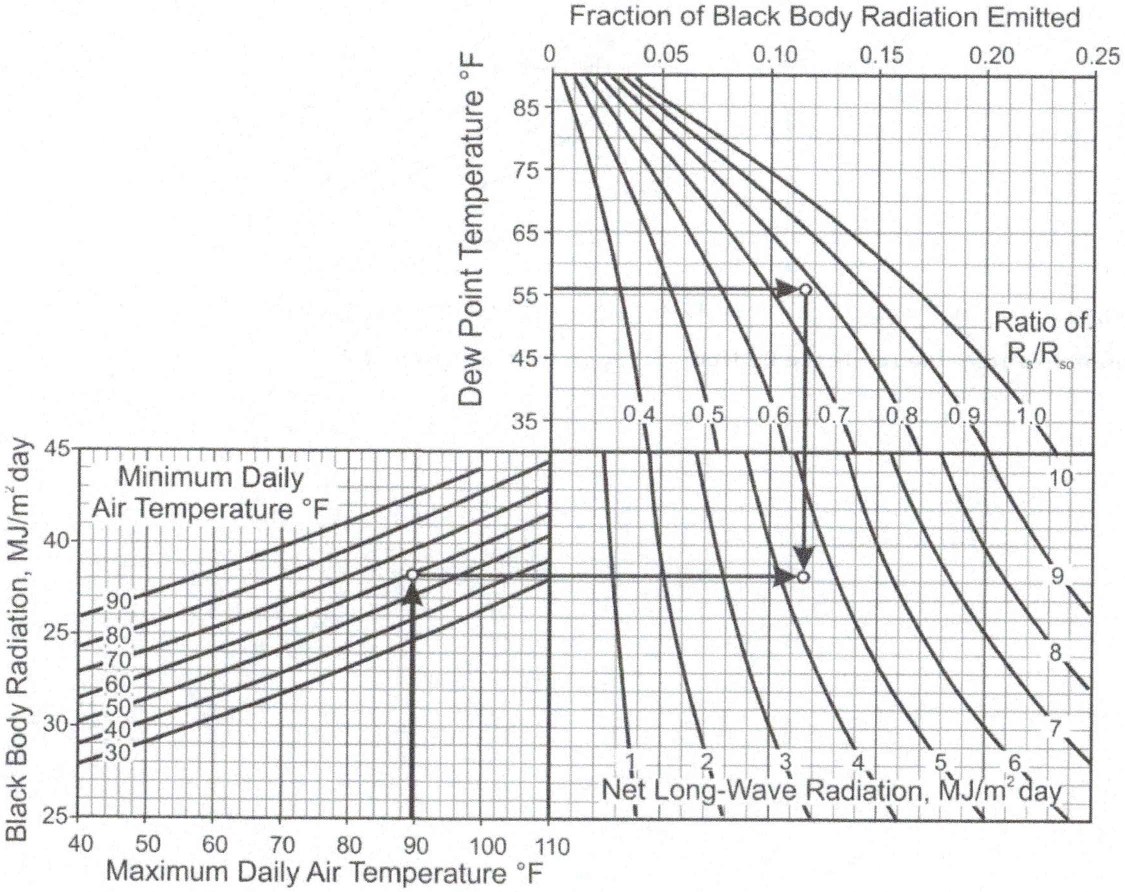

Figure 4.10. Diagram to determine net outgoing long-wave radiation.

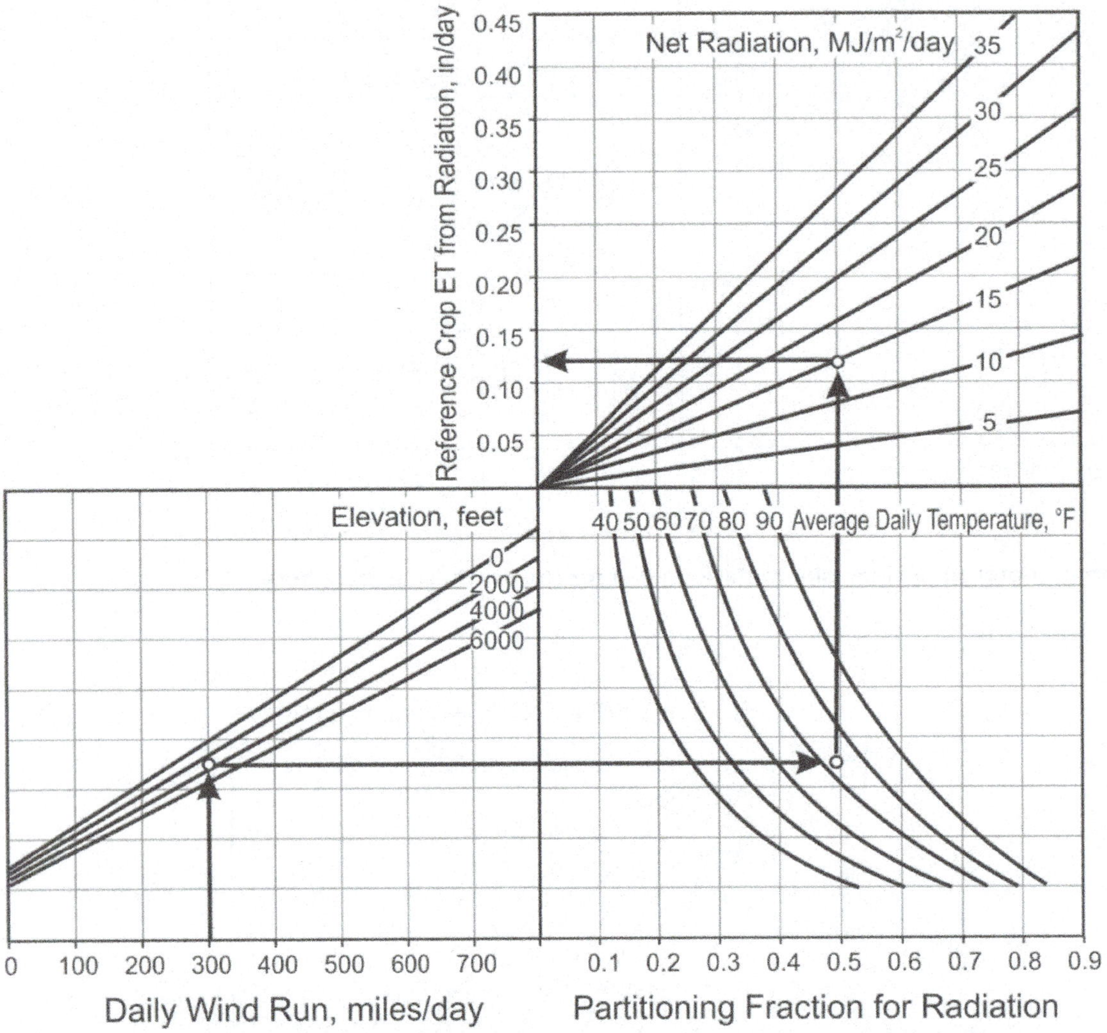

Figure 4.11. Diagram to determine reference ET from radiation from Equation 4.4.

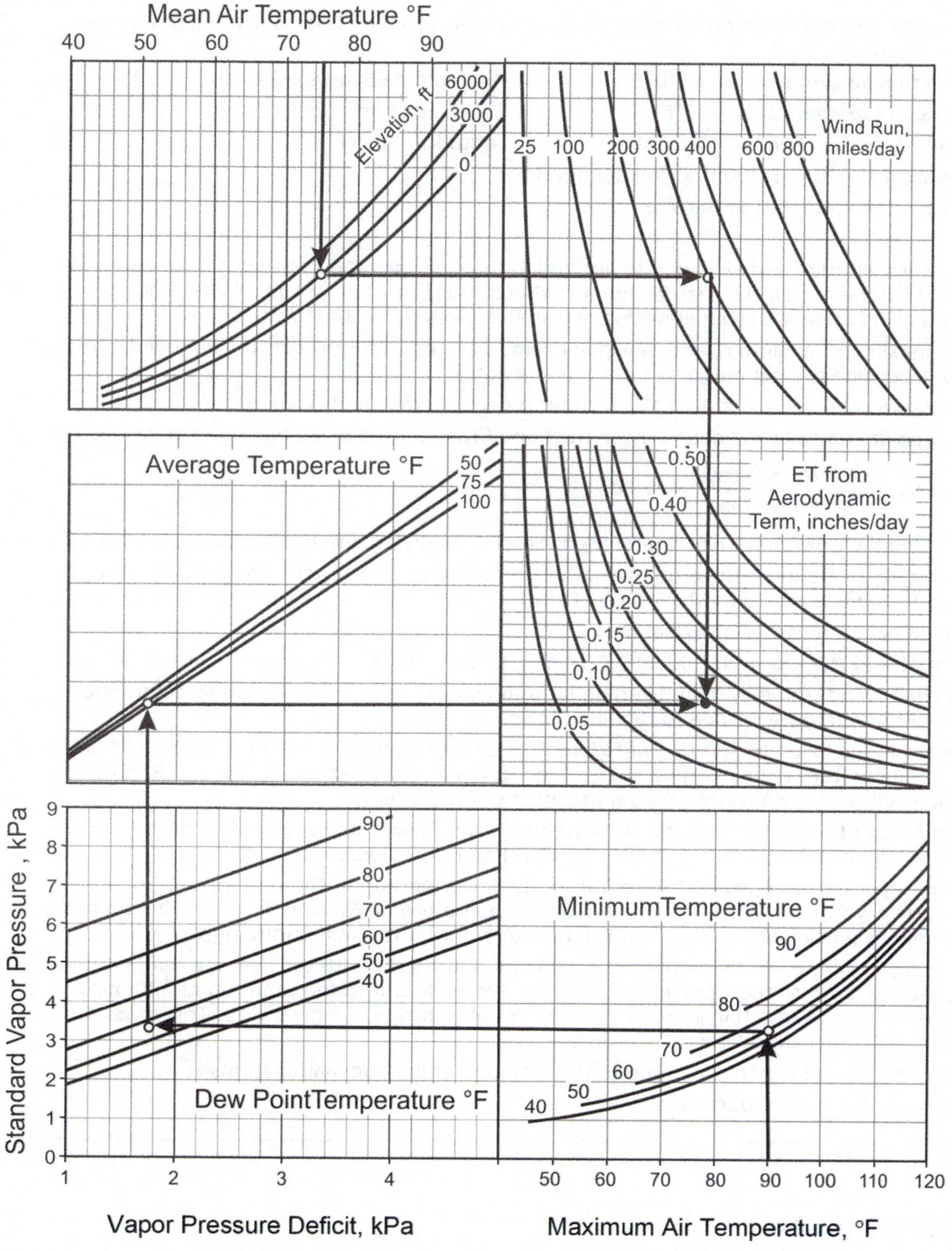

Figure 4.12. Diagram for determining reference ET from aerodynamic term for Equation 4.4.

Example 4.1

How much water might a reference crop us for a typical day in June? Data is listed below for an average day in June.

Given: Maximum air temperature = 90°F
Minimum daily air temperature = 60°F
Dew point temperature = 56°F
Solar radiation = 25 MJ/m²/d
Wind run at 2 m height = 300 mi/d
Latitude = 40°N
Elevation at site = 3,000 ft above sea level

Find: The ET_o for June 15th at the site.

Solution:

1. Use the date, latitude and elevation to determine the clear-sky radiation (R_{so}) from Figure 4.9: On June 15th the extraterrestrial radiation is about 42 MJ/m²/d at latitude 40°N. Tracing horizontally in Figure 4.9 to an elevation of 3,000 ft gives R_{so} = 32 MJ/m²/d.
2. Use the maximum, minimum, and dew point temperatures with the solar radiation in Figure 4.10 to determine the net outgoing long-wave radiation:

   Long-wave radiation is about 38.3 MJ/m²/d for a perfect black-body radiator.

   Use dew point and solar radiation for the fraction of the emitted black-body radiation:

$$R_s / R_{so} = \frac{25 \text{ MJ/m}^2\text{/day}}{32 \text{ MJ/m}^2\text{/day}} = 0.78$$

   Follow along the 56°F dew point line to the ratio of 0.78 giving emitted fraction as 11.7%.

   Proceeding down from the upper right and horizontally from the lower right gives the net outgoing long-wave radiation (R_{nl}) of 4.5 MJ/m²/d.
3. Determine net radiation using Equation 4.6:

$$R_n = (1 - \alpha)\, R_s - R_{nl} = (1 - 0.23)\, 25 - 4.5 = 14.8 \text{ MJ/m}^2\text{/d}$$
4. Use the daily wind run, elevation, average air temperature, and net radiation to determine the amount of reference ET due to radiation term:

   Go upward in the lower left portion of Figure 4.11 to an elevation of 3,000 ft, then right to the average temperature ([90 + 60] / 2 = 75), then upward to the net radiation of 14.8 MJ/m²/d, and finally left to the reference ET from radiation of 0.12 in/d.
5. Use Figure 4.12 for the ET from the aerodynamic term. Enter the diagram at 2 locations and determine where the lines in the middle right portion of the figure intersect.

   First, the vapor pressure deficit ($e_s - e_a$) from the lower portion of Figure 4.12 is 1.76 kPa. Going vertically from that point to the average temperature of 75°F provides a swing point on the left middle portion of the diagram. Follow that point horizontally to the right.

   For the second line, enter the upper left portion of Figure 4.12 with the average temperature (75°F) and go down to the elevation of 3,000 ft. Proceed right to the wind run of 300 mi/d, then down to the intersection point in the middle right portion of Figure 4.12. This gives the ET from the aerodynamic term as 0.19 in/d.
6. The total reference ET is the sum due to the radiation and aerodynamic terms:

$$ET_o = 0.12 + 0.19 = 0.31 \text{ in/d}$$

4.6 Crop Coefficients

Evapotranspiration from crops depends on the type of crop, stage of growth, water content of the soil, and the amount of energy available to evaporate water. The reference crop evapotranspiration rate (ET_o) is used to represent the amount of energy available. The ET for crops is computed relative to the ET_o using the ***crop coefficient*** (K_c):

$$ET = K_c\, ET_o \tag{4.7}$$

The crop coefficient consists of the basal crop coefficient (K_{co}) which represents crops with adequate soil water to maintain transpiration tempered by a stress factor (K_s) to account for water stress and a factor (K_w) to adjust for increased evaporation from a wet soil surface. The crop coefficient is calculated by:

$$K_c = K_{co}\, K_s + K_w \quad (4.8)$$

where: K_{co} = basal crop coefficient for unstressed crops with a dry soil surface,
K_s = stress factor to account for effects of water stress on ET, and
K_w = soil wetness factor to account for increased evaporation from wet soils.

The crop coefficient depends on the growth and development of the crop canopy. A measure of crop canopy development is the leaf area index (LAI). The LAI is the ratio of the amount of leaf area relative to the underlying land area. For example, if the total surface area of one side of the leaves is 2,600 in^2 for a 3-ft square area of a field (i.e., 1,296 in^2), then the LAI is about 2. The maximum LAI for many irrigated crops often exceeds 5 but depends on the crop variety, plant density and geographical location of the field. An example of the LAI for an annual crop during the year is illustrated in Figure 4.13.

The basal crop coefficient (K_{co}) resembles the LAI curve during the season (Figure 4.13). Early in the growing season, the basal crop coefficient is small for an annual crop. As the crop sprouts and seedlings start to grow, transpiration contributes a larger portion of daily water use, thus, the crop coefficient increases with canopy development. At some point the canopy develops sufficiently so that the crop coefficient reaches a maximum value. This time is referred to as the effective cover date. After effective cover, the crop coefficient is essentially constant for a period even though the plant canopy continues to expand. The crop coefficient decreases as the crop matures and leaves senesce. For crops that are harvested before senescence, the crop coefficient may remain at or near the peak value until harvest. Some perennial crops, such as citrus, maintain a near constant canopy from one season to the next year. Conversely, some perennial crops, such as fruit trees and grasses, emerge from dormancy and develop vegetation during the initial periods of growth. The initial crop coefficient for a crop breaking dormancy is often higher than for annual plants.

When the plant canopy is small the soil surface is not completely shaded and evaporation from wet soil contributes significantly to ET. When the soil surface is dry the rate of evaporation is small. Following a rain or irrigation the evaporation rate increases. Therefore, the crop coefficient increases immediately following a rain or irrigation (Figure 4.13). As the soil dries the crop coefficient decreases back to the rate for dry soil surfaces. As the canopy expands, the crop shades larger portions of

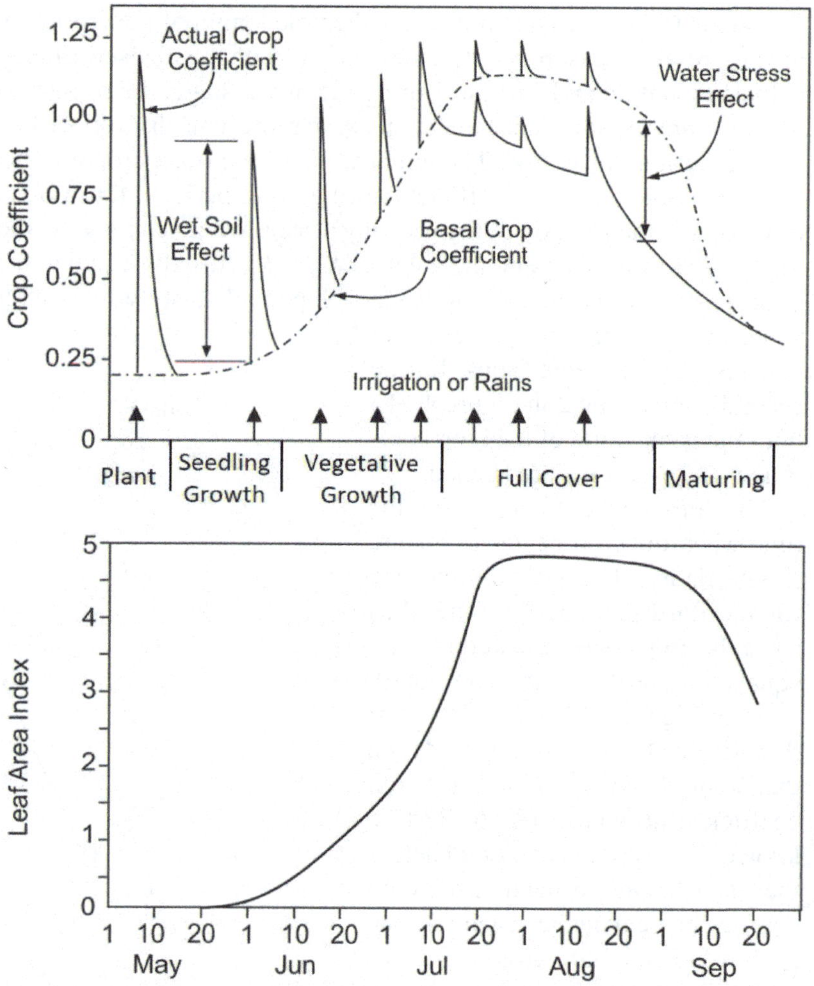

Figure 4.13. General shape of crop coefficient curve for an annual crop and the relationship to leaf area index.

the soil surface and absorbs energy that earlier would have been used to evaporate water from the soil. The effect of evaporation from wet soil, therefore, decreases as the canopy develops.

The crop ET rate decreases when plants are stressed by lack of water. Processes involved in reducing ET are complex but relate to the increased difficulty for the plant to extract soil water. For computing irrigation water requirements, the effect of water stress on ET can be estimated by decreasing the crop coefficient as in Figure 4.13.

4.6.1 Basal Crop Coefficients

The crop coefficient system presented by Allen, et al. (1998) provides a comprehensive list of basal crop coefficients. The basal coefficients represent water use of a healthy, well-watered crop where the soil surface is dry. Allen's basal crop coefficients are matched to the short reference crop used in this chapter. With this method the growing season is divided into four stages:

(1) initial stage: the period from planting through seedling growth when the soil is minimally shaded by the crop (ground shade < 10%).

(2) vegetative: the period from the initial stage to the time that the crop effectively shades the soil surface (ground shade ≅70 to 80%).

(3) midseason: period from full cover until the start of maturation when leaves begin to change color or senesce.

(4) maturing: the period from end of midseason until physiological maturity or harvest.

The progression of the basal crop coefficient during the season is illustrated in Figure 4.14 for field corn at an example site. The fraction of the growing season method developed by Stegman (1988) is used to normalize the time length of the crop growing season. The fraction of the growing season is defined as the ratio of the elapsed time since planting to the time between planting and harvest. During the initial stage, the primary water loss is due to evaporation from the soil. Since the basal curve represents dry soil surfaces, it has a constant value of 0.15 during this period. The initial value of the basal crop coefficient is denoted by K_{ci}.

To compute the crop coefficient during other periods, four points need to be defined. The first point is the fraction of the growing season where canopy development begins (point 1 in Figure 4.14). At this point, the value of $K_{co} = K_{ci}$ (usually equal to 0.15) is known. The second point occurs when the canopy has developed adequately to provide effective cover. This is when the basal crop coefficient reaches its peak value. Thus, for the second point (point 2 in Figure 4.14), both the peak value of the crop coefficient (K_{cp}') and F_{S2} are needed.

The third point in Figure 4.14 is the time when the crop begins to mature (loses vitality). The only value needed for the third point is the time (F_{S3}) since the crop coefficient at that point equals the peak value. The fourth point in Figure 4.14 represents crops that senesce before harvest. To define this point, the value of the basal crop coefficient at harvest (K_{cm}') must be known. If the crop is harvested before the plant begins to mature, the crop coefficient remains constant at the peak value until harvest.

The five factors needed to compute the basal crop coefficient (F_{S1}, F_{S2},

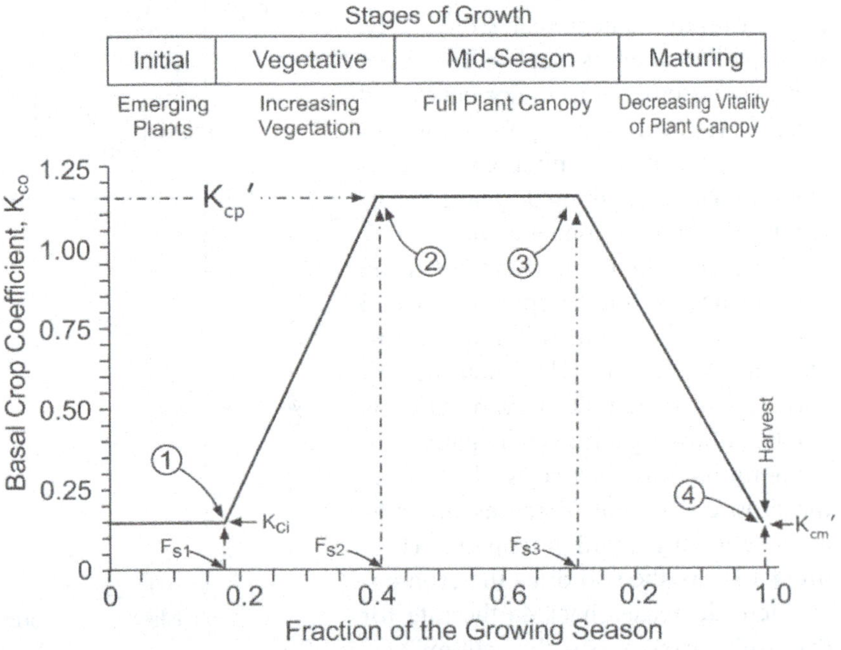

Figure 4.14. Development of the basal crop coefficient throughout the growing season for field corn.

F_{S3}, K_{cp}', K_{cm}') are labeled in Figure 4.14. The values presented in Figure 4.14 are $F_{S1} = 0.18$, $F_{S2} = 0.41$, $F_{S3} = 0.71$, $K_{cp}' = 1.15$ and $K_{cm}' = 0.15$.

Factors needed to compute basal crop coefficients for some crops are summarized in Table 4.3. Factors for crops not shown in Table 4.3 are reported by Allen, et al. (1998) or Doorenbos and Pruitt (1977). Locally developed crop coefficients can be used when available and reliable.

Doorenbos and Pruitt (1977) stress that "crop coefficient values relate to ET of a disease-free crop grown in large fields under optimum soil water and fertility conditions and achieving

Table 4.3. Basal crop coefficient information for selected crops (adapted from Allen, et al. 1998).

Crop	Crop coefficients			Fraction of growing season			Soil water stress threshold, f_{rc}[a]
	K_{ci}	K_{cp}'	K_{cm}'	F_{S1}	F_{S2}	F_{S3}	
Alfalfa, first cuttings	0.30	1.15	1.10	0.13	0.53	0.87	0.45
Alfalfa, later cuttings	0.30	1.15	1.10	0.11	0.56	0.78	0.45
Beans, dry	0.15	1.00	0.80	0.25	0.50	0.80	0.55
Beans, green	0.15	1.00	0.80	0.22	0.56	0.89	0.55
Carrot	0.15	0.95	0.85	0.17	0.42	0.83	0.65
Corn, field	0.15	1.15	0.15	0.18	0.41	0.71	0.45
Corn, sweet	0.15	1.10	1.00	0.30	0.60	0.90	0.50
Cotton	0.15	1.10	0.50	0.17	0.44	0.75	0.35
Cucumber	0.15	0.95	0.70	0.19	0.48	0.86	0.50
Grapes, table	0.15	0.80	0.40	0.10	0.34	0.71	0.65
Grapes, wine	0.15	0.65	0.40	0.10	0.34	0.71	0.55
Hay, Bermuda grass	0.50	0.95	0.80	0.07	0.19	0.74	0.45
Hay, rye grass	0.85	1.00	0.95	0.07	0.19	0.74	0.40
Lentil	0.15	1.05	0.20	0.15	0.35	0.76	0.50
Lettuce	0.15	0.90	0.90	0.29	0.67	0.90	0.70
Pepper, bell	0.15	1.00	0.80	0.14	0.33	0.86	0.70
Potato	0.15	1.10	0.65	0.27	0.45	0.88	0.65
Rice	1.00	1.15	0.55	0.20	0.40	0.80	0.80
Sorghum, grain	0.15	1.00	0.35	0.16	0.44	0.76	0.45
Soybeans	0.15	1.10	0.30	0.14	0.39	0.82	0.50
Sugar beet	0.15	1.15	0.90	0.28	0.50	0.78	0.45
Sunflower	0.15	1.10	0.25	0.19	0.46	0.81	0.55
Tomato	0.15	1.10	0.70	0.23	0.48	0.81	0.60
Watermelon	0.15	0.95	0.70	0.18	0.45	0.73	0.60
Wheat, spring	0.15	1.10	0.15	0.15	0.33	0.78	0.45
Wheat, winter[b]	0.15 / 0.50	1.10	0.15	0.48	0.70	0.93	0.45
Pasture, rotated grazing	0.30	0.90	0.80	0.05	0.15	1.00	0.40
Pasture, continuous grazing	0.30	0.70	0.70	0.05	0.15	1.00	0.40
Citrus, no ground cover							
70% canopy	0.65	0.60	0.65	0.16	0.41	0.74	0.50
50% canopy	0.60	0.55	0.60	0.16	0.41	0.74	0.50
20% canopy	0.45	0.40	0.50	0.16	0.41	0.74	0.50
Citrus, with ground cover							
70% canopy	0.75	0.70	0.75	0.16	0.41	0.74	0.50
50% canopy	0.75	0.75	0.75	0.16	0.41	0.74	0.50
20% canopy	0.80	0.80	0.85	0.16	0.41	0.74	0.50
Apples, cherries, pears							
No ground cover, killing frost	0.35	0.90	0.65	0.13	0.33	0.88	0.50
No ground cover, no frost	0.50	0.90	0.70	0.13	0.33	0.88	0.50
Ground cover, killing frost	0.45	1.15	0.90	0.13	0.33	0.88	0.50
Ground cover, no frost	0.75	1.15	0.80	0.13	0.33	0.88	0.50

[a] The critical fraction remaining (f_{rc}) is discussed in section 4.6.2.
[b] Larger value for initial period is when fallow wheat provides full ground cover, but the soil is not frozen.

full production under the given growing environment." Crops not meeting these provisions generally use less water unless raised in small fields where the effects of field boundaries can cause ET to be significantly different.

The crop coefficient depends upon the prevailing climatic conditions. The ET of tall crops, such as trees, is affected more by wind than short crops such as grass. This effect is amplified in arid climates. Therefore, Allen, et al. (1998) recommended that the basal crop coefficient be adjusted based on wind speed and humidity. The basal crop coefficient is computed from the values listed in Table 4.3 plus the adjustment factor in Table 4.4.

The initial value K_{ci} is not modified; however, K_{cp}' and K_{cm}' are adjusted according to the following equations:

$$K_{cp} = K_{cp}' + K_{cf}$$

and

$$K_{cm} = K_{cm}' + K_{cf} \tag{4.9}$$

where K_{cp} and K_{cm} are the adjusted coefficients, K_{cp}' and K_{cm}' are the tabular values of the coefficients (Table 4.3), and K_{cf} is the crop coefficient adjustment factor for crop height and wind speed (Table 4.4).

Table 4.4. Crop coefficient adjustment factor (K_{cf}) for wind speeds and relative humidity.

Wind Run	Average Minimum Relative Humidity (%)						
(mi/d)	20	30	40	50	60	70	80
			Crop height, 2 ft				
50	0.03	0.01	-0.02	-0.04	-0.06	-0.09	-0.11
100	0.06	0.03	0.01	-0.02	-0.04	-0.07	-0.09
150	0.08	0.05	0.03	0.00	-0.02	-0.05	-0.07
200	0.10	0.08	0.05	0.03	0.00	-0.02	-0.05
250	0.12	0.10	0.07	0.05	0.02	0.00	-0.03
300	0.15	0.12	0.10	0.07	0.05	0.02	0.00
350	0.17	0.14	0.12	0.09	0.07	0.04	0.02
			Crop Height, 4 ft				
50	0.04	0.01	-0.02	-0.05	-0.08	-0.11	-0.14
100	0.07	0.04	0.01	-0.02	-0.05	-0.08	-0.11
150	0.10	0.07	0.04	0.01	-0.02	-0.06	-0.09
200	0.12	0.09	0.06	0.03	0.00	-0.03	-0.06
250	0.15	0.12	0.09	0.06	0.03	0.00	-0.03
300	0.18	0.15	0.12	0.09	0.06	0.03	0.00
350	0.21	0.18	0.15	0.11	0.08	0.05	0.02
			Crop Height, 6 ft				
50	0.05	0.01	-0.02	-0.06	-0.09	-0.12	-0.16
100	0.08	0.04	0.01	-0.02	-0.06	-0.09	-0.13
150	0.11	0.08	0.04	0.01	-0.03	-0.06	-0.10
200	0.14	0.11	0.07	0.04	0.00	-0.03	-0.07
250	0.17	0.14	0.10	0.07	0.03	0.00	-0.04
300	0.20	0.17	0.13	0.10	0.06	0.03	0.00
350	0.23	0.20	0.16	0.13	0.10	0.06	0.03
			Crop Height, 8 ft				
50	0.05	0.01	-0.02	-0.06	-0.10	-0.14	-0.17
100	0.09	0.05	0.01	-0.03	-0.06	-0.10	-0.14
150	0.12	0.08	0.04	0.01	-0.03	-0.07	-0.11
200	0.15	0.12	0.08	0.04	0.00	-0.03	-0.07
250	0.19	0.15	0.11	0.07	0.04	0.00	-0.04
300	0.22	0.18	0.15	0.11	0.07	0.03	-0.01
350	0.25	0.22	0.18	0.14	0.10	0.07	0.03
		Multiplier for Other Crop Heights (multiply times values for 6-ft crop)					
Crop Height, ft		10	12	14	16	20	25
Multiplier		1.17	1.23	1.29	1.34	1.44	1.53

The climatic data used to adjust the crop coefficient are average values for the appropriate time of year for a specific region. Daily measured climatic conditions are not used to make the adjustment. The minimum relative humidity used in Table 4.4 can be computed by:

$$RH_{min} = 100 \frac{e^o(T_{dew})}{e^o(T_{max})} \tag{4.10}$$

Crop coefficient varies linearly with the fraction of the growing season during the vegetative and maturing growth stages. During the vegetative stage, the crop coefficient is computed with the following equation:

$$K_{co} = K_{ci} + \left(K_{cp} - K_{ci}\right)\left(\frac{F_s - F_{s1}}{F_{s2} - F_{s1}}\right) \tag{4.11}$$

During the maturing stage, the crop coefficient is computed with:

$$K_{co} = K_{cp} - \left(K_{cp} - K_{cm}\right)\left(\frac{F_s - F_{s3}}{1 - F_{s3}}\right) \tag{4.12}$$

Example 4.2

Basal crop coefficients are fundamental to using ET in irrigation management. The procedure depends on the midseason and harvest values of the basal crop coefficient. Determine the peak and harvest values of the basal crop coefficient for corn at the site described below.

Given: Field corn with the following conditions:

Average Conditions	Midseason	Harvest
Maximum air temperature	90°F	50°F
Dew point temperature	65°F	40°F
Wind run	200 miles per day	150 miles per day

Solution:

Corn will be about 8 ft tall.

From Table 4.3: $K_{cp}' = 1.15$ and $K_{cm}' = 0.15$

Determine minimum relative humidity using Equation 4.10 and Table 4.1:

at midseason:

$$RH_{min} = 100\left(\frac{e^o(65)}{e^o(90)}\right) = 100\left(\frac{2.11 \text{ kPa}}{4.81 \text{ kPa}}\right) = 44\%$$

at harvest:

$$RH_{min} = 100\left(\frac{e^o(40)}{e^o(50)}\right) = 100\left(\frac{0.84 \text{ kPa}}{1.23 \text{ kPa}}\right) = 68\%$$

From Table 4.4 the adjustment factors are:

at midseason: K_{cf} is about 0.06

at harvest: K_{cf} is about −0.07.

The adjusted basal crop coefficients would then be:

$K_{cp} = K_{cp}' + K_{cf} = 1.15 + 0.06 = 1.21$

$K_{cm} = K_{cm}' + K_{cf} = 0.15 - 0.07 = 0.08$

Example 4.3

Ultimately computing crop ET depends on the value of the basal crop coefficient during the season. To illustrate the process compute the basal crop coefficient on the 15th of May, June, July, August and September for corn at the site in Example 4.2.

Given: Corn grown for grain planted on May 1 is harvested on September 30. Use the basal crop coefficients for K_{cp} and K_{cm} from Example 4.2.

Solution:

Use the elapsed time since planting to describe canopy development.

Determine the time from planting to harvest:

31 days in May + 30 in June + 31 in July + 31 in August + 30 in September = 153 day growing season

Determine the fraction of the growing season for each date:

Date	Elapsed Time Since Planting	Fraction of the Growing Season
May 15	15	15 / 153 = 0.10
June 15	46	46 / 153 = 0.30
July 15	76	76 / 153 = 0.50
August 15	107	107 / 153 = 0.70
September 15	138	138 / 153 = 0.90

From Table 4.3: F_{s1} = 0.18, F_{s2} = 0.41, F_{s3} = 0.71

From Example 4.2: K_{cp} = 1.21 and K_{cm} = 0.08

On May 15: F_s = 0.1, which is between 0 and F_{s1}, so $K_{co} = K_{ci}$ = 0.15

On June 15: F_s = 0.30, which is between F_{s1} and F_{s2}, so

$$K_{co} = 0.15 + (K_{cp} - 0.15)\left[\frac{F_s - F_{s1}}{F_{s2} - F_{s1}}\right] \qquad \text{(Eq. 4.11)}$$

$$K_{co} = 0.15 + (1.21 - 0.15)\left[\frac{0.3 - 0.18}{0.41 - 0.18}\right] = 0.70$$

On July 15: F_s = 0.50, which is between F_{s2} and F_{s3}, so $K_{co} = K_{cp}$ = 1.21

On August 15: F_s = 0.70, which is between F_{s2} and F_{s3}, so $K_{co} = K_{cp}$ = 1.21

On September 15: F_s = 0.90, which is between F_{s3} and 1.0, so:

$$K_{co} = 1.21 - (1.21 - 0.08)\left[\frac{0.90 - 0.71}{1.00 - 0.71}\right] = 0.47 \qquad \text{(Eq. 4.12)}$$

4.6.2 Water Stress Effects

If management or water supply limitations restrict irrigation, the effect of water stress on ET should be considered. For managing irrigation, the effect of water stress on ET can be described using a stress factor K_s which is based on soil water content. A linear function (Figure 4.15) has been used by Hanks (1974) and Ritchie (1973). With this method the stress factor is based on the fraction of the available soil water that is stored in the crop root zone. The stress factor (K_s) is computed as:

$$K_s = \frac{f_r}{f_{rc}} \text{ for } f_r < f_{rc}$$

$$= 1 \text{ for } f_r \geq f_{rc} \qquad (4.10)$$

where: K_s = stress factor,

f_r = fraction of the available soil water that remains, and

f_{rc} = critical threshold of f_r when stress begins (Table 4.3).

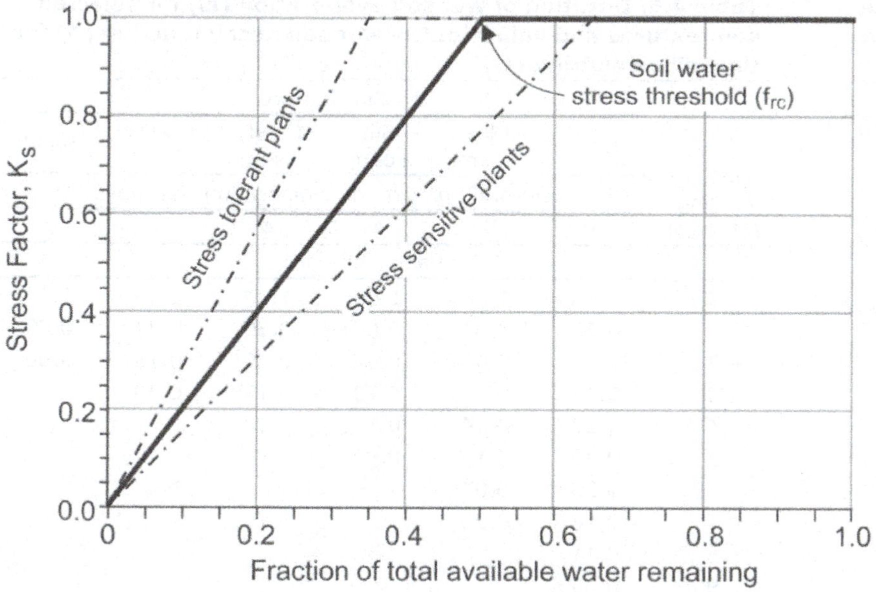

Figure 4.15. Relationship of the soil water stress factor (Ks) to available soil water.

Crops vary in the ability to withstand soil water stress. Some crops are tolerant and maintain ET rates under relatively dry conditions. Other crops are sensitive and ET rates decrease when soil is wetter (Figure 4.15). Values for the soil water stress threshold are in Table 4.3. Threshold values for other crops are available from Allen, et al. (1998).

> **Example 4.4**
>
> Stress reduces the rate of crop water use. Utilize the water stress factor method to determine the ET for the soil water conditions listed below.
>
> Given: Current volumetric water content of a sandy loam soil = 0.15
> Volumetric water content at field capacity = 0.24
> Volumetric water content at the wilting point = 0.12
> ET_o = 0.30 in/day
> K_{co} = 1.20
> Critical soil water content (f_{rc}) = 0.45
>
> Solution:
>
> Assume K_w = 0
>
> $$f_r = \frac{\theta_v - \theta_{wp}}{\theta_{fc} - \theta_{wp}} \qquad \text{(Eq 2.10)}$$
>
> $$f_r = \left(\frac{0.15 - 0.12}{0.24 - 0.12}\right) = \frac{0.03}{0.12} = 0.25$$
>
> $$K_s = \frac{f_r}{f_{rc}} = \frac{0.25}{0.45} = 0.56$$
>
> $ET = K_s\ K_{co}\ ET_o = 0.56 \times 1.20 \times 0.30$ in/day = 0.2 in/day

4.6.3 Wet Soil Evaporation

The increased rate of evaporation due to a wet soil surface is influenced by the amount of canopy development, the energy available to evaporate water, and the hydraulic properties of the soil. The factor (K_w) can be used to predict the amount of wet soil evaporation. The total amount of evaporation from a wet soil should be less than the amount of water received by rain or irrigation.

The method of Wright (1982) has been adapted to account for wet soil evaporation:

$$K_w = F_w\,(K_{cmax} - K_{co})\,f_t$$
$$K_{cmax} = 1.2 + K_{cf} \qquad (4.11)$$
$$f_t = 1 - \sqrt{\frac{t}{t_d}}$$

where: K_w = wet soil evaporation factor,
K_{cmax} = maximum crop coefficient for wet soil evaporation,

Table 4.5. Fraction of the soil surface wetted for various types of irrigation systems.

Wetting Method	F_w
Rain	1.0
Sprinkler irrigation:	
Above canopy sprinklers	1.0
In-canopy sprinklers	0.75
LEPA systems (alternate furrows wetted)	0.5
Borders and basin irrigation	1.0
Furrow irrigation:	
Large application depth	1.0
Small application depth	0.5
Alternate furrows irrigated	0.5
Surface trickle or drip irrigation	0.25
Subsurface drip irrigation:	
Large applications	0.1
Normal applications	0.0

Table 4.6. Duration of wet soil evaporation (t_d) for selected soil textures and values of the wet soil decay function (f_t) for time since wetting (t).

Time Since Wetting, (t), days	Soil Texture					
	Clay	Clay Loam	Silt Loam	Sandy Loam	Loamy Sand	Sand
	Duration of wet soil evaporation (t_d), days					
	10	7	5	4	3	2
	Values of f_t					
0	1.00	1.00	1.00	1.00	1.00	1.00
1	0.68	0.62	0.55	0.50	0.42	0.29
2	0.55	0.47	0.37	0.29	0.18	0.00
3	0.45	0.35	0.23	0.13	0.00	
4	0.37	0.24	0.11	0.00		
5	0.29	0.15	0.00			
6	0.23	0.07				
7	0.16	0.00				
8	0.11					
9	0.05					
10	0.00					

K_{cf} = crop coefficient adjustment factor (Table 4.4),
F_w = fraction of surface wetted,
t = time since last wetting of soil surface (d),
t_d = duration of wet soil evaporation (d), and
f_t = wet soil decay function.

The fraction of the surface wetted depends on the type of irrigation system (Table 4.5). The duration of wet soil evaporation depends on the type of soil. Sandy soils dry quicker than fine-textured soils. Representative values of the drying duration are given in Table 4.6. Local observations can also be used to determine values for the drying duration.

Example 4.5

You manage two adjacent fields with the same soil properties and crop conditions but different soil water status. Compute the evapotranspiration rate for each field for the following information.

Given: Soil water was measured in each field with the following results:

Field A: Soil water content is 0.22 Field B: Soil water content is 0.12

Volumetric water content at field capacity and permanent wilting are 0.25 and 0.10.

The crop root zone is 4 ft deep in both fields.

The reference ET rate is 0.30 in/d and the basal crop coefficient is 1.1 at this growth stage.

The crop grown at the sites has a critical soil water threshold (f_{rc}) of 0.35.

The soil has not been wetted during the last week.

Solution: Wet soil evaporation is insignificant since there was no irrigation or rain recently, so K_w = 0.

1. Compute f_r for each field: $f_r = \dfrac{\theta_v - \theta_{wp}}{\theta_{fc} - \theta_{wp}}$ (Eq. 2.10)

   Field A: $f_r = \left(\dfrac{0.22-0.10}{0.25-0.10}\right) = 0.80$ Field B: $f_r = \left(\dfrac{0.12-0.10}{0.25-0.10}\right) = 0.13$

2. Compute the ET for crop in field A:

   From Eq. 4.13 since $f_r \geq f_{rc}$ (i.e., 0.80 ≥ 0.35), then, K_s = 1.0

   and the ET rate is $ET = K_s\, K_{co}\, ET_o = 1.0 \times 1.1 \times 0.3 = 0.33$ in/d

3. Compute the ET for field B:

   In Field B, $f_r < f_{rc}$ (0.13 < 0.35), so $K_s = \dfrac{f_r}{f_{rc}} = \dfrac{0.13}{0.35} = 0.37$ (Eq. 4.13)

   and $ET = K_s\, K_{co}\, ET_o = 0.37 \times 1.1 \times 0.3 = 0.12$ in/d

4.6.4 Methods to Describe Canopy Development

Every year the weather is different causing the rate of crop growth to vary even for the same planting date. Methods are needed to ensure that the predicted rate of canopy development is accurate. The elapsed time (days) since planting and the cumulative growing degree days (sometimes called heat units) since planting are often used as the basis to estimate crop growth. The elapsed time since planting is easier to use; however, some of the annual variation of canopy development can be accounted for using growing degree days.

The definition for growing degree days is:

$$GDD_n = \sum_{i}^{n} (T_i - T_{base}) \tag{4.12}$$

where: GDD_n = cumulative growing degree days on the nth day after planting,
n = total number of days since planting,
T_i = average air temperature [$0.5 \times (T_{high} + T_{low})$] on day i, (°F),
T_{base} = base temperature at which crop photosynthesis and growth begins,
T_{high} = the smaller of the daily maximum temperature and 86°F, and
T_{low} = the larger of the minimum temperature and T_{base}

The base temperature depends on the crop species. The base temperature for warm weather crops such as corn is typically 50°F, while 40°F is commonly used for cool season crops such as wheat and barley. Because of local variations, the base temperature for specific crops at a location should be determined from regional information.

Growing degree days can be used to determine the fraction of the growing season for computing the crop coefficient:

$$F_s = \frac{GDD_n}{GDD_m} \tag{4.13}$$

where: GDD_n = the cumulative growing degree days from planting to day n and
GDD_m = the cumulative growing degree days needed to reach maturity.

4.7 Intercropping

Some irrigated fields are divided into more than one area for crop rotations. In these cropping systems only one crop is irrigated at a time, so only the ET for that species is relevant for managing that sector. However, interest has grown recently in various forms of intercropping. Intercropping involves growing two or more crops simultaneously juxtaposed within parts of the field. Intercropping includes various forms (Figure 4.16). Mixed intercropping is a complete mixture of multiple species in the same area. Row intercropping involves growing two or more crops at the same time within crop rows. This is common in developing countries and small holdings where an upper story crop—often corn—is first planted and then a shorter crop such as beans is planted in the furrow between crop rows. The crops occupy the same space but may have different growth schedules so the composition of the vegetation changes throughout the season. Alley and strip cropping involves alternating strips of single crops. Strip cropping generally involves paths of equal width off alternating crops. Alley cropping is frequently a form of agroforestry where tree lines are planted beside strips of crops. The width of crop strips in strip and alley cropping is usually some multiple of farming equipment width. Relay and/or cover cropping involves starting a second crop before the first crop is mature or harvested. Crop establishment can be difficult for the second crop in the series.

Computation of water use for intercropped systems is difficult because crops in the mixture have different characteristics. The distribution of leaf area usually involves some shading of

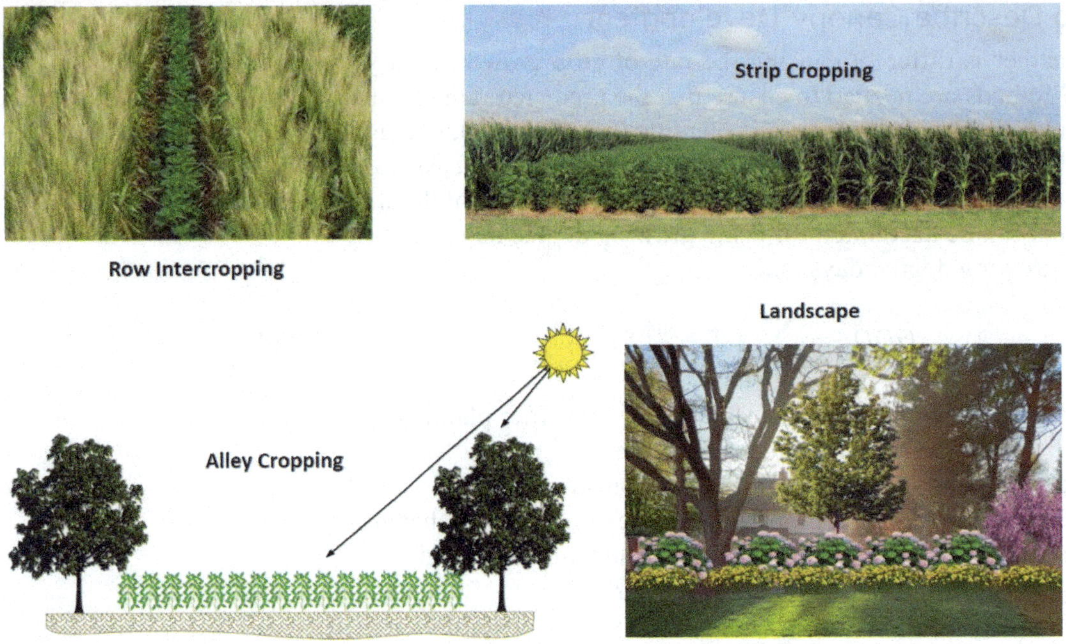

Figure 4.16. Examples of intercropping (upper left photo courtesy of USDA-NRCS).

lower crops as illustrated for alley cropping in Figure 4.16. Determining the capture of radiation requires information of leaf area distribution vertically and horizontally. Multiple stories of vegetation alter wind patterns within and above crops in the mixture. Rooting characteristics may be quite different leading to dissimilar levels of water stress. The development of the canopy for crops in the mixture may progresses at different rates and the plant density of species in the mixture may vary considerably from field to field. The leaf area, plant geometry and general crop health may be quite different than for single crop fields represented by crop coefficients—especially for small holdings in developing countries. These complexities require more elaborate procedures than simple crop coefficients. Methods are presented by Allen et al. (1998) to estimate compound crop coefficients. Computer models are also available for simulating micrometeorological processes in complex canopies. Methods to estimate intercropped ET are multifaceted and beyond procedures present in this text.

Irrigation of intercropped systems, especially strip and alley cropping, is difficult to achieve efficiently. Water needs of one crop may differ from requirements of the adjacent crop and some irrigation systems are incapable of applying water in that configuration. The soil water along the boundary between crops is often different than in the middle of the strip or alley. This dissimilarity amplifies the distribution of water need within the strip and confounds soil water monitoring. Soil water monitoring can be effective in row intercropping system like shown in Figure 4.16.

Landscapes contain a mixture of vegetation that is irrigated at the same time, so the composite ET is needed (Figure 4.16). It is difficult to measure ET for such plantings because of the interactions occurring in the landscape and due to the variability of species in landscapes. Planting densities vary considerably among landscapes. Young landscapes contain less leaf area than mature plantings and are less capable of absorbing radiation; thus, mature landscapes usually have higher transpiration rates. A landscape of trees with underlying shrubs or groundcover captures more radiation and will require more water than trees underlain with mulch. Many landscapes include a range of microclimates varying from shaded or protected areas to hot, sunny, and windy areas. These variations influence ET in ways not representative of large areas of homogeneous vegetation inherent in crop coefficients.

Costello and Jones (2014) provide updates to a method to estimate ET using landscape coefficients for multiple species:

$$ET = K_L \times ET_o \tag{4.14}$$

Table 4.7 Range of plant species, density, and microclimate factors for landscape coefficients (adapted from Costello and Jones, 2014).

	Species Factor (K_P)			Density Factor (K_D)			Microclimate Factor (K_M)		
Type of Vegetation	High	Avg.	Low	High	Avg.	Low	High	Avg.	Low
Trees	0.9	0.5	0.2	1.3	1.0	0.5	1.4	1.0	0.5
Shrubs	0.7	0.5	0.2	1.1	1.0	0.5	1.3	1.0	0.5
Groundcover	0.7	0.5	0.2	1.1	1.0	0.5	1.2	1.0	0.5
Mixed: trees, shrubs, and groundcover	0.9	0.5	0.2	1.3	1.1	0.6	1.4	1.0	0.5
Turf grass	0.8	0.7	0.6	1.0	1.0	0.6	1.2	1.0	0.8

where K_L is the landscape coefficient. The amount of ET for a landscape varies as a function of the species planted, the density of vegetation, and microclimate conditions. Assigning numerical values for these factors enables estimation of the landscape coefficient:

$$K_L = K_P \times K_D \times K_M \tag{4.15}$$

where K_P is the plant species factor, K_D is the density factor and K_M is the microclimate factor. The range of values for each factor for types of vegetation are presented in Table 4.7.

The landscape coefficient procedure differs from the crop coefficient procedure regarding the adequacy of water. Crop coefficients approximate water use for crops under well-watered conditions intended to maximize production. Landscape coefficients approximate the water needed to maintain the aesthetic or functional acceptability of a landscape. Rather than a measure of how much water can be lost from an area, the landscape coefficient is an estimate of the water needed to maintain landscape quality.

Species factors (K_p) for five types of vegetation are included in Table 4.7. Three levels are included for each type of vegetation depending on the water use characteristics of the plants included in the landscape. Mixed species plantings have a range of water use like those of tree, shrub, and groundcover species. The values presented in Table 4.7 represent the range assumed for individual species. Costello and Jones (2014) provide species for a very large number of specific plant species to develop an integrated species factor for the landscape.

The density of vegetation within a landscape varies considerably. Even though individual plants in a sparsely planted landscape may use more water for a given leaf area than individual plants in a dense landscape, water lost from the entirety of the dense planting will likely be greater than for the sparse landscape. To account for these differences, the density factor varies from a low of 0.5 to a high of 1.3. The density factor involves estimating the percent ground cover for a portion of the landscape. Canopy cover is defined as the percentage of ground shaded. A 50% ground cover will shade half of the land area in the landscape. With a canopy cover less than 60% a reduction in K_D is appropriate. Trees with a canopy cover of 25% or less should have a density factor of 0.5.

An upward adjustment of K_D is warranted when trees are the prevailing vegetation, but shrubs and groundcover also occur. Essentially, the groundcover or shrub represents another tier of vegetation where water loss occurs. Total water use would be expected to be greater for multiple tiers than for a single tier. Shrubs and groundcover are equivalent in K_D values. A complete or nearly complete cover (about 90%) with either shrubs or groundcover represents the average condition for these vegetation types and has a density factor 1.0. Higher density values may result when plantings are predominately groundcover or shrubs, but another vegetation type also occurs. Density values for high-density mixed plantings are greater than for other three vegetation types. High density plantings with three vegetation types would be assigned a maximum density factor of 1.3. Low-density mixed plantings may also occur and a commensurate reduction in the density factor is appropriate. Young or widely spaced plantings also qualify for a low-density value.

Figure 4.17. Landscapes with varying plant density and microclimate factors.

Environmental conditions vary significantly within a landscape. Buildings and other structures and paving typical of urban landscapes strongly influence foliar and air temperatures, wind, and humidity. For example, trees in parking lots are subject to higher temperature and lower humidity than trees in parks. Areas within a landscape that have different environmental conditions are called microclimates. Microclimates must be considered in estimating water needs. The microclimate factor accounts for such differences.

The microclimate factors are relatively easy to set. An average microclimate condition is where buildings, pavement, slopes, and reflective surfaces do not influence the microclimate. Essentially, this condition is like that for the reference ET conditions. For these conditions, the microclimate climate factor (K_M) is set to 1.0.

In a "high" microclimate condition, features increase the evaporative condition in the irrigation zone. Landscape surrounded by heat-absorbing surfaces or reflective surfaces or those exposed to particularly windy conditions would be assigned high microclimate factors. For example, medians, parking lots, west sides of buildings, west and south sides of slopes, and wind tunnel areas would be assigned a higher climate factor. Such areas might have a microclimate climate value between 1.0 and 1.4. See Figure 4.17 for examples of high and low microclimate factors.

"Low" microclimate conditions are as common as high microclimate conditions. Plantings that are shaded by buildings or other landscape features for part or most of the day, or that are protected from winds, would be assigned low microclimate values. Examples of conditions that should receive low microclimate factors include areas on the north sides of buildings, courtyards, under wide building overhangs, and the north side of slopes. Such situations would be assigned microclimate values between 0.5 and 1.0 (Figure 4.17).

Application of the landscape methodology is very well developed by Costello and Jones (2014) but is complicated. That publication should be utilized for specific applications. The method may also offer a basis for estimating ET for other intercropped systems.

4.8 Accessing Climatic Information

The rate plants use water determines irrigation schedules and ultimately the depth of irrigation water to apply. Without this information it is difficult to efficiently manage irrigation systems. The methods in this chapter rely on data for reference crop conditions with the Penman-Monteith equation. This involves accurate measurement of several climatic variables. Most irrigators will not measure these variables at their field.

Fortunately, weather data networks have been established across the United States to provide data for the Penman-Monteith method. In most situations, networks also compute the reference crop ET. Care must be taken to ensure that the reference ET provided by the service is for a short reference crop (i.e., grass clipped to a height of 5 inches). Providers of the climatic data may also compute water use rates for crops grown in the local vicinity. While these calculations should be carefully monitored for accuracy and reliability, the computed values

Table 4.8. Example weather, reference crop ET and growing degree day data from High Plains Regional Climate Center.

		Air Temperature, °F				Wind Speed	Solar Radiation	Rain	ET_o	Growing Degree
Month	Day	Max.	Min.	Average	Dew Point	(mi/hr)	(Lang/d)	(in)	(in/d)	Units
7	2	89.5	63.9	76.7	70.4	8.2	559	0.16	0.26	25
7	3	87.5	63.1	75.3	65.9	5.4	651	0	0.27	25
7	4	88.4	64.6	76.5	68.3	4.1	668	0	0.26	25
7	5	86.7	64.3	75.5	67.8	3.0	542	0	0.21	25
7	6	91.1	65.6	78.4	70.2	4.5	600	0	0.25	26
7	7	91.3	67.4	79.4	69.6	9.5	659	0	0.34	27
7	8	92.2	62.6	77.4	66.0	11.4	643	0.62	0.38	24
7	9	83.8	59.9	71.9	65.6	5.0	638	0	0.24	22
7	10	88.7	62.5	75.6	68.8	6.7	597	0.01	0.26	24
7	11	85.0	62.0	73.5	66.0	5.8	647	0	0.26	24
7	12	85.1	58.1	71.6	64.0	4.5	548	0	0.22	22
7	13	91.8	63.1	77.5	68.0	8.7	636	0.14	0.33	25
7	14	77.8	63.4	70.6	65.3	6.4	407	0.01	0.16	21
7	15	81.4	60.8	71.1	63.3	4.2	584	0	0.22	21

can often be used directly for managing irrigation. Readers should refer to the local Extension Service at their university for assistance in locating climatic data for their location. An example of data provided from the High Plains Regional Climate Center is provided in Table 4.8. These data can be used in irrigation scheduling and other applications.

Smith (1992) developed a decision support program to estimate crop water requirements for a wide range of crops. The program can utilize local weather data or historical climatic information to develop average water requirements for planning purposes. The program can be used for irrigation scheduling and is useful for managing whole-farm irrigation systems.

Data are also becoming available from analysis of Landsat and other remote sensing systems. Techniques have been developed to predict ET and crop coefficients. There are several current and emerging techniques along with new satellite capabilities that promise future opportunities for irrigation planning and management. The example from Barker, et al. (2018) shows that remote sensing can accurately predict crop coefficients periodically throughout the growing season. Methods like that by Barker et al. (2018) also provide methods to estimate ET and crop coefficients between the days that satellites pass over the specific locations. Currently, these methods are still being developed but promise substantial opportunity for real-time irrigation management in the future.

4.9 Summary

Management of irrigation systems depends on knowledge of the rate that plants use water. This chapter presents methods to compute the evapotranspiration rate for field crops. The methods are based on the Penman-Monteith equation used to compute the ET of a healthy and well-watered grass reference crop that is approximately five inches tall. Climatic data for air temperature, relative humidity, solar radiation, and wind speed are needed to compute the reference ET. The water use of crops and vegetables is computed by multiplying the reference ET times a crop coefficient. The crop coefficient represents the effect of canopy development as well as plant water stress and increased evaporation from wet soils.

Questions

1. Explain what a grass reference crop is and why this concept is used in estimating crop water use.
2. Describe the sources of energy that affect ET. Which sources are most important in semiarid locations?
3. An irrigation district must establish a schedule for water delivery to the 1,000 producers that they serve. Describe and explain the procedure you would use to develop the schedule.
4. Why does a cotton crop have a higher ET rate than field beans when they both completely shade the soil (i.e., after effective cover)?
5. List and explain three ways data on the rate of crop ET are used in irrigation management.
6. Compute the basal crop coefficient on July 1 for tomatoes planted on May 1 in the Central Valley of California.
7. Given the following information:
   - Maximum daily air temperature = 95°F
   - Minimum daily air temperature = 72°F
   - Dew point temperature = 68°F
   - Daily wind run = 200 mi/d
   - Daily solar radiation = 28 $MJ/m^2/d$
   - Elevation above sea level = 1,500 ft
   - Latitude = 30°N
   - Date is July 25

   a. Compute the daily reference ET using the Penman-Monteith method.

   b. What fraction of the total ET_o is due to the aerodynamic term?

   c. What fraction of ET_o is caused by the radiation term?
8. The reference ET is 0.30 in/d, the basal crop coefficient is 0.6, and it has been two days since a thorough rain occurred. The soil water depletion is about 20% of the available water holding capacity for the silt loam soil. How much crop ET occurs for these conditions if $f_{rc} = 0.5$? Assume that $K_{cf} = 0$.
9. A crop is irrigated with a subsurface drip irrigation system. The reference ET is 0.25 in/d and the basal crop coefficient is 1.05. The soil water depletion is 50% before irrigating, and $f_{rc} = 0.5$. The root depth is 4 ft, and the soil is a sandy loam. How many days will a 2-inch irrigation last if $K_{cmax} = 1.2$ and $K_{cf} = 0.0$?
10. How many growing degree days would accumulate in one day if the maximum and minimum air temperatures were 90° and 65°F, respectively and the base temperature was 50°F?
11. Given the following data:
    - Corn planted on May 1.
    - Effective cover date July 10 (end of week 10).
    - Maturity date is September 10 (end of week 19).
    - Irrigated to prevent soil water stress.
    - The crop is irrigated, or it rains once a week.
    - The soil is silt loam.

    a. Compute the average daily ET for corn for each week of the season.

    b. What is the total seasonal ET?

Weeks after May 1	Average ET_o (in/d)	Weeks after May 1	Average ET_o (in/d)
1	0.21	11	0.26
2	0.22	12	0.26
3	0.23	13	0.25
4	0.24	14	0.25
5	0.24	15	0.21
6	0.24	16	0.23
7	0.25	17	0.22
8	0.25	18	0.20
9	0.26	19	0.18
10	0.26	20	0.17

References

Allen, R. G., Pereira, L. S., Raes, D., & Smith, M. (1998). Crop evapotranspiration: Guidelines for computing crop water requirements. Irrigation and Drainage Paper No. 56. 300(9), D05109. Rome, Italy: Food and Agricultural Organization of the United Nations.

Allen, R. G., Howell, T. A., Pruitt, W. O., Walter, I. A., & Jensen, M. E. (1991). Lysimeters for evapotranspiration and environmental measurements. *Proc. International Symposium on Lysimetry.* ASCE.

ASCE-EWRI. (2004). The ASCE standardized reference evapotranspiration equation. Standardization of Reference Evapotranspiration Task Committee Final Report. Reston, VA: ASCE Environmental and Water Resources Institute.

Barker, J. B., Neale, C. M., Heeren, D. M., & Suyker, A. E. (2018). Evaluation of a hybrid reflectance-based crop coefficient and energy balance evapotranspiration model for irrigation management. *Trans. ASABE, 61*(2), 533-548.

Costello, L. R., & Jones, K. S. (2014). WUCOLS IV: Water use classification of landscape species. California Center for Urban Horticulture, University of California, Davis. http://ucanr. edu/sites/WUCOLS.

Doorenbos, J., & Pruitt, W. O. (1977). Guidelines for predicting crop water requirements. Irrigation and Drainage Paper No. 24. Rome, Italy: Food and Agricultural Organization of the United Nations.

Hanks, R. J. (1974). Model for predicting plant yield as influenced by water use. *Agron. J., 66*, 660-665.

Hatfield, J. L. (1983). Evaporation obtained from remote sensing methods. In D. Hillel (Ed.), *Advances in irrigation* (Vol. 2, pp. 395-415). New York, NY: Academic Press.

Jackson, R. D. (1982). Canopy temperature and crop water stress. In D. Hillel (Ed.), *Advances in irrigation* (Vol. 1, pp. 43-81). New York, NY: Academic Press.

Jensen, M.E., & Allen, R. G. 2016. Evaporation, evapotranspiration, and irrigation water requirements, ASCE Manuals and Reports on Engineering Practice No. 70. 2nd ed. Reston, VA : American Society of Civil Engineers.

Marek, T. H., Schneider, A. D., Howell, T. A., & Ebeling, L. L. (1988). Design and construction of large weighing monolithic lysimeters. *Trans. ASAE, 31*(2), 477-484.

Ritchie, J. T. (1973). Influence of soil water status and meteorological conditions on evaporation from a corn canopy. *Agron. J., 65*, 893-897.

Smith, M. (1992) *CROPWAT: A computer program for irrigation planning and management.* Irrigation and Drainage Paper No. 46. Rome, Italy: Food and Agricultural Organization of the United Nations.

Stegman, E. C. (1988). Corn crop curve comparisons for the central and northern plains of the U.S. *Appl. Eng. Agric., 4*(3), 226-233.

USDA-SCS. (1993). *National engineering handbook.* Part 623, Ch. 2. Irrigation water requirements. Washington, DC: USDA-SCS.

Wright, J. L. (1982). New evapotranspiration crop coefficients. *J. Irrig. Drain. Div., ASCE, 108*(IR2), 57-74.

Wright, J. L., & Jensen, M. E. (1972). Peak water requirements of crops in southern Idaho. *J. Irrig. Drain. Div. ASCE, 98*(IR2), 93-201.

Chapter 5
Irrigation System Performance

5.1 Introduction

Management of irrigation systems should be based on the desired objectives or outcomes consistent with economic, energy, environmental, labor, water, and resource constraints. Goals can vary from maximizing profit, producing a contracted yield, optimizing water resource use, maintaining the quality of produce, or assuring an attractive landscape. Managers cannot achieve these goals without considering the performance of the irrigation system.

This chapter discusses the basic characteristics of various irrigation systems, defines terms that quantify performance, describes basic requirements all systems must provide, gives a range of attributes for systems, and discusses how water supply requirements are governed by ET and system characteristics. Detailed characteristics of specific systems are presented in later chapters. The key here is to understand the basic systems and their relative performance.

5.2 Types of Systems

There are three general types of irrigation systems: (1) ***sprinkler irrigation***; (2) ***surface irrigation***; and (3) ***microirrigation***, including drip, trickle, and spray. All have advantages and disadvantages in given situations.

5.2.1 Sprinkler Irrigation

Sprinkler irrigation systems are used for agricultural or horticultural production and for landscape or turf applications. The principles of operation are the same for all applications even though the management objectives may differ. Sprinkler systems can be divided into four basic types: single-sprinkler, solid-set, moved lateral, and moving lateral systems. Figure 5.1 illustrates two types of sprinkler systems.

Single-sprinkler systems are designed to irrigate an entire area with only one sprinkler that is moved periodically or automatically moves across the area. Examples range from the single lawn sprinkler that is placed throughout the yard, to automatically moving systems equipped with a big gun sprinkler that throws water hundreds of feet (traveler irrigation system). The performance of single sprinkler systems depends on placing the sprinkler at the proper location for the correct amount of time. A disadvantage is that the systems generally apply water beyond the irrigated area to ensure that the targeted land is adequately watered. However, a significant advantage is that the single sprinkler system is quite versatile and widely used for irregularly shaped land areas.

A step up in complexity from the single-sprinkler system is the system with multiple sprinklers placed along a pipe called a lateral. The basic components of lateral-based sprinkler systems are the mainline and one or more laterals. The mainline is a pipe network designed to carry water from the water source to the laterals. The sprinkler devices are located on the lateral pipelines. Most lateral-based systems consist of multiple laterals. When the laterals are placed permanently in one location in the field, the system is called a ***solid-set*** system. Generally, the laterals and mainline of solid-set systems are installed under the soil surface and the sprinklers are mounted above ground with pipes called risers or the sprinklers are specially designed to pop up above the soil when water pressure builds in the lateral. Solid-set systems are commonly used on lawns, landscapes, golf courses, and some agricultural and horticultural applications. This type of system can be very efficient since each sprinkler in the system is only used in the area it was designed to irrigate. The systems are easily automated and can apply any depth desired.

To reduce investment costs, a single lateral could be set to water a portion of an irrigated area and then moved to multiple locations. The earliest and simplest of these ***moved lateral*** systems is carried by hand and is called a hand move system. The lateral can also be moved by pulling the lateral across the field. This type is called a tow line or towed sprinkler system. Laterals can be mounted on wheels that suspend the pipeline above the crop. These systems are called side roll systems because the wheels are rolled across the field to reposition the lateral. Because of the labor requirement, the moved laterals are usually left in one location for 8, 12, or even 24 hr. Thus, the systems usually apply large depths of water each irrigation.

(a)

(b)

Figure 5.1. (a) Center pivot sprinkler system used for agriculture, and (b) underground sprinkler system in turfgrass.

Automated systems have been developed to move the lateral across the field. Examples of ***moving lateral*** systems include center pivots and linear or lateral move systems. All of these systems use one lateral to irrigate a large area, but since the lateral moves at a controlled speed, the depth of water applied can be varied over a wide range.

5.2.2 Surface Irrigation

Several types of surface irrigation, including ***basins***, ***borders***, and ***furrows*** (Figure 5.2), are used depending on topography, soil texture, and the types of crops grown. Surface irrigation systems are used on agricultural or orchard crops and landscapes that have moderate slopes. With surface irrigation the water is distributed across the field as it flows over the soil surface. Surface irrigation methods generally have lower pressure requirements than sprinkler irrigation, and therefore are less expensive to operate per unit of water applied. The installation costs of surface systems may be lower than for sprinklers if land leveling is not necessary.

Three common problems occur with surface irrigation. To irrigate uniformly, water must advance across the field quickly. This means that some water will run off of the field. Some states have regulations that prohibit irrigation water from running off the field. The runoff problem is largely overcome if a runoff recovery system or ***return flow system*** is a component of the surface system. The second problem is that surface irrigation is labor-intensive. Irrigators are generally unwilling or unable to invest the time needed to irrigate efficiently. This results in excessive applications leading to water losses in the form of runoff or deep percolation. Deep percolation resulting from nonuniform distribution of infiltration is a third common problem with surface irrigation.

Figure 5.2. Furrow irrigation with gated pipe; one type of surface irrigation. (Photo courtesy of Steve Melvin, Nebraska Extension.)

A surface irrigation system consists of some type of water supply mechanism, similar to a mainline for sprinkler systems. This supply mechanism may be a "head" ditch, gated pipe, or buried pipelines with valves at the surface. A variation is the use of siphon tubes to deliver water from a supply ditch.

Whatever water supply device is used, water will flow across a constrained portion of the field. This area of the field may be constrained by small dikes in a border irrigated field or furrows in furrow irrigation. Sometimes an area is leveled and surrounded by small dikes. This type of system is called basin irrigation. If the field is nearly level in both the direction of flow and the transverse direction, the water that would run off the field may be blocked and forced to stay on the field.

5.2.3 Microirrigation

Microirrigation systems consist of laterals containing emitters (***drip irrigation***) or ***microsprinklers***, or laterals with outflow continuously along their lengths (soaker hose). Drip irrigation on the soil surface, also known as ***trickle irrigation***, is illustrated in Figure 5.3. Microirrigation is unique in that the discharge devices are intended to irrigate individual or groups of plants and not the entire soil surface. In landscape applications the flow rate from each emitter may be quite small, while in orchard applications several devices may be required to apply the needed irrigation. Microsystems are usually permanently installed and can be expensive. Labor requirements are minimal although maintenance may be high for situations where the water requires filtration.

Microirrigation systems are popular on high-value crops in locations where water is expensive, in short supply, or of degraded quality. Emitters and microsprinklers have very small orifices or outlets. Since the orifices are small, it is necessary to prevent plugging by soil particles or microorganisms such as bacteria.

Figure 5.3. Surface drip irrigation system in India. (Photo courtesy of IDE-India.)

Microsystems are among the most expensive methods of irrigation, primarily because of the expensive piping system and filtration requirements. They are generally not applicable to row crop production due to the expense and the need to remove the system each season. The latter problem is overcome by burying the laterals beneath the tillage zone, a practice called ***subsurface drip irrigation***

(SDI). Microirrigation is used extensively for landscape applications, especially for trees, shrubs, and gardens. Advantages of these systems include: (1) high efficiency, because evaporation loss is small since the whole plant area is not wetted; (2) water is applied at very low rates so runoff is negligible even for steep slopes; and (3) systems are easily automated to minimize labor.

5.3 Performance Measures

Achieving management objectives requires that water be applied at the proper time, rate and quantity, and in the desired location. However, irrigation systems are not perfect which results in some areas receiving more water than others while some water is simply lost to evaporation. How should an irrigator respond to inefficiency and nonuniformity? How does a management change affect operation and performance? To address these questions, relationships have been developed to quantify performance.

5.3.1 Efficiency

Irrigation systems are never 100% efficient. The major ways water can be "lost" from an irrigated field are illustrated in Figure 5.4. Water is never truly lost, but not all applied water is beneficially used. For irrigation systems such as sprinklers that throw water into the air while irrigating, some ***evaporation*** occurs while the droplets are in the air or once they reach the crop or soil surface. Research suggests that there is little evaporation of the drop while in the air. Losses to evaporation are usually significantly less than 10% of the applied water. If wind blows, droplets may be blown outside of the land to be irrigated. This is called drift. Drift losses may be important and are often significantly higher than evaporation losses.

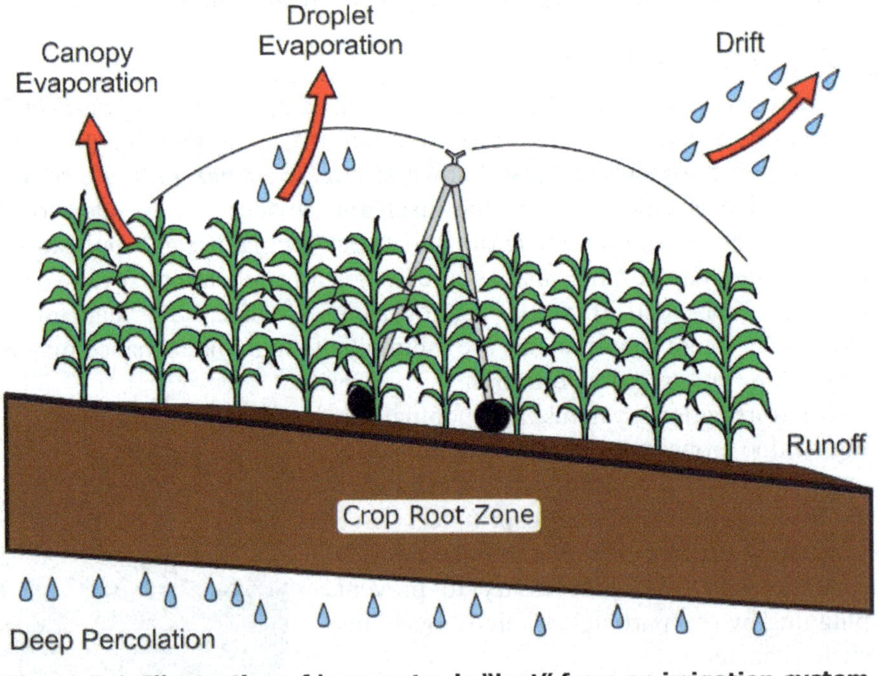

Figure 5.4. Illustration of how water is "lost" from an irrigation system.

When water is applied at a rate that exceeds the infiltration rate of the soil, water begins to accumulate on the soil surface. If the water builds up sufficiently it will begin to run off the soil surface where applied or off of the field. The ***runoff*** water could also infiltrate at a lower elevation in the field leading to poor uniformity of infiltration. When water is applied to the field, in excess of the soil water depletion (SWD), the excess water may percolate past the root zone, a quantity called ***deep percolation***. Irrigation water that remains in the soil at the end of the growing season may also be lost if off-season rains would have replenished the root zone anyway. Thus, there are many ways applied water can be lost from the plant root zone. The manager must minimize losses where possible, yet invariably some losses will occur. In this case, the manager should know how much water might typically be lost so that applications can be adjusted to meet plant needs. ***Application efficiency*** (E_a) is usually defined as the fraction of the applied water that is stored in the root zone and is available for crop water use. The water stored in the root zone is often called ***net irrigation*** and the total amount applied to the field is termed ***gross irrigation***. Thus, the application efficiency is defined as:

$$E_a = 100\%\left(\frac{d_n}{d_a}\right) \tag{5.1}$$

where: E_a = application efficiency,
d_n = net irrigation depth, and
d_a = gross or applied irrigation depth.

The E_a can be expressed as either a decimal fraction (i.e., ranging from 0 to 1.0) or a percentage (ranging from 0 to 100%). The applied depth refers to the volume applied from the water source divided by the area irrigated by that water. The E_a is the result of system characteristics, management, soil and crop conditions, and the weather--especially rainfall. Therefore, there is a broad range of application efficiencies.

This chapter focuses on irrigation water use in terms of the performance of the irrigation system (e.g., application efficiency, application uniformity). Water use can also be evaluated in terms of the yield of the irrigated crop, with the idea of increasing the ratio of crop production to water use. This has been called water use efficiency (Irmak et al., 2011) or water productivity (Trout and DeJonge, 2017; Giordano et al., 2017). In general, advancements in irrigation technology can improve both application efficiency and water productivity (Evett et al., 2020).

5.3.2 Application Uniformity

Irrigation systems are not capable of applying exactly the same depth of water to every location in the field. The distribution of applied water varies because of factors such as wind drift, improper pipeline pressure, poor design, and inappropriate system management. For many irrigation systems, the depth of water applied at a point is nearly the same as the depth entering the soil (infiltration) at the point. Thus, nonuniform applications lead to nonuniform depths of infiltration and ultimately to varying amounts of soil water in the root zone. This nonuniformity adversely affects plant performance so information about the uniformity of application is needed to manage irrigation systems effectively. Illustrations of the effects of poor water distribution on plant health are shown in Figure 5.5. The center pivot pictures (Figures 5.5a and 5.5b) are in Nebraska soybean fields during a drought year (August 2012), which exacerbated the effect of poor uniformity. Further, nonuniform application leads to more deep percolation which results in lower application efficiencies and sometimes to chemical leaching.

Uniformity can be measured for all irrigation systems. For sprinkler systems collection containers (catch cans) or rain gauges are placed in a grid pattern in the field. The irrigation system is then operated for a period of time and the depth of water caught in each container is measured. For microirrigation systems, the volume of water emitted in a given time is measured for all emitters on a lateral. For surface irrigation, experiments can be conducted to determine the depth of water that infiltrates at various points within the field.

To evaluate uniformity, a method is needed to compute a performance value from field test data. The two most commonly used methods are the ***distribution uniformity*** (DU) and the Christiansen uniformity coefficient.

The DU is a relatively simple method where:

$$DU = \frac{d_{LQ}}{d_z} \tag{5.2}$$

where: d_{LQ} = average low-quarter depth of water infiltrated, and
d_z = mean depth infiltrated for all observations.

The value of d_{LQ} is the average depth of application for the lowest one-quarter of all measured values when each value represents an equal area of the field. You can determine the low-quarter depth by ranking observed depths and computing the average for the smallest 25% of the values. Since DU is a ratio with the value of the denominator always being larger than the numerator,

Figure 5.5. Irrigation system having poor water distribution: (a) center pivot irrigation system with large leaks, (b) center pivot with end gun providing a larger application depth than the rest of the system, (c) furrow irrigation, and (d) underground sprinkler system for turfgrass. (Photos a and b courtesy of Gary Zoubek, Nebraska Extension; photo c courtesy of Richard Ferguson, Nebraska Extension.)

DU is always between 0 and 1. The larger the value of DU, the better the uniformity.

The ***Christiansen uniformity coefficient*** (CU) is another index to indicate application uniformity. When each observation represents the same area, the CU is determined as:

$$CU = 100\% \left(1 - \sum_{i=1}^{n} \frac{|d_i - d_z|}{n d_z} \right) \tag{5.3}$$

where: d_i = depth of observation i,
d_z = mean depth infiltrated for all observations, and
n = number of observations.

The calculated value is multiplied by 100 to provide an index value between 0 and 100. Note that $\sum_{i=1}^{n} \frac{|d_i - d_z|}{n}$ is the average deviation from the mean. Thus, another way to write Equation 5.3 is: 100% (1 – average deviation ÷ mean depth infiltrated).

Equation 5.3 was developed to interpret data collected with catch cans placed under sprinkler irrigation system. Typically, water depths in the equation are amounts caught in the cans, not infiltrated water. Since the distribution of infiltration is really what is of interest, the depth of water caught in the can used in Equation 5.3 will indicate infiltrated water only if no surface runoff occurs.

Example 5.1

Given: A sprinkler system was evaluated using 20 catch can containers. The depth caught in each container is given below.

#	d_i (in)	#	d_i (in)	#	d_i (in)	#	d_i (in)
1	1.2	6	1.7	11	2.1	16	2.0
2	2.6	7	2.9	12	1.7	17	1.6
3	1.8	8	2.7	13	1.9	18	2.3
4	2.1	9	1.6	14	1.4	19	1.8
5	2.2	10	2.0	15	2.4	20	2.0

Find: Compute the distribution uniformity (DU) and Christiansen's uniformity coefficient (CU).

Solution: Rank the data in descending order, compute d_z, and then calculate d_{LQ}.

#	d_i (in)	$\lvert d_i - d_z \rvert$	#	d_i (in)	$\lvert d_i - d_z \rvert$	#	d_i (in)	$\lvert d_i - d_z \rvert$	#	d_i (in)	$\lvert d_i - d_z \rvert$
1	2.9	0.9	6	2.2	0.2	11	2.0	0.0	16	1.7	0.3
2	2.7	0.7	7	2.1	0.1	12	1.9	0.1	17	1.6	0.4
3	2.6	0.6	8	2.1	0.1	13	1.8	0.2	18	1.6	0.4
4	2.4	0.4	9	2.0	0.0	14	1.8	0.2	19	1.4	0.6
5	2.3	0.3	10	2.0	0.0	15	1.7	0.3	20	1.2	0.8

d_{LQ} = average of #16 to 20 = 1.5 in

d_z = average of #1 to 20 = 2.0 in

Then compute the individual deviations $\lvert d_i - d_z \rvert$ and the sum of deviations $\sum \lvert d_i - d_z \rvert = 6.6$

Then: $DU = \dfrac{d_{LQ}}{d_z} \qquad DU = \dfrac{1.5}{2.0} = 0.75$ (Eq. 5.2)

$$CU = 100\% \left(1 - \sum_{i=1}^{n} \frac{\lvert d_i - d_z \rvert}{n\, d_z} \right) \quad \text{(Eq. 5.3)}$$

$$CU = 100\% \left(1 - \frac{6.6}{20 \times 2.0} \right) = 84\%$$

Typically, CU values are used for sprinkler and microirrigation systems while DU has become more popular for surface systems. However, some organizations use DU exclusively for all irrigation systems.

Methods used to measure the uniformity of center pivot irrigation systems are unique and a modified CU is normally used. The uniformity of a center pivot is measured by placing containers along two radial lines. The cans are usually placed with uniform spacing from 5 to 15 ft apart along each line. Then the pivot is operated so that the lateral passes over the containers. Since the pivot operates in a circular fashion, a container located far from the pivot point represents more area than one close to the pivot point. Therefore, the Heermann and Hein coefficient of uniformity (CU_H) is ordinarily used for pivots (Heermann and Hein, 1968):

$$CU_H = 100\% \left(1 - \frac{\sum_{i=1}^{n} \lvert d_i - d_z^* \rvert S_i}{\sum_{i=1}^{n} d_i S_i} \right) \quad (5.4)$$

where: S_i = distance from the pivot point to the container, and
d_z^* = weighted mean infiltration, which is equal to:

$$d_z^* = \frac{\sum_{i=1}^{n} d_i S_i}{\sum_{i=1}^{n} S_i} \quad (5.5)$$

Uniformity values are not used like efficiency terms; rather they provide an index of performance. The optimal value of CU or DU depends on the price of irrigation water, the value of the irrigated crop, the costs of drainage or water quality impacts on the environment, and the cost of system renovation and/or management changes. Guidelines to judge whether uniformity is acceptable have been established. For moved lateral sprinkler systems, a CU of 80 (or DU of 0.7) is commonly the lowest acceptable uniformity. For center pivots, a $CU_H = 90$ is often achieved. For furrow systems, a DU of 0.6 is frequently the lowest acceptable value. The DU for microirrigation systems (also known as emission uniformity) should be at least 0.8.

5.3.3 Adequacy of Irrigation

How should an irrigator react to nonuniformity? If the d_z equals the average SWD for each irrigation, then about half of the field will receive more water than needed to refill the crop root zone and deep percolation will ultimately occur. The other half of the field will not receive enough water to refill the root zone and plant water stress may occur. The irrigation manager is continually faced with this tradeoff between excessive deep percolation and plant water stress. The management decision affects profits and E_a. In this context, an important variable is the adequacy of irrigation.

Adequacy of irrigation is the percent of the field that receives the desired depth, or more, of water. It can most easily be evaluated by plotting a frequency distribution of infiltration depth as shown in Figure 5.6. Figure 5.6 is based on the data in Example 5.1 and assumes that each data point represents 5% of the field area. The curve is developed by grouping field measurements of infiltration depth in descending order and computing the percent of the field area that receives at least a given depth of water. The point where the curve intersects the desired depth indicates the percent of the field that is being adequately irrigated. In example 5.1, 5% of the area receives 2.9 in or more while 100% of the area receives 1.2 in or more. Assuming a desired depth of infiltration of 1.6 in, from Figure 5.6 we find that 90% of the land received the desired depth of infiltration or more. Thus, 90% of the area is adequately irrigated. The remaining 10% of the field experienced some plant water stress. Well designed and managed irrigation systems should adequately irrigate at least 80 to 90% of the field. The appropriate adequacy of irrigation depends on many factors and probably varies during the growing season. With an existing irrigation system, the manager can vary the average depth of application to change the adequacy. This amounts to a proportional change to the distribution curve in Figure 5.6, with the distribution curve retaining the original shape. To change the shape of the distribution curve for sprinkler and microirrigation systems may require system modification, which is usually impractical during the season. With

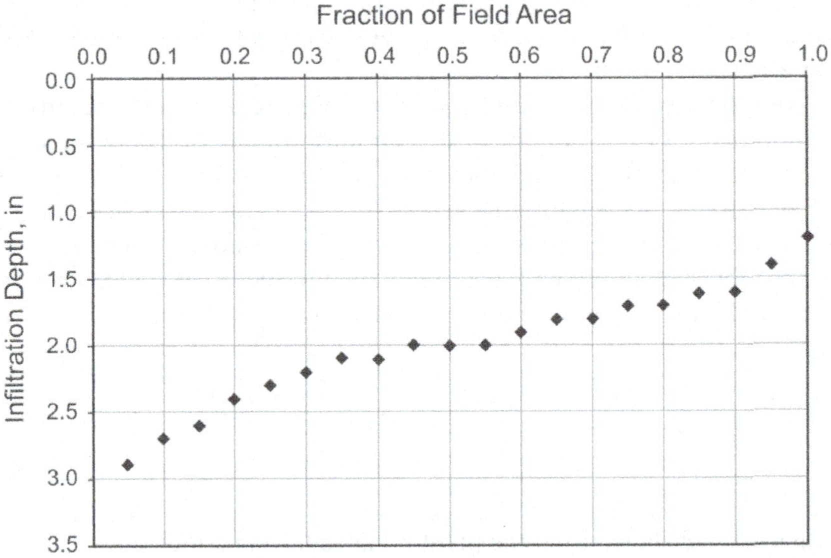

Figure 5.6. Distribution of infiltration based on data from Example 5.1.

surface irrigation, the shape of the distribution curve can be changed through system management as will be discussed in Chapter 10. Of course, if an irrigator increases the average depth applied, more deep percolation will occur. There is a direct link between E_a and uniformity.

5.3.4 Application Efficiency of the Low Quarter: Unification of Efficiency and Uniformity

It is important that all water "losses" during application be considered in an efficiency calculation. These losses shown in Figure 5.4 include:

- evaporation and drift,
- runoff,
- deep percolation due to nonuniform infiltration, and
- deep percolation due to excessive application.

Deep percolation occurs whenever infiltration exceeds the SWD. Excess infiltration can be caused by both the nonuniformity of application and excessive application. Non-uniformity of application is usually a result of a problem with the system for sprinkler and microirrigation, while excessive application is a result of system management. With surface irrigation, non-uniformity of application can also be a result of system management, e.g., if the flow rate in furrows is too low. Percolation caused by the nonuniformity occurs because the manager must decide how much of the field should be adequately irrigated. A common, albeit somewhat arbitrary, approach is to use the average low-quarter depth as the "management depth." Managing according to the average low-quarter depth results in approximately 90% of the field being adequately irrigated and potentially about 10% of the field being under irrigated.

Conservation of mass requires that the following water balance equation holds when conveyance losses (discussed later) are ignored:

$$d_g = d_z + d_r + d_{ev} \tag{5.6}$$

where: d_g = gross depth applied,
d_z = average depth infiltrated,
d_r = depth of runoff, and
d_{ev} = depth of evaporation and drift.

Rearranging Equation 5.6 results in:

$$d_z = d_g - d_r - d_{ev} \tag{5.7}$$

Note that Equation 5.7 accounts for above-ground losses, but the d_z includes both water that will be stored in the root zone and deep percolation. Rearranging Equation 5.2 yields:

$$d_{LQ} = (DU)(d_z) \tag{5.8}$$

The effectiveness of d_{LQ} depends upon the quantity of infiltration relative to the SWD. The ***effective depth*** (d_e) is the irrigation water that remains in the root zone for plant use, accounting for SWD and assuming that any irrigation depth in excess of the d_{LQ} will be lost to deep percolation (i.e., assuming a 90% adequacy of irrigation). The d_e, a managed term, is the amount of water that will be used in irrigation scheduling; its utility will be illustrated in Chapter 6. Figure 5.7 illustrates the concept of d_e with four scenarios. In 5.7a, the infiltrated water is perfectly uniform (DU = 1.0) and equal to SWD. No deep percolation would occur in this scenario. In this case, $d_{LQ} = d_z = d_e$.

In Figure 5.7b, the infiltrated water is perfectly uniform, but, due to excessive application, infiltration exceeds SWD. In this case, $d_{LQ} = d_z$ and $d_e = SWD$. The excessive application can be caused by irrigating too frequently or operating the system too long for the existing SWD. The interval between irrigations can be increased as long as SWD does not exceed the ***allowable depletion*** (AD)–a concept discussed in Chapter 6.

Nonuniform infiltration is illustrated in 5.7c. Here, the $d_{LQ} = SWD = d_e$. In this case, deep percolation is not due to excessive application caused by applying too much water or applying water too frequently but is due to the nonuniformity of the infiltration. The majority of the

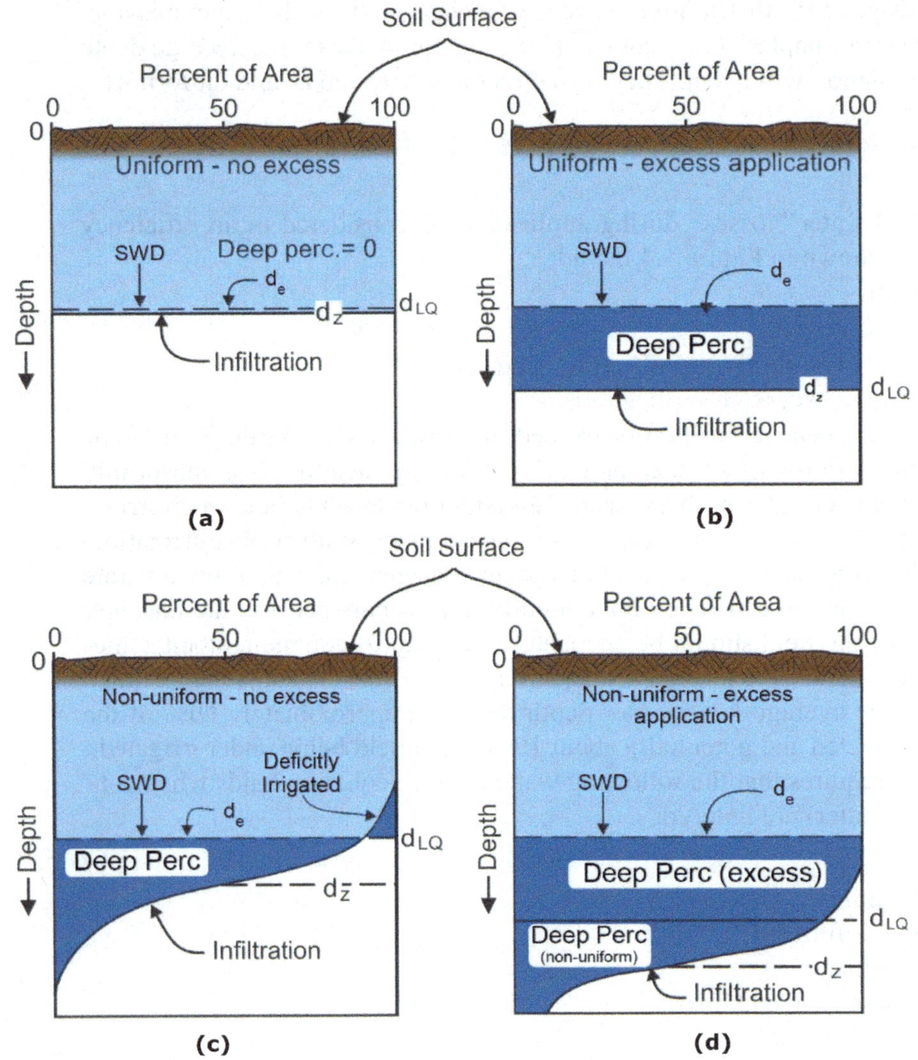

Figure 5.7. Distribution of infiltrated irrigation water and deep percolation under four scenarios.

field (approximately 90%) experiences deep percolation because of the management decision to only allow about 10% of the field to be under irrigated.

Figure 5.7d illustrates the case where there are deep percolation losses due to both excess application and nonuniform infiltration. The figure illustrates the division of the two losses. In this case, $d_e = SWD$.

Figure 5.7 can be summarized by the following equations:

$$\text{If } d_{LQ} \leq SWD, \text{ then } d_e = d_{LQ} \tag{5.9}$$

$$\text{If } d_{LQ} > SWD, \text{ then } d_e = SWD \tag{5.10}$$

Finally, the concepts of uniformity (irrigation adequacy), d_{LQ}, and d_e can be incorporated into the definition of application efficiency. The ***application efficiency of the low-quarter*** (E_{LQ}), discussed by Burt et al. (1997), is defined as:

$$E_{LQ} = 100\% \left(\frac{d_e}{d_a} \right) \tag{5.11}$$

where: E_{LQ} = application efficiency of the low-quarter (%), and
d_a = depth applied from the original source.

Determination of the depth of water from the original source is straightforward except when runoff recovery is part of the system. Either Equation 3.1 or 3.3 can be used for the calculation of d_a. Without runoff recovery, d_a and d_g are equal; d_a is always equal to the volume of water taken from the original source, such as a well, divided by the total land area irrigated. Runoff recovery, discussed in detail in Chapter 10, is a common practice in surface irrigation. If conveyance losses are ignored, the relationship between d_a and d_g for a closed runoff recovery system (runoff water reapplied on the same field) is:

$$d_a = d_g - d_r R_t$$

$$d_a = d_g (1 - R_r R_t) \qquad (5.12a)$$

while, for an open runoff recovery system (runoff water reapplied on different field):

$$d_a = \frac{d_g}{1 + R_r R_t} \qquad (5.12b)$$

where: d_g = gross depth applied which includes the volume applied from the runoff recovery system,
d_r = depth of runoff,
R_r = runoff ratio (d_r / d_g), and
R_t = return ratio, the depth of water returned (reused) divided by the depth of runoff.

Example 5.2

In Example 5.1, the DU was 0.75 and d_z equaled 2.0 in. If d_a = 2.2 in, runoff is zero, and SWD = 1.6 in, determine the system's E_{LQ} and d_{ev}.

Given: d_z = 2.0 in
d_a = 2.2 in
d_r = 0
SWD = 1.6 in
DU = 0.75

Find: d_{ev}
E_{LQ}

Solution:

Rearranging Equation 5.6

$d_{ev} = d_g - d_z - d_r$ (Eq. 5.6)

d_{ev} = 22 in − 20 in − 0 = 0.2 in

Using Equations 5.8, 5.9, and 5.11, you will find that

$d_{LQ} = (DU)(d_z)$ (Eq. 5.8)

d_{LQ} = (0.75)(2.0 in) = 1.5 in

Since d_{LQ} < SWD, d_e = 1.5 in, according to the criteria in Equation 5.9.

Since d_r = 0, $d_a = d_g$ = 2.2 in

$$E_{LQ} = \left(\frac{d_e}{d_a}\right) \times 100\% \qquad \text{(Equation 5.11)}$$

$$\text{Thus, } E_{LQ} = \left(\frac{1.5\text{ in}}{2.2\text{ in}}\right) \times 100\% = 68\%$$

Example 5.3

Repeat Example 5.2 if SWD equaled 1.2 in.

Solution:

Now, d_{LQ} > SWD, thus, Equation 5.10 applies and d_e = SWD = 1.2 in

Thus, E_{LQ} = (1.2 in)/(2.2 in) × 100% = 55%

5.3.5 The Scheduling Coefficient

Another term that is an index of irrigation uniformity and efficiency is the scheduling coefficient (Solomon, 1988). It is commonly used for a description of turf sprinkler systems. It is used to calculate how long a system needs to apply water with the realization that the water application will not be perfectly uniform. For example, if the goal is to apply 0.5 in of water and the sprinkler system applies 0.25 in/hr, it would take 2 h to apply the desired depth *if the water were distributed uniformly across the irrigated area.* However, it usually is not! Thus, to adequately irrigate the desired proportion of the lawn, the sprinkler must be run longer than 2 hr.

Assuming that 90% adequacy is the goal, the ***scheduling coefficient*** (SC) is calculated as:

$$SC = \frac{d_z}{d_{LQ}} \quad (5.13)$$

Example 5.4

A sod farm sprinkler system was tested and shown to have a DU of 0.80. If the average depth caught in the cans (d_z) was 1.5 in and the sprinkler had been running for 5 h, determine the scheduling coefficient (SC), the d_{LQ}, and the number of hours the sprinkler would need to run to achieve the same result if the pattern had been perfectly uniform.

Find: d_{LQ} and SC

Time if uniformity had been perfect

Solution:

$$d_{LQ} = (DU)(d_z) \quad \text{(Eq. 5.8)}$$

$$d_{LQ} = (0.8)\ (1.5\ \text{in}) = 1.2\ \text{in}$$

$$SC = \frac{d_z}{d_{LQ}} \quad \text{(Eq. 5.13)}$$

$$SC = \frac{1.5}{1.2} = 1.25$$

An SC of 1.25 indicates that the sprinkler had to run 25% longer because of uneven distribution. Thus, with perfect uniformity, the time to operate would have been:

Time = 5 h/1.25 = 4 h

As you can see, SC is simply the inverse of DU. The SC indicates how much longer an irrigation system will need to run in order to account for non-uniformity.

5.3.6 Chemical Leaching Losses

Deep percolation losses not only decrease irrigation efficiency, but also result in chemical movement or loss below the root zone. The volume of deep percolating water due to nonuniformity can be designated V_{dp1}. For an adequacy of 90% and a normally distributed (in a statistical sense) water application depth, the V_{dp1} is given by:

$$V_{dp1} = V_z\,(1 - F_1) \quad (5.14)$$

where: $V_z = d_z A$ = volume of water infiltrated,

d_z = average depth of water infiltrated,

A = total irrigated area, and

F_1 = factor (Table 5.1).

Table 5.1. Relationship between CU and F1 for a 90% adequacy of irrigation.

CU	F1	CU	F1
70	0.46	83	0.69
71	0.48	84	0.71
72	0.49	85	0.73
73	0.51	86	0.75
74	0.53	87	0.77
75	0.55	88	0.78
76	0.57	89	0.80
77	0.58	90	0.82
78	0.60	92	0.86
79	0.62	94	0.89
80	0.64	96	0.93
81	0.66	98	0.96
82	0.67		

Deep percolation due to excessive average irrigation depths and/or irrigating too frequently (excessive application) is denoted V_{dp2} and:

$$\text{If } d_{LQ} \le SWD, \text{ then } V_{dp2} = 0$$

$$\text{If } d_{LQ} > SWD, \text{ then } V_{dp2} \approx 0.95\,A\left(d_{LQ} - SWD\right) \quad (5.15)$$

Total deep percolation, V_{dp}, is given by:

$$V_{dp} = V_{dp1} + V_{dp2} \quad (5.16)$$

The depth of deep percolation, d_p, is:

$$d_p = \frac{V_{dp}}{A} \quad (5.17)$$

The amount of chemical lost with the leachate can be calculated by:

$$C_l = 0.226\, C\, d_p \qquad (5.18)$$

where: C_l = chemical loss (lb/ac),
C = concentration of the chemical in the leachate (deep percolation) (ppm), and
d_p = depth of deep percolation (in).

Example 5.5

Find the nitrate leached (lb/ac) for the field illustrated in Example 5.1 if the average concentration of nitrate-nitrogen in leachate is 20 ppm and SWD = 1.2 in.

Find: Determine the amount of nitrate-nitrogen leached from the field during each irrigation.

Solution:

Since we need to calculate this in lb/ac, assume that A = 1 ac.

From Table 5.1, F_1 = 0.71 for a CU of 84%.

Using Equations 5.14, 5.15, 5.16, 5.17, and 5.18:

$$V_{dp1} = d_z\, A\, (1 - F_1) \qquad \text{(Eq. 5.14)}$$

$$V_{dp1} = (2.0\text{ in})\,(1\text{ ac})\,(1 - 0.71)$$

$$V_{dp1} = 0.58\text{ ac-in}$$

$$V_{dp2} = 0.95\, A\, (d_{LQ} - SWD) \qquad \text{(Eq. 5.15)}$$

$$V_{dp2} = (0.95)\,(1\text{ ac})\,(1.5\text{ in} - 1.2\text{ in})$$

$$V_{dp2} = 0.29\text{ ac-in}$$

$$V_{dp} = V_{dp1} + V_{dp2} \qquad \text{(Eq. 5.16)}$$

$$V_{dp} = 0.58\text{ ac-in} + 0.29\text{ ac-in} = 0.87\text{ ac-in}$$

$$d_p = \frac{V_{dp}}{A} \qquad \text{(Eq. 5.17)}$$

$$d_p = \frac{0.87\text{ ac-in}}{1\text{ ac}} = 0.87\text{ in}$$

$$C_l = 0.226\, C\, d_p \qquad \text{(Eq. 5.18)}$$

$$C_l = 0.226\,(20)\,(0.87) = 3.9\text{ lb/ac}$$

Thus, 3.9 lb/ac of nitrate-nitrogen are lost to leaching for each irrigation.

Another approach for finding the average d_p, if data from a uniformity test is available, is to determine the d_p at each irrigation catch can and then averaging. From Example 5.1, the d_p in Can No. 1 is 0 in (1.2 in caught – 1.2 in SWD). For Can No. 20, it is 0.8 in (2.0 – 1.2). For the 20 cans in Example 5.1, the d_p is:

Can No.	Deep Perc. (d_p) (in)	Can No.	Deep Perc. (d_p) (in)
1	0.0	11	0.9
2	1.4	12	0.5
3	0.6	13	0.7
4	0.9	14	1.2
5	1.0	15	1.2
6	0.5	16	0.8
7	1.7	17	0.4
8	1.5	18	1.1
9	0.4	19	0.6
10	0.8	20	0.8

Averaging the 20 depths, we get an average d_p of 0.85 in, which compares well with the 0.87 in calculated in Example 5.5.

5.3.7 Conveyance Efficiency

Water can also be lost in delivering the water from its origin to the irrigation system. Losses are most significant for unlined canals, field laterals, or ditch systems that convey water over long distances through permeable soils. Water can be lost due to seepage from the canal or other conduit, by evaporation from exposed water surfaces, and by evapotranspiration from phreatophytes along the conveyance system. Water can also be lost because of operational problems in moving water through complex delivery systems. If an irrigator originally requested water delivery but later decided not to take the full supply, some water might "spill" from the system. Alternatively, a few irrigators might request water, but the canal may not be able to deliver water with such small flows. Thus, excess flow would be required to supply the requested amount.

The conveyance efficiency (E_c) is used to describe the ability of the delivery system to deliver the requested amount. The E_c is defined as the amount of water delivered to the irrigated area and applied divided by the total amount of water supplied or diverted from the supply (either reservoirs, rivers, or groundwater):

$$E_c = 100\% \left(\frac{d_a}{d_s} \right) \tag{5.19}$$

where: E_c = conveyance efficiency (%),
d_a = gross depth of irrigation water applied, and
d_s = depth of water diverted from the source.

The conveyance efficiency can be reported as either a decimal fraction or a percentage.

Measuring water losses in canals and other delivery systems is difficult and expensive, and for most management purposes, the E_c can be estimated. Several efficiency terms have been used depending on where the delivery system is located. Doorenbos and Pruitt (1977) divide the efficiency of an irrigation project into three components: supply conveyance efficiency (E_c), field canal efficiency (E_b), and field application efficiency (E_a). Conveyance efficiency and field canal efficiency are sometimes combined and called the distribution efficiency (E_d), where $E_d = E_c \times E_b$. The combination of the field canal and application efficiencies is often called the farm efficiency (E_f), where $E_f = E_a \times E_b$. Field application efficiency can be estimated from the methods described earlier in this section (e.g., Equation 5.11).

Factors affecting E_c include: the size of the irrigated area, type of schedule used to deliver water, types of crops, canal lining material, and the capabilities of the water supplies. The field canal conveyance efficiency is primarily affected by the method and control of operation, the type of soils, the canal transects, the length of the canal, and the size of the irrigated block and fields. The farm efficiency is very dependent on the operation of the supply system relative to the supply required on the farm. Doorenbos and Pruitt (1977) present approximate efficiencies for various conditions as summarized in Table 5.2.

A procedure used in the USDA-SCS *Washington State Irrigation Guide* (1985) can also be used to estimate seepage losses. The method gives a range of expected seepage losses depending on the type of material lining the delivery system and the amount of fines in the material (Figure 5.8). In addition to these guidelines, the following losses may be expected:

- Ditch side vegetation: 0.5 to 1.0% loss per mile
- Buried pipeline: 0.01 to 0.15 $ft^3/ft^2/d$ depending on the age and type of pipe.

An example calculation of the season water loss from an earthen ditch follows.

Table 5.2. Conveyance, field, and distribution efficiencies for various types of systems (adapted from Doorenbos and Pruitt, 1977).

Project Characteristics	Conveyance Efficiency
Continuous supply with no substantial change in flow	90%
Rotational supply for projects with 7,000 to 15,000 ac and rotational areas of 150 to 800 ac and effective management	80%
Rotational supply for large projects (> 25,000 ac) and small projects (< 2,500 ac) with problematic communication and less effective management:	
• based on predetermined delivery schedules	70%
• based on arranged delivery schedules	65%
Field Size and Canal Characteristics	**Field Canal Efficiency**
Irrigated blocks bigger than 50 ac with:	
• unlined canals	80%
• lined canals or pipelines	90%
Irrigated blocks smaller than 50 ac with:	
• unlined canals	70%
• lined canals or pipelines	80%
For rotational delivery systems with management and communication adequacies of:	
• adequate	65%
• sufficient	55%
• insufficient	40%
• poor	30%

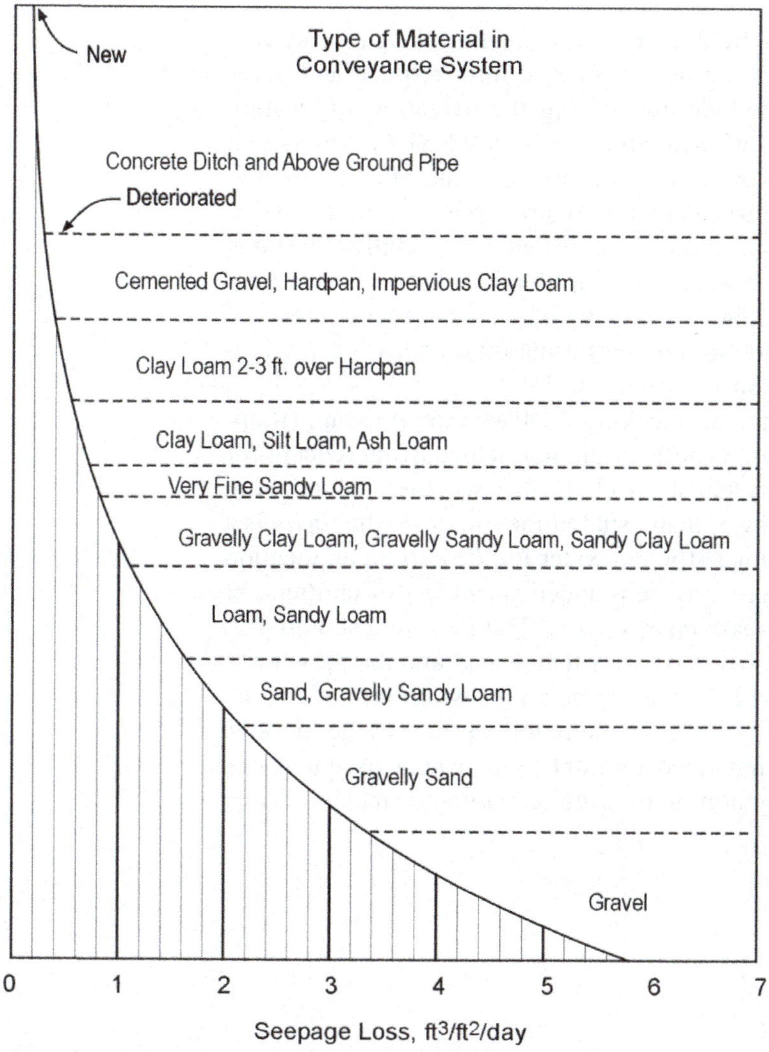

Figure 5.8. Method to estimate seepage losses from irrigation delivery systems (adapted from USDA-SCS, 1985).

Example 5.6
An unlined field ditch is 1,320 ft long, transports 2.5 cfs with a flow contact area (wetted perimeter) of 2.5 ft^2 per ft of length for 180 d/yr. The ditch traverses through loam soil.

Find: Total conveyance loss in ac-ft/yr

Solution:

Figure 5.8 shows the seepage loss of a loam soil to be about 1.4 ft^3/ft^2/d

$$\text{Seepage loss} = \frac{\text{Flow Area} \times \text{Length} \times \text{Seepage Loss Rate} \times \text{Length of Irrigation}}{43{,}560\ \text{ft}^2/\text{ac}}$$

$$\text{Seepage loss} = \frac{(2.5\ \text{ft}^2/\text{ft})\ (1{,}320\ \text{ft})\ (1.4\ \text{ft}^3/\text{ft}^2/\text{d})\ (180\ \text{d})}{43{,}560\ \text{ft}^2/\text{ac}} = 19\ \text{ac-ft}$$

Assuming vegetation loss at 1% of the total flow for the period per mile, then:

Vegetative loss =

$$\left[\left(\frac{1\%}{100\%}\right)(2.5\ \text{cfs})\left(\frac{1{,}320\ \text{ft}}{5{,}280\ \text{ft}}\right)\right] \times \left(\frac{1\ \text{ac-in/h}}{1\ \text{cfs}}\right) \times \left(\frac{24\ \text{h}}{1\ \text{d}}\right) = 0.15\ \text{ac-in/d}$$

$$0.15\ \text{ac-in/d} \times \left(\frac{1\ \text{ft}}{12\ \text{in}}\right) \times 180\ \text{d/yr} = 2.3\ \text{ac-ft}$$

Total conveyance loss = seepage loss + vegetation loss = 19 + 2.3 = 21.3 ac-ft/yr

5.4 System Evaluation

It is important to do a ***system evaluation*** at the field site regularly to check irrigation system performance. Activities for a system evaluation can be categorized into frequent and occasional activities. Occasional activities would include quantifying the irrigation application uniformity (CU or DU). Standard procedures are available, such as ANSI/ASAE S436.2 (2020) for mechanized irrigation systems. Data from the uniformity test can often be used to determine the d_z and d_e, from which the E_{LQ} can be calculated. If a pump is used in the irrigation system, the performance of the pumping plant should be checked occasionally (Martin et al., 2017). Pumps can be a significant source of energy consumption for a farming operation, so maintaining a high pump efficiency can result in cost savings. The energy requirements of pumps are discussed in detail in Chapter 8. More thorough information on occasional irrigation performance audits is presented in Thompson and Ross (2011).

Activities for frequent system evaluations include checking for flow rate, pressure (if applicable), leaks, and runoff (Heeren et al., 2020). Runoff should not be occurring (except for surface irrigation systems). For pressurized systems, check to see whether the pressure matches the design pressure. If the pressure is lower than usual, it may indicate that there is a leak in the system or that the pump is not pumping sufficient water for the current application system. If the pressure is higher than usual, there may be plugged sprinklers or emitters, or the system is set up improperly, which can increase energy costs. The flow rate should also be compared to the design flow rate. If your flow rate is lower than usual, and the pressure is lower than usual, this may indicate a problem with the well or pump. Possibilities include the screen (clogged or crusted over), declining water table, or the pump speed may be too low. For an above-ground system, binoculars or an unmanned aircraft (drone) can be used to check for leaks or plugged nozzles. Cloud-based irrigation monitoring technologies make it easier to frequently check system performance.

5.5 Irrigation System Capacity

In addition to meeting the cumulative seasonal irrigation requirement, irrigation systems must be able to supply enough water to prevent crop water stress during short time periods when plant water requirements are at their highest. The system capacity is the rate of water supply that the irrigation system must provide to prevent this water stress during peak demand. The system capacity must account for peak crop need and the efficiency of the irrigation system. The net system capacity (C_n) is determined by the supply rate needed to maintain the soil water balance above a specified level that will reduce or minimize water stress. The gross system capacity (C_g) is the combined effect of crop needs and system inefficiency. Net and gross capacity are related by the application efficiency and the percent downtime (D_t) for the system:

$$C_g = \frac{C_n}{\frac{E_{LQ}}{100\%}\left(1 - \frac{D_t}{100\%}\right)} \qquad (5.20)$$

where: C_g = gross system capacity,
C_n = net system capacity,
E_{LQ} = application efficiency of low quarter (%), and
D_t = irrigation system downtime (%).

Here, system capacity can be expressed as depth per unit of time, e.g., in/d, or flow rate per unit area, e.g., gpm/ac. For the latter case gross system flow rate is determined by multiplying C_g by the irrigated area. A useful conversion is 18.86 gpm/ac = 1 in/d.

5.6 Determining System Capacity Requirements

Determining the C_n is difficult. Irrigation systems must supply enough water over prolonged periods to satisfy the difference between ET demands and rainfall. Water stored in the crop root zone can supply part of the crop demand. However, the volume of water that can be extracted from the soil should not exceed the amount that will induce crop water stress and likely yield loss. A careful accounting of the soil water status is required if stored soil water is used to supply crop water needs during periods when the crop ET demands are larger than the C_n plus any rainfall. Some irrigation designs have been developed to completely meet peak ET without reliance on either rain or stored soil water. Other techniques intentionally rely on stored soil water to meet peak crop requirements to reduce the required capacity, which decreases the initial cost of the irrigation system.

The most conservative method is to provide enough capacity to meet the maximum expected or "peak" ET rate of the crop. In this case, rain and stored soil water are not considered in selecting the C_n. This design procedure relies on determining the distribution of crop ET during the year. The ET during the season varies from year to year (USDA-SCS, 1993). With the peak ET method, the maximum daily ET for each year is determined. Then the annual maximum daily ET rates are ranked and plotted. The C_n required to meet peak daily ET 70% of the time (i.e., in 7 of 10 yr) is normally taken as the acceptable capacity when using this method.

A method to predict the daily peak period ET rate for general conditions was presented by the USDA-SCS (1970) as shown in Table 5.3. This relationship should only be used for general estimates and only if more localized peak data are not available.

Table 5.3. Peak daily crop ET rates as related to maximum monthly ET for the crop during the season and the net depth applied per irrigation (i.e., allowable depletion).

Allowable Depletion (in)	Maximum Monthly Crop Evapotranspiration (in/mo)													
	5	6	6.5	7	7.5	8	8.5	9	9.5	10	10.5	11	11.5	12
	Peak Daily Evapotranspiration (ETd) in/d													
1.0	.20	.24	.26	.28	.31	.33	.35	.37	.40	.42	.44	.46	.49	.51
1.5	.19	.23	.25	.27	.29	.32	.34	.36	.38	.41	.43	.45	.47	.50
2.0	.18	.23	.25	.27	.29	.31	.33	.35	.37	.39	.41	.44	.46	.48
2.5	.18	.22	.24	.26	.28	.30	.32	.34	.36	.39	.41	.43	.45	.47
3.0	.18	.22	.24	.26	.28	.30	.32	.34	.36	.38	.40	.42	.44	.46
3.5	.18	.21	.23	.25	.27	.29	.31	.33	.35	.37	.39	.41	.44	.46
4.0	.17	.21	.23	.25	.27	.29	.31	.33	.35	.37	.39	.41	.43	.45
4.5	.17	.21	.23	.25	.27	.29	.31	.33	.35	.37	.39	.41	.43	.45
5.0	.17	.21	.23	.25	.26	.28	.30	.32	.34	.36	.38	.40	.42	.44
5.5	.17	.21	.22	.24	.26	.28	.30	.32	.34	.36	.38	.40	.42	.44
6.0	.17	.20	.22	.24	.26	.28	.30	.32	.34	.36	.38	.40	.41	.43

The peak ET method is based on selecting a C_n that can supply water at a rate equal to the peak ET for a period. However, it is unlikely that several periods with water requirements equal to the peak ET will occur consecutively. The crop water use during the combined time period can come from the irrigation supply or from rain and stored soil water. Therefore, the capacity could be reduced if rain is likely or if stored soil water can contribute part of the ET demand.

Relying on soil water can reduce capacity requirements in two ways. First, the soil water can supply water for short periods of time when climatic demands exceed the capacity. The soil water used during the short period can be stored prior to its need or be replaced to some extent during the subsequent period when the ET demand decreases. When the C_n is less than the peak ET rate, there will be periods of shortage when crop water use must come from the soil or rain (Figure 5.9). However, during other periods, the capacity may exceed the ET and the water supplied during the surplus period can replenish some of the depleted soil water (Figure 5.9).

The second way soil water can contribute to reduced capacity requirements is through allowable depletion (AD). This is the amount of water that can be depleted from the soil before crop stress occurs. The minimum capacity that maintains soil water above the AD during critical periods of the season can be used to design the irrigation system. An example of the effect of C_g on soil water mining and the magnitude of SWD during the season are shown in Figure 5.10.

The positive bars in Figure 5.10 represent the amount of rainfall and ET during 10-d periods. After mid-May ET exceeds rain. The deficit bars represent the difference between ET and rain. The largest 10-d deficit occurs in mid-July. Without considering the use

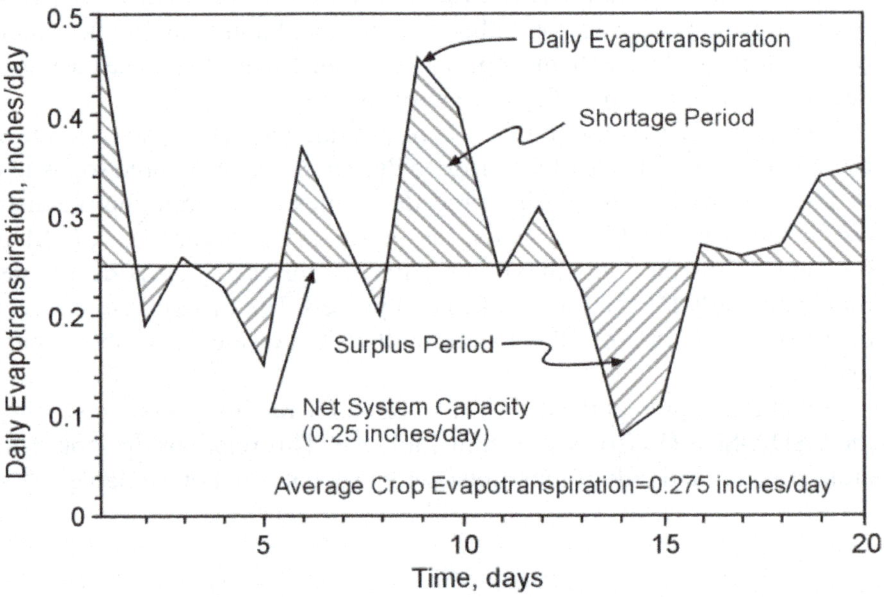

Figure 5.9. An example of the shortage and surplus periods for a system where the net system capacity is less than the average ET during a peak water use period.

of soil water, the irrigation system would have to supply all of the deficit in that period. The peak 10-d irrigation requirement would be 3.3 in per 10 d (or 6.24 gpm/ac). For the 130-ac field shown in Figure 5.10, the C_n for the peak 10-d period would be 810 gpm, and, using an 85% E_{LQ}, the C_g requirement would be approximately 950 gpm.

The amount of water that a 500 gpm capacity system, with an 85% E_{LQ} and assuming no D_t, can supply is also shown in Figure 5.10. The C_n for this system is:

$$C_n = 500 \text{ gpm} \times \frac{1 \text{ ac-in/hr}}{450 \text{ gpm}} \times \frac{24 \text{ hr}}{\text{day}} \times \frac{1}{130 \text{ ac}} \times 0.85 = 0.17 \text{ in/day}$$

The 500 gpm capacity (1.7 in/d in Figure 5.10) falls short of meeting the ET in late June and soil water would be depleted. The 500 gpm capacity continues to fall short of the 10-d deficit from early July through late August, resulting in a cumulative depletion of 4 in.

Suppose that the AD before stress occurs is 3 in for the crop and soil in Figure 5.10. With the 500 gpm capacity system the soil water would be depleted below the allowable level in late July and the crop would suffer yield reduction. Obviously, 500 gpm is inadequate for maximum yield at this site.

The C_n for a 700 gpm system is also shown in Figure 5.10. Here the system can supply the 10-d deficit for all but 20 d in late July. The cumulative soil water deficit for the 700 gpm system would be about 1.25 in with proper management. That depletion is well above the AD and should not reduce crop yield.

This example shows that the maximum cumulative soil water depletion would be approximately 4, 1.25, and 0 in for gross capacities of 500, 700, and 950 gpm, respectively. Clearly the opportunity to utilize available soil water substantially reduces the required system capacity.

Simulation programs using daily time steps to predict the soil water content have been used to determine the C_n when soil water is intentionally depleted. Some models such as by Heermann et al. (1974) and Bergsrud et al. (1982) use the soil water balance equation to predict daily soil water content. von Bernuth et al. (1984) and Howell et al. (1989) used crop simulation models to predict the C_n to maintain soil water above the specified AD or the C_n needed to maintain yields above a specified percentage of the maximum crop yield. University extension services have also created guides for determining C_n and C_g (e.g., Kranz et al., 2008).

The capacities determined using soil water and/or crop yield simulation are usually very dependent on the available water capacity (AWC) of the soil. An example from the results of Heermann et al. (1974) is shown in Figure 5.11 and is illustrated in the subsequent example problem for a sandy loam soil. To use this procedure, the

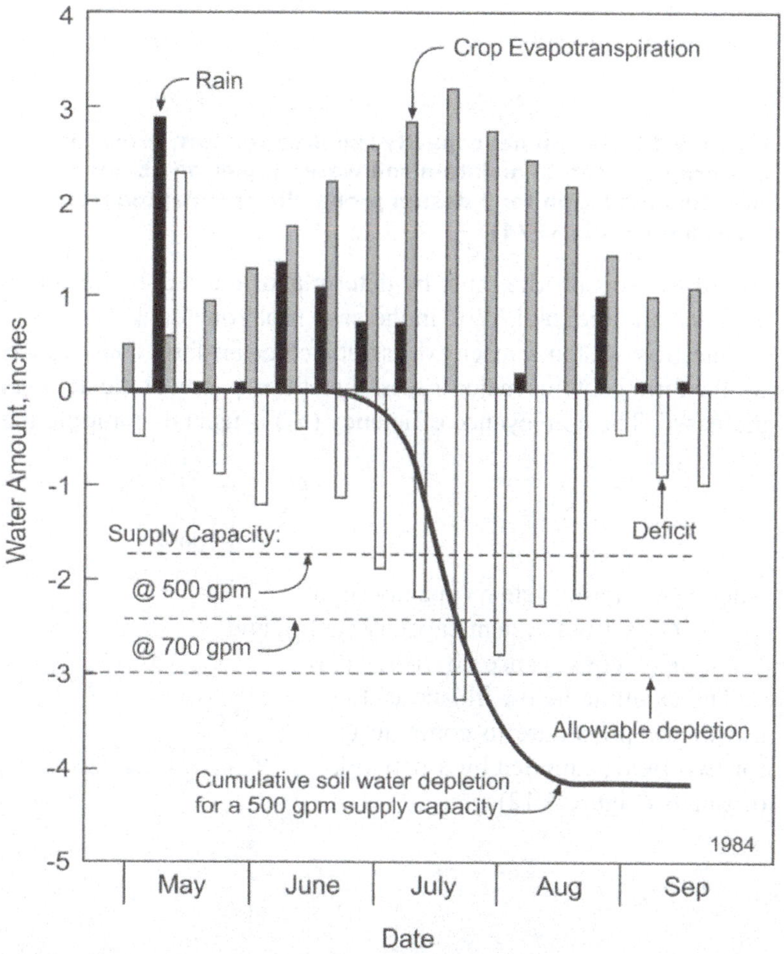

Figure 5.10. Diagram of the 10-d ET, rain and the corresponding water deficit, plus the soil water depletion pattern over a growing season as affected by gross system capacity. Based on a 130-ac field and 85% E_{LQ}.

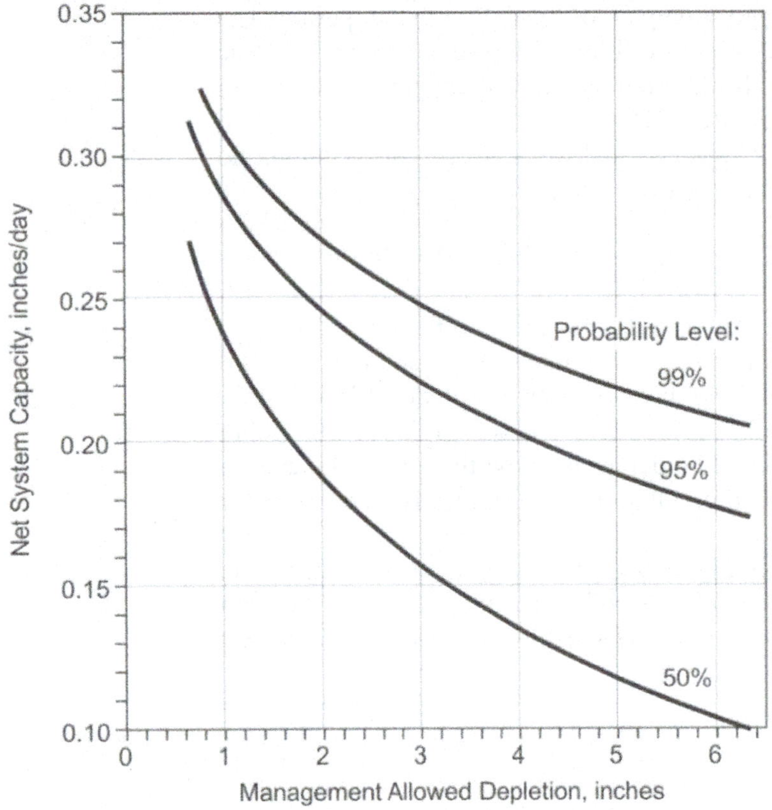

Figure 5.11. Design net capacity required for corn grown in Eastern Colorado to maintain soil water depletion above a specified depletion for 3 design probabilities (adapted from Heermann et al., 1974).

> **Example 5.7**
> Given:
> A sandy loam soil that holds 1.5 in of available water per ft of soil depth.
> Corn root zone depth of 4 ft.
> Allowable fraction depleted = 0.50.
>
> Find:
> The net system capacity needed at a 95% probability level.
>
> Solution:
> The allowable depletion is computed as:
> 1.5 in/ft × 4 ft × 0.5 = 3.0 in
>
> From Figure 5.11, the C_n is approximately 0.22 in/d.

AD of the soil profile must be determined; the AD is the product of the allowable fraction depleted and the total AWC in the crop root zone.

The gross system capacity does not include on-farm conveyance losses. If the delivery system for the farm contains major losses, then the capacity at the delivery point on the farm should be increased. The conveyance efficiency (E_c) is used to compute the farm capacity (Q_f):

$$Q_f = \frac{Q_g}{\left(\frac{E_c}{100\%}\right)} \tag{5.21}$$

where: Q_f = farm system capacity (gpm)
Q_g = gross system capacity (gpm), and
E_c = conveyance efficiency (%).

The example below illustrates the use of the procedure to compute Q_f for two fields supplied by a network of canals (Figure 5.12).

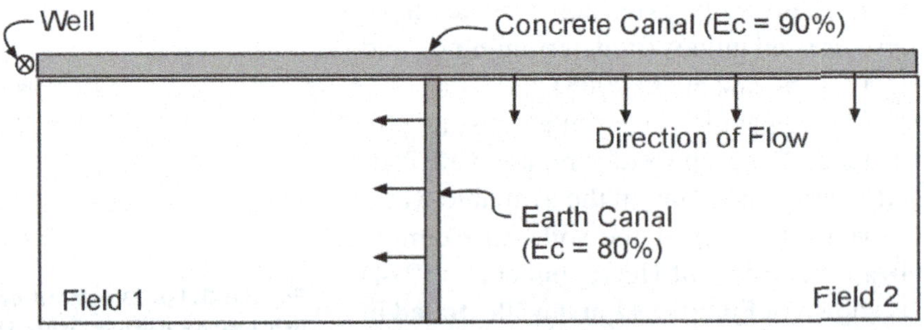

Figure 5.12. Example of a farm layout with seepage losses between the source of the water and delivery to the field.

Example 5.8

Given: A farm has an irrigation system (Figure 5.12) with a net capacity of 0.3 in/d. Each field is 80 ac, and both are furrow-irrigated with siphon tubes. The E_{LQ} is 65% for both fields. The system is shut down about 10% of the time

Find: Determine the discharge needed from the well.

Solution:

1. Net capacity for the farm is expressed in in/d, so convert to flow rate per unit area (gpm/ac):

$$C_n = 0.30 \text{ in/d} \times \frac{452 \text{ gpm}}{1 \text{ ac-in/hr}} \times \frac{1 \text{ d}}{24 \text{ hr}} = 5.7 \text{ gpm/ac}$$

2. The gross capacity for each field is:

$$C_g = \frac{5.7 \text{ gpm/ac}}{0.65\,(1-0.1)} = 9.7 \text{ gpm/ac}$$

3. System capacity is then:

$$Q_g = C_g \times \text{area}$$
$$Q_g = 9.7 \text{ gpm/ac} \times 80 \text{ ac} = 780 \text{ gpm}$$

4. However, the losses in the conveyance system must also be supplied by the pump.

Discharge needed for Field 1 is: $Q_{f1} = 780 \text{ gpm}/0.8 = 975 \text{ gpm}$

Discharge for Field 2 would be: $Q_{f1} = 780 \text{ gpm}/0.9 = 867 \text{ gpm}$

The well must supply the flow to each field plus the loss in the main supply canal:

$$Q_f = (975 + 867)/0.9 = 2{,}047 \text{ gpm, or about } 2{,}050 \text{ gpm}$$

5.7 Operational Factors

An irrigated area is often subdivided into tracts of land called sets or stations. A ***set*** or ***station*** is the smallest subdivision of the total area that can be irrigated separately. The term set is often used for agricultural systems. The set is the area of the field that is irrigated at one time or by a terminal section of the delivery system. For example, for a moved lateral sprinkler system, the land area irrigated while the lateral is stationary would be a set. The block of furrows supplied water at one time would be a set for a furrow system. In landscape and turf applications, the total area is divided into stations. The term "station" comes from the use of controllers that have "stations." The plumbing of the sprinkler or microirrigation systems is such that the station is irrigated at one time. The size of the stations may vary considerably depending on the geometry of the landscape.

The length of time that water is applied to a set is called the ***application time***. The time between starting successive sets in the field is called the ***set time***. The application time and the set time may be the same if the irrigation system is not stopped to change sets. Some systems require that the laterals drain before they are moved. Then the set time is longer than the actual application time. To apply the desired depth of water the application time must be correct. For automated systems the set time can vary for each set or station depending on the water requirement. For manually moved systems the set time may be less flexible. It is common that the set time is adjusted to fit the labor schedule. For example, a 12-h set time is very common for furrow or moved lateral sprinkler systems even though less water may be required at certain times of the season. An inflexible set time can lead to over irrigation and deep percolation if adjustments in flow rate are not made.

The amount of time between starting successive irrigations is called the ***cycle time*** or ***irrigation interval***. For example, suppose a furrow irrigated field is irrigated once per week. The cycle time would be 7 d. The time during the irrigation interval that the irrigation system is not operated is called the ***idle time***. Suppose that the furrow field just mentioned could be irrigated in 5 d. The idle time would then be 2 d. Idle time is very similar to the downtime used to determine system capacity. They would be the same if the application time and the set time are the same. If some time is needed to change sets, then the downtime will be larger than the idle time.

When systems are supplied by an irrigation district, you will often hear the terms duration and rotation used. The ***duration*** is the time that water is provided to the farm. The ***rotation*** is time between the start of times when the water is provided. If the whole field is irrigated each time water is provided, the rotation time is the same as the cycle time. For example, an irrigator might receive water for 4 consecutive days and then be without water for 10 d. In this case, the duration would be 4 d and the rotation would be 14 d.

5.8 System Characteristics

Characteristics of irrigation systems are listed in Table 5.4. The values listed in this guide are average quantities for the respective systems. The table is useful in the preliminary stages of developing and managing irrigation systems. The actual value of the various parameters can vary considerably depending on both design and management.

There has been much written and said about the selection of irrigation systems to fit specific properties of a site. Some factors affecting the selection of a water application method are listed in Table 5.5. The reader should consider these criteria to be general. Since this text deals with managing irrigation systems, it is important to operate the system as efficiently as possible. The practitioner will find that many systems have been installed and operated quite economically even though they do not conform to traditionally defined limits on the irrigation method. An *Irrigation Consumer Bill of Rights*™ has been developed which provides several questions to ask when discussing the selection of an irrigation system with a dealer (ITRC, 2019).

Table 5.4. Typical characteristics of various irrigation systems.

System Type	Maximum Slope (%)	Pressure Required (psi)	Labor Required (hr/ac/irrig)	E_{LQ} (%)	Nominal Application Depth (in)
Surface:					
Furrow gated pipe without reuse	2	0.5–10	0.5–1.0	40–70	2.0–6.0
Furrow gated pipe with reuse	2	0.5–10	0.5–1.0	60–85	2.0–6.0
Furrow siphon tube	2	0	1.0–1.5	35–65	2.0–6.0
Graded border	2–4	0–10	0.2–1.0	50–85	1.5–6.0
Level basin	0	0–10	0.05–0.5	70–85	1.5–6.0
Sprinkler:					
Hand move	20	50–70	0.5–1.5	60–80	1.0–6.0
Solid-set	No limit	50–70	0.05–0.1	60–85	0.5–4.0
Side roll & towline	10	50–70	0.1–0.3	60–80	1.0–6.0
Boom	5	60–80	0.2–0.5	55–75	1.5–4.0
Traveler	5–15	70–100	0.1–0.3	55–75	1.5–4.0
Center pivot	10–20	20–70	0.05–0.15	75–90	0.25–2.0
Pivot with corner system	10–20	30–70	0.05–0.2	70–85	0.25–2.0
Linear move	5–8	20–50	0.1–0.3	75–90	0.2–2.5
Micro, drip, trickle:					
Point source	No limit	20–50	0.05–0.2	70–90	Small
Lateral (continuous) source	No limit	20–50	0.05–0.2	70–90	Small

Table 5.5. Factors affecting the selection of a water application method.

Water Application Method	Factors Affecting Selection			
	Land Slope	Water Intake Rate of Soil	Water Tolerance of Crop	Wind Action
Sprinkler	Adaptable to both level and sloping ground surfaces.	Adaptable to any soil intake rate.	Adaptable to most crops. Typical systems may promote fungi and disease on foliage and fruit.	Wind may affect application efficiency and uniformity.
Surface	Land area must be leveled or graded to slopes less than 2% for most systems. It issometimes possible to flood steeper slopes that are sodded.	Not recommended for soils with high intake rates of more than 2.5 in/hr or with extremely low intake rates such as peats or mucks.	Adaptable to most crops. May be harmful to root crops and to plants which cannot tolerate water standing on roots.	No effects.
Trickle/drip micro	Adaptable to all land slopes.	Adaptable to any soil intake rate.	No problems.	No effects.
Subsurface drip irrigation	Adaptable to all land slopes.	Best adapted to medium and fine-textured soils with moderate to good capillary movement.	Adaptable to most crops. Saline water tables limit application.	No effects.
Below surface subirrigation	Land area must be level or contoured.	Adaptable only to soils which have an impervious layer below the root zone, or a high, controllable water table.	Adaptable to most crops. Saline water tables limit application.	No effects.

5.9 Safety with Irrigation Systems

Irrigation systems can pose several potential hazards, so safety should always be a priority. Hazards from mechanized irrigation systems include missing driveshaft covers, possible falls from ladders and towers, numerous moving parts, and lightning. Drowning is a concern with canals and water storage ponds. Some micro and sprinkler irrigation systems are used to apply chemicals which can be toxic. A very important safety concern is electrical safety, since many irrigation systems use a high voltage (480 V) power supply to pump water and/or to run motors which move the system. The combination of metal structure and wet environment results in a risk of electrocution. Irrigation managers should always be cautious when working or irrigating near overhead power lines. It is the responsibility of producers, service technicians, and others working around irrigation systems to be aware of hazards and safety practices. Anyone designing or constructing an irrigation system must follow the applicable laws, codes, and engineering standards. More thorough information on electrical safety related to irrigation systems is presented in ANSI/ASAE S397.4 (2018), ANSI/ASAE S362.2 (2014), Nolletti (2011), and Marek and Porter (2018).

5.10 Irrigation Efficiency and Water Resources Sustainability

The performance measures discussed in Section 5.3 are all related to the more general term irrigation efficiency. Irrigation efficiency is the ratio of the irrigation water that is beneficially used to the depth of water applied or delivered. Irrigation technologies that improve irrigation efficiency can reduce pumping and the associated energy costs, and in some cases can reduce labor. Reduced pumping often improves the water quality of water resources: reduced deep percolation reduces the leaching of nitrates and other solutes from the root zone to aquifers, and reduced runoff reduces the transport of sediment, nutrients, and pesticides to surface water bodies.

Often it is incorrectly assumed that ***water conservation*** at the watershed scale will automatically follow an improvement in irrigation efficiency at the farm scale. Whether or not liquid water is actually conserved depends upon what led to improved irrigation efficiency in the first place. If efficiency is increased by reducing evaporative losses, liquid water will certainly be conserved. However, if efficiency is improved by reducing deep percolation in a groundwater irrigated region, water may not be conserved since the percolating water may recharge the aquifer from where it originated. In that case, the water is simply being recycled. While the deep percolation could be causing water quality degradation and increased energy expenditures, reducing deep percolation to increase irrigation efficiency may not actually conserve liquid water. A similar example can be developed for surface runoff of irrigation water. Downstream irrigators often depend on the water "losses" or waste from upstream irrigators. A good discussion of this topic is presented by CAST (1988).

Hydrological conservation is needed when water must be conserved to sustain a fresh water supply or to meet a downstream demand for fresh water. From a watershed-scale perspective, "consumptive use" is a helpful concept. ***Consumptive use*** is defined as water that is diverted for use and is not returned to the water resource system. A coal power plant that diverts stream water for cooling returns that water to the stream; this is not a consumptive use and the water is available to downstream users. In agricultural watersheds, the largest consumptive use of water is ET. For example, over long time scales, if groundwater levels remain constant, outflow from a watershed is approximately equal to the difference between the precipitation and ET (Figure 5.13). To reduce aquifer depletion and/or increase stream flow, consumptive use must be decreased. In some situations, water allocations may be required to reduce yield-producing ET. Many irrigation technologies help at the farm scale and help with water quality but don't reduce consumptive use (Grafton et al., 2018).

Since the term irrigation efficiency does not identify the disposition of unused water, Perry et al. (2009) encourage the use of alternative terms when hydrological conservation, not irrigation system performance, is the consideration. Key terms that they suggest are consumed fraction, recoverable fraction, and non-recoverable fraction. The ***consumed fraction*** includes both beneficial consumptive use (transpiration resulting in yield) and non-beneficial consumptive use (soil evaporation, transpiration from weeds). The ***recoverable fraction*** is water

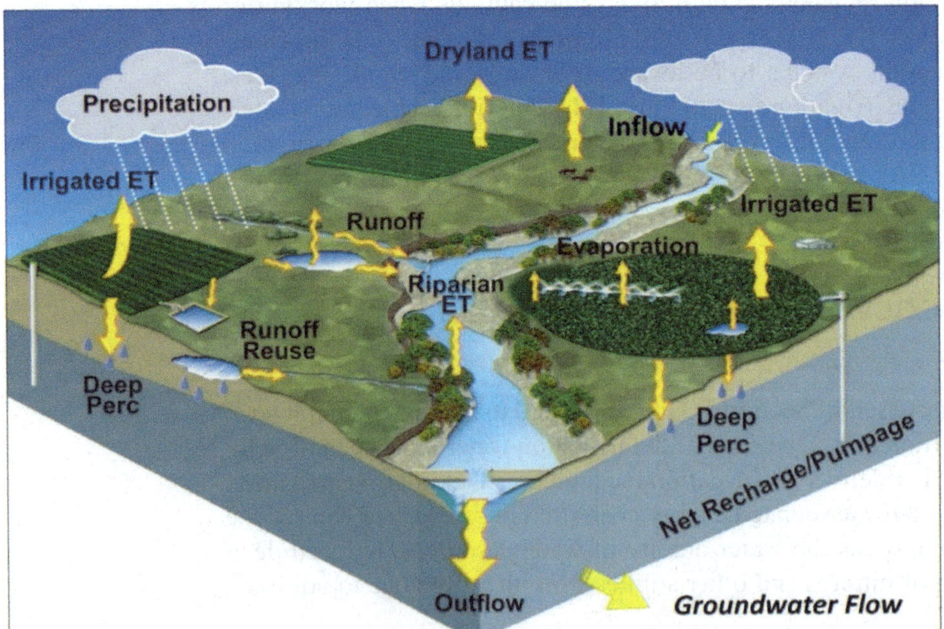

Figure 5.13. Watershed-scale water balance.

that can be reused, such as deep percolation to an aquifer or return flows to a river. The ***non-recoverable fraction*** is not consumed but also is not available for further use, e.g., water that drains from an irrigated region into a saline system, or deep percolation to a very deep aquifer (from which it is too expensive to pump the water). Watershed-scale conservation programs should target reduction of the consumed fraction and/or the non-recoverable fraction.

5.11 Summary

Irrigation systems can be classified into three general categories: Surface, sprinkler, and micro. While the characteristics of each of these systems differ, none of them apply water perfectly to an irrigated area. Water is never uniformly distributed across the land, and some water goes to evaporation, runoff and deep percolation rather than being used by plants. Common terms can be used to describe how efficiently irrigation systems apply water. Distribution uniformity (DU) and Christiansen's Uniformity Coefficient (CU) are used as indices of water application uniformity. Application efficiency (E_a) and application efficiency of the low-quarter (E_{LQ}) are used to describe what proportion of the applied water is stored in the soil and available to plants.

Deep percolation is an important loss in irrigation because, not only does it result in larger applications of water than needed, but also chemicals can be leached with the percolating water. The amount of chemical leaching loss can be quantified by knowing the deep percolation losses and the concentration of the chemical in the leachate.

Water can also be lost to seepage and evaporation during conveyance. Seepage losses can be significant in unlined ditches and canals. It is important to consider losses at both the field scale and the watershed scale. Irrigation technologies that increase application efficiency often do not conserve water at the watershed scale, particularly if the technology does not reduce consumptive use of water.

The amount of water needed to meet irrigation needs is called the system capacity requirement. System capacity is determined by knowing land area, plant needs, E_{LQ}, and downtime or system operation time.

Questions

1. Consider a sprinkler-irrigated sports field where the depth of water applied from the original source is 0.90 in, the soil water deficit (SWD) prior to irrigation is 0.8 in and the depth of water lost to runoff, evaporation, and drift is 0.05 in. Determine the application efficiency of the low-quarter (E_{LQ}) for the following three conditions: (a) the infiltrated water is perfectly uniform and d_z exceeds SWD, (b) the average depth of water infiltrating in the low quarter of the field is 0.70 in, and (c) the average depth of water infiltrating the lowest quarter of the turf area is 0.80 in.
2. For the three conditions described in Question 1, calculate the distribution uniformity (DU).
3. If you had sufficient funds and were irrigating an apple orchard, which irrigation system would you choose and why? If funds were limited and the apple orchard was nearly level, which system would you select? Why?
4. Which irrigation system would you install in your area to irrigate a golf course? Why?
5. If a turf field needs 1.2 in of water, the scheduling coefficient is 1.25, and the sprinkler system applies 0.5 in/hr, how many hours of irrigation are required to be sure that 90% of it is adequately irrigated?

6. Calculate the distribution uniformity and Christiansen's coefficient of uniformity for a lateral move sprinkler system with the depths of water collected in the following 16 catch can containers.

Can No.	Depth (in)	Can No.	Depth (in)	Can No.	Depth (in)
1	1.2	7	1.4	13	1.0
2	1.1	8	0.8	14	0.9
3	1.3	9	0.7	15	0.9
4	0.9	10	0.9	16	1.2
5	1.0	11	0.9		
6	1.0	12	0.8		

7. If one million gallons of water are applied to three holes of a golf course and 0.8 million gallons of this application are stored in the root zone, what is the application efficiency?

8. Calculate Christiansen's coefficient of uniformity for a center pivot system with the following catch can container data.

	Water Depth in Can (in)	
Distance from Pivot Point (ft)	Radial Line #1	Radial Line #2
15	0.9	1.0
30	1.0	1.0
45	1.1	1.1
60	0.8	1.0
75	1.0	0.9
90	1.0	0.9
105	1.0	1.0
120	0.9	1.0
135	1.0	1.0
150	1.0	1.0
165	1.1	1.1
180	1.0	1.0
195	0.9	1.0
210	1.1	1.1
225	0.9	0.9
240	0.9	0.9
255	1.1	1.0
270	1.0	1.0
285	0.9	0.9
300	1.0	

9. If an irrigation system has a distribution uniformity of 0.85 and a total depth of 2.0 in was applied, d_z equaled 1.9 in, and the SWD was 1.7 in, determine the system's loss of water due to evaporation, drift, and runoff.

10. Calculate the annual seepage loss for a new concrete-lined ditch that is 10 miles long, carries water for 200 d each year, and has a flow area of 3 ft^2/ft. Report your answer in ac-ft/yr.

11. Determine the gross system capacity (Q_g) for a golf course if the application efficiency for the low-quarter is 75%, the system is inoperable no more than 10% of the time, and the net system capacity is 20 million gal/d.

References

ANSI/ASAE S362.2. (2014). Wiring and equipment for electrically driven or controlled irrigation machines (R2014). St. Joseph, MI: ASABE.

ANSI/ASAE S397.4. (2018). Electrical service and equipment for irrigation. St. Joseph, MI: ASABE.

ANSI/ASAE S436.2. (2020). Field test procedure for determining irrigation water distribution uniformity of center pivot and lateral move systems. St. Joseph, MI: ASABE.

Bergsrud, F. G., Wright, J. A., Werner, H. D., & Spoden, G. J. (1982). Irrigation system design capacities for west central Minnesota as related to the available water holding capacity and irrigation management. Tech. Paper No. NCR 82-101. St. Joseph, MI: ASAE.

Burt, C. M., Clemmens, A. J., Strelkoff, T. S., Solomon, K. H., Bliesner, R. D., Hardy, L. A., Howell, T. A., & Eisenhauer, D. E. (1997). Irrigation performance measures: Efficiency and uniformity. *J. Irrig. Drain. Eng., 123*(6), 423-442.

CAST. (1988). Effective use of water in irrigated agriculture. Report No. 113. Ames, IA: Council for Agricultural Science and Technology.

Doorenbos, J., & Pruitt, W. O. (1977). Guidelines for predicting crop water requirements. Irrigation and Drainage Paper No. 24. Rome, Italy: United Nations FAO.

Evett, S. R., Colaizzi, P. D., Lamm, F. R., O'Shaughnessy, S. A., Heeren, D. M., Trout, T. J., Kranz, W. L., & Lin, X. (2020). Past, present, and future of irrigation on the U.S. Great Plains. *Trans. ASABE, 63*(3), 703-729.

Giordano, M., Turral, H., Scheierling, S. M., Treguer, D. O., & McCornick, P. G. (2017). Beyond "more crop per drop": Evolving thinking on agricultural water productivity. IWMI Research Report 169. Colombo, Sri Lanka: International Water Management Institute.

Grafton, R. Q., Williams, J., Perry, C. J., Molle, F., Ringler, C., Steduto, P., Udall, B., Wheeler, S. A., Wang, Y., Garrick, D., & Allen, R. G. (2018). The paradox of irrigation efficiency. *Science, 361*(6404), 748-750.

Heeren, D. M., Melvin, S. R., Nygren, A., & Wilkening, E. (2020). Now is the time of year to check pivot performance. Online extension article, UNL Water.

Heermann, D. F., & Hein, P. R. (1968). Performance characteristics of self-propelled center-pivot sprinkler irrigation system. *Trans. ASAE, 11*(1), 11-15.

Heermann, D. F., Shull, H. H., & Mickelson, R. H. (1974). Center pivot design capacities in eastern Colorado. *J. Irrig. Drain. Divi., 100*(2), 127-141.

Howell, T. A., Copeland, K. S., Schneider, A. D., & Dusek, D. A. (1989). Sprinkler irrigation management for corn-Southern Great Plains. *Trans. ASAE, 32*(1), 147-154.

Irmak, S., Odhiambo, L. O., Kranz, W. L., & Eisenhauer, D. E. (2011). Irrigation efficiency and uniformity, and crop water use efficiency. Ext. Circular EC732. Lincoln: Nebraska Ext.

ITRC. (2019). Irrigation Consumer Bill of Rights. ITRC Report No. R 19-002. San Luis Obispo, CA: Irrigation Training and Research Center, California Polytechnic State University.

Kranz, W. L., Martin, D. L., Irmak, S., van Donk. S., J., & Yonts, C. D. (2008). Minimum center pivot design capacities in Nebraska. NebGuide G1851. Lincoln: Nebraska Ext.

Marek, T., & Porter, D. (2018). Think you understand center pivot safety and maintenance—really? *Proc. Central Plains Irrigation Conf.* Colby, KS: Central Plains Irrigation Association.

Martin, D. L., Kranz, W. L., Irmak, S., Rudnick, D. R., Burr, C., & Melvin, S. R. (2017). Pumping plant performance. *Proc. 29th Annual Central Plains Irrigation Conf.* Colby, KS: Central Plains Irrigation Association.

Nolletti, N. (2011). Electricity for Irrigation. In L. E. Stetson, & B. Q. Mechan (Eds.), *Irrigation* (6th ed.). Falls Church, VA: Irrigation Association.

Perry, C., Steduto, P., Allen, R. G., & Burt, C. (2009). Increasing productivity in irrigated agriculture: Agronomic constraints and hydrological realities. *Agric. Water Manag. 96,* 1517-1524.

Soloman, K. H. (1988). A new way to view sprinkler patterns. Irrigation Notes: CATI Publ. 880802. Fresno: Center for Irrigation Technology, California State University.

Thompson, K. K., & Ross, G. A. (2011). Performance audits. In L. E. Stetson, & B. Q. Mechan (Eds.), *Irrigation* (6th ed.). Falls Church, VA: Irrigation Association.

Trout, T. J., & DeJonge, K. C. (2017). Water productivity of maize in the U.S. High Plains. *Irrig. Sci., 35*(3), 251-266.

USDA-SCS. (1970). *Irrigation water requirements*. Technical Release No. 21. (Rev. 2). Washington, DC: United States Department of Agriculture-Soil Conservation Service.

USDA-SCS. (1985). *Washington State Irrigation Guide.* Part WA-686. Farm Distribution Systems. Washington, DC: United States Department of Agriculture-Soil Conservation Service.

USDA-SCS. (1993). Irrigation water requirements. In *National engineering handbook.* Part 623, Chapter 2. Washington, DC: United States Department of Agriculture-Soil Conservation Service.

von Bernuth, R. D., Martin, D. L., Gilley, J. R., & Watts, D. G. (1984). Irrigation system capacities for corn production. *Trans. ASAE, 27*(2), 419-424.

Chapter 6 Irrigation Scheduling

6.1 Introduction

In Chapter 4 we discussed plant water use (ET). In Chapter 2 we discussed how the soil-plant root zone serves as a reservoir. But how should we manage this reservoir? The ET demand of a crop is supplied from three sources of water: (1) rainfall that occurs during the growing season, (2) precipitation that was stored in the plant root zone during the dormant or off season and (3) irrigation. For example, in central Nebraska, the growing season ET for corn averages about 26 inches per year. Approximately 12 inches of this water would come from growing season rainfall, 4 inches from the stored soil water, and the balance 10 inches from irrigation (net). Good managers of irrigation water strive to maximize the use of precipitation while minimizing deep percolation. Proper irrigation scheduling will help reach these goals.

Irrigation scheduling includes determining how often to apply water and how much water to apply. Irrigation scheduling is imperative for good water management. In this chapter we discuss irrigation scheduling and how system efficiency, available water capacity, plant root zone, and evapotranspiration affect the frequency and amount of water application, i.e., we build on what was covered in Chapters 2, 3, and 4.

In practice irrigation scheduling is often based on the irrigator's personal experience, plant appearance, watching the neighbor, or just simply irrigating whenever water is available. Sometimes the irrigation controller is set and never adjusted for rainfall or changes in ET. As water resources become limited and as more concern is raised about the effects of irrigation on water quality, the need for improved methods for scheduling will increase. Three general approaches or philosophies for scheduling irrigations are:

- maintain soil water content within desired limits,
- use plant status indicators to signal the need for water, and
- irrigate according to calendar date or other fixed schedule.

The soil water approach uses the plant root zone as a reservoir for storing and releasing water, as we discussed in Chapter 2. This reservoir can be managed using irrigation scheduling. The soil water status can be determined by checkbook accounting or by direct measurement of soil water. Checkbook accounting uses estimates of ET, rainfall, and irrigation to calculate the soil water level. Plant status indicators range from simply observing the stage of plant development to more sophisticated methods, such as measuring plant leaf temperature and plant water potential. Fixed schedules are often associated with irrigation water supply districts that lack the flexibility to deliver water on demand.

No one scheduling philosophy is correct in itself. The emphasis in this chapter is on managing the soil water reservoir, but it also includes a short discussion on plant status indicators. It is likely that in the future a combination of soil water maintenance and plant status will be the most appropriate choice.

6.2 Plant Response to ET and Soil Water

The relationship between crop yield and transpiration and ET is illustrated in Figure 6.1. As illustrated in Figure 6.1b, once soil water evaporation is satisfied, there is a linear increase in yield as evapotranspiration increases until maximum yield is reached. In this book most of the discussion relating to crop or forage production will center on managing water for maximum, or near maximum, yield. More advanced books and papers discuss deficit or limited irrigation, where yield is reduced because less irrigation water is applied than necessary to meet full crop water requirements (English et al., 1990, Trout et al., 2020). With deficit irrigation, ET is less than the crop ET necessary for maximum yield. Figure 6.2 shows the relationship between growth or yield and f_r (fraction of available soil water remaining). If soil moisture is maintained above certain limits, maximum, or near maximum, yield is achieved. The minimum fraction of available water remaining that should occur to avoid plant stress and a yield reduction is the critical fraction remaining (f_{rc}) a term presented in Chapter 4. When the available soil water is maintained equal to or above f_{rc}, maximum yields are attainable because the plants are able to extract adequate water from the soil.

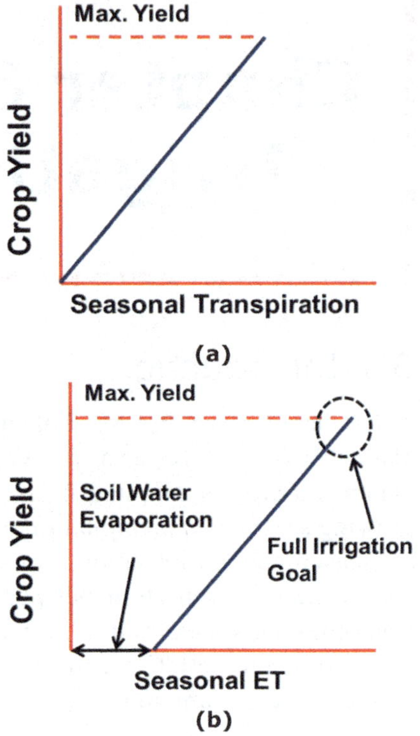

Figure 6.1. Relationship between yield, T, and ET.

The f_{rc} is related to f_{dc}, the maximum allowable fraction depletion of the available soil moisture, by the following equation:

$$f_{dc} = 1 - f_{rc} \tag{6.1}$$

f_{dc} is dependent on the plant species and genotype and on weather conditions. Weather influences the maximum ET each day. According to Doorenbos and Kassam (1979), f_{dc} ranges from 0.18 to 0.88, depending upon how plants respond to soil water deficits and on the maximum ET for a given day. Data for various conditions are given in Table 6.1. From the table you can see that for corn with a maximum evapotranspiration of 0.28 inches per day, f_{dc} is 0.5 and f_{rc} is 0.50 (from Equation 6.1). A crop, such as onions, grown under the same environment or weather conditions, will have a f_{dc} of about 0.23 (f_{rc} = 0.77). Thus, the criteria for management depends on the crop and the environmental conditions. If the weather is relatively cool (low ET), a high percentage of the soil water can be depleted before stress occurs. Conversely, on hot days (high ET), less soil water depletion is allowed before plants undergo stress.

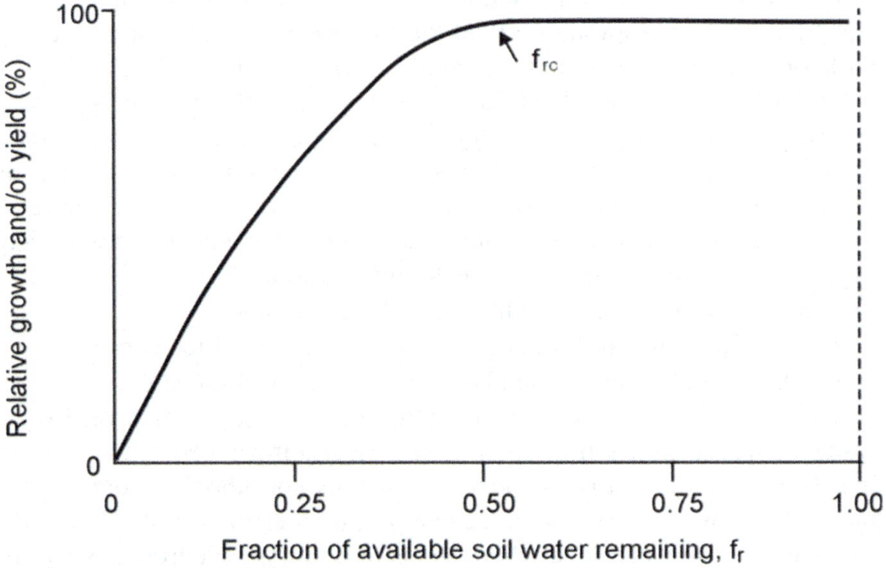

Figure 6.2. Relationship between available water remaining and yield (adopted from Stegman, 1983).

A common level of f_{dc} is 0.50. This is an average value and can be used where more appropriate data, such as that shown in Table 6.1, is not available.

The above discussion has implied that the management goal is to produce maximum (or near maximum) yield or biomass. This may not be the case for landscaping plants and

Table 6.1. Estimated maximum allowable fraction depletion (f_{dc}) to maintain maximum yields of crops grouped according to sensitivity (modified from Doorenbos and Kassam, 1979).

Crop Group	
1	onion, pepper, potato
2	banana, cabbage, pea, tomato
3	alfalfa, bean, citrus, groundnut, pineapple, sunflower, watermelon, wheat
4	cotton, sorghum, olive, grape, safflower, corn, soybean, sugarbeet, tobacco

Crop Group	Maximum ET (in/day)								
	0.08	0.12	0.16	0.20	0.24	0.28	0.31	0.35	0.39
	f_{dc} to Maintain Maximum Evapotranspiration Rates								
1	0.50	0.43	0.35	0.30	0.25	0.23	0.20	0.20	0.18
2	0.68	0.58	0.45	0.40	0.35	0.33	0.28	0.25	0.23
3	0.80	0.70	0.60	0.50	0.45	0.43	0.38	0.35	0.30
4	0.88	0.80	0.70	0.60	0.55	0.50	0.45	0.43	0.40

turfgrass. With these plants the goal is satisfactory plant appearance and/or adequate functional quality. To maintain a high-quality golf green will require more water than is required to satisfy the needs of a low-maintenance utility turf.

The management objective must be defined for any irrigation scheduling procedure. Possible objectives include:

- maximum yield or biomass production,
- maximum economic return,
- functional value of the plants (e.g., an athletic field),
- aesthetic value (i.e., keeping the plants healthy), and
- maintaining plant life.

6.3 Capacity of the Soil Water Reservoir

As discussed in Chapter 2 and illustrated in Figure 2.8, the plant root zone can be viewed as a reservoir for storing water for use by plants. For non-layered soils, total available water capacity, or TAW, is:

$$TAW = (R_d)(AWC) \tag{6.2}$$

where: R_d = root zone depth and
AWC = available water capacity

For layered soils:

$$TAW = (AWC_1)(t_1) + (AWC_2)(t_2) + ... + (AWC_n)\,[R_d - (t_1 + t_2 + ... + t_{n-1})] \tag{6.3}$$

where: $AWC_1, AWC_2, ...$ = available water capacity in layers 1, 2, etc.,
$t_1, t_2, ...$ = thickness of soil layer 1, soil layer 2, etc., and
n = number of soil layers.

So, the size of the reservoir is dependent on both the soil and the root zone depth.

Allowable depletion, an important irrigation management term, is the amount of available water that can be removed from a root zone before plants undergo moisture stress. The allowable depletion, AD, is:

$$AD = f_{dc}\,(TAW) \tag{6.4}$$

AD is expressed as a depth of water. Likewise, the minimum allowable available soil moisture, or minimum balance (MB) is:

$$MB = f_{rc}\,(TAW) \tag{6.5}$$

Note that $MB = TAW - AD$ or $TAW = MB + AD$.

Example 6.1

The root zone depth is 3 ft, AWC = 0.17 in/in, and f_{dc} = 0.45. What is TAW, AD, and MB?

Given: AWC = 0.17 in/in

f_{dc} = 0.45

Find: TAW, AD, and MB

Solution:

$TAW = (R_d)(AWC)$ (Eq. 6.2)

TAW = (36 in)(0.17 in/in) = 6.1 in

$AD = f_{dc}\ (TAW)$ (Eq.6.4)

AD = (0.45)(6.1 in) = 2.7 in

$MB = f_{rc}\ (TAW)$ (Eq.6.5)

MB = (0.55)(6.1 in = 3.4 in

In Example 6.1, irrigation should be applied when or before 2.7 inches of soil water have been depleted. If AD is reached, i.e., if SWD = AD at the time of irrigation, the maximum amount of water that the root zone would hold without exceeding field capacity is 2.7 inches. One goal is to keep infiltration less than or equal to the soil water deficit (SWD). As we discussed in Section 2.9, whenever infiltration exceeds SWD, deep percolation will occur.

6.3.1 Plant Root Zone

Established perennial plants such as alfalfa, grasses, trees, and shrubs have relatively constant root zone depths. The maximum effective root depth depends on several environmental, crop, and soil factors. The range of maximum effective root zone depths for various crops is summarized in Table 6.2. The maximum effective depth used for scheduling, which is usually less than the maximum depth where roots are found, represents the depth of the soil profile that has enough rooting density for extraction of available water. Values in Table 6.2 should be used cautiously and adjusted for local soil and climatic conditions.For annual crops the root depth prior to the date of maximum rooting is described by:

$$R_d = R_{dMIN} + (R_{dMAX} - R_{dMIN})R_f \quad (6.6)$$

where: R_d = root depth

R_{dMIN} = minimum root depth for young plants,

R_{dMAX} = maximum effective root depth, and

R_f = root growth factor.

The development of a corn root zone during the season is illustrated in Figure 6.3.

The minimum root depth for seedlings is normally considered to be 4 to 6 inches. The actual initial depth may deviate slightly from this value, but an error on the minimum root depth will have very little effect on the soil water balance or irrigation scheduling.

The root growth factor, which describes the rate

Table 6.2. Range of maximum effective rooting depths for fully grown plants (from Martin et al., 1990).

Crop	Maximum Effective Depth (ft)	Crop	Maximum Effective Depth[a] (ft)
Alfalfa	3.3–10	Olives	2.6–6.6
Banana	1.3–2.6	Onions	2.6–6.6
Barley	3.3–4.3	Palm trees	2.3–3.6
Beans	1.3–2.6	Peas	2.0–3.3
Cabbage	2.0–3.3	Peppers	1.7–3.3
Carrots	1.6–3.3	Pineapple	1.0–2.0
Celery	1.0–1.7	Potatoes	1.3–2.6
Citrus	3.3–5.9	Safflower	3.3–6.6
Clover	2.0–3.0	Sorghum	3.3–6.6
Corn	3.3–6.6	Soybeans	2.6–5.0
Cotton	3.3–6.6	Spinach	1.0–1.7
Cucumber	2.3–4.0	Strawberries	0.7–1.0
Dates	5.0–8.3	Sugar beet	2.6–6.6
Flax	3.3–5.0	Sugarcane	4.0–6.6
Grapes	3.3–6.6	Sunflower	3.3–8.3
Grass	1.7–5.0	Sweet potatoes	3.3–5.0
Groundnuts	1.7–3.3		
Lettuce	1.0–1.7	Tobacco	1.7–3.3
Maize	3.3–6.6	Tomatoes	2.3–5.0
Melons	3.3–5.0	Wheat	3.3–5.0

[a] The maximum values given represent the full expression of the genetic potential for root growth and are only found in uniform, fertile soils of low resistance to root penetration.

of root zone expansion during the season, can be computed as:

$$R_f = \frac{D_{ag}}{D_{tm}} \tag{6.7}$$

where: D_{ag} = days after germination and

D_{tm} = days from germination to maximum effective depth.

The time required for roots to reach the maximum effective depth varies considerably for different environments, crops, and varieties. Local values and individual experience must be used to determine these values. Root zone depths for various stages of crop development are given in Table 6.3.

Plants do not extract water uniformly throughout their rooting depth. Usually there is more water extracted from shallow depths and less from deeper depths. An approximation of the extraction pattern is the 4-3-2-1 rule, i.e., 40% of the water comes from the top 25% of the root zone, 30% from the second 25%, and so forth. This conceptual approximation is illustrated in Figure 6.4. If for example, the root zone depth is 4 feet, and the plants extract 2 inches of water between irrigations, 0.8 inches would be obtained from the first foot, 0.6 inches from the second foot, 0.4 inches from the third foot, and 0.2 inches from the fourth foot. This concept applies only when the root zone is refilled, or nearly refilled, following irrigation. If the root zone is not completely refilled to field capacity during irrigation, then more water will be obtained from the shallower depths. Under these conditions, there is usually a sandwiched layer of dryer soil between the upper part of the root zone and the lower part.

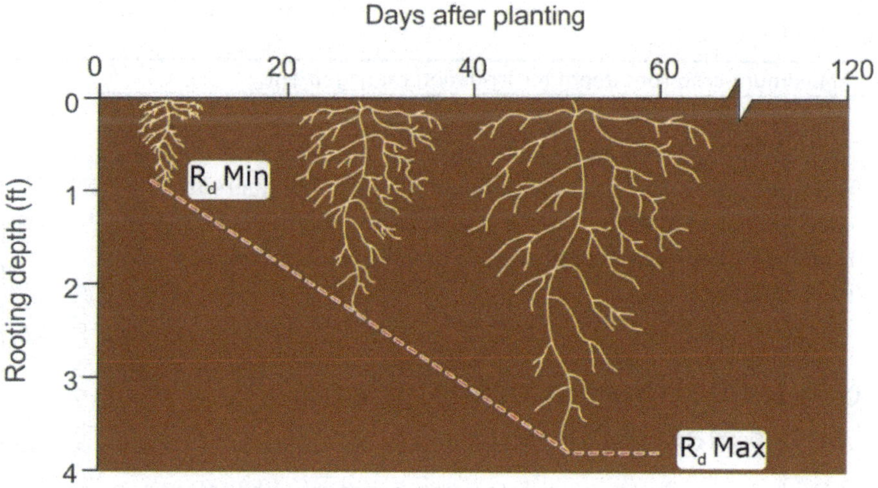

Figure 6.3. Development of a corn plant's root zone.

Example 6.2

Determine the root zone depth for corn at early tassel assuming that depth at germination is 6 in, maximum rootding depth is 4 ft, full depth occurs 90 d after germination, and early tassel occurs 50 d after germination.

Given: D_{ag} = 50 d

D_{tm} = 90 d

R_{dMIN} = 0.5 ft

R_{dMAX} = 4.0 ft

Find: R_d at early tassel

Solution:

$$R_f = \frac{D_{ag}}{D_{tm}} \quad \text{(Eq. 6.7)}$$

$$R_f = \frac{50 \text{ days}}{90 \text{ days}} = 0.56$$

$$R_d = R_{dMIN} + (R_{dMAX} - R_{dMIN})R_f \quad \text{(Eq. 6.6)}$$

$$R_d = 0.5 \text{ ft} + (4.0 \text{ ft} - 0.5 \text{ ft})\ 0.56 = 2.5 \text{ ft}$$

Table 6.3. Example root zone information for various annual crops grown in Nebraska (adapted from Melvin and Yonts, 2009).

Assumed Root Depth (ft)	Corn (3)[a]	Grain Sorghum (3)[a]	Soybean (3)[a]	Dry Beans (2.5)[a]	Sugar Beets (3)[a]	Winter Wheat (4)[a]	Alfalfa (4)[a]
1.0	vegetative	vegetative	vegetative	vegetative			
1.5				initial flowering pod set			
2.0	12 leaf		early bloom	beginning pod fill	June 1	fall growth	
2.5	16 leaf	flag leaf	full bloom	full seed fill	July 1	spring growth	
3.0	silking	boot	pod elongation		July 15	joint	
3.5	blister	bloom			August 1	boot	
4.0	beginning dent	dough	full seed fill		Sept. 1	dough	
5.0							
6.0							established stand

[a] Maximum crop root depth for irrigation management.

Even though root zone depths exceed 3 feet by midseason for many crops, to be on the safe side, many managers use a 3-foot root zone until late in the season. As maturity approaches, the plants are allowed to extract water from the entire root zone.

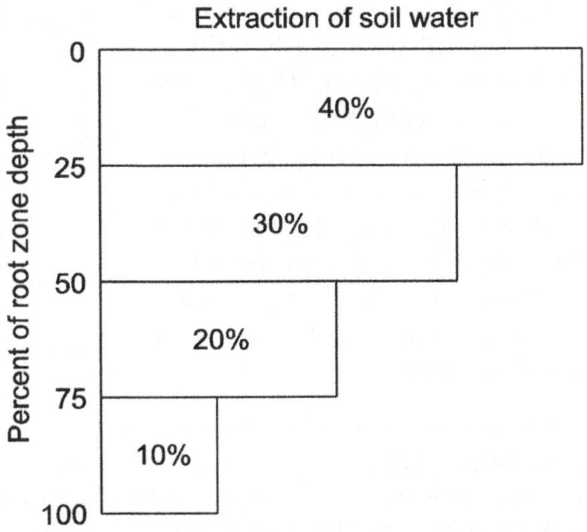

Figure 6.4. Average water extraction pattern from the plant root zone, the 4-3-2-1 rule.

6.4 Irrigation Scheduling for Soil Water Maintenance

With the soil water maintenance approach, the plant's needs for water are assumed to be met as long as the soil water is maintained between TAW and MB. As shown earlier, f_{rc} and MB are dependent on the plant's microclimate, specifically the atmospheric demand for water.

An important variable in irrigation is the allowed depletion (AD). The interval between irrigations is controlled by the AD and the evapotranspiration. The maximum time interval between irrigations, T_{MAX}, is as follows:

$$T_{MAX} = \frac{AD}{ET} \tag{6.8}$$

where: T_{MAX} = maximum time interval between irrigations and
ET = average daily evapotranspiration.

In Example 6.1, AD was 2.7 inches. What is T_{MAX} if ET = 0.3 of an inch a day? The answer is 9 days. This suggests that if water is not applied until AD is reached, then the appropriate maximum time between irrigation is 9 days. And, if irrigation is withheld for 9 days and limited to 2.7 inches, deep percolation is avoided.

Using Equations 6.4 and 6.8 you can determine how the root zone depth, evapotranspiration, and the available water capacity of the soil all influence the frequency and the amount of irrigation. A shallow root zone requires more frequent irrigations but lighter applications.

A coarse-textured soil with a lower available water capacity requires lighter and more frequent irrigations. Medium-textured soils combined with deep root zones allow for less frequent irrigations and larger water applications.

The irrigation interval does not have to equal T_{MAX}; it can be less. It is controlled by ET and the effective depth of water application, i.e.:

$$T = \frac{d_e}{ET} \tag{6.9}$$

where: T = the time interval between irrigations and

d_e = the effective water applied per irrigation.

Many of the modern irrigation systems are managed to apply light, frequent irrigations even when root zones are deep and the AWC is large. For example, a center pivot irrigation system might be managed to apply an effective depth of 0.9 inches even if AD is much larger. Suppose that SWD = AD = 2.7 inches on the day of irrigation. The effective application of 0.9 inches is okay as long as the irrigation frequency is adjusted accordingly. Using our earlier example where ET = 0.3 inches per day, the appropriate interval between irrigations would be:

$$T = \frac{0.9\text{ in}}{0.3\text{ in/d}} = 3\text{ days}$$

The basic goals of irrigation management are that the deficit not exceed AD before water is applied and that infiltration not exceed the SWD. To avoid exceeding AD, irrigation should occur on or before the latest date (LD). LD is calculated as:

$$LD = \frac{AD - SWD}{ET_f} \tag{6.10}$$

or using the balance approach:

$$LD = \frac{AW - MB}{ET_f} \tag{6.11}$$

where: AW = available water (defined below),

MB = minimum allowable balance, and

ET_f = forecasted daily ET.

The LD concept is illustrated in Figure 6.5.

For non-layered soils,

$$AW = (\theta_v - \theta_{wp})R_d \tag{6.12}$$

or

$$AW = f_r(AWC)R_d \tag{6.12b}$$

or

$$AW = f_r(TAW) \tag{6.12c}$$

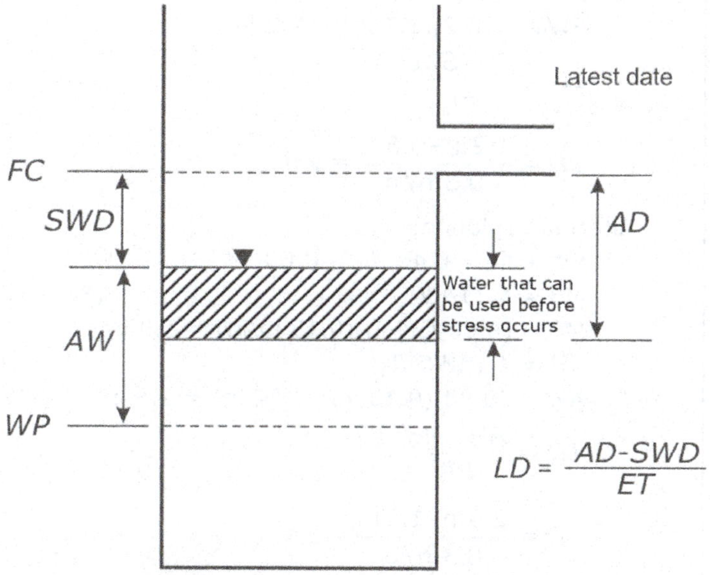

Figure 6.5. Illustration of latest day (LD) concept.

For layered soils,

$$AW = (\theta_{v1} - \theta_{wp1})(t_1) + (\theta_{v2} - \theta_{wp2})t_2 + \ldots + (\theta_{vn} - \theta_{wpn})\,[R_d - (t_1 + t_2 + \ldots + t_{n-1})] \quad (6.13)$$

where: θ_{v1}, θ_{v2}, θ_{vn}, θ_{wp1}, θ_{wp2}, θ_{wpn} = volumetric water content of soil layer 1, 2, and n, respectively, numbered from the surface layer down;

t_1, t_2, t_{n-1} = thickness of soil layers 1, 2, and n – 1, respectively; and

n = the number of soil layers that contain roots.

It is convenient to combine Equations 2.12 and 2.14 to obtain:

$$SWD = f_d(TAW) \quad (6.14)$$

Another useful conversion is that

$$TAW = AW + SWD \quad (6.15)$$

In Example 6.3, the irrigation system should water this location in the irrigated area within 2 days to prevent plant stress. If it will take 3 days to get there, irrigation will be 1 day late. Usually, a beginning or start position and an ending or stop position is designated within the irrigated area. A record should be kept of each position so that irrigation occurs before AD is exceeded at either position. An example of the starting and ending position for a center pivot system is illustrated in Figure 6.6.

Example 6.3

Field beans (Crop Group 3) are being grown in a fine sandy loam soil (AWC = 0.13 in/in). The feel and appearance method for determining soil water revealed that the average f_r = 0.80 in the root zone. Determine the latest date for irrigatin. Assume that the root zone depth is 24 in, and ET of the unstressed crop is 0.3 in/d.

Given: AWC = 0.13 in/in $\quad$ R_d = 2 ft = 24 in

Current f_d = 0.20 $\quad$ ET_f = 0.3 in/d

Find: LD

Solution:

$f_{dc} = 0.40$ (Table 6.1)

$TAW = (R_d)(AWC)$ (Eq. 6.2)

TAW = (24 in)(0.13 in/in) = 3.1 in

$AD = f_{dc}\,(TAW)$ (Eq. 6.4)

AD = (0.40)(3.1 in) = 1.2 in

$SWD = f_d\,(TAW)$ (Eq. 6.14)

SWD = (0.2)(3.1 in) = 1.2 in

$$LD = \frac{AD - SWD}{ET_f} \quad \text{(Eq. 6.10)}$$

$$LD = \frac{1.2\text{ in} - 0.6\text{ in}}{0.3\text{ in/d}} = 2\text{ d}$$

Alternate solution:

Since f_{dc} = 0.40, f_{rc} = 0.60

$MB = f_{rc}\,(TAW)$ (Eq. 6.4)

MB = (0.60)(24 in)(0.13 in/in) = 1.9 in

$AW = f_r\,(AWC)R_d$ (Eq. 6.12b)

AW = (0.80)(0.13 in/in)(24 in) = 2.5 in

$$LD = \frac{AW - MB}{ET_f} \quad \text{(Eq. 6.11)}$$

$$LD = \frac{2.5\text{ in} - 1.9\text{ in}}{0.3\text{ in/d}} = 2\text{ d}$$

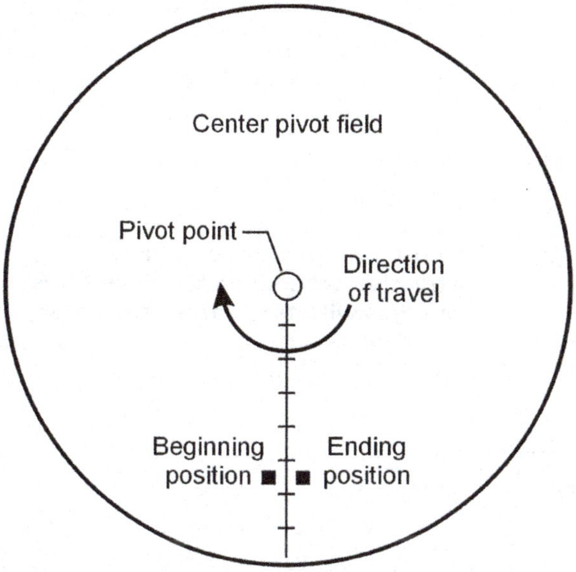

Figure 6.6. Location of beginning and ending positions for a center pivot irrigation system.

Example 6.4

Suppose in Example 6.3 that d_e = 0.5 in, r_a = 0.4 in, and SWD = 0.6 in. Find the earliest date you should irrigate.

Given: d_e = 0.5 in

r_a = 0.4 in

SWD = 0.6 in

Find: ED

Solution:

$$ED = \frac{r_a + d_{ep} - SWD}{ET_f} \quad \text{(Eq. 6.16)}$$

$$ED = \frac{0.4\,\text{in} + 0.5\,\text{in} - 0.6\,\text{in}}{0.3\,\text{in/d}} = 1\,\text{d}$$

Since the LD date was 2 d, irrigation should occur either 1 or 2 days from now.

Another goal is not to irrigate too soon, i.e., not before the earliest date (ED). To avoid deep percolation there must be room in the root zone to store the planned effective depth of water, d_{ep}. In addition, in humid and semi-humid regions, it is good management to allow room in the soil profile for storing rainfall that might occur immediately following the irrigation. This is called the rainfall allowance, r_a. ED is calculated as:

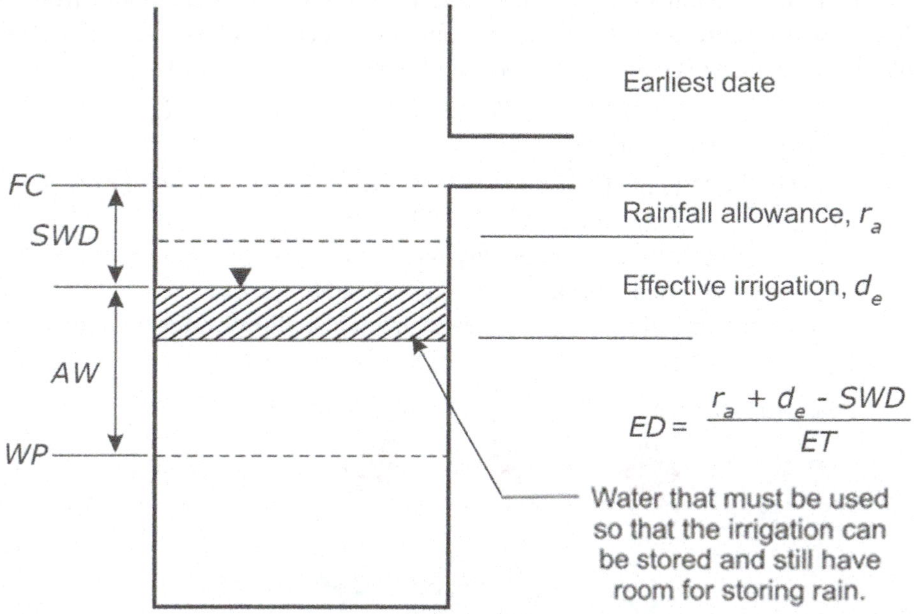

Figure 6.7. Illustration of the earliest date (ED) for irrigation concept.

$$ED = \frac{r_a + d_{ep} - SWD}{ET_f} \quad (6.16)$$

$$ED = \frac{r_a + d_{ep} - (TAW - AW)}{ET_f} \quad (6.17)$$

The ED concept is illustrated in Figure 6.7.

The concepts of TAW, MB, AW, AD, and SWD and how they change with time for an annual crop are shown in Figure 6.8. Note that one goal of irrigation scheduling is to keep the AW between TAW and MB.

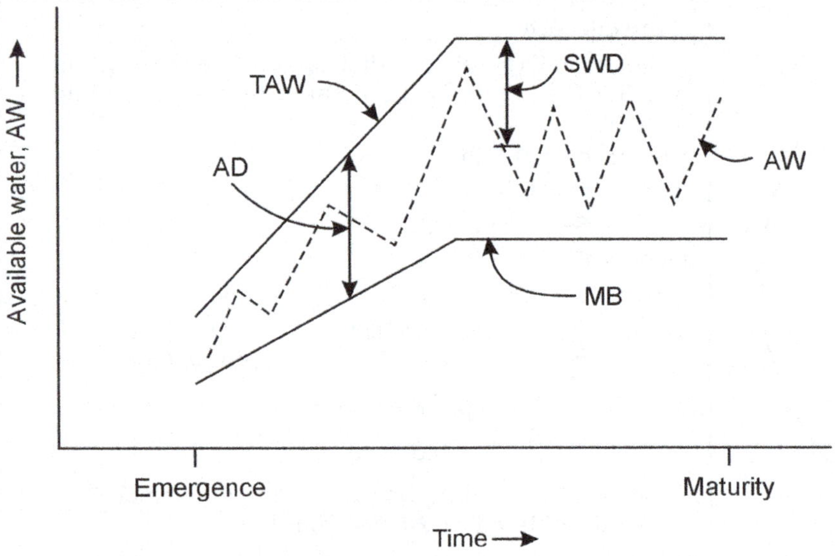

Figure 6.8. Illustration of key irrigation scheduling terms and their changes with time for annual crops.

6.4.1 Checkbook Accounting Method

The checkbook accounting or water balance approach can be used to schedule irrigations. This approach accounts for all of the additions and withdrawals to and from the root zone as illustrated in Figure 6.9. The checkbook method keeps track of the soil water deficit (SWD) on a daily basis. SWD on a given day can be calculated as:

$$SWD_i = SWD_{i\text{-}1} + ET_{i\text{-}1} - d_{e\,i\text{-}1} - P_{e\,i\text{-}1} - U_{f\,i\text{-}1} \quad (6.18)$$

where: SWD_i = SWD on day i,
$SWD_{i\text{-}1}$ = SWD on day i-1
$ET_{i\text{-}1}$ = evapotranspiration on day i-1
$d_{e\,i\text{-}1}$ = effective irrigation on day i-1,
$P_{e\,i\text{-}1}$ = effective precipitation on day i-1, and
$U_{f\,i\text{-}1}$ = upward flow of groundwater from a shallow water table on day i-1.

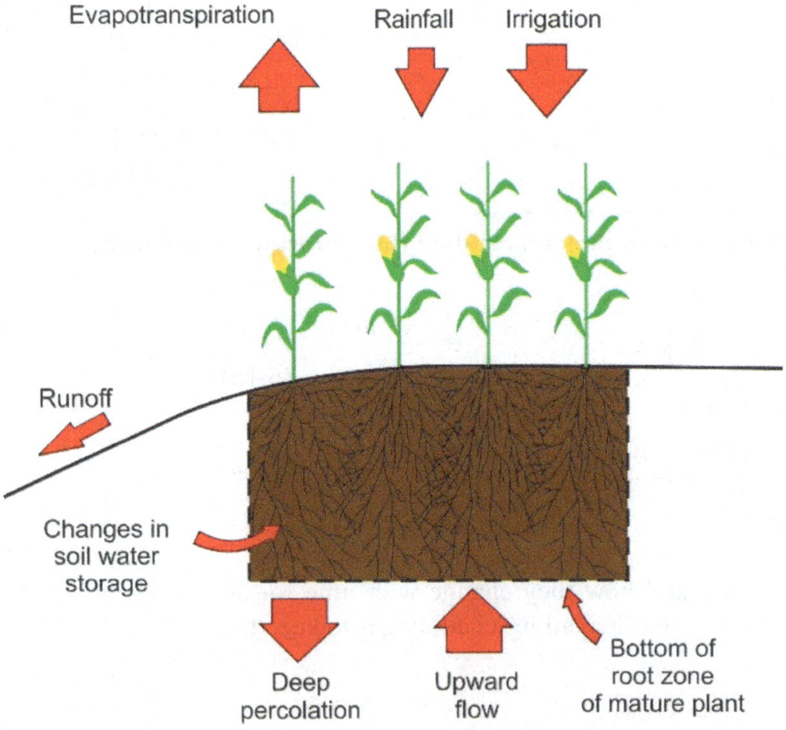

Figure 6.9. Additions and subtractions from the plant root zone (adapted from Cassel, 1984).

In available water balance form, Equation 6.18 is:

$$AW_i = AW_{i-1} - ET_{i-1} + d_{ei-1} + P_{ei-1} + U_{fi-1} \quad (6.19)$$

where: AW_i = water balance on day i and

AW_{i-1} = water balance on day $i - 1$.

Note that runoff and deep percolation are not considered in Equations 6.18 and 6.19. This is because we have used the terms effective precipitation and effective irrigation. Methods were presented in Chapter 5 to determine effective irrigation depths. If the infiltrated depth of water in the low quarter from precipitation and irrigation exceeds SWD, then the effective depth equals the SWD. In mathematical terms:

if $d_{LQ\,i-1}$ (infiltrated irrigation depth) $\leq SWD_{i-1}$ then $d_{ei-1} = d_{LQi-1}$ (6.20)

if $d_{LQ\,i-1} > SWD_{i-1}$ then $d_{ei-1} = SWD_{i-1}$ (6.21)

The same equations can be used for rainfall infiltration to determine effective precipitation.

Using Equation 6.19 is analogous to keeping the balance in your checkbook. AW is the balance; irrigation, rainfall, and upward flow are the deposits; and ET is a withdrawal. If the AW becomes lower than the MB, a penalty is paid, such as a reduction in crop yield.

To use Equation 6.18 or 6.19, a starting or initial estimation of SWD or AW is needed. This can be done by using one of the soil water measurement techniques discussed in Chapter 2. Another approach is to begin the checkbook accounting procedure following a wet period or following a thorough irrigation when the soils can be assumed to be at or near field capacity.

The ET in Equations 6.18 and 6.19 can be calculated from weather data using the approaches given in Chapter 4. A question that often arises is, what should be used as the forecast ET for the LD and ED calculations? Equations 6.18 and 6.19 use ET as determined by the weather that has already occurred. The forecast ET can be based on long-term average weather conditions for a region, such as illustrated in Table 6.4.

Table 6.4. Example of long-term water use data (ET) for crops in Nebraska (adapted from Melvin and Yonts, 2009).

Water Use Rate (in/day)	Corn	Grain Sorghum	Soybeans	Dry Beans	Sugar Beets	Winter Wheat	Alfalfa[a]
0.18					June 15	spring growth	
0.22			full bloom		July 1		
0.24	12 leaf			rapid vegetative growth		joint	
0.26		flag leaf	begin pod				
0.28	early tassel	boot					June 15
0.30	silking	half bloom	full pod	flowering and pod development	July 15	boot	July 1
0.28							August 1
0.26	blister kernel	soft dough			August 1		
0.24	milk		seed fill				August 15
0.22						dough	September 1
0.20	begin dent						
0.18	full dent	hard dough		pod fill and maturation			
0.19							

[a] Alfalfa water use rates should be multiplied by 0.50 during the first 10 days following cutting and by 0.75 from the 10th to 20th day following cutting.

Another approach is to predict ahead based on what occurred during the past few days. If the weather is forecasted to be similar to what has just occurred, then it can be assumed that the forecasted ET is equal to ET of the prior few days.

When a water table exists close to the root zone, crops may extract water from the capillary fringe, or water may flow upward into the crop root zone. Water tables that are within 3 feet of the bottom of the root zone can provide a substantial fraction of the ET even for saline groundwater if the crop is relatively salt tolerant.

The rate of upward groundwater flow depends on the depth to the water table and the soil type. Shallow water tables supply water more rapidly than deep water tables. The soil type has two influences. First, the capillarity of the soil provides the energy or potential for upward movement. Second, the hydraulic conductivity of the soil determines the rate of upward flow. Sandy soils have a high conductivity when nearly saturated, but the conductivity drops very quickly with distance above the water table as the soil becomes unsaturated. Sandy soils are usually irrigated to prevent large soil water potentials; they provide less energy for upward flow. Therefore, sandy soils usually have small rates of upward flow. Clay soils can produce large potentials for upward flow; however, their low hydraulic conductivity limits the rate of upward flow. Upward flow is generally most significant for medium-textured soils where the soil water potential and conductivity together produce significant flow rates.

A simplified method of estimating upward flow from Doorenbos and Pruitt (1977) is shown in Figure 6.10. More detailed analysis has been presented by Skaggs et al. (1981) for use with combined drainage and subsurface irrigation systems.

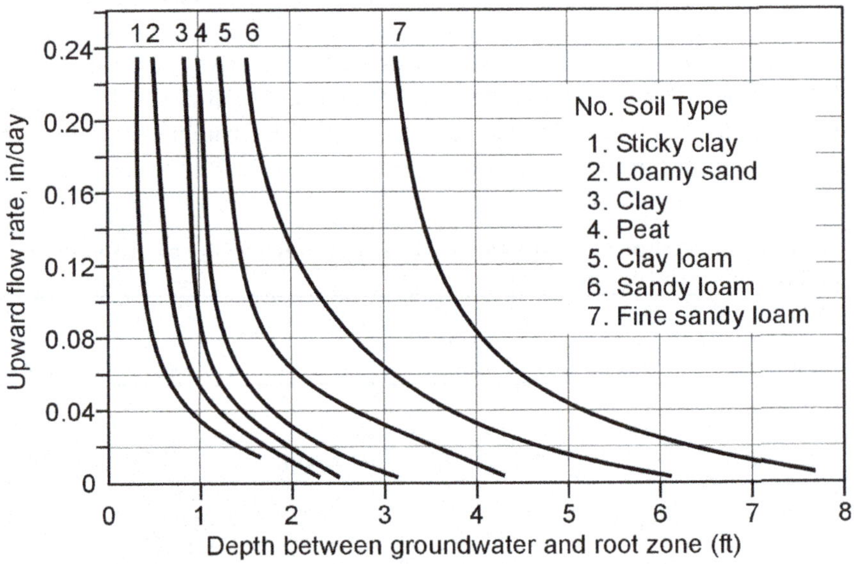

Figure 6.10. Upward flow of water from a groundwater table (modified from Dorrenbos and Pruitt, 1977).

For annual crops where the root zone depth is expanding with time, the SWD and AW calculations should consider the soil water conditions that the roots are growing into. For example, if the roots are growing into a soil with soil water levels less than FC, then the ED will decrease because of the extra room for water storage provided by the root zone expansion. The LD might increase or it might decrease depending upon the SWD in the new portion of the root zone and on f_{dc}.

Equations 6.18 and 6.19 each could have a component that accounts for root zone expansion. The root zone expansion can be treated as a continuum or in discrete steps.

If the root zone expansion is treated in discrete steps, Equation 6.18 is modified as follows:

$$SWD_i = SWD_{i-1} + ET_{i-1} - d_{e\,i-1} - P_{e\,i-1} - U_{f\,i-1} + \Delta SWD_{i-1} \quad (6.22)$$

$$\Delta SWD_{i-1} = (AWC)\left(\frac{f_{do}}{100\%}\right)(\Delta R_d) \quad (6.23)$$

where: ΔSWD_{i-1} = change in SWD due to the additional root depth (ΔR_d) and

f_{do} = initial f_d in the new layer of soil explored by roots.

The modified Equation 6.19 is:

$$AW_i = AW_{i-1} - ET_{i-1} + d_{ei-1} + P_{ei-1} + U_{fi-1} + \Delta AW_{i-1} \qquad (6.24)$$

$$\Delta AW_{i-1} = (AWC)\left(\frac{f_{ro}}{100\%}\right)(\Delta R_d) \qquad (6.25)$$

where: ΔAW_{i-1} = added available water due to the root zone expansion and
f_{ro} = initial f_r in the new layer of soil explored by roots.

The use of the water balance method is illustrated in Example 6.5. In Example 6.5, Location 1 could be irrigated on June 29. The soil could store the effective depth applied and yet there would be room for storing a 0.5-inch rainfall. At most, irrigation could be delayed until July 6 (8 days after June 28). At Location 2, irrigation is not required until July 7 but it would be allowable to irrigate in 3 days (July 1), based on the ED calculation.

The results of using checkbook accounting are shown graphically in Figures 6.11 and 6.12. Figure 6.11 shows an example where f_r is maintained between 0.40 and 0.70 throughout the growing season. In Figure 6.12 you see an example where the soil water is allowed to gradually deplete to below 0.40 f_r at plant maturity (PM). Figure 6.12 illustrates an important concept that can be followed in semiarid and subhumid regions. The evolution of the soil water is the result of water applications which, by design, only replace a fraction of ET. This concept, called programmed soil moisture depletion (Fischbach and Somerhalder, 1973), depletes the soil water reservoir to low, yet safe, levels late in the growing season. By depleting soil water, there is room in the soil for storing precipitation during the offseason. Storing offseason precipitation is an effective way of reducing irrigation requirements.

Example 6.5

Corn is grown on a silt loam soil at two different locaitons. The pertinent site conditions are:

Given: f_{dc} = 0.45 $\quad R_d$ = 2.5 ft = 30 in

d_e = 1.1 in $\quad AWC$ = 0.2 in/in

r_a = rainfall allowance = 0.5 in

Depth to water table = 10 ft

The SWDs at the start of June 25 were 2.2 and 0.80 inches for Locations 1 and 2 in the irrigated area, respectively. The ET and P_e for June 25 to June 28 are given in the table below.

Find: Determine the *LD* and *ED* for each location for June 25 to 28

Solution:

Use Equations 6.2, 6.4, 6.10, 6.16, and 6.18:

TAW = (R_d)(AWC) (Eq. 6.2)

TAW = (30 in)(0.2 in/in) = 6 in

AD = f_{dc} (*TAW*) (Eq. 6.4)

AD = (0.45)(6 in) = 2.7 in

The results of the calculations using Equations 6.10, 6.16, and 6.18 are shown in the bold italics in the table below.

Date	Actual ET (in/day)	Forecast ET (in/day)	P_e (in)	U_f (in)	Location 1 SWD (in)	Location 1 d_e (in)	Location 1 ED (days)	Location 1 LD (days)	Location 2 SWD (in)	Location 2 d_e (in)	Location 2 ED (days)	Location 2 LD (days)
June 25					2.20		-	***3***	0.80		***4***	***11***
	0.20	0.18	0.0	0.0		0				0		
26					***2.40***		-	***2***	***1.00***		***3***	***9***
	0.21	0.18	0.0	0.0		0				0		
27					***2.61***		-	***1***	***1.21***		***2***	***8***
	0.13	0.18	0.3	0.0		1.1				0		
28					***1.34***		***1***	***8***	***1.04***		***3***	***9***
	0.17	0.18	0.0	0.0		0				0		

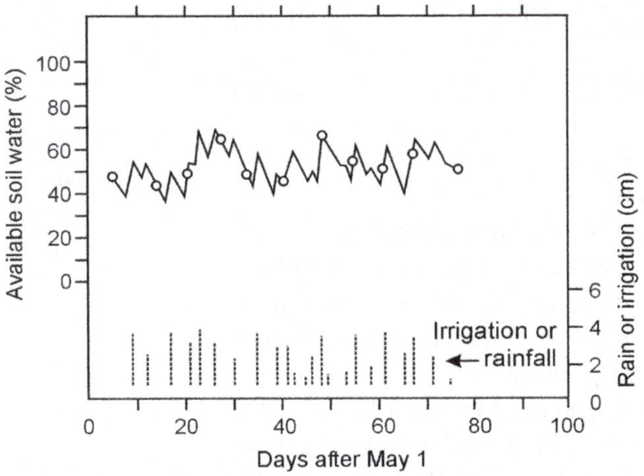

Figure 6.11. Graphical results of soil checkbook accounting method (adapted from Stegman, 1983). Soil water levels are kept between 40 and 70%.

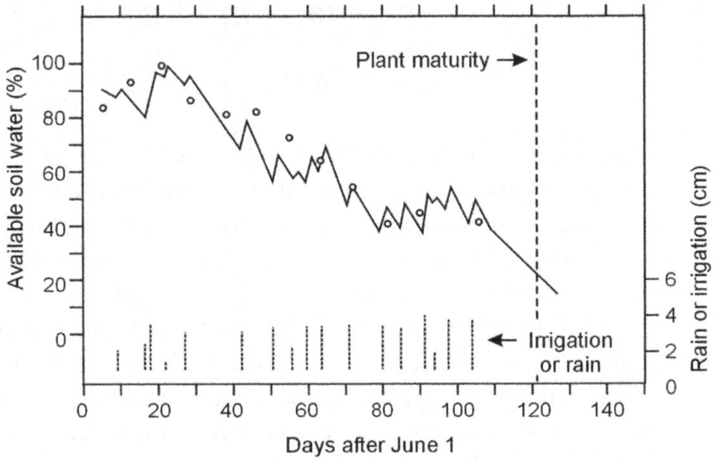

Figure 6.12. Graphical results of checkbook accounting method where soil water was managed to deplete slowly (adapted from Stegman, 1983). Soil water levels were allowed to gradually deplete.

6.4.2 Simplified Checkbook Method

A major limitation to the checkbook accounting procedure has been the lack of reliable real time ET data. This has largely been overcome with the advent of automated weather station networks. Weather data from these stations are used to calculate crop ET on a continuous basis. Now with the easier availability of ET data, the biggest problem is estimation of effective irrigation depths. Water measurement is key. Once the water is measured, effective depths can be determined using the techniques described in Chapter 5.

The checkbook accounting procedure requires daily record keeping. This has slowed its acceptance. Computer software eliminates the need to manually perform the daily calculations, but keeping records is still necessary except when all sensors are electronic and data can be transmitted from remote locations directly to computers or hand-held smart devices.

One way that irrigators apply the ET data without daily recording is to simply irrigate when the effective depth has been consumed (Equation 6.9). For example, if ET is 0.25 inches per day and the effective depth is 0.75 inches per application, irrigation must be applied every 3 days. Thus, the water manager is reacting to the amount of water applied and ET. To adjust for rainfall, irrigation can be delayed in accordance with how long it will take to consume the rainfall. If a 0.5-inch rainfall occurs, the irrigation schedule should be delayed 2 days (assuming that you weren't behind with irrigation before the rainfall occurred and all of the rainfall was effective in satisfying ET).

Another simple checkbook accounting approach is to adjust the effective depth of application according to the amount of ET and rainfall that has occurred over some pre-established time interval. Suppose weekly irrigations are desired. Then the ET and effective precipitation is summed for the time interval. The effective irrigation needed is then equal to the accumulated ET minus the accumulated effective precipitation or

$$d_e = \sum ET - \sum P_e \qquad (6.26)$$

Example 6.6

The daily ET for a potato field and the effective precipitation (P_e) for a week period is given below. How much effective irrigation water is needed to make up the balance between ET and rainfall?

Given:

Day	ET (in)	P_e (in)
Sunday	0.20	--
Monday	0.30	0.50
Tuesday	0.15	--
Wednesday	0.25	0.20
Thursday	0.20	--
Friday	0.25	--
Saturday	0.25	--
Total (Σ)	1.60	0.70

Find: Effective irrigation required

Solution:

$$d_e = \sum ET - \sum P_e$$

$$d_e = 1.60 - 0.70 = 0.90 \text{ in}$$

A drawback to the two "simpler" checkbook accounting approaches discussed above is that the water applications lag behind the time of the water use. On soils with low AWC and/or for plants with shallow root zones, the lag may cause some water stress before water is applied.

6.4.3 Soil Water Measurement Method

An alternative, or supplement, to the checkbook accounting method is to measure soil water directly for irrigation scheduling. In concept, it is quite simple. Rather than predicting or calculating SWD, the SWD is inferred from measures of f_r, f_d, θ_m, θ_v, or soil water tension. Once SWD or AW is determined, then Equations 6.10 or 6.11 and 6.16 or 6.17 are used to calculate the LD and ED for the location where the measurements were taken. The soil water content must be measured throughout the entire plant root zone. Samples or measurements at 1-foot intervals are usually adequate. If a 3-foot root zone is to be sampled, then sensors could be placed at 6, 18, and 30 inches, respectively, and each sensor would represent a 1-foot interval.

Techniques such as feel and appearance, gravimetric sampling, neutron scattering, and TDR measure water content directly (Chapter 2). Water contents can be used in the LD and ED calculations, just as was done by checkbook accounting in Example 6.5. When soil water potential (soil water tension) is measured, such as with tensiometers, granular matrix sensors, or electrical resistance blocks, a soil water release curve is needed to convert tension to volumetric water content. This is essentially a local calibration. The soil water release curve is not easily determined. Land-grant universities and government agencies, such as the Natural Resources Conservation Service, can provide release curves that represent the soils in question. An example of data used for converting tension to SWD for general soil texture classifications in Nebraska is shown in Table 6.5. More detailed data are provided by Irmak et al. (2016) and Melvin and Martin (2018). The use of soil water sensing to schedule irrigations is illustrated in Example 6.7.

Table 6.5. Example SWD versus tension for selected soil textures in Nebraska.

Tension (cb)	Fine Sand	Loamy Sand	Sandy Loam	Fine Sandy Loam
	Fraction Depleted (in/in)			
0	0.000	0.000	0.000	0.000
10	0.008	0.000	0.000	0.000
20	0.025	0.025	0.025	0.017
30	0.042	0.033	0.042	0.042
40	0.050	0.042	0.050	0.058
50	0.054	0.050	0.058	0.067
60	0.058	0.058	0.067	0.083
70	0.067	0.067	0.071	0.092
80	-	-	0.075	0.100
AWC (in/in)	0.083	0.092	0.117	0.150

An alternative to converting soil water tension to water content is to monitor the soil water sensors frequently and irrigate when the soil water tension has reached a "threshold level" (Irmak et al, 2016) In fact, manufacturers of soil water sensing equipment often provide the users with guidelines for these threshold levels for various crops and soil textures. They are usually based on sensing near the vertical center of the root zone. Example thresholds are given in Table 6.6. One problem with this approach is that it is difficult to predict ahead to determine the LD. This is overcome by more frequent monitoring. Graphical extrapolation, as shown in Figure 6.13, can be used to lessen the frequency of monitoring. The graphical method provides a good visual record of soil water variations during the season. A limitation of the threshold level method is that the irrigator does not know how much water the soil can hold during each irrigation.

Table 6.6. Example threshold soil water tensions for irrigation scheduling based on f_{dc} = 0.35.

Texture	Threshold Tension (cb)
Fine sand	20
Loamy sand	25
Sandy loam	35
Fine sandy loam	45
Silt loam	80
Clay loam	80

A list of questions to consider when selecting a soil water monitoring system, including sensors, communications, and data storage, has been provided by ITRC (2019).

An important, and often frustrating, consideration is the number of locations that must be sampled to reliably estimate the average soil water condition within the area of interest. You must not only consider the spatial variability of the soil itself, but also the spatial variability of water application from the irrigation system. A minimum of four locations should be sampled in a large, irrigated area that has "relatively" uniform soils and slopes. It is often good to sample stress-prone areas (low AWC and/or shallower root zone), areas where infiltration is low (steeper slopes, etc.), areas where water applications are low due to the inherent nature of the system (e.g., the downstream end of furrow irrigated fields), or where ET is the highest (e.g., nonshaded and wind exposed areas within a landscape). Do this only if the stress-prone area represents a "significant" portion of the irrigated area. Using checkbook accounting in conjunction with the soil water sensing method reduces the number of locations that must be sampled. Another method, which can greatly reduce the uncertainty of θ_v data from sensors, is to monitor and manage trends in SWD instead of θ_v or AW (Singh et al., 2020). In this method, SWD and AD should be calculated using the observational FC (FC_{obs}) for the specific sensor and location, which is determined from the trend in the sensor data when the field is approaching FC conditions (e.g., at the beginning of the season after the profile becomes saturated and ET is low). In this way, much of the uncertainty from the sensor and spatial variability cancels out during the calculation:

$$SWD = (\theta_{FC,obs} - \theta_v)R_d$$

Example 6.7

Assume a lettuce field with a root zone depth of 12 in. The soil is a silt loam with an AWC of 0.17 in/in. Assume f_{dc} = 0.50. The feel and appearance method was used to measure soil water. The f_r was 0.60 in Location 1 and 0.80 in Location 2. ET (forecasted) is 0.2 in/d. The d_e = 0.5 in and r_a = 0.3 in. Determine the LD and ED dates for the two locations that were sampled.

Given: $ET = 0.2$ in/d $AWC = 0.17$ in/in
$R_d = 12$ in $d_e = 0.5$ in
$r_a = 0.3$ in

Find: *LD* and *ED*

Solution:

$$TAW = (R_d)(AWC) \quad \text{(Equation 6.2)}$$

$$TAW = (12\text{ in})(0.17\text{ in/in}) = 2.0\text{ in}$$

$$AD = f_{dc}(TAW) \quad \text{(Equation 6.4)}$$

$$AD = (0.50)(2.0\text{ in}) = 1.0\text{ in}$$

Calculations for Location 1:

$$f_d = 1 - f_r \quad \text{(Equation 2.10a)}$$

$$f_d = 1 - 0.60 = 0.40$$

$$SWD = f_d(TAW) \quad \text{(Equation 6.13)}$$

$$SWD = (0.4)(2.0\text{ in}) = 0.8\text{ in}$$

$$LD = \frac{AD - SWD}{ET_f} \quad \text{(Equation 6.10)}$$

$$LD = \frac{1.0\text{ in} - 0.8\text{ in}}{0.2\text{ in/d}} = 1\text{ d}$$

$$ED = \frac{r_a + d_e - SWD}{ET_f}$$

$$ED = \frac{0.3\text{ in} + 0.5\text{ in} - 0.8\text{ in}}{0.2\text{ in/d}} = 0\text{ d}$$

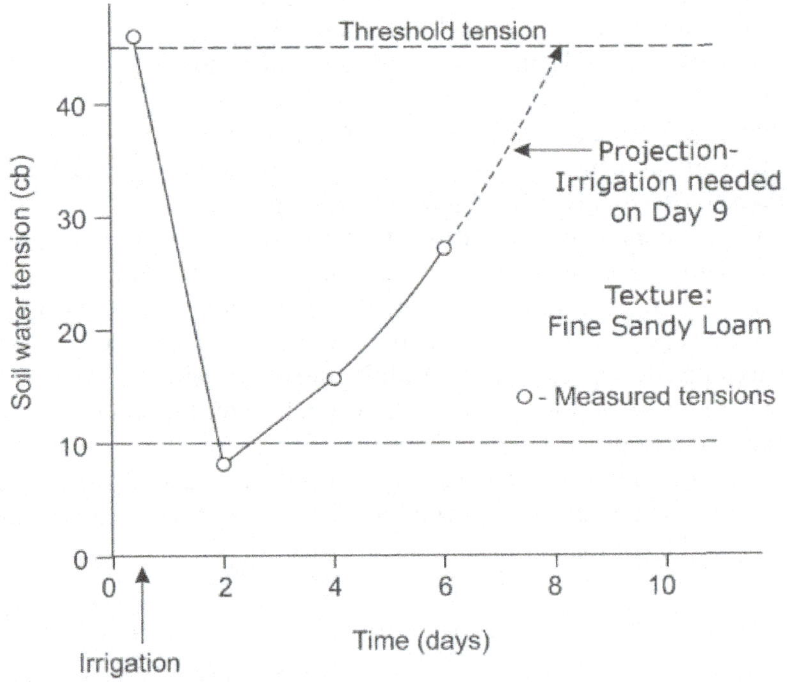

Figure 6.13. Graphical method of predicted date of irrigation.

6.5 Scheduling Using Plant Status Indicators

The emphasis of this chapter has been on management of the soil water reservoir. While soil water management has been successfully employed for irrigation scheduling, it does not directly evaluate all of the factors influencing plant response to water. As pointed out by Jones (2004) plants actually respond directly to change in water status in the plant tissues, whether in the roots or in other tissues, rather than to changes in soil water status. The plant response to a given soil water content varies as evaporative demand varies. Plant status indicators integrate all of the important factors, i.e., soil water conditions, atmospheric demand for water, and plant characteristics. All three factors are taken into account to a degree by selecting the appropriate f_{dc} as suggested by Doorenbos and Kassam (1979) and presented in Table 6.1. Direct measures or indicators of plant water status are often suggested for irrigation scheduling. Jones (2004) presents a good overview of plant-based methods for irrigation scheduling. A few of these are discussed below.

6.5.1 Leaf Water Potential

Turner (1990) and Stegman (1983) discussed the use of plant water status indicators for irrigation scheduling. Leaf water potential, a measure of the energy status of the water in a plant leaf, is an indicator of the water status of a plant. Figure 6.14 shows the relationship between the various potentials in the soil plant atmosphere continuum. Instruments used to measure leaf water include pressure chambers and thermocouple psychrometers. Stegman (1983) found that threshold levels of leaf water potential ranged from –12 to –12.5 bars for corn (midafternoon readings). The thresholds were dependent on ambient temperatures much like f_{dc} is dependent on ET.

While leaf water potential is a direct measure of plant water status, using it as a scheduling tool has some limitations. Threshold levels must be developed for the plants in question. Also, like using threshold levels of soil water potential, it lacks in predictability. Third, measuring leaf water potential is time-consuming and must be done during a narrow time window during midday. A large number of samples are necessary for an accurate estimation of the mean.

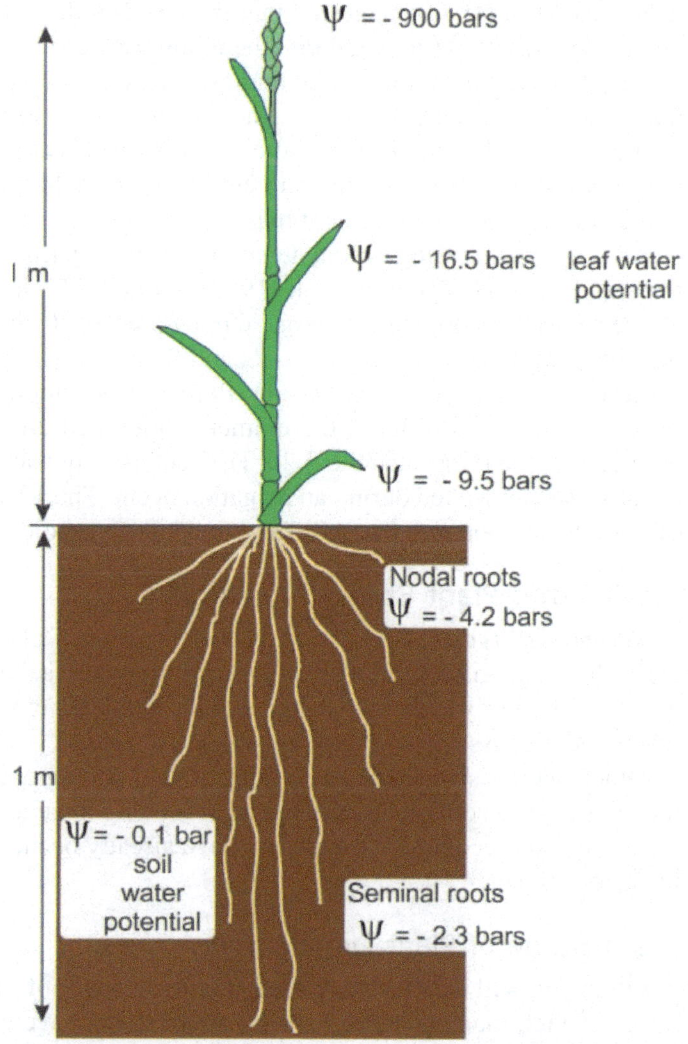

Figure 6.14. Water potentials expected at different points of the pathway for water transport through a wheat plant growing in soil at a potential of -0.1 bars and in an atmosphere with a potential of -900 bars (adapted from Turner and Burch, 1983).

6.5.2 Plant Canopy Temperature

Since evaporation of water is a cooling process, the foliage of well-watered plants is usually cooler than the surrounding air, especially in arid climates. Plants that are experiencing water stress will have higher leaf temperatures than well-watered plants. With the advent of infrared thermometry, it is relatively easy to measure canopy temperature (Lo et al., 2018; Figure 6.15). In general, two approaches have been developed to use canopy temperature in irrigation scheduling, the crop water stress index method (CWSI) and the time-temperature-threshold (TTT) method.

The temperature difference (DT) between the air and the plant canopy depends on both the plant water status and the vapor pressure deficit (VPD). Jackson (1982) presented an excellent overview of plant water stress response to DT and VPD. Jackson et al. (1981) developed the CWSI method for quantifying plant stress. It relies on baseline values of canopy temperature of non-transpiring reference and a non-stressed canopy. Crop yield can then be related to CWSI as illustrated by Irmak et al. (2000). To apply this technique the base-line or reference lines must be established. Jackson (1982) suggested that the CWSI will be a very useful tool for irrigation scheduling because it is easy to use handheld infrared thermometers for measuring canopy temperature. He points out some of the problems, such as the effects of bare soil in the field of view and the establishment of the threshold stress indicators for various crops or plants. As pointed out by Stegman (1983), the effect of wind and cloud cover on the interpretation of DT data and how it relates to irrigation management must be considered.

Figure 6.15. Diagram of a hand-held infrared thermometer.

Another way of using canopy temperature in irrigation scheduling is the time-temperature-threshold method. Wanjura et al. (1995) defined TTT as "the amount of time accumulated above a specific temperature in one day by a crop." If the time-temperature value exceeds a threshold, a temperature stress exists and irrigation is needed. Peters and Evett (2008) successfully automated the irrigation scheduling of a center pivot irrigation system using the TTT method. They used infrared thermometers mounted on the center-pivot to monitor canopy temperatures as it irrigated the field. The center-pivot was equipped with LEPA drops thus the canopy was not wetted during an irrigation event. The TTT method requires the establishment of both the temperature threshold and the time threshold.

6.5.3 Other Plant Status Indicators

Turner (1990) discussed various other measures of plant water status indicators, including leaf color. With many crops, stressed plants often turn a darker color if soil water stress occurs. This is particularly evident in turfgrass. Bluegrass, for example, will turn to a blue-green color when under stress.

Other plant responses to stress include leaf rolling and wilting. While all of these visual techniques are useful, many often appear too late to be useful for water management. Significant yield and economic losses may have already occurred. Prediction is still a problem for these techniques.

6.5.4 Stage of Plant Development

Often you will hear that irrigating at critical stages of plant growth is a good way of scheduling. In fact, many of the Extension publications have been written on this concept. While the method has merit, local calibration is necessary to account for soil crop climate conditions.

6.6 Variable Rate Irrigation Management

Variable rate irrigation (VRI) or precision irrigation technology allows for spatial management of soil water. The irrigation prescription map determines the application depth throughout the field, which can be varied spatially to account for spatial variability in soils, ET, and topography.

One option for VRI is to utilize the stored soil water in heavier soils by not irrigating them at the beginning of the season (Miller et al., 2018). Soils with lower AWC will need to be irrigated before soils with a larger AWC if all parts of the field begin the season at the same soil water depletion level (e.g., if the field starts at field capacity). An example prescription map based on this approach is shown in Figure 6.16. This particular prescription map would need to be used twice in order to mine the water from the heavier soils. After that, the soil water deficit would equal AD in each irrigation management zone, and uniform irrigation could be used to replace ET. One study estimated that 13% of center pivot irrigated fields in Nebraska could reduce pumping by at least one inch by using VRI to account for spatial variation in AWC (Lo et al., 2016). It is recommended that soil water sensors be placed in each irrigation management zone.

VRI can also be used to manage problems associated with topography, e.g. applying less water in a low spot that tends to be wet from accumulated runoff. Prescription maps for VRI are often based on soil maps and yield maps (Kranz et al., 2014). Soil properties can be cor-

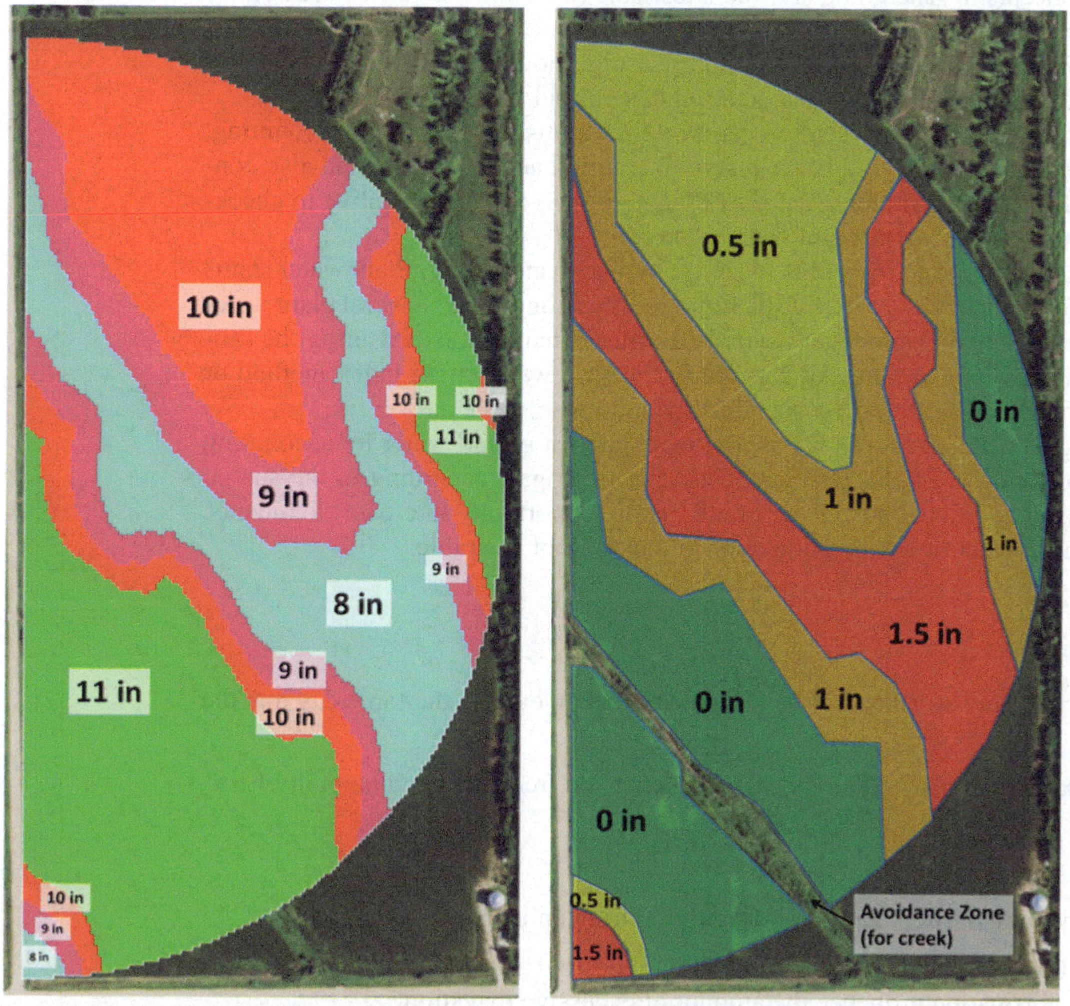

Figure 6.16. Map of spatial variability in total available water (left) and the corresponding irrigation prescription map (right).

related to apparent electrical conductivity (EC_a), elevation, or other topographic indices. Prescription maps can include an "avoidance zone" for waterways (Figure 6.16), other non-cropped areas, or low spots where the pivot tends to get stuck. This feature allows producers to utilize chemigation on fields with a waterway, where regulation would prevent application of chemical on a surface water body.

Remote sensing data, whether from satellite or unmanned aircraft, can also be used to inform variable rate irrigation management (Chavez et al., 2020). Data may indicate an area of the field that is under stress and requires special attention. Remote sensing can also be used to quantify the spatial variability of ET in a field. Ongoing development in sensors, big data, communications (e.g., internet of things), decision support systems, and computational intelligence will allow for more advanced management of irrigation systems as a part of precision agriculture (Evett et al., 2020).

6.7 Summary

Irrigation scheduling refers to the timing and amount of irrigation water applications. By accounting for or measuring soil water and by knowing plant water needs, the goals of good irrigation scheduling can be accomplished; production goals can be met with minimum water. By minimizing water applications, deep percolation of water and chemicals is minimized and energy is saved.

Two important concepts in scheduling are: the latest date (LD) and the earliest date (ED). By irrigating on or before the LD, plant water stress is avoided. By waiting, at least until the ED, deep percolation losses are avoided or minimized. Built into the ED is an allowance for storing rainfall in the soil, an important consideration in semiarid and subhumid regions.

Scheduling according to soil water content can be achieved using checkbook accounting, which considers the deposits to the soil water reservoir (rainfall and irrigation) and also considers withdrawals from the soil water reservoir (ET and drainage). An alternative to checkbook accounting is to directly measure soil water content.

Another scheduling option is to irrigate in response to plant water status. Plant water status is an integrator of soil water, plant characteristics, and weather conditions. Several plant water status methods are available including measuring leaf water potential and measuring the canopy temperature. Canopy temperature can be used in the crop water stress index method or the threshold-time-temperature method to schedule irrigations.

Variable rate irrigation (VRI) allows for spatial management of soil water by accounting for variability in soils within a field. Besides being able to irrigate according to the spatial distribution of soils, an additional advantage of VRI is the opportunity to create "avoidance zones" in a field where water or chemigation applications are not desirable.

Questions

1. Explain why irrigation water does not have to be applied on exactly the same day that the AD is reached.
2. How do soil texture and plant rooting depth influence the frequency and amount of irrigation?
3. Explain why f_{dc} is affected by ET.
4. What is the maximum desired depth of infiltration during an irrigation? Why?
5. Explain why infiltration often exceeds the maximum desired amount of infiltration. Does the uniformity of irrigation influence the amount of excessive irrigation?

6. Develop the water balance (AW) equivalent to Example 6.5.

7. Which do you prefer, the SWD approach or AW approach to the checkbook accounting method? Why?

8. Tensiometers where used in a field for estimating soil water depletion. The beginning position and the ending position of the irrigation system were sampled.

   a. Determine the latest and the earliest dates for both positions given the following information:

Layer (in)	Tensiometer Depth (in)	Tension	
		Beginning Position (cb)	Ending Position (cb)
0–12	6	40	35
12–24	18	45	10
24–36	30	25	35

Root depth = 3 ft
Forecast ET_c = 0.30 in/d
Soil type = fine sandy loam
AWC = 0.15 in/in
Effective irrigation depth = 1 in
Rainfall allowance = 0.5 in
f_{dc} = 0.55

   b. If the irrigation cycle time is 3 days, when would you recommend that the system be started and why?

9. The layers of a Hastings silt loam (Hc) are described below (taken from Soil Survey of Clay County).

Depth (in)	Texture	AWC (in/in)	Bulk Density (g/cm^3)
0–10	silt loam	0.22–0.24	1.20–1.40
10–38	silty clay loam	0.11–0.20	1.30–1.40
38–60	silt loam	0.18–0.22	1.20–1.40

   If tomato roots are 48 inches deep, determine:

   a. Total available water (TAW) in inches

   b. AD (in inches) assuming that ET_c = 0.24 in/d

   c. MB (in inches) assuming that ET_c = 0.24 in/d

   d. The maximum interval between irrigation (T_{MAX}) in days

10. Tensiometers are used in a loamy sand. Readings are taken as follows:

	Tension (cb)
July 2	10
July 4	12
July 6	18

   a. Using the graphical procedure (Figure 6.13), predict the date when irrigation will be needed.

   b. If the root zone is 36 inches deep, at what depth should the tensiometer(s) be placed?

   c. Discuss the pros and cons to this approach of scheduling.

11. Rework Example 6.6 given the following data:

Day	Crop ET (in/d)	P_e (in)
Sunday	0.25	-
Monday	0.35	0.70
Tuesday	0.15	0.10
Wednesday	0.30	-
Thursday	0.15	-
Friday	0.30	-
Saturday	0.20	-

12. If crop ET for the previous four days was 0.35, 0.20, 0.30, and 0.15 in/d and the effective depth of water applied per irrigation is 1.25 inches, how often should you irrigate? If a 0.75-inch rainfall occurs during this schedule, how many days can the system be left idle before it is restarted?

13. Site, crop, weather, and irrigation data are for a field area given below. Using the checkbook accounting approach, develop an irrigation schedule for July 1-14 for both the start position and the stop position. Do the ED and LD calculations on July 1 and July 8. For this 2-week period, tell us when the beginning and ending positions would be irrigated (keep in mind how fast the system can get to each point). Assume that on July 1, the system is in the start position.

Crop: Corn
Emergence date: 4/28
Effective cover date: 7/10
Date of maximum root depth: 8/1
Maximum allowable depletion f_{dMAX}: 0.50
AWC: 0.18 in/in
System capacity: 850 gpm
Forecasted ET: 0.25 in/d

Irrigated area: 136 ac
Gross depth applied: 1.0 in
Cycle time (T_c): 3 d
E_{LQ}: 85%
Allowance for storing rainfall: 0.5 in
f_r on 7/1 in the root zone: 0.70 (both positions)
f_r below the root zone: 1.0

Date	Crop ET (in/d)	Rain (in)	Root Depth (in)
7/1	0.31		
2	0.18	0.15	30
3	0.21		
4	0.20		
5	0.25		
6	0.36		
7	0.22		
8	0.25		33
9	0.23		
10	0.20	0.78	
11	0.26		
12	0.22		
13	0.35		
14	0.23		

References

Cassel, D. K. (1984). Irrigation scheduling I. Crop and soil parameters. *Crops Soils (Feb.)*, 16-20.

Chavez, J. L., Torres-Rua, A. F., Woldt, W. E., Zhang, H., Robertson, C. C., Marek, G. W., Wang, D., Heeren, D. M., Taghvaeian, S., & Neale, C. M. U. (2020). A decade of unmanned aerial systems in irrigated agriculture in the Western U.S. *Appl. Eng. Agric., 36*(4), 423-436.

Dorrenbos, J., & Kassam, A. H. (1979). Yield response to water. FAO Irrigation and Drainage Paper. *33*, 1-193. Rome, Italy: United Nations FAO.

Dorrenbos, J., & Pruitt, W. J. (1977). Guidelines for predicting crop water requirements. FAO Irrigation and Drainage Paper. *24*, 1-144. Rome, Italy: United Nations FAO.

English, M. J., Musick, J. T., & Murtz, V. V. (1990). Deficit irrigation (Ch. 17). In G. J. Hoffman, T. A. Howell, & K. H. Solomon (Eds.), *Management of farm irrigation systems.* St. Joseph, MI: ASAE.

Evett, S. R., O'Shaughnessy, S. A., Andrade, M. A., Kustas, W. P., Anderson, M. C., Schomberg, H. S., & Thompson, A. (2020). Precision agriculture and irrigation: Current U.S. perspectives. *Trans. ASABE, 63*(1), 57-67.

Fischbach, P., & Somerhalder, B. (1973). Programmed soil moisture depletion. NebGuide G73-58. Nebraska Cooperative Ext. Service, University of Nebraska-Lincoln.

Irmak, S., Haman, D. Z., & Bastug, R. (2000). Determination of crop water stress index for irrigation timing and yield estimation of corn. *Agron. J., 92*(6), 1221-1227.

Irmak, S., Payero, J., VanDeWalle, B., Rees, J., Zoubek, G., Martin, D., Kranz, W., Eisenhauer, D., & Leininger, D. (2016). Principles and operational characteristics of Watermark granular matrix sensor to measure soil water status and its practical applications for irrigation management in various soil textures. EC 783. University of Nebraska-Lincoln Ext.

ITRC. (2019). Irrigation Consumer Bill of Rights: Soil and plant moisture monitoring systems. ITRC Report No. R 19-004. San Luis Obispo, CA: Irrigation Training and Research Center, California Polytechnic State University.

Jackson, R. D. (1982). Canopy temperature and crop water stress. In D. Hillel (Ed.), *Advances in irrigation* (Vol. 1, pp. 43-85). New York, NY: Academic Press.

Jackson, R. D., Idso, S. B., Reginato, R. J., & Pinter Jr., P. J. (1981). Canopy temperature as a crop water stress indicator. *Water Resour. Res., 17*(4), 1133-1138.

Jones, H. G. (2004). Irrigation scheduling: advantages and pitfalls of plant-based methods. *J. Exp. Bot., 55*(407), 2427-2436.

Kranz, W. L., Irmak, S., Martin, D. L., Shaver, T. M., & van Donk, S. J. (2014). Variable rate application of irrigation water with center pivots. University of Nebraska-Lincoln Ext. Circular EC2000.

Lo, T., Heeren, D. M., Martin, D. L., Mateos, L., Luck, J. D., & Eisenhauer, D. E. (2016). Pumpage reduction by using variable-rate irrigation to mine undepleted soil water. *Trans. ASABE, 59*(5), 1285-1298.

Lo, T., Rudnick, D. R., Ge, Y., Heeren, D. M., Irmak, S., Barker, J. B., Qiao, X., & Shaver, T. M. (2018). Ground-based thermal sensing of field crops and its relevance to irrigation management. NebGuide G2301, University of Nebraska-Lincoln Ext.

Martin, D. L., Stegman, E. C., & Fererer, E. (1990). Irrigation scheduling principles. In G. J. Hoffman, T. A. Howell, & K. H. Soloman (Eds.), *Management of farm irrigation systems. ASAE Monograph.* St. Joseph, MI: ASAE.

Melvin, S. R., & Martin, D. L. (2018). Irrigation scheduling strategies when using soil water data. EC 3036. University of Nebraska-Lincoln Ext.

Melvin, S. R., & Yonts, C. D. (2009). Irrigation scheduling: Checkbook method. EC 709. University of Nebraska-Lincoln Ext.

Miller, K. A., Luck, J. D., Heeren, D. M., Lo, T., Martin, D. L., & Barker, J. B. (2018). A geospatial variable rate irrigation control scenario evaluation methodology based on mining root zone available water capacity. *Precis. Agric., 19*(4), 666-683.

Peters, R. T., & Evett, S. R. (2008). Automation of a center pivot using the temperature-time-threshold method of irrigation scheduling. *J. Irrig. Drain. Eng., 134*(3), 286-291.

Singh, J., Heeren, D. M., Rudnick, D. R., Woldt, W. E., Bai, G., Ge, Y., & Luck, J. D. (2020). Soil structure and texture effects on the precision of soil water content measurements with a capacitance- based electromagnetic sensor. *Trans. ASABE, 63*(1), 141-152.

Skaggs, R. W., Fausey, N. R., & Nolte, B. H. (1981). Water management model evaluation for North Central Ohio. *Trans. ASAE, 24*(4), 922-928.

Stegman, E. C. (1983). Irrigation scheduling: Applied timing criteria. In D. Hillel (Ed.), *Advances in irrigation* (Vol. 2, pp. 1-30). New York, NY: Academic Press.

Trout, T. J., Howell, T. A., English, M. J., & Martin, D. L. (2020). Deficit irrigation strategies for the Western U.S. *Trans. ASABE, 63*(6), 1813-1825.

Turner, N. C. (1990). Plant water relations and irrigation management. *Agric. Water Manag., 17*(1), 59-73.

Turner, N. C., & Burch., G. J. (1983). The role of water in plants. In I. D. Teare, & M. M. Peet (Eds.), *Crop water relations* (pp. 73-125). New York, NY: John Wiley & Sons.

Wanjura, D. F., Upchurch, D. R., & Mahan, J. R. (1995). Control of irrigation scheduling using temperature-time thresholds. *Trans. ASAE, 38*(2), 403-409.

Chapter 7
Salinity Management

7.1 Introduction

Salinity is frequently a threat to irrigated agriculture. Have you ever wondered, What is the impact of salinity on crop production or the environment? How and where would you measure salinity? There are various types of salinity problems which can affect crops and soil in different ways. How can a producer best cope with the threats of excess salts? These questions are answered in the following sections.

All soils and irrigation waters contain salt. In humid areas—soils, surface waters, and groundwaters—are normally low in salinity. Salt concentrations are minimal because rainfall typically exceeds crop water requirements, which results in dilution of any salts in the soil. The excess water normally percolates through the soil flushing salts below the crop root zone. In dry climates potential evapotranspiration exceeds rainfall. Thus, small amounts of water percolate through the soil to remove salts. With time, the salt content of soils in arid regions may increase and crop yields decrease. When crop productivity is reduced by the presence of excess salt, the soil is said to be salt affected. Documented occurrences of salt-affected soils are illustrated in Figure 7.1. Estimates of the amount of irrigated land impacted by salination are given in Table 7.1 for the world and five selected countries. For detailed information on salinity management refer to Tanji (1990) and Wallender and Tanji (2012).

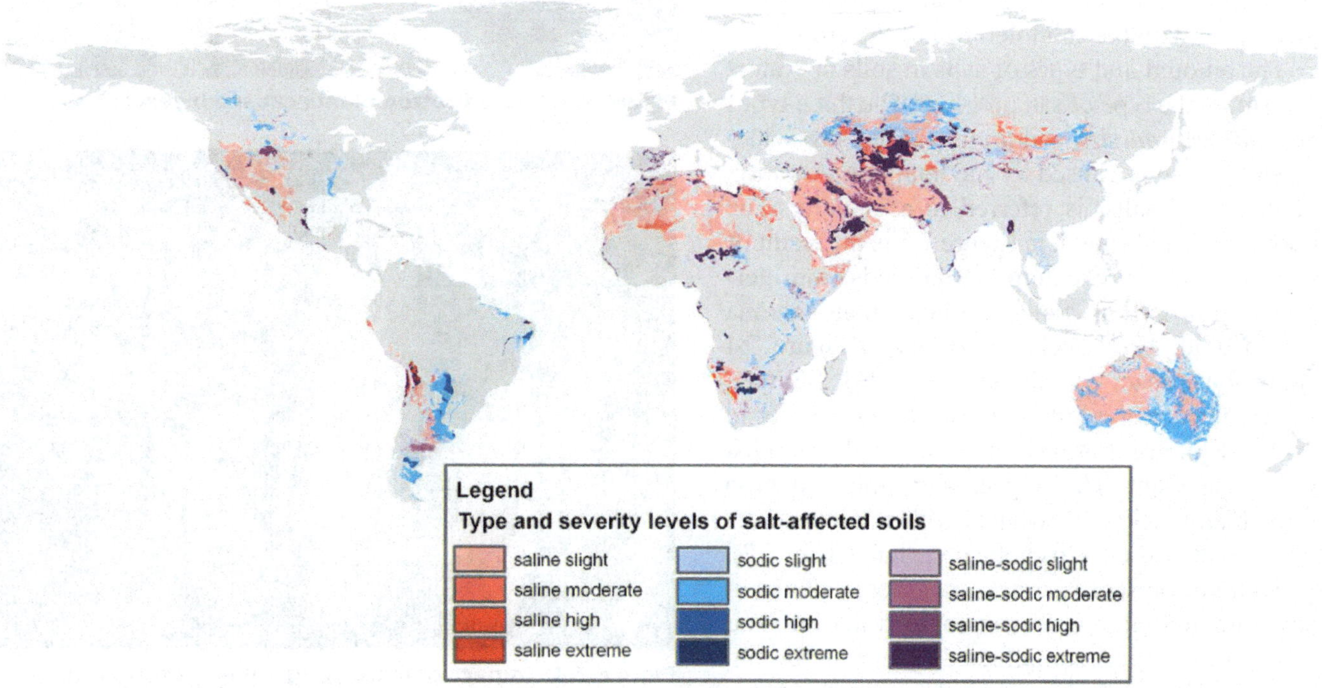

Figure 7.1. Salt-affected soils across the world (Reproduced from Wicke et al., 2011 with permission from the Royal Society of Chemistry.)

Table 7.1. Estimate of irrigated land damaged by salination during the mid-1980s for the top five irrigated countries and the world (adapted from Postel, 1990).

Country	Area Damaged (million ac)	Share of Irrigated Land Damaged (%)
India	50	36
China	18	15
United States	13	27
Pakistan	8	20
Former Soviet Union	6	12
World	150	24

The primary cause of increasing salt content in soils is evapotranspiration. As water is removed from the soil by plant roots or evaporates from the soil surface, salts are left behind. If salt concentrations become so high that they can no longer be held in solution, precipitation of salt occurs. Precipitation is the chemical process whereby dissolved salts change to their solid form. In the field, precipitated salts appear as a white to gray crust on the soil surface. Figure 7.2 shows salt precipitated on the soil surface in the Imperial Valley of California. Within the soil, nodules or layers of precipitated salts, called caliche, are found in some salt-affected soils. Within the crop root zone the salt concentration is controlled by the ability of the crop's root system to extract water. This ability is associated with the salt tolerance of the crop.

The amount and types of salts in soils or waters determine the type of salt problem. The three types are ***salinity, sodicity,*** and ***toxicity***. The most widespread problem, caused by the total concentration of dissolved salts, is referred to as salinity. The poor crop stand shown in Figure 7.3 is the result of excess salinity preventing cotton seeds from germinating or killing young seedlings in an Arizona cotton field. The bunches of celery in Figure 7.4 show the impact of salt on crop yield. Salinity, nearly zero on the left, increases progressively to high salt concentrations on the right that almost killed the plant. The impact of salinity on crop growth and yield is associated with osmotic stress, which is measured as osmotic potential (Chapter 2).

Sodium, present in excess, deteriorates the soil structure and inhibits movement of water into and through the soil. A soil affected by excess sodium is referred to as a sodic soil or, the outmoded term, alkaline soil. Figure 7.5 shows the effects of excess

Figure 7.2. Two bands of salt precipitated out of solution along the top of furrow irrigation beds in California.

Figure 7.3. Loss of cotton plants caused by excess salinity.

Figure 7.4. Impact of increasing salinity (from left to right) on the size of a bunch of celery.

sodium in a corn field in Idaho. The white chunks are precipitated salts and the black areas are organic matter released when the excess sodium destroyed the soil structure. Some crops are sensitive to specific ions such as chloride, boron, sodium, and certain heavy metals in relatively low concentrations. Trees and other woody crops, in particular, are sensitive to specific ions. In these circumstances, excessive concentrations of specific ions are toxic. Figure 7.6 illustrates the toxic effects of three specific ions with considerable potential to damage or adversely affect plants: sodium (Na^+), chloride (Cl^-), and boron (B).

Figure 7.5. Effect of excess sodium in a corn field. Note the white colored salts and the black organic matter.

The major impact of salinity is a reduction in the osmotic potential caused by the salt concentration of the soil water. As the osmotic potential of the soil water external to the plant decreases (becomes more negative), the difference between the water potential internal and external to the plant roots is reduced. The smaller the difference between internal and external water potentials the higher the degree of difficulty for the roots to extract water from the soil. This phenomenon is frequently noted as a reduction in the availability of water to the plant. As soil water becomes more saline, plants must use more energy to extract soil water. This utilization of energy means less energy for plant growth and the plant becomes smaller and yield is reduced.

Figure 7.6. Leaf burn on trees caused by specific ion toxicity; (a) indicates minimal leaf damage, (b) shows moderate damage, and (c) illustrates significant damage (adapted with permission from Tanji et al., 2007).

7.2 Origin of Salt in Soils

Salt-affected soils are part of the geochemical processes that have continued since ancient geologic time. Soluble salts originate from the weathering of primary minerals in rocks forming the continents. The types of soluble salts depend on the composition of the weathered rock. Normally, salts move from sites of weathering into the groundwater system, eventually moving into streams and then into oceans. The present-day location of salts is dominated by the amount of water that has passed through each point of the hydrologic system. If rainfall is high, as in humid climates, most salts have been transported into oceans or to deep groundwater systems. In arid environments where rainfall is limited, salts are frequently still present in the soil.

Salts accumulate in landscapes having certain relief and geologic conditions. Salt moves with water; thus, saline conditions are linked to lowlands or depressions where water naturally drains and accumulates. Often this situation is associated with restricted internal drainage of the soil, which is conducive to high water table conditions. Salts frequently accumulate in these low areas. Low-lying lands may be relatively small areas in fields or they may be as large as the Great Basin of Utah and Nevada. Drainage water collects at some terminus in closed basins and evaporates. Water in these terminals increases in salt content and, eventually, may lose biological value and become less attractive for recreation.

In addition to weathering, secondary deposits are a major source of saline soils. Throughout geologic history, large portions of the continents have been covered by saline seas. Marine sediments deposited during extended periods of inundation serve as parent material for large areas now devoted to agriculture. These secondary deposits include shales, sandstones, mudstones, and conglomerates. Saline marine shales, for example, are notorious sources of salt. A prime example is the Mancos shale formation that occurs extensively in the upper Colorado River Basin of Colorado, Wyoming, and Utah.

When new lands are developed and brought under irrigation, soils that are prone to salt accumulation are frequently very saline. Before crop production is economically feasible, these salt-affected soils must be reclaimed. The reclamation process, whether it be for saline, sodic, or toxic soils, requires copious amounts of nonsaline water to flush the salts from the intended crop root zone. Frequently, man-made drainage systems are required to augment natural drainage to remove the extra water applied to flush salts from the soil.

Once irrigated lands are in production, the primary source of salt is the irrigation water. The salt introduced into the crop root zone by irrigation is additive to any salt already present in the soil.

Example 7.1

An irrigation source contains 500 mg of dissolved salt per liter of water. How much salt is applied to a 50-acre corn field if 15 inches of irrigation water are applied?

Given: Salt concentration of irrigation water (C) = 500 mg/L = 500 ppm

Depth of irrigation water (d_a) = 15 in

Field size (A) = 50 ac

Find: Amount of salt applied (W)

Solution:

$$C_1 = 0.226\ C\ d_a \qquad \text{(Eq. 5.18)}$$

$$C_1 = 0.226\ (500\ \text{ppm})(15\ \text{in})$$

$$C_1 = 1695\ \text{lb/ac}$$

$$W = 1695\ \text{lb/ac}\ (50\ \text{ac})\left(\frac{1\ \text{ton}}{2{,}000\ \text{lb}}\right)$$

$$W = 42\ \text{tons}$$

7.3 Measurement of Salinity

Electrical conductivity (EC) is used frequently to estimate the salt concentration of solutions. This method is based upon the fact that salts dissociate into charged ions in water and can conduct an electric current. As the concentration of salts increases, the capacity of the solution to conduct an electrical current, called electrical conductivity, increases. Electrical conductivity is expressed in units of Siemens per meter (S/m). For most natural systems, the EC unit of dS/m (10^{-1} S/m) is convenient and is equal numerically to millimhos/cm, an outmoded unit. The approximate relationship between osmotic potential (Ψ_o, bars) and electrical conductivity (EC, dS/m) is:

$$\Psi_o = 0.36\ EC \qquad (7.1)$$

The relationship between salt concentration (C) in units of mg/L and EC is approximated by:

$$C = 640\ EC \qquad (7.2)$$

It is important to remember that electrical conductivity is sensitive to the temperature of the solution. Between the temperatures of 15° and 35°C, a one degree increase in temperature increases EC by about 2%. A solution at 35°C that measures an EC of 5 dS/m will have an EC of 4 dS/m if the temperature of the solution is decreased to 25°C. For consistency, ECs are normally reported at a temperature of 25°C.

Ideally, soil salinity should be measured at the soil water content found in the field. This is not easily done although several methods are now available that operate at field water contents. The most common method of determining soil salinity is by extracting a solution from a soil sample that has been saturated. The procedure begins by taking a soil sample in the field. The sample is brought to saturation in a laboratory by adding distilled water and then a sample of the saturated soil solution is extracted by vacuum filtration. The electrical conductivity of this saturated extract (EC_e) is then measured.

7.4 Crop Salt Tolerance

The salt tolerance of a plant is defined as the plant's capacity to endure the effects of salt. Crop salt tolerance is not an exact value because it depends on many factors. Although salt tolerance cannot be stated in absolute terms, relative crop response to known salt concentrations under typical conditions can be predicted. For a more complete reference on crop salt tolerance see Maas and Hoffman (1977).

The salt tolerance of a crop can be described by plotting relative crop yield as a continuous function of soil salinity (Figure 7.7). For most crops, this response function follows a sigmoidal relationship where crop yield is not reduced significantly as salinity initially begins to increase but, as salinity increases further, yield is reduced rather rapidly. Then, as salinity reaches high levels, crop yields, although low, do not decrease as rapidly as at moderate concentrations. For practical purposes this sigmoidal relationship for crop salt tolerance can be represented by two straight lines, one line is a tolerance plateau with a slope of zero and the other line is concentration dependent and its slope indicates the yield reduction per unit increase in salinity.

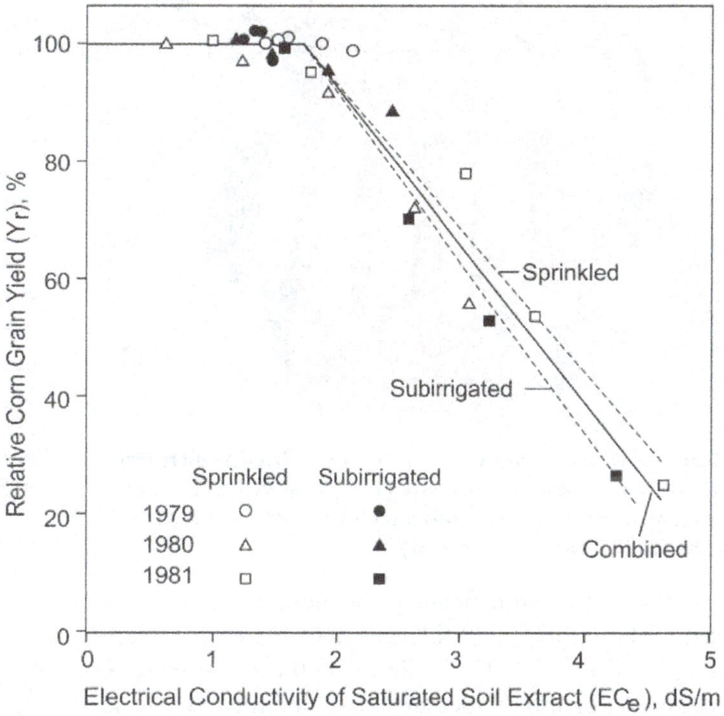

Figure 7.7. Relative grain yield of corn grown in the Sacramento-San Joaquin Delta of California as a function of soil salinity (adapted from Hoffman et al., 1983).

Figure 7.7 shows the "two straight lines" model fitted to actual field data for corn grain yield. The point at which the two straight lines intersect designates the salt tolerance threshold which is the maximum soil salinity that does not reduce yield appreciably below that achieved under nonsaline conditions. For soil salinities exceeding the threshold, relative yield (Y_r) in percent can be estimated from:

$$Y_r = 100 - S(EC_e - \mathrm{T}) \quad \text{for } EC_e > \mathrm{T} \tag{7.3}$$

where: T = salt tolerance threshold expressed in EC_e units of dS/m,

S = slope expressed in % per dS/m, and

EC_e = the mean salt concentration in units of electrical conductivity of saturated soil extracts taken from the crop root zone.

The threshold and slope values provide general guidelines about salt tolerance for crop management decisions. Irrigators need to know the level of soil salinity that initiates yield reduction (T, threshold) and the rate at which yield is reduced at salt levels greater than the threshold (S, slope).

Typical ears of corn from the experimental results plotted in Figure 7.7 are shown in Figure 7.8. The top row of ears were grown using nonsaline irrigation water; the bottom row with irrigation water having an EC of 8 dS/m.

Figure 7.8. Example ears of corn produced with irrigation water having no salt (top) and with salt concentrations equal to one-fourth the salt concentrations of sea water (bottom).

Crops differ significantly in tolerance to soil salinity. The relative salt tolerances of major crops are given in Table 7.2. The table gives the salt tolerance threshold (T) and the percent yield decline (S). These two values can be inserted into the salt tolerance equation (Equation 7.3) to predict relative crop yield (Y_r). Qualitative ratings for ease in comparisons among crops are also given in Table 7.2. The qualitative salt tolerance ratings are sensitive (s), moderately sensitive (ms), moderately tolerant (mt), and tolerant (t). These qualitative ratings can be seen in Figure 7.9.

A handy guide to classify potential crop damage from increasing salt levels in irrigation waters is given in Table 7.3. The reader is cautioned, however, that the

Table 7.3. Classification guide for saline irrigation water.

Irrigation Water	Salt Concentration (ppm or mg/L)	Electrical Conductivity (dS/m)	Crop Problems
Fresh	< 300	< 0.5	none
Slightly saline	300–600	0.5–1	rare
Moderately saline	600–1,200	1–2	occasional
Saline	1,200–2,400	2–4	common
Highly saline	2,400–4,800	4–7	severe

Table 7.2. Salt tolerance of major crops (adapted from Maas and Hoffman, 1977).

Crop	Salt Tolerance Threshold, T (dS/m)	Percent Yield Decline, S %/(dS/m)	Qualitative Salt Tolerance Rating[a]
Grain Crops			
Barley	8.0	5.0	t
Corn	1.7	12	ms
Cowpea	4.9	12	mt
Rice	3.0	12	s
Sorghum	6.8	16	mt
Soybean	5.0	20	mt
Wheat	6.0	7.1	mt
Fiber, Sugar and Oil Crops			
Cotton	7.7	5.2	t
Flax	1.7	12	ms
Peanut	3.2	29	ms
Sugar beet	7.0	5.9	t
Sugar cane	1.7	5.9	ms
Grasses and Forage Crops			
Alfalfa	2.0	7.3	ms
Bermuda grass	6.9	6.4	t
Clover	1.5	12	ms
Fescue	3.9	5.3	mt
Orchard grass	1.5	6.2	ms
Ryegrass	5.6	7.6	mt
Trefoil, birdsfoot	5.0	10	mt
Vegetables and Fruit Crops			
Asparagus	4.1	2.0	t
Bean	1.0	19	s
Cabbage	1.8	9.7	ms
Carrot	1.0	14	s
Celery	1.8	6.2	ms
Corn, sweet	1.7	12	ms
Lettuce	1.3	13	ms
Potato	1.7	12	ms
Strawberry	1.0	33	s
Sweet potato	1.5	11	ms
Tomato	2.5	9.9	ms
Woody Crops			
Almond	1.5	19	s
Apricot	1.6	24	s
Blackberry	1.5	22	s
Date palm	4.0	3.6	t
Grape	1.5	9.6	ms
Grapefruit	1.8	16	s
Guayule	15	13	t
Orange	1.7	16	s
Peach	1.7	21	s
Plum	2.6	31	s

[a] s = sensitive; ms = moderately sensitive; t = tolerant; mt = moderately tolerant

use of saline water depends upon the crop, soil, climate, geology, and management practices. Thus, this classification is only a rough guide.

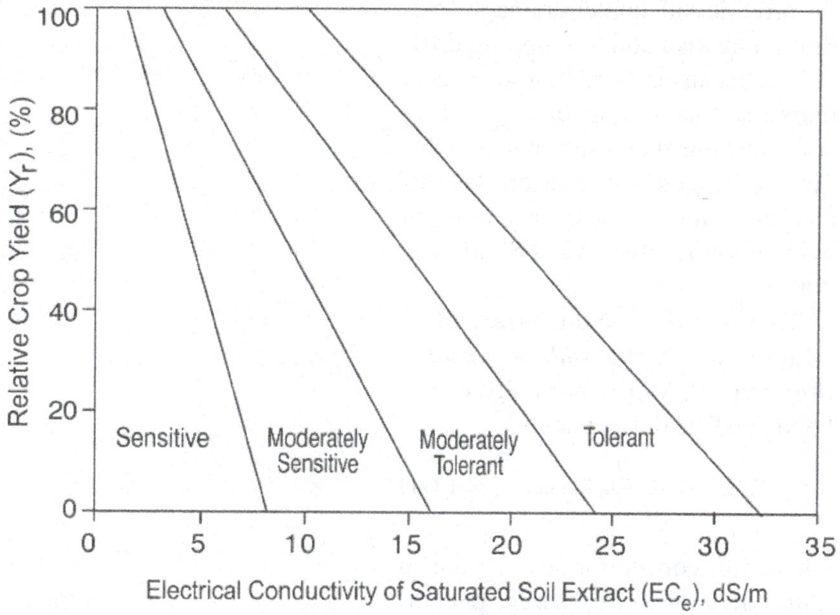

Figure 7.9. Division boundaries for qualitative salt tolerance ratings of crops (adapted from Maas and Hoffman, 1977).

Example 7.2

A saline area of a field has an average salt concentration of 3,000 mg/L. Calculate the relative yield of corn in this salt-affected soil. If the nonsaline portion of the field produces 180 bushels per acre, what is the actual yield of the saline area in the field?

Given: C = 3,000 mg/L

Nonsaline corn yield = 180 bu/ac

S = 12% / (dS/m) and T = 1.7 dS/m (Table 7.2 for corn)

Find: Relative (Y_r) and actual (Y_a) corn yields in the saline area

Solution:

$$EC = \frac{C}{640} = \frac{3{,}000}{640} = 4.7\ \mathrm{dS/m} \quad \text{(Eq. 7.2)}$$

$$Y_r = 100 - S\,(EC_e - T) \quad \text{(Eq. 7.3)}$$

$$Y_r = 100 - 12\,(4.7 - 1.7)$$

$$Y_r = 64\%$$

$$Y_a = Y_r \times Y_{max}$$

$$Y_a = 0.64\,(180\ \mathrm{bu/ac})$$

$$Y_a = 115\ \mathrm{bu/ac}$$

7.5 Sodicity

If sodium is the predominate cation adsorbed in the soil, the clay particles in the soil swell and soil aggregates disperse. This deterioration leads to reduced penetration of water into and through the soil. When calcium and magnesium are the predominate cations, the soil tends to have a granular structure that is easily tilled and readily permeable. Excess sodium becomes a concern when the rate of infiltration is reduced to the point that the crop cannot be adequately supplied with water or when the hydraulic conductivity of the soil profile is too low to provide adequate drainage. Sodium may also add to cropping difficulties because of crusting seed beds; temporary saturation of the surface soil; and the increased potential for disease, weeds, soil erosion, lack of oxygen, and inadequate nutrient availability (Hoffman and Shalhevet, 2007).

To assess the sodium hazard of irrigation water, the ***sodium absorption ratio*** (SAR) is normally calculated. SAR is defined as:

$$SAR = \frac{C_{Na}}{\sqrt{C_{Ca} + C_{Mg}}} \quad (7.4)$$

where ion concentrations (C) are in units of moles of charge per m^3 (mol_c/m^3) for sodium (Na), calcium (Ca), and magnesium (Mg). Equation 7.4 is valid for soil water under steady-state conditions where the SAR of the irrigation water approximates the SAR of the soil water. The SAR for the soil water under nonsteady-state conditions needs to be adjusted. Figure 7.10 can be used to determine whether an irrigation water will lead to a sodicity problem. If the relationship between the SAR of the irrigation water and its salinity results in a point to the left in Figure 7.10, a sodicity hazard is likely to occur. If the point is between the two lines, a slight to moderate sodicity hazard is likely.

Ionic concentrations are sometimes reported in units of milliequivalents per liter (meq/L). The relationship between the two units frequently used to report ionic concentrations is:

$$mol_c/m^3 = \frac{\text{meq/L}}{\text{valence of ion}} \quad (7.5)$$

where the valence of the ion can be one or more. Recall from chemistry that the valence of sodium is positive one and the valence of calcium and magnesium is positive two.

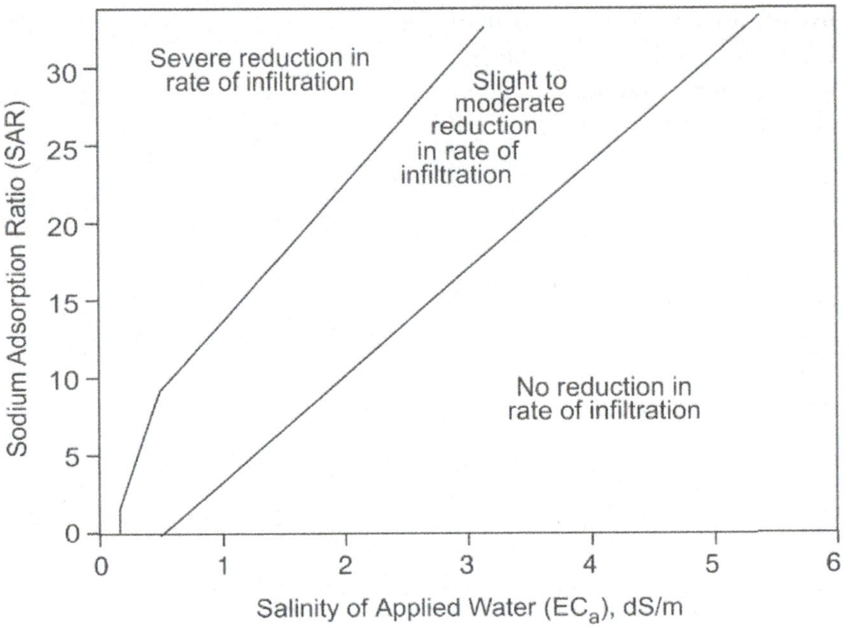

Figure 7.10. Division of waters that cause inadequate water penetration because of chemical conditions (adapted from Rhoades, 1982).

Example 7.3

Water from an irrigation well in Arizona has an electrical conductivity of 0.4 dS/m at 25°C and the concentration of sodium is 33 meq/L. The concentrations of calcium and magnesium are 24 and 8 meq/L, respectively. Determine whether this irrigation water will create a sodicity hazard.

Given: C_{Na} = 33 meq/L
C_{Ca} = 24 meq/L
C_{Mg} = 8 meq/L
EC = 0.4 dS/m

Find: Will water cause a sodicity hazard?

Solution:

$$C_{Na} = \frac{(33\ \text{meq/L})}{1} = 33\ \text{mol}_c/\text{m}^3$$

$$C_{Ca} = \frac{(24\ \text{meq/L})}{2} = 12\ \text{mol}_c/\text{m}^3$$

$$C_{Mg} = \frac{(8\ \text{meq/L})}{2} = 4\ \text{mol}_c/\text{m}^3$$

$$SAR = \frac{C_{Na}}{\sqrt{C_{Ca} + C_{Mg}}} = \frac{33}{\sqrt{12 + 4}}$$

$$SAR = 8.2\left(\text{mol}_c/\text{m}^3\right)^{1/2}$$

From Figure 7.10, the intersection of lines extended from a SAR of 8.2 and an EC of 0.4 dS/m indicates that water penetration will probably be decreased due to excess sodium.

7.6 Toxicity

Toxicity occurs as the result of the uptake and accumulation of certain ions within plant tissue. The toxicity of any ion is highly dependent upon the crop. Specific ions that may be toxic include boron, chloride, and sodium. Some ions, like chloride, can be absorbed directly into the leaves when moistened during sprinkler irrigation. Foliar damage from sprinkling is particularly acute during periods of high temperature and low humidity. Many trace elements, such as cadmium and lithium, are also toxic to plants at very low concentrations. Suggested maximum concentrations for many trace elements are given by Pratt (1973). Fortunately, most irrigation supplies contain insignificant concentrations of these potentially toxic trace elements and are generally not a problem.

7.7 Leaching

Salinity in the crop root zone can be controlled if the quality of the irrigation water is satisfactory and the flow of water through the soil is sufficient. Leaching, the net downward movement of soil water and solutes, is the key to successful irrigation where salts are a hazard. As the salinity of the irrigation water increases or if more salt sensitive crops are grown, leaching must be increased to maintain high crop yields. This chapter presents general guidelines for leaching that can be applied to various types of irrigation systems; guidelines specific to drip irrigation are presented in Hanson and May (2011).

The simplest general expression describing the actual amount of leaching is:

$$L_f = \frac{d_p}{d_t} = \frac{C_a}{C_d} \tag{7.6a}$$

$$d_t = d_z + d_r \tag{7.6b}$$

where: L_f = actual ***leaching fraction***,

d_p = depth of water draining below the crop root zone (deep percolation),

d_t = total depth of infiltrated water,

C_a = weighted mean salt concentration of the applied water,

C_d = salt concentration of the draining water,

d_z = mean depth of infiltrated irrigation, and

d_r = depth of infiltrated rainfall.

The weighted mean salt concentration of the applied water can be calculated from:

$$C_a = \frac{C_i\, d_z + C_r\, d_r}{d_z + d_r} \tag{7.7}$$

where: C_i = concentration of irrigation water

C_r = concentration of rain water

The salt concentration of rainfall is so low that it is considered to be zero. Thus, the term $C_r\, d_r$ in Equation 7.7 is zero.

> **Example 7.4**
>
> What is the annual leaching fraction of a soil if 300 mm of rain fell during the year, with 250 mm infiltrating into the soil, and 300 mm of irrigation infiltrated? The electrical conductivity of the irrigation water was measured to be 0.5 dS/m at 25° and the electrical conductivity of the soil water draining below the root zone was found to be 2.5 dS/m.
>
> Given: C_d = 2.5 dS/m C_i = 0.5 dS/m
>
> d_z = 300 mm d_r = 250 mm
>
> Find: The leaching fraction (L_f) for this condition.
>
> Solution:
>
> $$C_a = \frac{C_i\, d_z + C_r\, d_r}{d_z + d_r}$$
>
> $$C_a = \frac{0.5(300) + 0(250)}{300 + 250}$$
>
> $$C_a = 0.27 \text{ dS/m}$$
>
> $$L_f = \frac{C_a}{C_d} = \frac{0.27 \text{ dS/m}}{2.5 \text{ dS/m}}$$
>
> $$L_f = 0.11$$

The ***leaching requirement*** (L_r) is the minimum leaching fraction that will prevent a reduction in crop yield. The L_r can be derived from Equation 7.6 as:

$$L_r = \frac{d_p^*}{d_t} = \frac{C_a}{C_d^*} = \frac{EC_a}{EC_d^*} \tag{7.8}$$

in which the superscript * distinguishes required from actual values. Because electrical conductivity (EC) is easily measured and is almost linearly related to the salt concentration of a relatively dilute salt solution, it is customary to substitute EC for C in these relationships.

Several mathematical models have been proposed to relate L_r to some readily available value of soil salinity that is indicative of the crop's leaching requirement. One such model is represented graphically in Figure 7.11. This graphical solution relates the salinity of the applied water, crop salt tolerance threshold, and leaching requirement.

The salt tolerance of many annual crops increases as the growing season progresses. This suggests that if soil salinity levels are low enough at the beginning of the season and adequate amounts of low salt water are applied, soil salinity can be permitted to increase gradually during the irrigation season. For the next crop, rainfall, either singly or in combination with dormant season or pre-plant irrigations, can replenish soil water and leach accumulated salts to permit irrigation the next season without the need for further leaching. An important exception to this procedure is perennial crops, like trees, that form their buds for the next year during the latter half of the irrigation season. High salinity levels during bud formation will be detrimental to fruit production the following season.

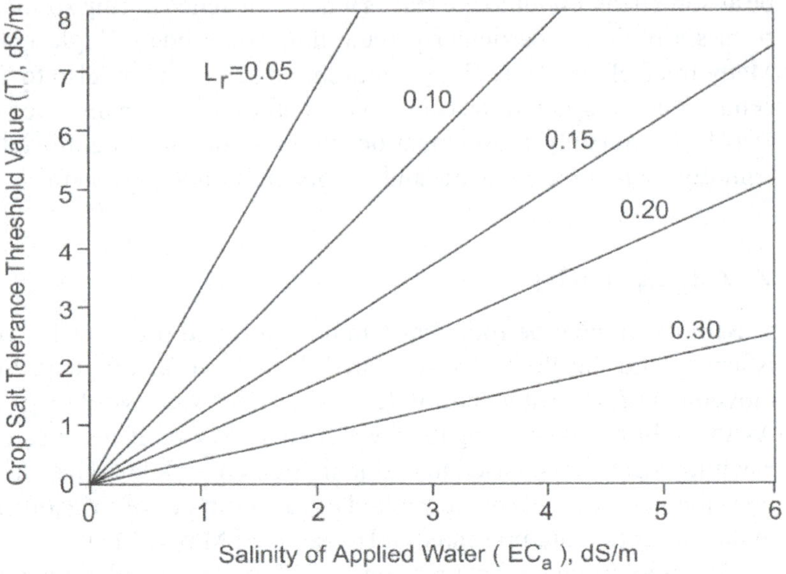

Figure 7.11. Graphical solution for the leaching requirement (L_r) as a function of the salinity of the applied water and the salt tolerant threshold value for the crop (adapted from Hoffman and van Genuchten, 1983).

If irrigation waters are saline, rainfall and out of season leaching may not be sufficient and leaching during the irrigation season will be required to prevent yield reduction. The key factor to remember is that leaching is not required until accumulated soil salinity surpasses the salt tolerance threshold for the crop. Leaching can be done each irrigation or less frequently, such as seasonally or at even longer periods, provided soil salinity is maintained below the salt tolerance threshold if yield losses are to be avoided.

Some irrigation systems are managed to apply copious amounts of water. Thus, in many cases, this excess amount of irrigation supplies water for leaching without a conscious effort by the irrigator. In some situations, the nonuniform applications of the irrigation system result in some areas of the field receiving water in excess of the crop water and leaching requirements, while underirrigated areas cause water and salt stress. This problem is best solved by an irrigation system that is more uniform in water application rather than applying more water to compensate for nonuniformity.

The leaching requirement model presented here assumes steady state conditions. In reality, steady state never occurs in the field. Several complex computer models have been developed which account for transient conditions that more closely represent field

Example 7.5

Determine the leaching requirement for tomatoes if the salinity of the irrigation water is 3 dS/m with 16 inches of irrigation water and 4 inches of rainfall contributing to the crop water requirement.

Given: EC_i = 3 dS/m

d_z = 16 in

d_r = 4 in

T for tomatoes is 2.5 dS/m from Table 7.2

Find: Leaching requirement for tomatoes under the specified conditions.

Solution:

$$EC_a = \frac{EC_i d_z + EC_r d_r}{d_z + d_r} \quad \text{(from Eq. 7.7)}$$

$$EC_a = \frac{3\ \text{dS/m}\ (16\ \text{in}) + 0\ (4\ \text{in})}{16\ \text{in} + 4\ \text{in}}$$

$$EC_a = \frac{48}{20} = 2.4\ \text{dS/m}$$

From Figure 7.11, for an EC_a of 2.4 dS/m and a salt tolerance threshold value of 2.5 dS/m, the leaching requirement is 0.17.

conditions (Minhas et al., 2020). These transient models predict that steady state models overestimate the leaching requirement (Letey et al., 2011; Corwin and Grattan, 2018). Unfortunately, these models require huge data sets, are not readily available to irrigators, and do not directly predict the leaching requirement. Nevertheless, the irrigator should be aware that the leaching requirements given in Figure 7.11 overestimate the amount of saline drainage water entering the environment.

7.8 Reclamation

Reclamation of salt-affected soils is frequently required when semiarid or arid lands are first brought into agricultural production; when saline groundwater persists near the soil surface; or when irrigation and rainfall have failed to meet the leaching requirement. The only proven method of reclaiming salt-affected soils is the leaching of accumulated salts down below the crop root zone. For reclaiming sodic soils, an amendment or deep tillage may be required before leaching is effective. Soils excessively high in boron are particularly difficult to reclaim because of the tenacity by which boron is held in the soil.

Adequate drainage is essential for reclamation. Natural internal drainage alone may be adequate, provided there is storage capacity in the profile for salt below the root zone or a permeable subsurface layer is present that drains to a suitable outlet. Where such natural drainage is lacking, an artificial system must be provided or reclamation will not be feasible.

7.8.1 Saline Soils

The amount of water that must leach through the soil profile to remove soluble salts depends primarily on the initial soil salinity level, the technique of applying water, and the soil type. Water suitable for irrigation is normally suitable for reclamation. The relationship between the fraction of the initial salt concentration (C_o) remaining in the soil profile (C_f/C_o) and the amount of water leached through the profile (d_L) per depth of soil (d_S) to be leached (d_L/d_S) when water is ponded continuously on the soil surface can be described by:

$$\left(\frac{C_f}{C_o}\right)\left(\frac{d_L}{d_s}\right) = K \qquad (7.9)$$

where K is a constant that differs with soil type. Equation 7.9 defines the curves in Figure 7.12 for organic (peat) soils where $K = 0.45$, for fine-textured (clay loam) soils where $K = 0.3$, and for coarse-textured (sandy loam) soils where $K = 0.1$. The initial offset at the top of each curve in Figure 7.12 is indicative of the amount of water that must be added to the profile before leaching commences.

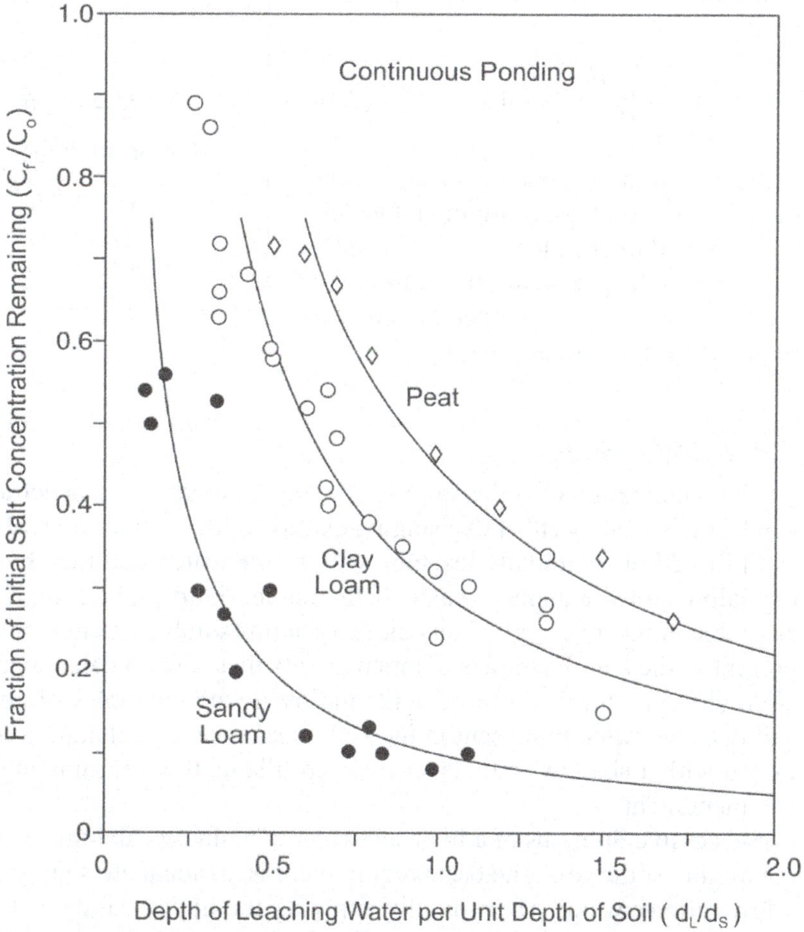

Figure 7.12. Depth of leaching water per unit depth of soil required to reclaim a saline soil by continuous ponding (adapted from Hoffman, 1986).

The amount of water required for leaching soluble salts, particularly for fine- textured soils, can be reduced by intermittent applications of ponded water or by sprinkling. The differences in leaching efficiency among the leaching methods (continuous ponding versus intermittent ponding or sprinkling to prevent water ponding on the soil surface) are caused by differences in dispersion and diffusion. The concept of soil pores is useful in visualizing these differences. The amount of solution retained in small soil pores is considerable for saturated soils, as for continuous ponding, and decreases with decreasing soil water content. Consequently, the drier the soil, as with intermittent ponding or sprinkling, the larger the fraction of water flowing through fine pores and the more efficiently the leaching water displaces the saline solution. The reclamation equation for intermittent ponding and sprinkling can be written as:

$$\left(\frac{C_f}{C_o}\right)\left(\frac{d_L}{d_s}\right) = 0.1 \qquad (7.10)$$

By intermittent ponding or sprinkling, the effect of soil type is minimal. One disadvantage of intermittent ponding or sprinkling is that the period of time required for reclamation may be extended beyond that required by continuous ponding.

Example 7.6

An irrigator has a saline field and wishes to reclaim to a soil depth of 3 feet. The EC_e of the soil now averages 10 dS/m and the irrigator desires the final EC_e to be 2 dS/m. How much water must be continuously ponded on the soil surface if the soil is clay loam? How much water is needed if the water is applied by sprinkling without saturating the soil?

Given: Clay loam soil, $K = 0.3$ for continuous ponding and $K = 0.1$ for sprinkling

$d_s = 3$ ft

$EC_o = 10$ dS/m

$EC_f = 2$ dS/m

Find: The depth of water to apply for reclaiming the soil by continuous ponding and sprinkling.

Solution:

$$\left(\frac{C_f}{C_o}\right)\left(\frac{d_L}{d_s}\right) = K$$

$$d_L = \frac{(K\, d_s\, C_o)}{C_f}$$

For continuous ponding:

$$d_L = \frac{(0.3 \times 3\text{ ft} \times 10\text{ dS/m})}{2\text{ dS/m}}$$

$$d_L = 4.5\text{ ft}$$

For sprinkling:

$$d_L = \frac{(0.1 \times 3\text{ ft} \times 10\text{ dS/m})}{2\text{ dS/m}}$$

$$d_L = 1.5\text{ ft}$$

7.8.2 Sodic Soils

The reclamation of sodic soil usually requires that water penetration into and through the soil be improved by either exchanging excess sodium in the soil with calcium, so that leaching can proceed or by initially leaching with saline water, and then by progressively decreasing the salinity of the applied water. If the choice is to replace sodium with calcium, then an amendment must be applied that either contains soluble calcium or dissolves calcium already present in the soil. Examples of amendments that contain calcium are gypsum, lime, and calcium chloride. Sulfur, sulfuric acid, and pyrite are examples of amendments that will react and dissolve calcium present in the soil. Occasionally, calcium present in the subsoil can be mixed with a shallow sodic layer by deep tillage, thus, eliminating or reducing the need for an amendment.

Successive dilutions of a high salt water containing calcium can be an effective method of reclaiming sodic soil. The basic requirement is an adequate supply of a high saline water and a low salinity water. After initially applying the highly saline water, this water is diluted in steps with the low salinity water until only the low salinity water is applied and the reclamation process is complete.

Tillage to create a rough, yet thoroughly disturbed, soil surface is a common practice for improving water infiltration. Typically, a sodic soil is tilled prior to each intermittent water application during reclamation.

7.9 Salinity and the Environment

Irrigation always degrades water quality and can cause a salinity hazard. Without proper management, the land can become waterlogged and salinized. Regardless of management, drainage water from irrigated lands carries salt that requires disposal. Questions arise as to whether salination is inevitable and if the environment is jeopardized.

Where salinity is a hazard, irrigation must have drainage. A net downward movement must occur through the soil profile to prevent the accumulation of soluble salts to a level detrimental to crops. Whether drainage is natural or man-made, the rate of movement of soil water must be sufficient to prevent salination. This drainage water must go somewhere. Depending on the geologic and hydrologic conditions, the need for drainage may become evident after only a few irrigations or after many decades.

Permanent irrigated agriculture frequently requires the sacrifice of some value elsewhere. An example is the Colorado River in the southwest corner of the United States. Lohman et.al. (1988) estimated damages from salinity for the period of 1976 to 1985 to be $311 million per year when based on a reference salinity of 500 mg/L, the Public Health Service standard for drinking water. Damages occurred to agriculture, households, water utilities, and industry. Of the figure quoted, $113 million reflect damages to agriculture.

Ultimately, saline drainage water must be transported out of the region, disposed of locally, or treated. It is technically feasible to treat saline water. Several desalination studies have evaluated reverse osmosis. The world's largest desalination plant was constructed near Yuma, Arizona, to remove salt from irrigation drainage water before it returns to the Colorado River. However, it's difficult to justify such an approach economically (van Schilfgaarde, 1982). An alternative to treating the water is to convey it to evaporation ponds. Experience in California indicates that 10 to 14% of the land area must be devoted to evaporation ponds. Loss of land, construction costs, and avoidance of leakage makes this alternative unattractive. Transporting saline water out of the region remains the primary means of disposal using natural or man-made water courses.

7.10 Summary

In regions where rainfall is not adequate to leach salts from the soil, water must be managed to avoid crop losses from excess salinity. Crops differ by nearly a factor of 10 in their sensitivity to salinity. With appropriate management to provide drainage and ensure downward movement of soil water through the crop root zone, crop productivity can be maintained even if salinity is a hazard.

The amount of water that must leach below the root zone to prevent yield loss depends on the salt content of the irrigation water and the salt tolerance of the crop. If both the soil and the irrigation water are low in salt concentration, no leaching may be required for several years, particularly if rainfall is significant.

Soils high in salinity can be made productive by applying copious amounts of water to leach the salt and reclaim the soil. The amount of water required to reclaim a saline soil depends on the soil type, the depth of soil to be reclaimed, and the method of applying the water.

Where salinity is a problem, salts must be flushed from the soil. The disposal of this salt can be detrimental to the receiving body of water, whether surface water or groundwater.

Questions

1. Describe saline and sodic soils.
2. Why do large imbalances occur in the distribution of salts in soils?
3. A saturated soil extract has an electrical conductivity of 5 dS/m at 20°C; what value of electrical conductivity should be reported and used in calculations?
4. What is the specific meaning of EC_e?
6. Calculate the yield reduction expected for sorghum produced on a soil having a salt concentration of 2,500 mg/L.
6. At a salt concentration of applied water of 4,000 mg/L, what would the L_r be for sorghum?
7. In Example 7.5, what would the leaching requirement for tomato be if rainfall were 10 inches rather than 4 inches?
8. Define sodium absorption ratio.
9. Distinguish between leaching fraction and leaching requirement. Discuss the field conditions that would exist when L_r is less than L_f, and when L_r is greater than L_f.
10. Explain the concept shown in Figure 7.11.
11. What benefits are derived from intermittent soil drying in a salt reclamation project?
12. For a clay loam soil how much water must be applied before any significant reclamation will occur if a soil depth of 3 ft is to be reclaimed by continuously ponding water on the soil surface?
13. How much water would be needed to reclaim the field in Example 7.6 if the soil was a sandy loam and the water was applied by sprinkling to prevent surface ponding?
14. Under what conditions would deep plowing aid in the reclamation of sodic soils?
15. If you had access to a large quantity of lime sulfur (9% Ca + 24% S), would it be useful to reclaim sodic soil? If useful, why?

References

Corwin, D. L., & Grattan, S. R. (2018). Are existing irrigation salinity leaching requirement guidelines overly conservative or obsolete? *J. Irrig. Drain. Eng., 144*(8), 02518001.

Hanson, B., & May, D. (2011). Drip irrigation salinity management for row crops. Agriculture and Natural Resources (ANR) Publ. 8447. University of California, Davis.

Hoffman, G. J. (1986). Guidelines for reclamation of salt-affected soils. *Appl. Agric. Res., 1*(12), 65-72.

Hoffman, G.J. & Shalhevet, J. (2007). Controlling salinity. *Design and operation of farm irrigation.* (2nd Ed.), (pp.160-207). St. Joseph, MI: ASABE

Hoffman, G. J., & van Genuchten, M. T. (1983). Water management for salinity control. In H. Taylor, W. Jordan, & T. Sinclair (Eds.), *Limitations to efficient water use in crop production* (pp. 73-85). American Society of Agronomy Monogr.

Hoffman, G. J., Maas, E. V., Prichard, T. L., & Meyer, J. L. (1983). Salt tolerance of corn in the Sacramento-San Joaquin Delta of California. *Irrig. Sci., 4*(1), 31-44.

Letey, J., Hoffman, G. J., Hopmans, J. W., Grattan, S. R., Suarez, D., Corwin, D. L., Oster, J. D., Wu, L., Amrhein, C. (2011). Evaluation of soil salinity leaching requirement guidelines. *Agric. Water Manag., 98*(4), 502-506.

Lohman, L. C., Milliken, J. G., Dorn, W. S., & Tucky, K. E. (1988). Estimating economic impacts of salinity of the Colorado River. U.S. Bureau of Reclamation Report. Denver, CO: Water Quality Office.

Maas, E. V., & Hoffman, G. J. (1977). Crop salt tolerance-current assessment. *J. Irrig. Drain. Div., 103*(2), 115-134.

Minhas, P. S., Ramos, T. B., Ben-Gal, A., & Pereira, L. S. (2020). Coping with salinity in irrigated agriculture: Crop evapotranspiration and water management issues. *Agric. Water Manag., 227*, 105832. doi: 10.1016/j.agwat.2019.105832.

Postal, S. (1990). Saving water for agriculture. In L.Starke (Ed.), *State of the world 1990* (pp. 39-58). New York, NY: W. M. Morton & Co.

Pratt, P. F. (1973). Quality criteria for trace elements in irrigation waters. California Agricultural Experiment Station Bull.

Rhoades, J. D. (1982). Reclamation and management of salt-affected soils after drainage. *Proc. of 1st Annual Western Provincial Conf., Rationalization of Water and Soil Research and Management*, 123-197.

Tanji, K. K. (1990). Agricultural salinity assessment and management. ASCE Manuals and Reports on Engineering Practice No. 71. New York, NY.

Tanji, K. K., Rollins, R., Suyama, P., & Farris, C. (2007). Salinity management guide. Accessed 4/29/2021. https://watereuse.org/salinity-management/ls/ls_3e.html.

van Schilfgaarde, J. (1982). The Welton Mohawk dilemma. *Int. J. Water Supply Manag. 6,* 115-127.

Wallender, W. W., & Tanji, K. K. (2012). Agricultural salinity assessment and management. ASCE Manuals and Reports on Engineering Practice No. 71. ASCE: Reston, VA.

Wicke, B., Smeets, E., Dornburg, V., Vashey, B., Gaiser, T., Turkenburg, W., Faaij, A. (2011). The global technical and economic potential of bioenergy from salt-affected soils. *Energy Environ. Sci.4*(8): 2669-2681.

Chapter 8
Pump and Pipeline Hydraulics

8.1 Introduction

We've all had the experience of carrying a bucket of water up a hill. It takes work to get it done, right? In a similar fashion it takes energy to move and distribute water for irrigation. Proper development and operation of irrigation requires that considerable attention be given to the hydraulics of the system. Knowledge of system hydraulics is necessary when selecting and sizing system components such as pipelines, valves, sprinklers, emitters, and pumps. Mistakes made in designing and installing the components of an irrigation system are often very expensive to correct, whereas the cost of appropriate planning to avoid errors is small. Pressure distribution in a pipe affects the distribution water discharge from sprinklers, gates, and emitters and hence the uniformity and efficiency of the application. Furthermore, pressurizing water requires energy. Thus, it is important to understand how to select a pump that efficiently matches the water supply and pressure requirements of the irrigation system.

8.2 Basic Hydraulics

There are two important physical laws that apply to hydraulics, conservation of energy and conservation of mass (continuity). The energy in the water will be in any of the following forms:

- kinetic energy due to the velocity of the water,
- potential energy due to the elevation of the water relative to an arbitrary reference elevation, and
- potential energy due to water pressure.

In this book, the energy in water is expressed either as energy per unit of weight of water (head) or energy per unit of volume of water (pressure). Since energy has the dimensions *FL* (force × length) and weight has the dimension *F*, energy per unit of weight has the dimension of length (*L*). Energy expressed in this manner is referred to as head. Energy per unit of volume has the dimension of FL/L^3 or F/L^2. A common unit is pounds per square inch (psi). The energy of water in an irrigation system includes velocity head, elevation head, and pressure head.

Kinetic energy is a result of the movement of the fluid. Velocity head (h_v) is given by:

$$h_v = \frac{V_m^2}{2g} \tag{8.1}$$

where: V_m = average velocity at a point in the pipe or channel, ft/s, and

g = gravitational constant, 32.2 ft/s^2.

In general, the maximum recommended average velocity of flow in an enclosed or pressurized pipeline is 5 feet per second (ft/s). When the velocity in a mainline exceeds 5 ft/s, there is potential to develop relatively high-pressure surges which may damage pipelines. Pressure surges are due to flow being stopped suddenly while the upstream water has a large amount of momentum. When the flow is stopped too quickly, the rapid change of momentum results in an impulsive force called ***water hammer***. The allowable maximum velocity may be higher than 5 ft/s if special precautions (pressure relief valves, surge tanks, etc.) are used to relieve possible pressure surges.

The potential energy due to elevation is a result of the location of the water relative to an arbitrary reference plane. Water at a higher elevation has more potential energy than water at a lower elevation. Consider water flowing downhill. Energy is the ability to do work, and work can be described as a force acting over a distance. As water flows downhill, the force is gravity, and the distance is the length over which the water flows. The water has the ability to do work as it flows downhill such as eroding the surface, generating power, etc. The potential energy of the water decreases as it flows downhill. The letter Z will be used to represent ***elevation head*** or gravitational head.

The potential energy due to the pressurization of water can be a very large component in an irrigation system. Pressure is the force per unit area exerted on the walls of a container. The pressure may be expressed as:

$$P = \gamma h \quad \text{or} \quad h = P/\gamma \tag{8.2}$$

where: P = pressure,

γ = weight of a unit volume of fluid (specific weight), and

h = pressure head.

For water, $\gamma = 62.4$ lb/ft³. Figure 8.1 illustrates how the pressure is related to the depth (head) of water in a container. The shape and volume of the container are not important when applying Equation 8.2.

In *USCS* units, the following conversions are convenient:

$$\gamma = 0.433 \frac{\text{psi}}{\text{ft}} \tag{8.3}$$

or

$$\frac{1}{\gamma} = 2.31 \frac{\text{ft}}{\text{psi}}$$

Because different fluids have different weights per unit volume (γ), Equation 8.3 is only valid for water.

In Example 8.1 the pressure is independent of the surface area of the columns, but realize of course that the forces on the container bottoms are different, one having 10 times the force as the other.

Example 8.1
Two columns of water are filled to a height of 10 feet with water. One column has a cross-sectional area of 1 in², the other 10 in². Find the pressure due to the fluid at the bottom of each column.

Given: h = 10 ft
γ = 0.433 psi/ft

Find: P (Figure 8.1)

Solution:
$P = \gamma h$ (Eq. 8.2)

$$P = 0.433 \frac{\text{psi}}{\text{ft}} (10 \text{ ft}) = 4.33 \text{ psi}$$

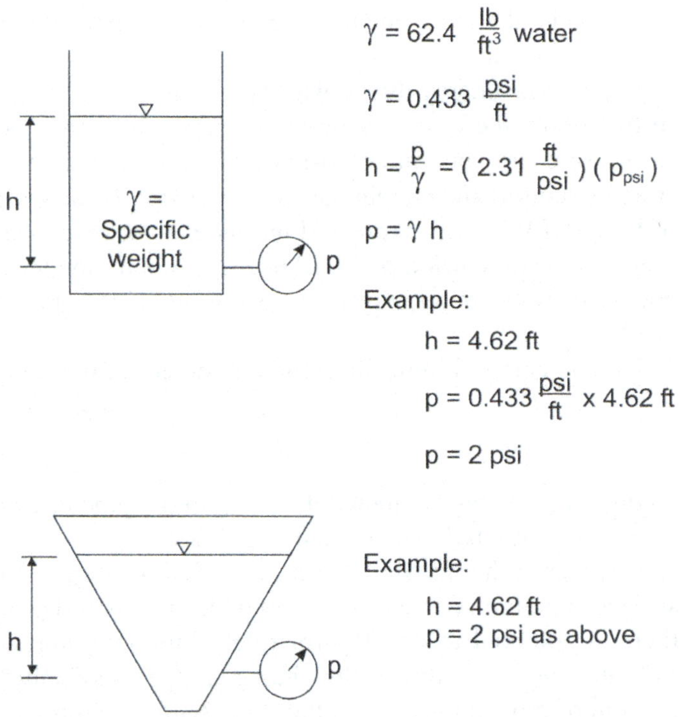

Figure 8.1. Pressure head for water in a vessel.

The sum of the three energy forms is the total energy per unit weight of water called total head (H). Total head is:

H = velocity head + elevation head + pressure head

or
$$H = \frac{V_m^2}{2g} + Z + h \qquad (8.4)$$

The sum of elevation head and pressure head is called hydraulic head. Figure 8.2 illustrates the components of hydraulic head for a pipeline with various orientations.

Another important concept of water flow is continuity. In a hydraulic system mass must be conserved. Therefore, for an incompressible fluid such as water, the volumetric flow rate (Q) must be the same for all points in a system with only one inlet and one outlet. The continuity equation for an incompressible fluid, such as water, may be expressed as:

$$Q = V_m A_f \qquad (8.5)$$

where: Q = volumetric flow rate or discharge,
V_m = average flow velocity, and
A_f = cross-sectional area of flow.

The laws of conservation of mass and energy are applied in Example 8.2. The conservation of mass states that the volumetric flow rate (Q) must be the same for all points in the system. Thus, the flow rate everywhere in the system shown in Example 8.2 must be 400 gallons per minute (gpm). By combining the continuity equation and the concept of mass flow, problems other than just calculating the total head at a point may be solved.

An important law of fluid mechanics is conservation of energy. Conservation of energy for irrigation systems is described by the Bernoulli Equation, which is expressed as:

$$H_1 = H_2 + h_L \qquad (8.6)$$

or
$$\frac{V_1^2}{2g} + Z_1 + h_1 = \frac{V_2^2}{2g} + Z_2 + h_2 + h_L \qquad (8.7)$$

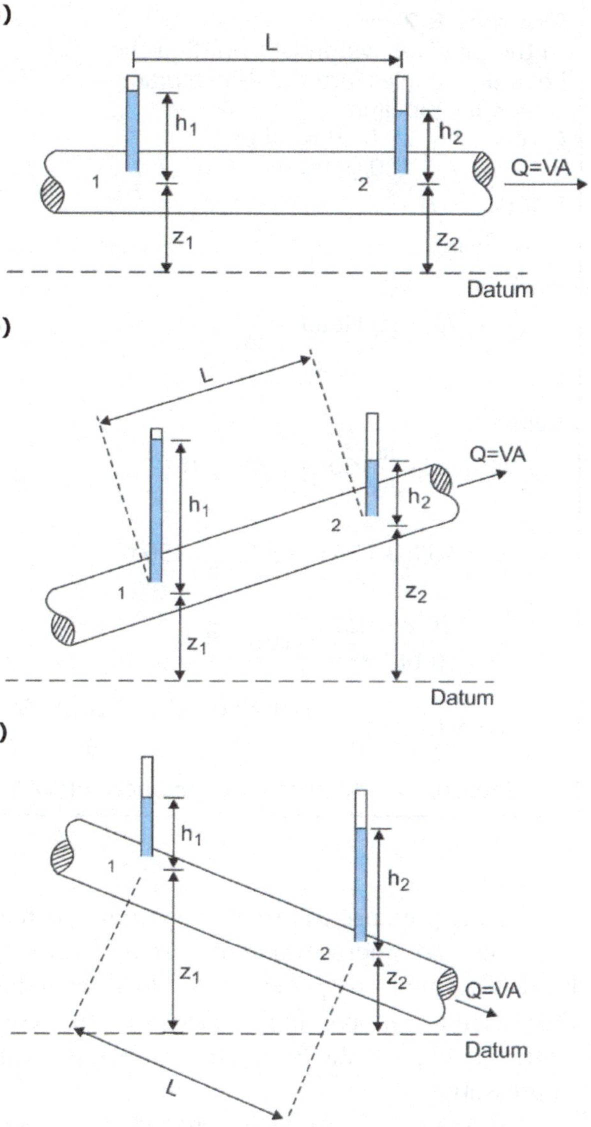

Figure 8.2. Components of hydraulic head for pipelines with various orientations (velocity heads not considered).

where: H_1 = total head at point 1 in a system,
H_2 = total head at point 2 in a system, and
h_L = head loss during flow from point 1 to point 2.

Velocity head (h_v) can be determined graphically using Figure 8.3.

The head loss from point 1 to point 2 is due to ***friction loss*** (h_f) from the resistance to flow along a pipeline and to ***minor losses*** (h_m) of energy through pipe fittings, etc. Thus,

$$h_L = h_f + h_m \qquad (8.8)$$

Expressed as pressure loss,

$$P_L = P_f + P_m \qquad (8.9)$$

where: P_L = pressure loss,
P_f = pressure loss due to friction, and
P_m = pressure loss due to minor losses.

Example 8.2

In the pipeline system shown, find the total head at the inlet into the 4-inch diameter (d = 4 in) pipeline.

Given: Z = 15 ft, P = 60 psi

Q = 400 gpm, d = 4 in

Find: h, H

$$V_m = \frac{Q}{A_f}$$

$$\text{Velocity Head} = \frac{V_m^2}{2g}$$

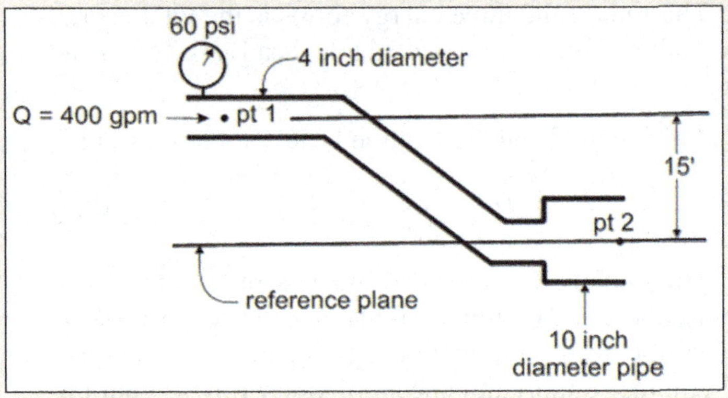

Solution:

$$h = 2.31\frac{\text{ft}}{\text{psi}}(60\text{ psi}) = 139\text{ ft}$$

$$A = \frac{\pi(4\text{ in})^2}{4} = 12.57\text{ in}^2\left(\frac{1\text{ ft}^2}{144\text{ in}^2}\right) = 0.087\text{ ft}^2$$

$$H = 15\text{ ft} + 139\text{ ft} + 1.6\text{ ft} = 156\text{ ft}$$

$$Q = \frac{400\text{ gpm}}{450\text{ gpm / cfs}} = 0.89\text{ cfs}$$

$$V = \frac{(0.89\text{ cfs})}{(0.087\text{ ft}^2)} = 10.23\text{ ft / s}$$

Since g = 32.2 ft/s², $2g$ = 64.4 ft/s²

$$\text{Velocity Head} = \frac{(10.23\text{ ft / s})^2}{2g} = \frac{(10.23\text{ ft / s})^2}{(64.4\text{ft / s}^2)} = 1.6\text{ ft}$$

Pressure is the primary component of total head in this example.

In many pressurized irrigation systems, such as sprinkler and micro-irrigation systems, velocity head is a minor component of the total head and thus it can be ignored. In this case, it is more convenient to express the Bernoulli equation in terms of pressure:

$$0.433Z_1 + P_1 = 0.433Z_2 + P_2 + P_L \qquad (8.10)$$

Application of this equation for level and sloping pipelines is shown in Figure 8.4. After studying Figure 8.4 you might ask yourself the question, "How do you apply Equation 8.10 when the pipeline goes up and over a hill?"

Example 8.3

What is the velocity head at point 2 in Example 8.2?

Given: $Q_2 = Q_1$ = 400 gpm

d_2 = 10 in

Find: $A_2 = \frac{\pi d_2^2}{4}$

V_m

Velocity head at point 2

Solution:

$$A_2 = \frac{\pi(10\text{ in})^2}{4} = 78.5\text{ in}^2 = 0.545\text{ ft}^2$$

$$V_2 = \frac{(0.89\text{ ft}^3/\text{s})}{(0.545\text{ ft}^2)} = 1.63\text{ ft/s}$$

$$\text{Velocity head} = \frac{(1.69\text{ ft / s})^2}{(64.4\text{ ft / s}^2)} = 0.04\text{ ft}$$

This means that the velocity head in the 10-in pipe is 0.025 times the velocity head in the 4-in pipe.

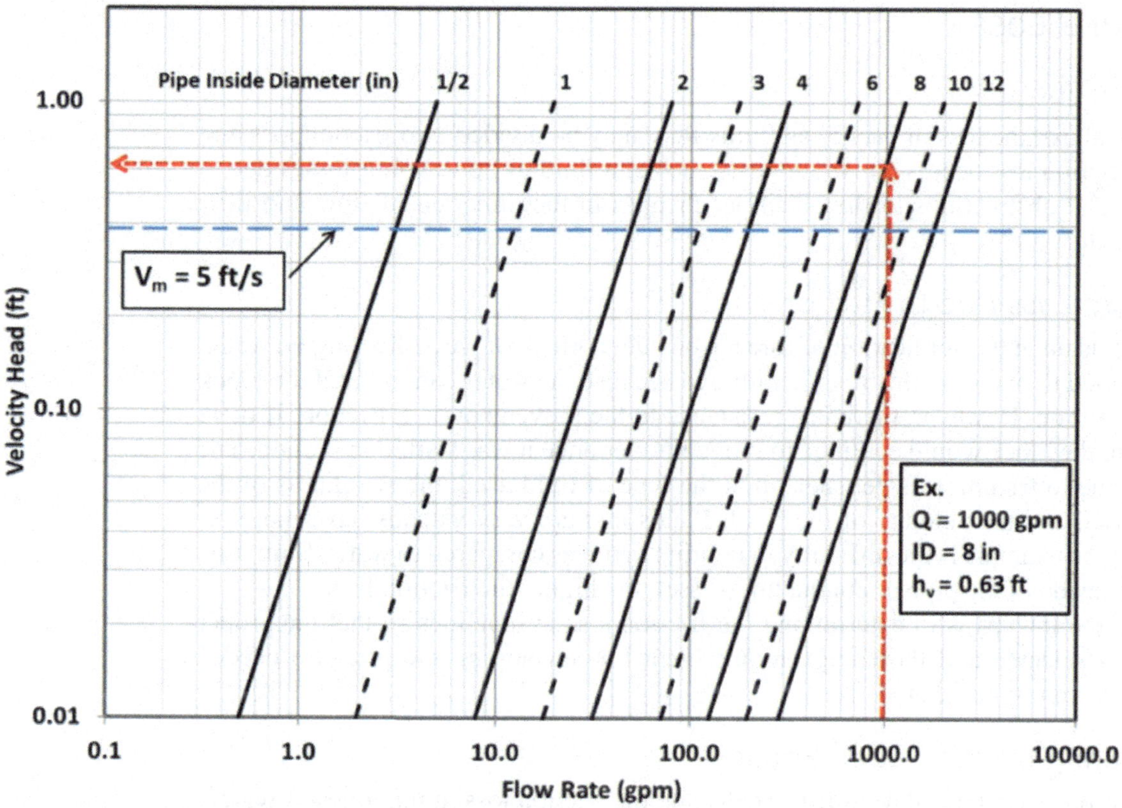

Figure 8.3. Graph for determining velocity head in pipelines.

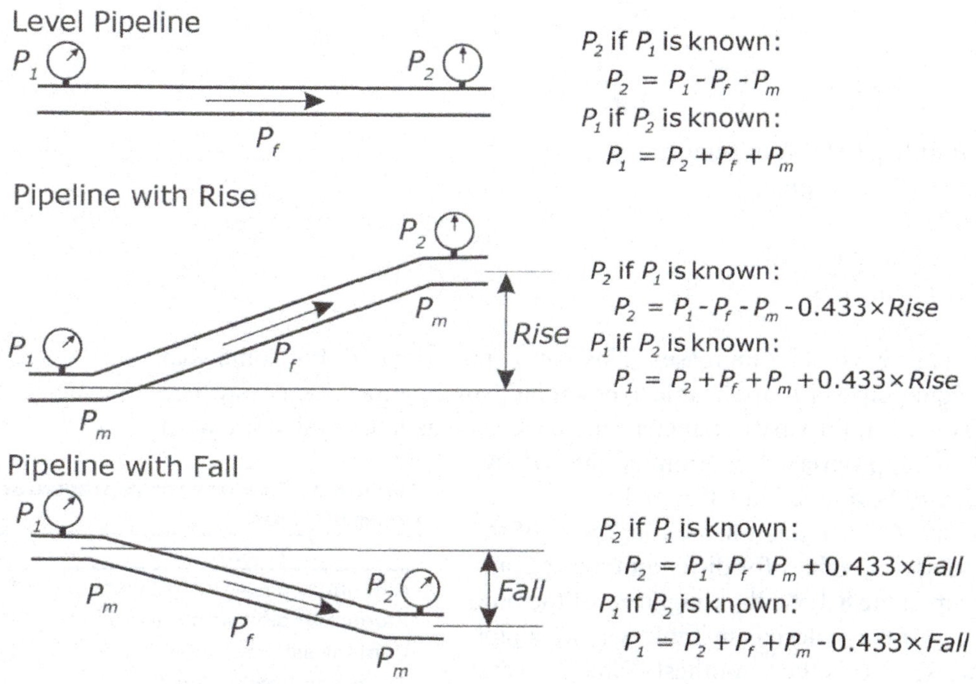

Figure 8.4. Application of pressure form of Bernoulli equation for level and sloping pipelines (velocity head changes assumed insignificant).

8.3 Pressure Loss

8.3.1 Introduction

As discussed above the loss of energy as fluids flow may be divided into friction loss and minor losses. Friction loss occurs due to the resistance of a fluid to flow. Minor losses due to turbulence occur at obstructions to flow such as changes in the direction of flow and flow through valves, etc.

8.3.2 Pressure Loss Due to Friction Loss

A fluid deforms upon the application of force. Consider a block of wood floating on water. If a force is applied to one side, the block will move because the water cannot hold the block in its original position. However, there is a resistance to the movement, i.e., friction. If there were no friction, the block would continue to move forever once it was started in motion. The loss of energy due to friction loss depends upon the type of fluid used, the roughness of the conducting vessel, and the velocity of the fluid. Fluids that are very viscous have more resistance to flow. An example is the difficulty in pouring syrup as compared to water. Similarly, the rougher the inside of the pipe or conducting vessel, the higher the friction loss.

In irrigation, the interest is in determining the friction loss in pipelines so that the proper pipe diameter is selected and the energy requirement for developing the pressure needed within the system can be calculated.

8.3.3 Computing Losses Due to Friction

Several equations have been developed to calculate the friction loss in pipelines. A widely used empirical method is the Hazen-Williams Equation. The Hazen-Williams Equation for circular pipes is given by:

$$h_f = 1054F\left(\frac{Q}{C}\right)^{1.852}\left(\frac{1}{d^{4.866}}\right) \tag{8.11a}$$

or

$$P_f = 456F\left(\frac{Q}{C}\right)^{1.852}\left(\frac{1}{d^{4.866}}\right) \tag{8.11b}$$

where: h_f = friction loss, ft of head/100 ft of pipe,
P_f = friction loss, psi/100 ft of pipe,
Q = flow rate (gpm),
d = inside diameter of the pipe (in),
C = roughness coefficient, and
F = outlet factor.

Friction loss increases as flow velocity increases. This fact is incorporated, but somewhat hidden in Equation 8.11. Equation 8.11 is applicable to essentially all pipelines used in surface and sprinkler irrigation. However, for small diameter pipelines, such as laterals that are used in microirrigation, a more appropriate equation is the Darcy-Weisbach equation which will be applied in Chapter 14.

The roughness coefficient, *C*, accounts for the roughness of the wall of the pipe. Representative *C* values for different types of pipe materials are summarized in Table 8.1. As the roughness of the pipe wall increases *C* decreases. Of the materials in Table 8.1, steel pipe is the roughest material while PVC is the smoothest. Table 8.2a and b contain pressure losses due to friction for selected pipe materials and diameters based on the Hazen-Williams equation.

Table 8.1. *C* values for representative types of pipes.

Material	*C*
Aluminum pipe with couplers	120
Aluminum pipe with gates	110
Cement asbestos pipe	140
Galvanized steel pipe	140
Standard steel pipe	100
PVC	150
PVC pipe with gates	130

Table 8.2a. Pressure loss due to friction for smaller diameter pipes (Hazen-Williams Formula). Bold font with shading represents region where velocity exceeds 5 ft/s.

	Aluminum Sprinkler Pipe, 150 psi Rating $C = 120$				PVC IPS Class 160 $C = 150$				
Nominal Diameter (in):	2	3	4	6	2	$2^1/_2$	3	4	6
Inside Diameter (in):	1.900	2.900	3.900	5.884	2.193	2.655	3.230	4.150	6.120
Q (gpm)	Pressure Loss Due to Friction (psi/100 ft)								
2	0.01								
4	0.04				0.01				
6	0.08				0.03				
8	0.13				0.04				
10	0.20	0.03			0.07	0.03			
15	0.43	0.05			0.14	0.06			
20	0.73	0.09			0.24	0.09	0.04		
25	1.10	0.14			0.36	0.14	0.06		
30	1.54	0.20			0.51	0.20	0.08		
35	2.05	0.26			0.68	0.27	0.10		
40	2.63	0.34			0.87	0.34	0.13		
45	**3.27**	0.42			1.08	0.42	0.16		
50	**3.97**	0.51	0.12		1.31	0.52	0.20	0.06	
55	**4.74**	0.61	0.14		1.56	0.62	0.24	0.07	
60	**5.57**	0.71	0.17		**1.83**	0.72	0.28	0.08	
65	**6.46**	0.83	0.20		**2.13**	0.84	0.32	0.10	
70	**7.41**	0.95	0.22		**2.44**	0.96	0.37	0.11	
75	**8.42**	1.08	0.25		**2.77**	1.09	0.42	0.12	
80	**9.49**	1.21	0.29		**3.12**	1.23	0.47	0.14	
85	**10.61**	1.36	0.32		**3.49**	1.38	0.53	0.16	0.02
90	**11.80**	1.51	0.36	0.05	**3.88**	**1.53**	0.59	0.17	0.03
100	**14.34**	1.83	0.43	0.06	**4.72**	**1.86**	0.72	0.21	0.03
110	**17.11**	**2.19**	0.52	0.07	**5.63**	**2.22**	0.86	0.25	0.04
120	**20.10**	**2.57**	0.61	0.08	**6.62**	**2.61**	1.01	0.30	0.04
140	**26.74**	**3.42**	0.81	0.11	**8.80**	**3.47**	**1.34**	0.40	0.06
150		**3.88**	0.92	0.12		**3.95**	**1.52**	0.45	0.07
160		**4.38**	1.03	0.14		**4.45**	**1.71**	0.51	0.08
170		**4.90**	1.16	0.16		**4.98**	**1.92**	0.57	0.09
180		**5.44**	1.29	0.17		**5.53**	**2.13**	0.63	0.10
190		**6.02**	**1.42**	0.19		**6.11**	**2.36**	0.70	0.11
200		**6.61**	**1.56**	0.21		**6.72**	**2.59**	0.76	0.12
220			**1.87**	0.25			**3.09**	**0.91**	0.14
240			**2.19**	0.30			**3.63**	**1.07**	0.16
260			**2.54**	0.34			**4.21**	**1.24**	0.19
280			**2.92**	0.39			**4.83**	**1.43**	0.22
300			**3.32**	0.45			**5.49**	**1.62**	0.24
320			**3.74**	0.51			**6.18**	**1.83**	0.28
340			**4.18**	0.57				**2.04**	0.31
360			**4.65**	0.63				**2.27**	0.34
380			**5.14**	0.69				**2.51**	0.38
400			**5.65**	0.76				**2.76**	0.42
420			**6.18**	0.84				**3.02**	0.46
440			**6.74**	**0.91**				**3.29**	0.50
460				**0.99**				**3.58**	**0.54**
480				**1.07**					**0.58**
500				**1.15**					**0.63**
550				**1.38**					**0.75**
600				**1.62**					**0.88**
650				**1.88**					**1.03**
700				**2.15**					**1.18**
750				**2.45**					**1.34**
800				**2.76**					**1.51**

Table 8.2b. Pressure loss due to friction for larger diameter pipes (Hazen-Williams Formula). Bold font with shading represents region where velocity exceeds 5 ft/s.

	Aluminum Gated Pipe, 0.051 Wall C = 110			PVC PIP Class 125 C = 150			
Nominal Diameter (in):	6	8	10	6	8	10	12
Inside Diameter (in):	5.898	7.898	9.898	5.766	7.658	9.572	11.486
Q (gpm)	Pressure Loss Due to Friction (psi/100 ft)						
240	0.34						
260	0.40			0.25			
280	0.46			0.29			
300	0.52			0.33			
320	0.59	0.14		0.37			
340	0.66	0.16		0.41			
360	0.73	0.18		0.46			
380	0.81	0.19		0.51	0.13		
400	0.89	0.21		0.56	0.14		
420	0.97	0.23		**0.61**	0.15	0.05	
440	**1.06**	0.26	0.09	**0.66**	0.17	0.06	
460	**1.15**	0.28	0.09	**0.72**	0.18	0.06	
480	**1.24**	0.30	0.10	**0.78**	0.20	0.07	
500	**1.34**	0.32	0.11	**0.84**	0.21	0.07	
550	**1.60**	0.39	0.13	**1.01**	0.25	0.09	
600	**1.88**	0.45	0.15	**1.18**	0.30	0.10	0.04
650	**2.18**	0.53	0.18	**1.37**	0.34	0.12	0.05
700	**2.50**	0.60	0.20	**1.57**	0.39	0.13	0.05
750	**2.84**	0.69	0.23	**1.79**	**0.45**	0.15	0.06
800	**3.20**	**0.77**	0.26	**2.01**	**0.51**	0.17	0.07
850	**3.58**	**0.86**	0.29	**2.25**	**0.57**	0.19	0.08
900	**3.98**	**0.96**	0.32	**2.50**	**0.63**	0.21	0.09
950	**4.40**	**1.06**	0.35	**2.77**	**0.70**	0.23	0.10
1000	**4.84**	**1.17**	0.39	**3.04**	**0.76**	0.26	0.11
1050	**5.30**	**1.28**	0.43	**3.33**	**0.84**	0.28	0.12
1100	**5.77**	**1.39**	0.46	**3.63**	**0.91**	0.31	0.13
1150	**6.27**	**1.51**	0.50	**3.94**	**0.99**	**0.33**	0.14
1200	**6.78**	**1.64**	**0.55**	**4.26**	**1.07**	**0.36**	0.15
1250		**1.77**	**0.59**		**1.16**	**0.39**	0.16
1300		**1.90**	**0.63**		**1.24**	**0.42**	0.17
1350		**2.04**	**0.68**		**1.33**	**0.45**	0.19
1400		**2.18**	**0.73**		**1.43**	**0.48**	0.20
1450		**2.33**	**0.78**		**1.52**	**0.51**	0.21
1500		**2.48**	**0.83**		**1.62**	**0.55**	0.23
1550			**0.88**			**0.58**	0.24
1600			**0.93**			**0.62**	0.25
1650			**0.99**			**0.65**	**0.27**
1700			**1.04**			**0.69**	**0.28**
1750			**1.10**			**0.73**	**0.30**
1800			**1.16**			**0.77**	**0.32**
1850			**1.22**			**0.81**	**0.33**
1900			**1.28**			**0.85**	**0.35**
1950			**1.34**			**0.89**	**0.37**
2000			**1.41**			**0.93**	**0.38**
2050							**0.40**
2100							**0.42**
2150							**0.44**
2200							**0.46**

Table 8.3. Multiple outlet factors for laterals with equally spread outlets of the same discharge (first outlet one full spacing from inlet to pipe). For center pivots see footnote.*

No. of Outlets	F	No. of Outlets	F
1	1.0	16	0.377
2	0.634	17	0.376
3	0.528	18	0.373
4	0.480	19	0.372
5	0.451	20	0.370
6	0.433	22	0.368
7	0.419	24	0.366
8	0.410	26	0.364
9	0.402	28	0.363
10	0.396	30	0.362
11	0.392	35	0.359
12	0.388	40	0.357
13	0.384	50	0.355
14	0.381	100	0.350
15	0.379	>100	0.345

* F = 0.54 for center pivots without end guns.
F = 0.56 for center pivots with end guns.

Example 8.4

A four-inch aluminum sprinkler lateral is 1280 feet long. Sprinklers are spaced at 40-foot intervals. The lateral goes up (rises) 12 feet in elevation along its length. Each sprinkler on the lateral discharges 5 gpm.

Given: L = 1280 ft
sprinkler spacing = 40 ft
rise = 12 ft
q = 5 gpm

Find: Pressure loss due to friction in the lateral in psi.
If the inlet pressure to the lateral is 60 psi, what is the pressure at the downstream end of the lateral? Ignore minor losses.

Solution:

There are 33 sprinklers on the lateral (1280/40).

The inlet flow rate is then 165 gpm (i.e., 5 gpm × 33).

From Table 8.3, the multiple outlet factor is 0.36.

Interpolating from Table 8.2a, the pressure loss due to friction is 1.1 psi/100 ft.

$$P_f = F \times (P_f/100\text{ ft}) \times L$$

$$P_f = \frac{0.36 \times 1.1 \times 1280\text{ ft}}{100\text{ ft}} = 5.1\text{ psi}$$

The pressure at the downstream end of the lateral can be determined using the concepts shown in Figure 8.4.

$$P_2 = P_1 - P_f - P_m - 0.433 \times Rise$$

$$P_2 = 60 - 5.1 - 0.433 \times 12 = 49.7\text{ psi}$$

A pipeline with outlets, such as a lateral where water is removed by sprinklers, gates, or emitters, has a lower friction loss than a conveyance pipe because the velocity decreases with distance along the pipe. To correct for the effect of the outlets a multiple outlet factor F is used. F = 1.0 for a pipeline without outlets. For laterals with constant spaced outlets, and nearly the same discharge per outlet, use Table 8.3. With center pivots, sprinkler discharge increases with distance from the pivot point. Outlet factors for pivots are given at the bottom of Table 8.3.

8.3.4 Minor Losses Due to Pipeline Fittings

Head or pressure losses also occur in the fittings used in the pipeline system. These head losses are due to friction in the fitting, plus losses resulting from turbulence and changes in the direction of flow. Head loss in fittings, valves, etc., can be described by:

$$h_m = K\left(\frac{V_m^2}{2g}\right) \quad (8.12)$$

where: h_m = head loss in fitting (ft),
K = resistance coefficient for fitting, and
V_m = velocity of flow (ft/s).

Resistance coefficients for various types of fittings and valves are given in Table 8.4.

Table 8.4. Resistance coefficient *K* for determining head losses in fittings and valves (USDA, 2016).

Fitting or Valve	Standard Pipe Nominal Diameter						
	3 in (76.2 mm)	4 in (101.6 mm)	5 in (127.0 mm)	6 in (152.4 mm)	7 in (177.8 mm)	8 in (203.2 mm)	10 in (254 mm)
Bends							
Return flanged	0.33	0.30	0.29	0.28	0.27	0.25	0.24
Return screwed	0.80	0.70					
Elbows							
Regular flanged 90°	0.34	0.31	0.30	0.28	0.27	0.26	0.25
Long radius flanged 90°	0.25	0.22	0.20	0.18	0.17	0.15	0.14
Long radius flanged 45°	0.19	0.18	0.18	0.17	0.17	0.17	0.16
Regular screwed 90°	0.80	0.70					
Long radius screwed 90°	0.30	0.23					
Regular screwed 45°	0.30	0.28					
Tees							
Flanged line flow	0.16	0.14	0.13	0.12	0.11	0.10	0.09
Flanged branch flow	0.73	0.68	0.65	0.60	0.58	0.56	0.52
Screwed line flow	0.90	0.90					
Screwed branch flow	1.20	1.10					
Valves							
Globe flanged	7.0	6.3	6.0	5.8	5.7	5.6	5.5
Globe screwed	6.0	5.7					
Gate flanged	0.21	0.16	0.13	0.11	0.09	0.075	0.06
Gate screwed	0.14	0.12					
Swing check flanged	2.0	2.0	2.0	2.0	2.0	2.0	2.0
Swing check screwed	2.1	2.0					
Angle flanged	2.2	2.1	2.0	2.0	2.0	2.0	2.0
Angle screwed	1.3	1.0					
Foot	0.80	0.80	0.80	0.80	0.80	0.80	0.80
Strainers (basket type)	1.25	1.05	0.95	0.85	0.80	0.75	0.67

8.4 Pipelines

Irrigation pipelines are made of many materials. Currently, the most common materials used for aboveground sprinkler systems and gated pipe systems are aluminum and ultraviolet radiation protected PVC (polyvinyl chloride plastic). Center pivot and lateral systems are the exception where it is common to use galvanized steel as the pipeline material. Above ground microirrigation laterals are usually made of polyethylene (PE) plastic. For pipelines that are buried below the ground, the most common material in agricultural applications is PVC, and in microirrigation systems it is PE.

Sizing mainline pipelines is usually based on a maximum of 5 to 6 ft/s average velocity. Table 8.5 shows the typical flow ranges for selected aluminum and PVC pipe at various nominal sizes and 5 ft/s flow velocities. For example, the recommended maximum flow rate for an 8-inch pipeline is in the range of 700 to 800 gpm.

Table 8.5. Maximum flow rates in pipelines rounded to nearest 5 gpm at V_m = 5 ft/s (based on Table 8.2a and b).

Nominal size (in)	Aluminum		PVC	
	Inside dia. (in)	Q (gpm)	Inside dia. (in)	Q (gpm)
	Sprinkler		IPS	
2	1.900	45	2.193	60
$2^1/_2$	-	-	2.655	85
3	2.900	105	3.230	130
4	3.900	185	4.154	210
6	5.884	425	6.120	460
	Gated		PIP	
6	5.898	425	5.776	405
8	7.898	765	7.658	720
10	9.898	1200	9.572	1120
12	-	-	11.486	1615

Example 8.5

A PVC PIP mainline will supply water to a drip irrigation system. The flow rate of the system is 700 gpm. The mainline is 600 feet long and drops (falls) 25 feet in its length. The pressure at the inlet to the pipe is 35 psi.

Given: P_1 = 35 psi
Fall = 25 ft
L = 600 ft
Q = 700 gpm

Find: The appropriate size pipe for this system
The pressure at the downstream end of the pipe, i.e., at the end of the mainline (ignore minor losses).

Solution:

Referring to Table 8.5, an 8-in pipe should be selected to keep the mean velocity below 5 ft/s.

From Table 8.2b, we find that the pressure loss due to friction is 0.39 psi/100 ft.

Using the concepts from Figure 8.4, we can solve for the downstream pressure:

$$P_2 = P_1 - P_f - P_m + 0.433 \times Fall$$

$$P_2 = 35 \text{ psi} - (0.39 \text{ psi}/100 \text{ ft}) \times 600 \text{ ft} - 0 + 0.433 \times 25 \text{ ft} = 43.5 \text{ psi}$$

In this example, pressure has increased with length in the pipeline because of the relatively steep fall.

Pipelines must be protected from excessive pressures and vacuums. It is also imperative that air is relieved from pipelines so that it is not compressed while filling the pipeline. At high points, it is important to relieve the air so that an air blockage to flow does not occur. Figure 8.5 shows the layout of valves which is required to adequately protect pipelines. To release air and relieve vacuums, a combination vacuum-air vent relief valve is used. These should be used at the entrance to the pipeline, at high points in the pipeline, and at the end of the pipeline. There should also be an air vent at 1,000-foot intervals along the pipeline. In addition to air and vacuum relief, pressure relief valves should be provided in case surges occur within the pipeline (Figures 8.5 and 8.6). These valves should be installed at the inlet and at dead ends of the pipeline. At the inlet to the pipeline, a check valve is suggested so that reverse flow will not occur when the pumping system stops. For pipelines connected to municipal water systems or when chemigation is used (Chapter 15), proper backflow prevention equipment must be installed. For pipelines that are buried shallower than the frost depth, drainage should be provided so that freezing water does not burst the pipeline. More information on pipeline hydraulics can be found in Colt Industries (1979) and Waller and Yitayew (2016).

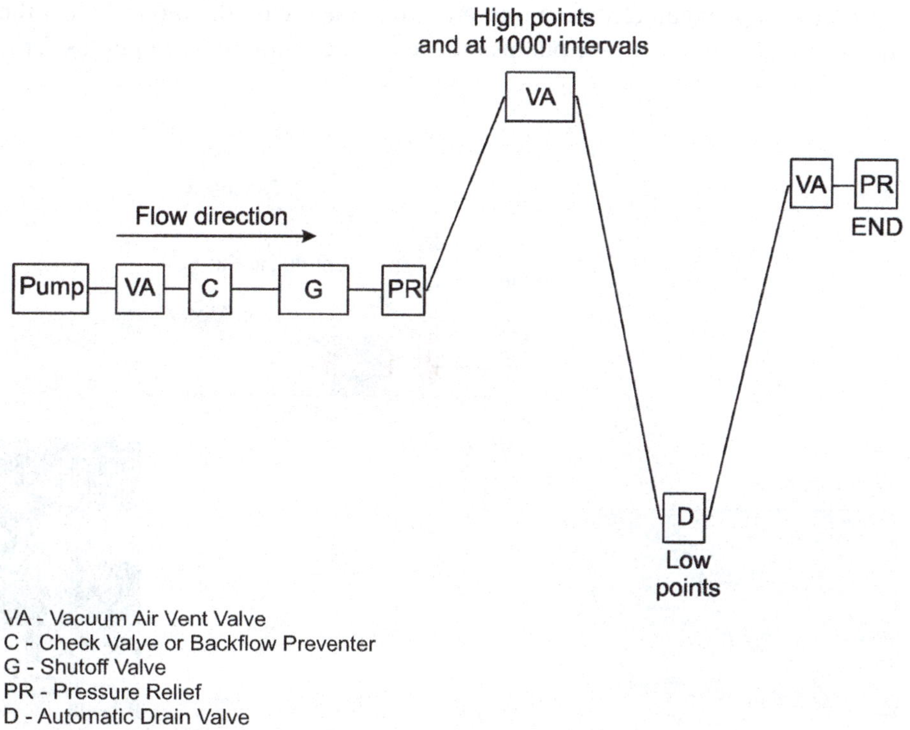

Figure 8.5. Suggested location of valves for buried pipelines.

Figure 8.6. Irrigation pipeline protection valves.

8.5 Pumps

Irrigation systems are designed to operate at specified pressures and flow rates. In order to develop the required pressure and to lift water from a reservoir or a well, it is often necessary to pump the water.

Pumps that lift and pressurize water in irrigation most commonly use the principal of centrifugal force to convert mechanical energy into hydraulic energy. This category includes horizontal centrifugal pumps and vertical turbine or submersible pumps. Horizontal ***centrifugal pumps*** are often used for pumping from an open water source (e.g., Figure 8.7) or for boosting the pressure in an irrigation pipeline. A ***vertical turbine pump*** has a vertical axle with the power source (motor or engine) above ground (e.g., Figure 8.8). A ***submersible pump*** is similar, except that both the pump and an electric motor are submersed, with the motor below the pump. The submersible and vertical turbine pumps are the most commonly used pumps for irrigation wells.

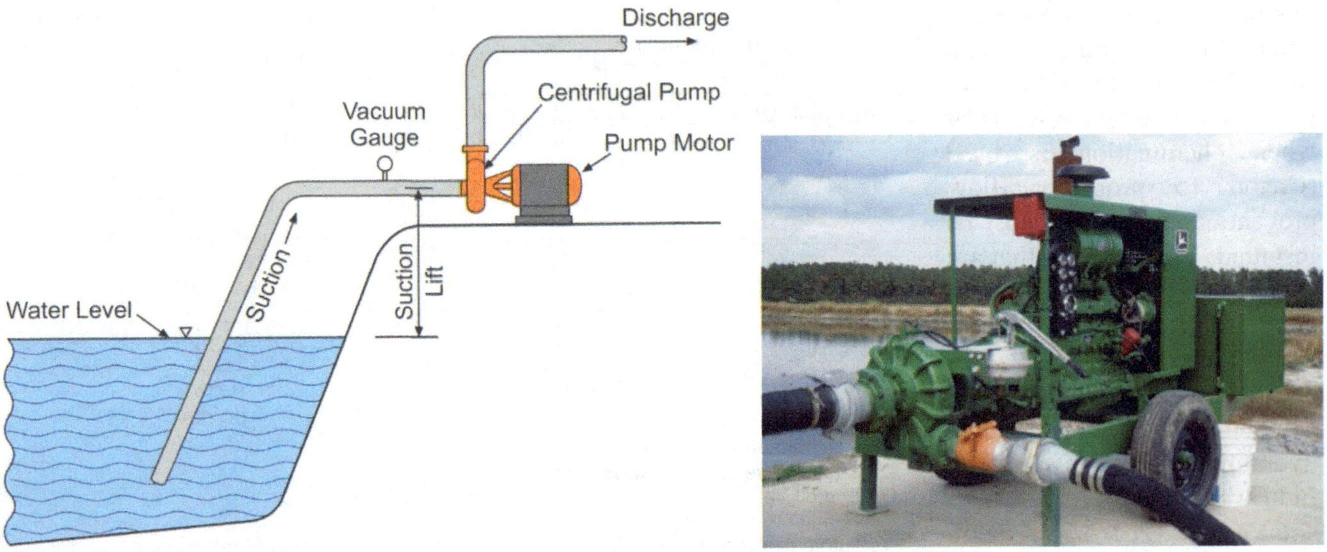

Figure 8.7. Application of horizontal centrifugal pump.

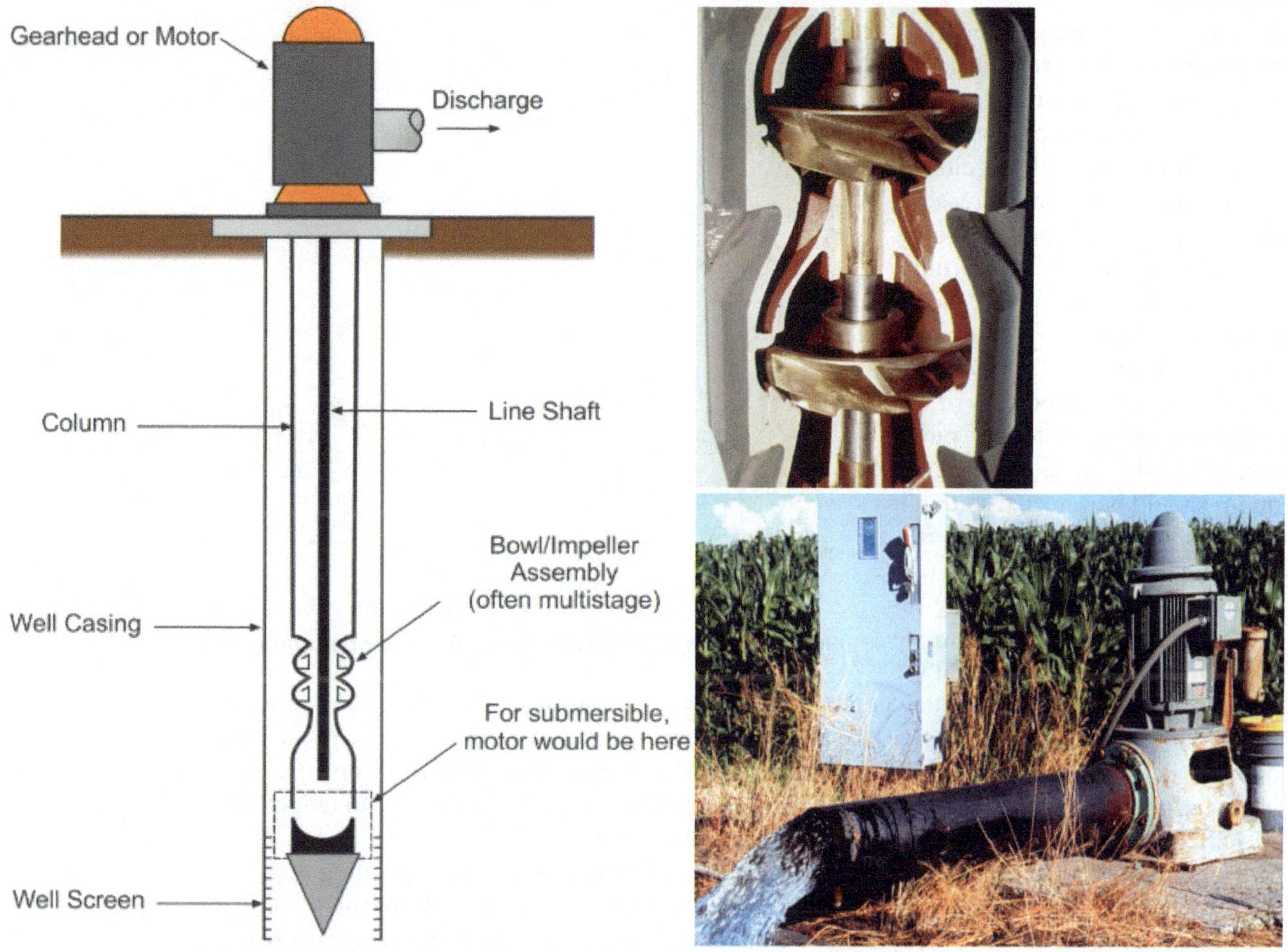

Figure 8.8. Vertical turbine pump installed in a well (left), cutaway of bowls with impellers in series (top right), and vertical turbine pump discharging to open ditch (bottom right).

The flow rate that is delivered by a pump is dependent upon the design of the impeller (the device that puts the energy into water), the diameter of the impeller, the speed of the impeller, and the total dynamic head that the impeller develops. Total dynamic head is the total head produced by the pump at a given flow rate. The total dynamic head (TDH) is the sum of the pressure head and elevation head (lift), i.e.

$$TDH = 2.31P + L \tag{8.13}$$

where: P = discharge pressure of the pump (psi) and

L = vertical distance water is moved from source to the pump discharge elevation (ft).

Solving for TDH in this manner is an approximation. We have ignored the velocity head and friction and minor losses required to move the water to the land surface. It is adequate for many, but not all, pumping conditions.

When a horizontal centrifugal pump is used as a booster pump the total dynamic head equation is

$$TDH = 2.31(P_{out} - P_{in}) \tag{8.14}$$

where: P_{out} = discharge pressure (psi)

P_{in} = inlet pressure to pump (psi)

A characteristic of a horizontal centrifugal pump is that as the total dynamic head increases the flow rate from the pump will decrease. Envision closing a valve downstream of the pump.

As the valve is closed, the flow rate decreases. If a pressure gauge were mounted upstream of the valve, it would indicate a rise in the pressure as the valve is closed. The pressure rise is an increase in the total dynamic head. The variable flow nature of centrifugal pumps is illustrated in the head-capacity relationship shown in Figure 8.9. Pump efficiency is a measure of the proportion of the energy transmitted to the pump that is transferred to the water. A pump should be selected so that it operates near its maximum efficiency at the desired flow rate (capacity) and the corresponding total dynamic head. In the example in Figure 8.9, it is evident that the pump reaches its peak efficiency at about 1,100 gpm and 190 feet of head. As you move to the left on the head-capacity curve, the pump efficiency goes down. As you move to the right of the peak efficiency point, the efficiency also goes down. Note that the peak pump efficiency is approximately 80% for the example shown. You can expect peak efficiencies to range from 55 to 82% for pump sizes most commonly used in irrigation.

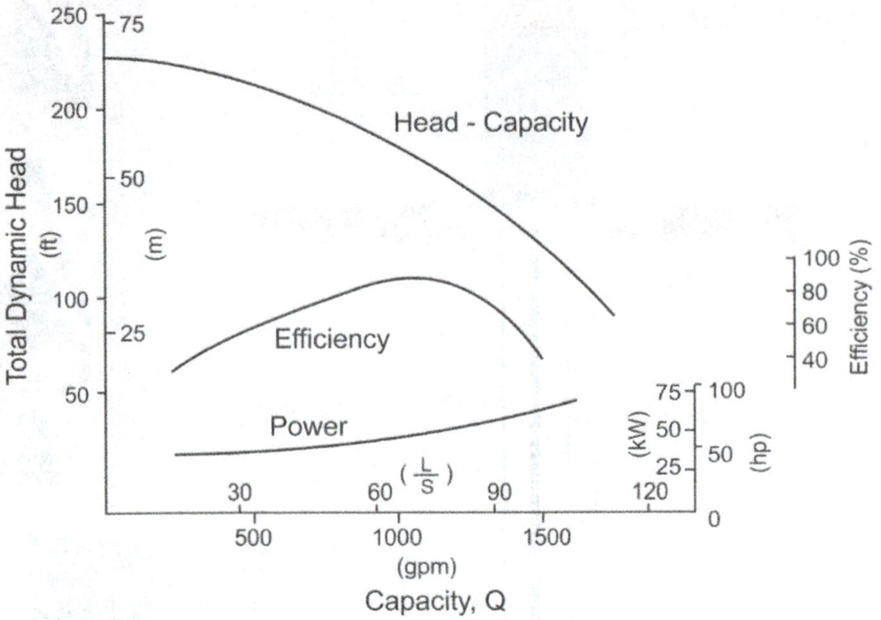

Figure 8.9. Head-capacity curve for a centrifugal pump.

The head discharge relationship shown in Figure 8.9 applies to a pump operating at a constant speed. If the pump speed is changed, the head discharge relation also changes. This is illustrated in Figure 8.10. As the speed of the pump decreases, its discharge pressure decreases at a given flow rate. Therefore, there is a different head discharge relationship for the slower speed. The slower speed head discharge curve is approximately parallel to the curve for the higher speed. Note that as the speed of the pump is lowered, the point for peak efficiency has shifted to the left, that is, to a lower flow rate.

Another factor affecting the head capacity relationship is the diameter of the impeller. Figure 8.11 illustrates what happens as an impeller is trimmed to reduce its diameter. Again, as the impeller diameter is decreased, the point of peak efficiency of the pump shifts to a lower flow rate, much like what happened when the speed was reduced.

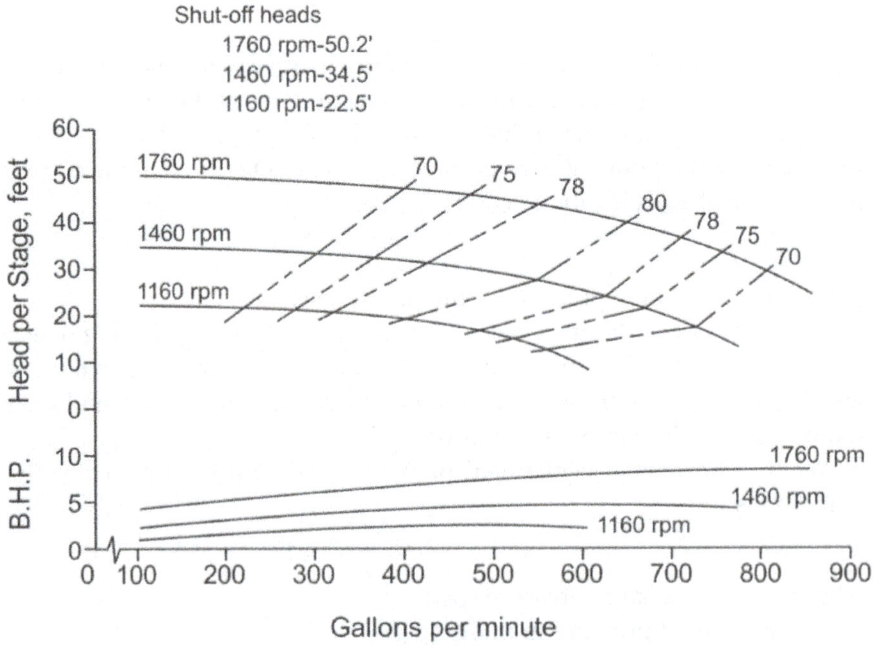

Figure 8.10. Head-capacity curve for centrifugal pump with various pump speeds.

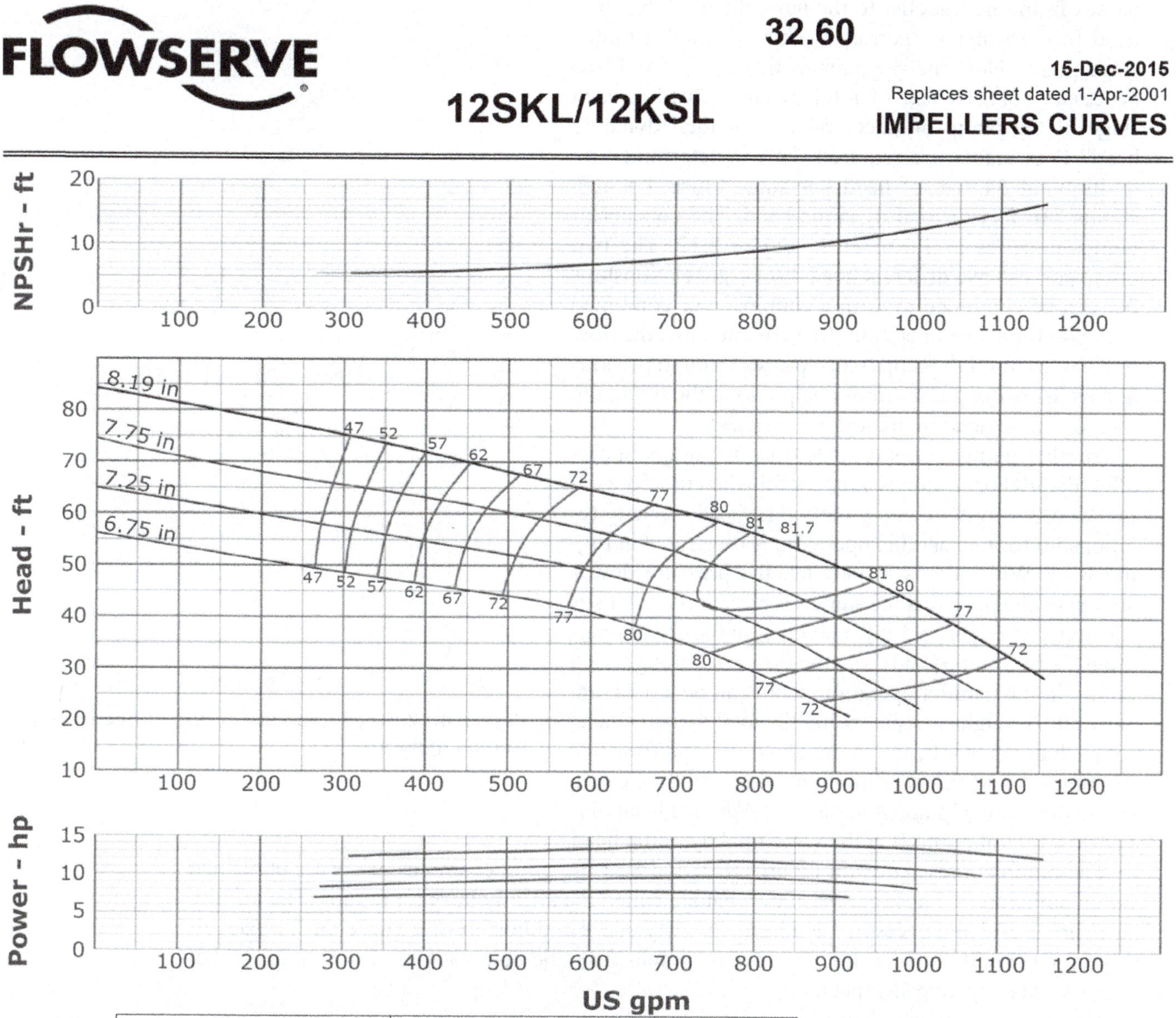

Pump / Curve Info				STG DERATES			
MODEL:	12SKL	PERFORMANCE BASED ON PUMPING CLEAR WATER BELOW 85°F. WITH LISTED MATLS. COLUMN LOSSES NOT INCLUDED		1	-3		
CURVE #:	EC-1926			2	-2	CURVE INPUT UNITS:	US UNITS
CRV DATE:	15/2/1996	1ST TRIM INCREMENT:	0.07	3	-1	CURVE INPUT SPEED	1770
FILENAME:	TVTP1422.05	TRIM INCREMENT:	0.0625	4	0	SPECIFIC SPEED:	2686
DEF UNITS:	US UNITS	THRUST K/STG:	11.80	5	0	SUCT SPECIFIC SPEED:	9138

Figure 8.11. Head-capacity curve for centrifugal pump with various pump diameters. (Figure credit: Flowserve.)

How are irrigation pumps selected? The key is to select a pump that is efficient at the system flow rate and total dynamic head. For example, a system having a flow rate of 800 gpm and lifting water out of a well a distance of 100 feet with a discharge pressure of 50 psi, the pump must be able to deliver the 800 gpm at a total dynamic head of 216 feet. Now, suppose the manufacturer has a pump that operates at 800 gpm, very efficiently, but the total dynamic head produced by that pump with a single impeller is only 54 feet. How can we develop the total dynamic head that is required for the irrigation system? One approach is to place the pump bowl and impeller assemblies in a series. With the vertical turbine pump (Figure 8.8) and submersible pump, several bowl and impeller assemblies are placed into a series. The same flow rate goes through each impeller, hence the concept of "in series". As water passes from one impeller to the next, the total dynamic head in the water is increased. This is called a multi-stage pump. How many stages of this pump would be necessary for 216 feet of total dynamic head if each stage of the pump produces 54 feet of total dynamic head? Four stages are required. This is determined by multiplying 54 feet of head per stage times 4 which equals 216 feet of total dynamic head. The concept of pumps-in-series is illustrated in Figure 8.12. The two pump curves have different head discharge relationships but can be combined to form a composite or combined curve for the series operation. Keep in mind that the flow that passes through pump A also passes through pump B and as the water passes from one pump to the other, the total dynamic head in the water increases.

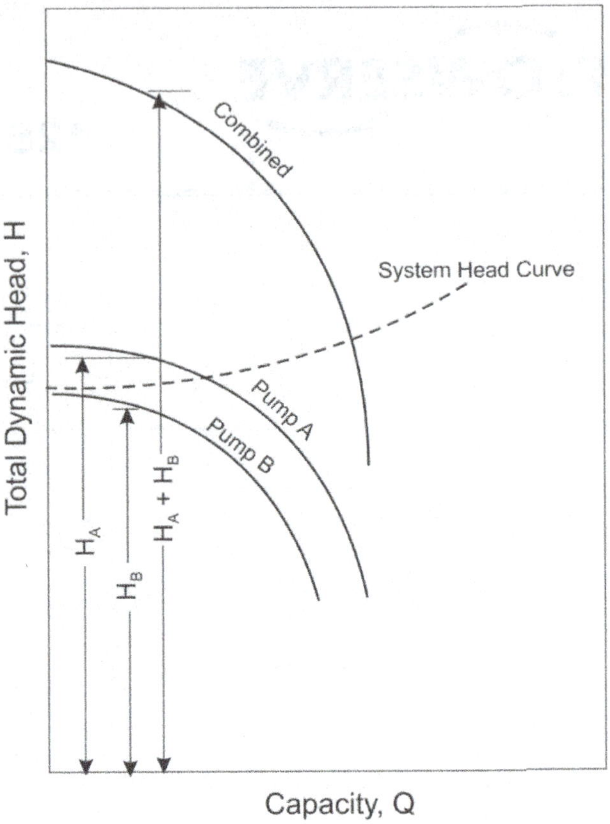

Figure 8.12. Head-capacity curves for centrifugal pumps in series.

Another pumping option is to operate pumps in parallel. Parallel operation is very useful when the flow demands of the system vary greatly. The head capacity relationship for this parallel operation is illustrated in Figure 8.13. With pumps in parallel, the pressure downstream of the pumps is the same for both pumps. This is illustrated in Figure 8.14. Remember, in the series operation the two pumps had the same flow rate through each pump. In the parallel operation, there can be a different flow rate through each pump, but the total dynamic head for each pump will be the same. Thus, the total flow rate of pumps A and B, operating in parallel will be the sum of the flow rate of pump A at the total dynamic head plus the flow rate of pump B at the same total dynamic head.

The pump curves shown in Figures 8.10 and 8.11 are good examples of curves published by manufacturers. These curves obey what are called the affinity laws for pumps. The affinity laws are useful and necessary if a head-capacity curve and horsepower curve must be developed for a condition that is not provided by graphs from the manufacturer. For example, what if you want to operate at a speed that is different than what is shown if Figure 8.10? Or, what if pump speed is fixed and the pump does not perfectly match expected pumping conditions, how much should the impeller be trimmed (reduced in diameter) to better match the expected conditions? In Figure 8.11, four trims are shown, but the most appropriate trim may not be shown on the graph. The affinity laws shown below are useful for determining appropriate pump speeds and impeller diameters:

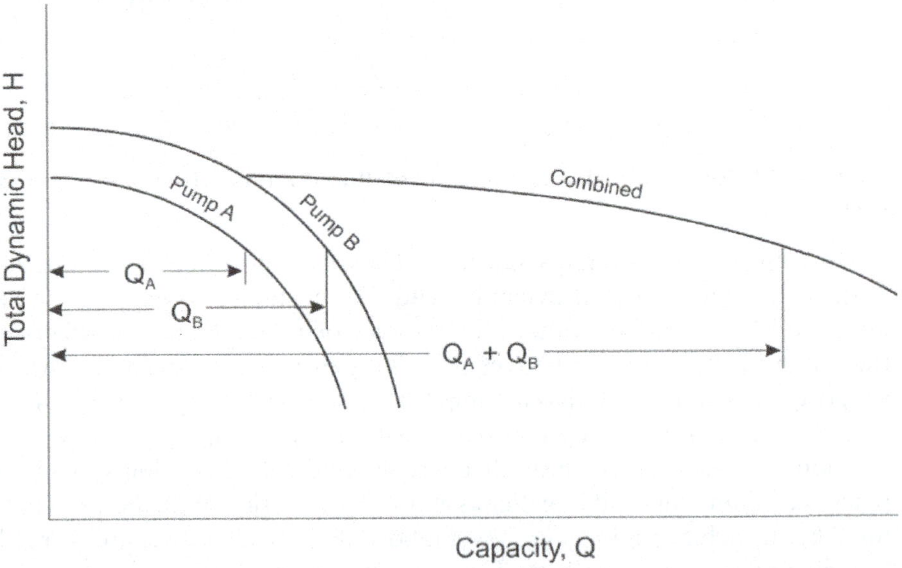

Figure 8.13. Head-capacity curve for centrifugal pumps in parallel.

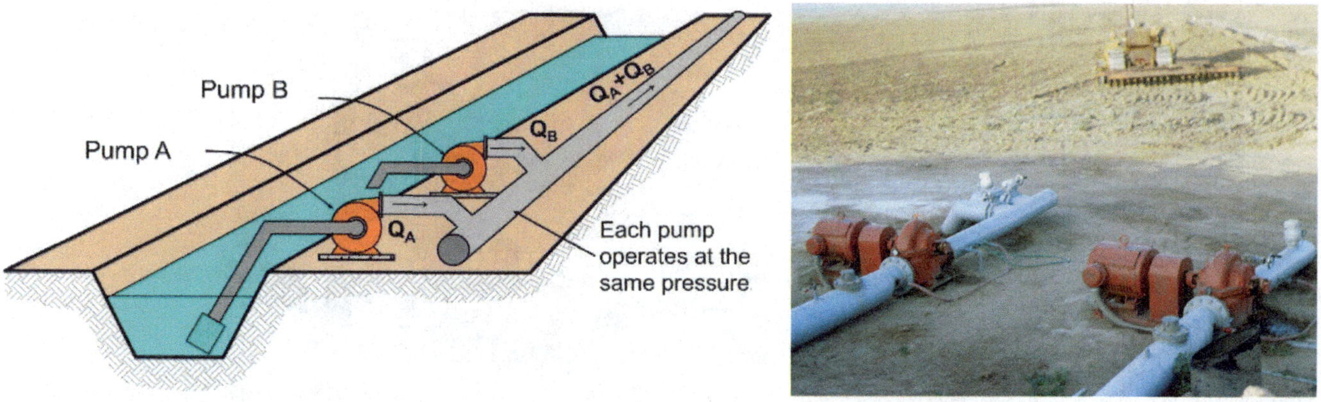

Figure 8.14. Centrifugal pumps connected in parallel.

$$\frac{Q_2}{Q_1} = \frac{RPM_2}{RPM_1} \qquad \frac{TDH_2}{TDH_1} = \left(\frac{RPM_2}{RPM_1}\right)^2 \qquad \frac{BHP_2}{BHP_1} = \left(\frac{RPM_2}{RPM_1}\right)^3 \qquad (8.15a)$$

$$\frac{Q_2}{Q_1} = \frac{DIA_2}{DIA_1} \qquad \frac{TDH_2}{TDH_1} = \left(\frac{DIA_2}{DIA_1}\right)^2 \qquad \frac{BHP_2}{BHP_1} = \left(\frac{DIA_2}{DIA_1}\right)^3 \qquad (8.15b)$$

where: RPM = pump speed in revolutions per minute,
DIA = impeller diameter,
BHP = brake horsepower (discussed in section 8.6 below), and
Subscripts 1 and 2 = current condition and new condition, respectively.

For example, Q_1 is the current flow rate and Q_2 is the future or predicted flow rate. The affinity laws can be used to generate head-capacity and horsepower curves based on known or current conditions. Pumps still operate efficiently if you change diameter or speed, and the affinity laws will be obeyed. Note how the laws behave. While flow rate is directly proportional to speed and diameter, TDH and power vary by the square and cube, respectively, of the speed and diameter.

An irrigation pumping system should be planned so that the pump operates at near peak efficiency. If the operating conditions change, the efficiency of the irrigation pump is likely to change at the same time. It is best to avoid undersizing or oversizing a pump; when pumps are oversized, they are sometimes throttled with a valve which leads to excess energy consumption. This concept and energy management are discussed further in Section 8.7.

8.6 Power Requirements

A pump transfers energy from an electric motor or engine to the water (Figure 8.15). Since a pump cannot be 100% efficient, ***pump efficiency*** (E_p) is used to account for the energy lost in pumping and is defined as:

$$E_p = \frac{\text{Output of energy or power}}{\text{Input of energy of power}} \qquad (8.16)$$

It is also necessary to determine how large of an engine or motor is required to pump the water. Horsepower (hp) is the typical unit of power in the USCS system and is defined as:

$$1 \text{ hp} = 33{,}000 \text{ ft-lb/min} \qquad (8.17)$$

Thus, to lift 33,000 pounds of water at the rate of 1 foot per minute, 1 horsepower would be required. A gallon of water weighs 8.33 pounds, so 1 horsepower would lift approximately 3,960 gallons of water at the rate of 1 foot per minute. The power required to pressurize and lift water (called ***water horsepower***) may be expressed by:

Figure 8.15. Pumping plants including a well with a vertical turbine pump and a power supply: electric motor (left) and internal combustion engine (right).

$$whp = \frac{(Q \times TDH)}{3960} \tag{8.18}$$

where: whp = water horsepower,
Q = flow rate (gpm), and
TDH = total dynamic head (ft).

Water horsepower is the power that is actually added to the water.

Since the pump has some inefficiency, the power input to the pump must be more than the water horsepower. The power input to the pump is called the ***brake horsepower*** (bhp) or ***pump horsepower*** and is determined by:

$$bhp = \frac{whp}{E_p} \tag{8.19}$$

Example 8.6

A pump operating at 80% efficiency lifts water form a reservoir a vertical distance of 100 feet and also develops a pressure of 50 psi. If the flow rate is 800 gpm, what is the water horsepower requirement? What is the brake horsepower requirement?

Given: Q = 800 gpm P = 50 psi
L = 100 ft E_p = 0.80

Find: wph, bhp

Solution:

$$TDH = 2.31\frac{\text{ft}}{\text{psi}}P + L \quad \text{(Equation 8.14)}$$

$$TDH = 2.31\frac{\text{ft}}{\text{psi}}(50\text{ psi}) + 100\text{ ft} = 216\text{ ft}$$

$$whp = \frac{(Q \times TDH)}{3960} \quad \text{(Equation 8.18)}$$

$$whp = \frac{(800\text{ gpm})(216\text{ ft})}{3960} = 44\text{ hp}$$

$$bhp = \frac{whp}{E_p} \quad \text{(Equation 8.19)}$$

$$bhp = \frac{44\text{ hp}}{0.80} = 55\text{ hp}$$

8.7 Energy Consumption

Pumping water for irrigation consumes energy; it takes energy to lift water and it takes energy to pressurize water. Below we discuss ways to determine energy consumption so that irrigation managers can appreciate the energy costs of operating irrigation systems.

To analyze the rate of energy consumption, we will use what is called the Nebraska Pumping Plant Performance Criteria (Kranz et al., 2012a; Martin et al., 2017). given in Table 8.6. To illustrate how this table was developed, consider a 1.34 horsepower motor attached to an irrigation pump (one kilowatt is equivalent to 1.34 horsepower). Electric motors are not 100% efficient. For motors from 5 to 250 horsepower, the fully-loaded efficiency will range from 83 to 94%. The Nebraska Performance Criteria were developed assuming a motor efficiency of 88%. Thus, the power produced by the motor would be 0.88 times 1.34 horsepower or 1.18 horsepower. Therefore, 12% of the energy is lost due to the inefficiencies of the motor. The next step is to consider the energy that is transmitted from the motor to the pump. Many electric motors are directly connected to the pump and there is no energy loss in transmission. Thus, we would say that the drive efficiency is 100%. If a V-belt or right-angled gear drive is used to transmit the power from a motor or engine to the pump, energy is lost to heat in the drive. Typically, about 5% of the energy is lost between the motor and the pump if a gear drive or belt drive is used to transmit the power. With electric motors, the Nebraska Performance Criteria assumes a direct connection between the pump and the motor and thus a drive efficiency of 100%. Therefore, it is assumed that 1.18 horsepower is transferred to the shaft of the pump. The next step is to consider the efficiency of the pump. Nebraska Performance Criteria assumes a reasonable pump efficiency of 75%. Remember, as stated earlier, the peak efficiency of the pumps can vary from approximately 55 to 82%, depending upon the size and design of the pump. So, how much power is in the water leaving the pump? The power out of the pump will be equal to 1.18 horsepower going to the shaft of the pump times 0.75, which equals 0.885 water horsepower. Again, water horsepower is the power that is actually added to the water. Keep in mind now that we started with 1 kilowatt of power entering the motor. Thus, we have produced 0.885 water horsepower per kilowatt of input power.

Power is the rate of consuming energy. If power is multiplied by time, the result is the amount of energy consumed. Referring to Table 8.6, the Nebraska Pumping Plant Performance Criteria are expressed as an energy output power unit of energy input. If the water horsepower is multiplied by hours and the kilowatts by hours, the result is water horsepower hours and kilowatt hours, respectively. Thus, the Nebraska Pumping Plant Performance Criteria for electric powered pumping plants is 0.885 water horsepower hours per kilowatt hour.

Table 8.6. Nebraska pumping plant performance criteria (from Dorn et al., 1981).

Energy Source	whp-hr/Unit of Energy[a]	Energy Unit
Diesel	12.5	gallon
Propane	6.89	gallon
Natural gas	61.7	1,000 ft^3 (mcf)
Electricity	0.885	kW-hr
Gasoline	8.66	gallon

[a] whp-hr (water horsepower-hours)/unit of energy is the performance of the pumping plant as a complete unit—power unit, drive, and pump. The values are based on a field pump efficiency of 75% and natural gas energy content 925 btu/mcf.

The procedure that we just illustrated for developing the performance criteria for electric powered pumps was also followed for gasoline, diesel, natural gas, and propane. The only differences are how the units of energy are expressed and the fact that the drive efficiency of internal combustion engines is assumed to be 95%, because belt drives or right-angle gear drives are used.

The Nebraska Pumping Plant Performance Criteria was developed with what are considered to be reasonable design objectives. We would expect well-designed and well-maintained pumping plants to perform at the level indicated. However, most pumping plants do not operate at this criteria. An index, called performance rating, is used to evaluate the performance and is calculated by:

$$\text{Performace Rating} = \frac{\text{Actual Performance}}{\text{Performance Criteria}} \tag{8.20}$$

Example 8.7

Given the following conditions, determine the performance rating of the irrigation pumping plant.

L = 100 ft

P = 50 psi

Q = 800 gpm

Measured diesel fuel consumption: 4 gal/hr

Find: Performance rating

Solution:

$$TDH = 2.31\frac{\text{ft}}{\text{psi}}(50\text{ psi}) + 100\text{ft} = 216\text{ ft}$$

$$whp = \frac{(800\text{ gpm})(216\text{ ft})}{3960} = 44\text{ hp}$$

$$\text{Performance} = \frac{44\text{ whp}}{4\text{ gal/hr}} = 11\frac{\text{whp-hr}}{\text{gal}}$$

$$\text{Performance Criteria} = 12.5\frac{\text{whp-hr}}{\text{gal}} \quad \text{(From Table 8.6)}$$

$$\text{Performance Rating} = \frac{11\frac{\text{whp-hr}}{\text{gal}}}{12.5\frac{\text{whp-hr}}{\text{gal}}} = 0.88$$

The example illustrates how pumping plants can be evaluated. By measuring the lift, discharge pressure, flow rate, and energy consumption, the actual performance of a system can be determined. This actual performance can then be compared to the Nebraska Pumping Plant Performance Criteria.

To calculate the energy use rate per hour of an irrigation pump, use Equation 8.21.

$$\text{Energy/hr} = \frac{whp}{(PC)(PR)} \quad (8.21)$$

where: PC = Nebraska Pumping Plant Performance Criteria and
PR = performance rating.

Example 8.8

How much diesel fuel would be used per hour if a pumping plant is operating at 100% of the Nebraska Performance Criteria? Assume the same conditions as in Example 8.7.

If the pumping plant were operating at the criteria, the performance rating would be 1.

Solution:

$$\text{Energy/hr} = \frac{44\text{ whp}}{\left(12.5\frac{\text{whp-hr}}{\text{gal}}\right)(1.00)} = 3.52\text{ gal/hr}$$

So, the pumping plant evaluated in Example 8.7 is using approximately one-half of a gallon per hour more diesel than needed according to the Nebraska Pumping Plant Performance Criteria.

Another equation that can be useful for determining the energy consumed per unit volume of water pumped is

$$E = \frac{TDH}{(8.75)(PC)(PR)} \tag{8.22}$$

where: E = energy consumed per ac-in of water.

Example 8.9
For the same conditions given in Example 8.7, determine the energy required per acre-inch of water pumped.

Solution:

$$E = \frac{216 \text{ ft}}{(8.75)\left(12.5\,\frac{\text{whp-hr}}{\text{gal}}\right)(0.88)} = 2.24 \text{ gal/ac-in}$$

Example 8.9 shows that the diesel pumping plant that has a performance rating of 0.88 would consume about 2.24 gallons of diesel per acre-inch of water pumped. Now, what would the consumption rate be if the pumping plant performance were improved to 1? To find the answer, refer to Example 8.8. Example 8.8 shows that the pump would be consuming about 2 gallons of diesel per acre-inch if performing at the Nebraska Criteria. This is about 10% less energy per acre-inch than when it is performing at its current rating of 0.88.

Example 8.10
Determine the energy consumption per acre-inch for the pump in Example 8.7 if the performance rating can be improved to 1.

Given: $PR = 1$

Find: E

Solution:

$$E = \frac{216 \text{ ft}}{(8.75)\left(12.5\,\frac{\text{whp-hr}}{\text{gal}}\right)(1.00)} = 1.97 \text{ gal/ac-in}$$

Equation 8.22 can be used by the irrigation manager to evaluate the costs of applying a known volume of water versus the expected return from that water. The equation can also be used to estimate the performance rating of an irrigation pumping plant if the manager knows the total dynamic head, the volume of water that is pumped in a year, and the energy consumed in that year. By using this equation, the manager can decide whether or not improvements need to be made to the irrigation pumping plant to improve its efficiency. The techniques for measuring water volumes were discussed in Chapter 3. Obviously, to determine total dynamic head, both lift and pump discharge pressure must be known. Investing in and maintaining accurate pressure gauges on an irrigation system is a must not only for energy management but also for managing and assessing the functionality of the irrigation system itself. For example, do the flow rate and system pressure agree with the original design? A distinction must be made here between pump discharge pressure, which as the name implies is measured immediately at the pump discharge, and system pressure which is the water pressure actually going into the irrigation system or mainline. Example 8.11 illustrates how Equation 8.22 can be used to assess energy management alternatives.

Example 8.11

Suppose we have a sprinkler irrigation system that requires 800 gpm at 40 psi pressure. The pumping lift, L, is 143 feet. A vertical hollow-shaft electric motor powers the pump at 1770 rpm. A 5-stage 12 SKL pump (curve shown in Figure 8.11) is installed with 8.19-inch diameter impellers.

Given: Q = 800 gpm
Design discharge pressure = 40 psi
L = 143 feet at 800 gpm

Find: Does the pump match the pumping requirements, or is it oversized?
If it does not match, what would be the correct impeller diameter?
What effect would a throttling valve have on the energy consumption of this system?

Solution:

Design $TDH = 2.31 \times 40 \text{ psi} + 143 = 235 \text{ ft}$

Actual TDH produced = 5 stages × 57 ft/stage = 285 ft

So, the pump is oversized for this pumping condition. The correct impeller would deliver 47 feet per stage based on the following computation:

TDH/stage = 235/5 = 47 ft per stage

According to Figure 8.11, the correct diameter would have been 7.75 in.

Given the 8.19-in diameter impeller, a throttling valve would have to dissipate 50 ft of head, or 22 psi of pressure. Thus, using a throttling valve will result in a discharge pressure of 62 psi (between the pump and the valve) and 40 psi downstream of the valve. Assuming that the performance rating of this pump is 1, we can calculate how much extra energy is being consumed per ac-in of water:

At TDH = 235 ft, $$E = \frac{235}{8.75 \times 0.885 \times 1} = 30.3 \text{ kWh/ac-in}$$

At TDH = 285 ft, $$E = \frac{285}{8.75 \times 0.885 \times 1} = 36.8 \text{ kWh/ac-in}$$

Thus, the system should consume 30.3 kWh/ac-in with the proper impeller trim but instead, with the throttling valve, it is consuming 36.8 kWh/ac-in, 21% more energy than needed. This energy is being burned up or lost in the throttling valve.

In Example 8.11, the appropriate impeller diameter would have been 7.75 inches, but an 8.19-inch impeller was incorrectly installed. Once installed, it can be very expensive to make the impeller diameter change. A useful alternative would be to use a ***variable frequency drive*** (VFD) on the electric motor so that pump speed could be changed. The VFD is a motor controller which can be set to control the speed so that the desired pressure, 40 psi in our example, is maintained. In Example 8.11 the correct pump speed to obtain the 40 psi is 1670 rpm. This is based on application of the affinity laws discussed earlier. Variable frequency drives have many other useful applications in irrigation, especially where pumping conditions are not constant with time. A good example is variable rate irrigation, or a center pivot with a corner arm. A panel for a VFD motor controller is shown in Figure 8.16.

Figure 8.16. Panel for a variable frequency drive (VFD) electric motor controller.

Measuring pumping lift is probably the most difficult of all the measurements needed for evaluating the performance of the pumping plant. Permanent installation of air lines on the irrigation well can be a useful addition to the system. This method is discussed by USGS (2010). A detailed procedure for evaluating pumping plant performance is provided in Kranz et al. (2012b).

8.8 Summary

Transporting water for irrigation in pipelines requires energy. Water moving in pipelines obeys the basic laws of physics, conservation of energy and conservation of mass. The components of energy in the water are made up of kinetic, pressure, and elevation energy. While water is moving, the forms of energy can exchange with one another and some energy will be lost due to friction in pipelines and minor losses in pipeline fittings such as elbows, valves, etc. Pipes are sized based on economics and limiting water velocities within reasonable limits. The latter consideration was emphasized in this chapter. Pipe materials used in irrigation are dominated by aluminum, steel, and plastic. Pipelines must be protected from excessive pressures, vacuums, and air locks and from damage by frost. Pumps are used to add head or pressure to water. This requires energy which is usually supplied by electricity or fossil fuels (diesel, propane, natural gas, or gasoline). Proper selection, operation and maintanance of pumping systems is imperative for minimizing energy consumption in irrigation.

Questions

1. Explain the three energy forms in irrigation system hydraulics.
2. List three things that will increase pressure loss due to friction and explain why they impact this loss of energy.
3. Discuss the components of total dynamic head and how they are impacted by the setting of the irrigation site, such as water source, etc.
4. A water surface elevation of mountain reservoir (lake) is 240 ft above and irrigated valley. A pipeline conveys water from the reservoir to the irrigated valley and the pressure loss due to friction and minor losses in the pipeline is 40 psi. If at least 60 psi is needed to operate the sprinklers in the valley, will there be adequate pressure without using a pump?
5. A furrow irrigated field uses 10-in gated pipe and the well flow rate is 1000 gpm. The last irrigation set in the field starts 900 ft from the well and there is a 9-ft elevation rise or gain from the well to the last set.
   a. If the discharge pressure at the well is 10 psi, what is the water pressure at the beginning of the last set?
   b. In the last set there are 50 gates open and the spacing between open gates is 5 ft. There is an additional 2.5-ft elevation rise or gain from the start of the set and the last gate. What is the water pressure at the last gate?
   c. If you inserted a clear plastic tube in the last gate and held it vertical, how high would the water rise in the tube?
6. An irrigated field has a highly variable demand for water. On some days 750 gpm are required, and on other days 1300 gpm are required. The water is stored in a nearby reservoir and the elevation of the water surface in the reservoir is 34 ft lower than the elevation of the field. It was decided to connect two pumps in parallel to meet this variable demand. The pressure required in the irrigation system is 40 psi. The pumps selected were the 12 SKL pump shown in Figure 8.11 and they will be powered by electric motors with speeds of 1770 rpm.
   a. If pump number one is to deliver 750 gpm and the impeller diameter is 7.25 inches, how many pump stages will be required?
   b. What will be the efficiency of this pump?
   c. If electric motors come in nominal sizes of 10, 15, 20, 25, 30, 40, and 50 hp, what size motor would you select for this pump?

d. The second pump must deliver 550 gpm to the system. It was decided to use 3 stages of the 12 SKL for this pump. What diameter impeller would you recommend?

e. What will be the efficiency of this pump?

f. What motor size would you recommend for the second pump?

7. Water is flowing at 500 gpm in a 6-in inside diameter pipeline.

a. What is the velocity head of the water (ft of head)?

b. If the water flows through a 90° regular flanged elbow, what is the head loss in ft (minor loss) in the elbow?

c. What is the pressure loss in the elbow in psi?

8. A farmer kept records of diesel fuel consumption and water applications for a center pivot irrigated field. The field conditions are:

Q = 800 gpm

Volume of water applied for the year was 1300 ac-in.

Discharge pressure at the pump was 50 psi.

Pumping lift from the well is 150 ft.

Amount of diesel full consumed for the year was 3900 gal.

a. Estimate the performance rating of this farmer's irrigation pumping plant.

b. Based on what you learned in Chapter 3, how many hours did the pump operate in a season in Question 8?

c. How many gallons of diesel fuel were consumed per hour?

d. If the performance rating were improved to 1.0, how many gallons of fuel would be consumed in a year?

References

Colt Industries. (1979). *Hydraulic handbook* (11th ed.). Kansas City, KS: Colt Industries, Fairbanks Morse Pump Division.

Dorn, T. W., Fischbach, P. E., Eisenhauer, D. E., & Gilley, J. R. (1981). It pays to test your irrigation pumping plant. EC 81-713. University of Nebraska Ext.

Kranz, W. L., Martin, D. L., Patterson, D., Hudgins, J., van Donk, S., & Yonts, D. (2012a). Updating the Nebraska pumping plant performance criteria. *Proc. 22nd Annual Central Plains Irrigation Conf.* Colby, KS: Central Plains Irrigation Association.

Kranz, W. L., Werner, H. D., Go, A., & Grosskopf, K. R. (2012b). Irrigation energy audit manual. Training manual for the Nebraska Department of Labor (NDOL). University of Nebraska-Lincoln.

Martin, D. L., Kranz, W. L., Irmak, S., Rudnick, D. R., Burr, C., & Melvin, S. R. (2017). Pumping plant performance. *Proc. 29th Annual Central Plains Irrigation Conf.* Colby, KS: Central Plains Irrigation Association.

USGS. (2010). GWPD 13-Measuring water levels by use of an air line, Ver. 2010. USGS. Retrieved from https://pubs.usgs.gov/tm/1a1/pdf/GWPD13.pdf

USDA-NRCS. (2016). Sprinkler irrigation. Chapter 11 in *Part 623 National engineering handbook.* Washington, DC: USDA-NRCS.

Waller, P., & Yitayew, M. (2016). *Irrigation and drainage engineering.* Springer.

Chapter 9
Water Supply Systems

9.1 Introduction

When you consider installing an irrigation system, there are several water supply questions that must be answered. First, where is a suitable supply of water? More specifically, will the water come from a reservoir, a water course, or a well? Second, will water rights need to be addressed? The third question is about the quality of the water: is the water saline or will it be reclaimed water? Such questions are addressed in this chapter.

Although the most plentiful substance on the earth's surface, water is frequently not available in sufficient quantity when and where it is needed. To overcome these deficiencies, water resources are frequently stored and then conveyed from the time and place of natural occurrence to the time and place of beneficial use. The demand for reliable water supplies continues to increase as world population grows and becomes more affluent. Depending on the location and climate, a large portion of water withdrawals are used for irrigation (Figure 9.1). Where the water supply is inadequate, competition arises between agricultural and urban water users and other users. In many locations, agricultural interests developed the water supply initially, thereby acquiring rights to the water through prior use. As urban water use increases, municipalities frequently can afford to pay higher water costs and can achieve a greater economic return per unit of water than agriculture. One solution to this dilemma is the purchase of agricultural land by municipalities to acquire the rights to the water. Another alternative is to seek rights to water through the legislative and judicial systems. Yet another potentially more attractive alternative is for the nonagricultural water users to pay for improved irrigation delivery or management systems with the water from reduced irrigation consumption going to the municipalities and agriculture being paid for the water (see also section 5.10). While this chapter focuses on water quantity for irrigation, the water quality of a water supply system should also be assessed for any potential negative impacts on the crop and soil (Chapter 7 and Suarez, 2012).

When considering the development of a water supply, water is categorized as surface water or groundwater. Surface water originates from precipitation on the landscape moving downslope to streams and rivers. A portion of the water in streams and rivers is from overland runoff. The balance is baseflow, streamflow that comes from groundwater. When flows are always ample to satisfy water demands, surface waters can be withdrawn directly from the natural water course. The flow of many water courses, however, fluctuates too widely over time to satisfy water demands. For many rivers, peak water demands occur at times of minimal flow. This situation requires the construction of reservoirs to store high flows to be released later for beneficial uses. Reservoirs are normally created by constructing a dam across a stream or river. In special situations reservoir sites are located off-stream. Surface storage may range in size from huge multipurpose reservoirs to small ponds.

Most precipitation that infiltrates and deep percolates beyond the plant root zone eventually reaches the groundwater table, called groundwater recharge. Groundwater is water beneath the earth's surface that occurs in saturated materials. A zone of saturation in a substratum capable of yielding enough water to satisfy a particular demand is referred to as an aquifer. A major challenge facing water managers is ensuring that withdrawals of groundwater plus baseflow requirements do not exceed recharge. In many areas where groundwater is the major water supply, withdrawals exceed replenishment, and a sustainable water supply is in jeopardy.

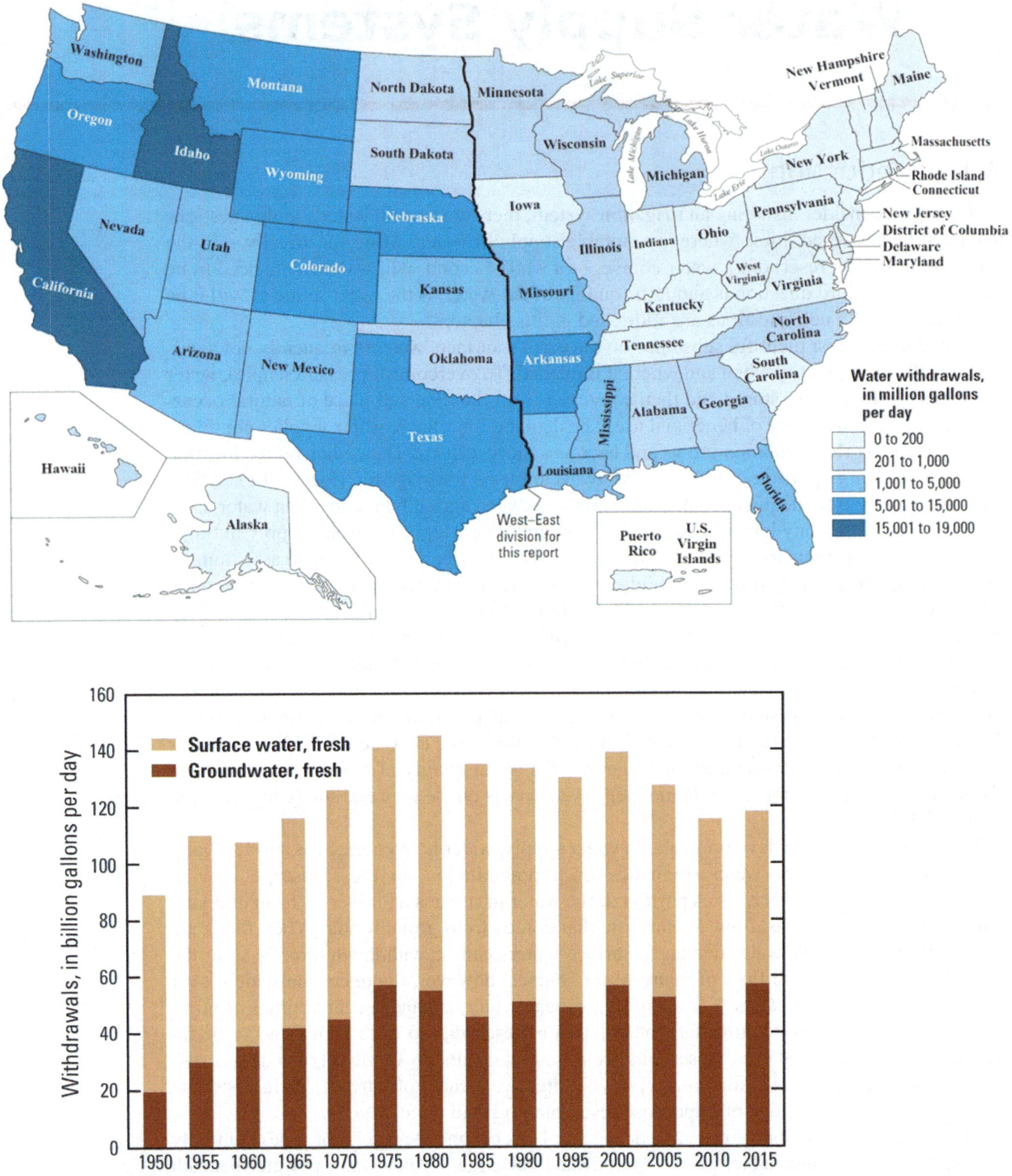

Figure 9.1. Irrigation withdrawals by state for 2015 (top) and irrigation withdrawals over time (bottom). (Both illustrations from Dieter et al., 2018.)

9.2 Water Rights and Laws

From a legal perspective in the U.S., water may be classified as diffused surface water, water in well-defined surface channels, water in well-defined aquifers, and underground percolating water. Diffused surface water is precipitation spread across the landscape. Diffused surface water and underground percolating water, because of their diverse nature, are normally regulated by common or civil law. In most states diffused surface water is considered the property of the landowner, who may use it without regard to the water supply of others. In many states, particularly those in the east, the law of diffused surface water has addressed who is responsible for the damage caused by diffused surface water. In some western states, diffused surface water is treated the same as water in well-defined channels.

9.2.1 Surface Water

In the U.S., the right to use surface water in natural watercourses is governed by two different doctrines: riparian and prior appropriation. In different states these doctrines are recognized either separately or in combination. In the future, adjudicated water rights based on highest-value use will become increasingly important.

The ***riparian doctrine*** (National Agricultural Law Center, 2020) recognizes the right of an owner of riparian land to make reasonable use of the stream's flow on the riparian land. Riparian land is contiguous to the stream or other body of surface water from which water is withdrawn. The right-of-land ownership includes the right of access to and use of the water. This right is not lost even if water is not used. Reasonable use of water generally implies that the landowner may use all the water needed for drinking, for household purposes, and for watering livestock. Where large herds of livestock are watered or where irrigation is practiced, the riparian owner is not permitted to exhaust the stream flow. Owners may only use their equitable share of the flow. This doctrine is used in many eastern states.

The ***doctrine of prior appropriation*** is based upon the priority of development and use. The first person to develop and put water to beneficial use has the right of continued use. The right of appropriation is generally acquired by filing a claim in accordance with state laws. If the use is beneficial, the appropriator has the right to all water required at the given time and place. This doctrine assumes that it is better to let individuals, prior in time, to take all the water rather than distribute inadequate amounts among all water users. Appropriated water rights are not limited to riparian land and may be lost by nonuse. This doctrine is recognized in most western states, although in some states it is in combination with the riparian doctrine. It is difficult to make generalizations because state laws on water rights differ on specific details and many change with time.

Today, almost all riparian states have moved towards allocating water through a permitting system. Using the same "reasonable use" criteria as common law, the states first determine whether a new use is reasonable. The permitting system allows the state to plan for and maximize water usage in the future. In many states, agricultural uses are exempt from permit requirements.

Some states, such as California and Oklahoma, have developed hybrid allocation systems. Hybrid systems combine aspects of both the riparian and the appropriation systems.

9.2.2 Groundwater

Most states in the U.S. have a different allocation system for groundwater than for surface water. Groundwater allocation systems often differentiate between on-tract and off-tract uses. On-tract use is where water is used on the tract where the well is located. Off-tract use is where water is transferred to another location.

Under the ***absolute dominion rule*** (National Agricultural Law Center, 2020 and Driscoll, 1986), a landowner may use as much groundwater as possible. The impact of the groundwater use on neighboring users is not taken into account. Although some states follow this doctrine with allowances for remedies for willful injury, most states have rejected this doctrine as malicious withdraws of water.

The ***correlative rights doctrine*** distributes water on an equitable basis among landowners and allows off-tract uses, although these are subordinate to on-tract uses. With the Correlative Rights Doctrine the landowners overlying the same aquifer are limited to a reasonable share of the aquifer supply.

Some of the western U.S. states apply the ***doctrine of prior appropriation*** (similar to its application to surface water), which gives the earlier water users priority over later users. The water use amount is limited to beneficial uses.

Another legal approach is to apply the ***rule of reasonable use.*** As the name implies the landowners have the right to use the groundwater beneath their land as long as it is deemed to be a reasonable and beneficial use.

In the U.S. these legal approaches, along with others, for groundwater use are applied on a state-to-state basis. For example, the State of Nebraska uses a unique blend of the Rule of Reasonable Use and the Correlative Rights Doctrine along with statutory preferences for use (Aiken, 1980).

9.3 Aquifers

Geologically, the loose and discontinuous layers of decayed rock debris overlying solid bedrock are termed regolith. Soil, where chemical and physical weathering are the most active, is the uppermost part of the regolith. The regolith is a potential storage medium for water. Above bedrock, which is essentially impermeable to water, the rock is fractured and frequently consists of gravels, sands, and soil particles. As illustrated in Figure 9.2 water can be contained in the pores (interstices) of soil, sand, gravel, and rock (Meinzer, 1923). The substrata containing interstitial water is divided into the unsaturated zone and the zone of saturation. Groundwater that can be successfully extracted for a water supply only occurs in the saturated zone. The boundary between the unsaturated and saturated zones is called the water table. The water table may be at or above the soil surface as in swamps, wetlands, and near lakes and continuously flowing streams.

Permeability or hydraulic conductivity is a measure of the ability of an aquifer to transmit water. The total porosity and permeability of an aquifer depends upon the size and shape of the pores. Table 9.1 provides approximate values for total porosity and relative permeability. The specific yield of an aquifer, the portion of the stored water that can be withdrawn for a water supply, is also given in Table 9.1. Except for clay, porosity is generally a good indicator of the amount that can be withdrawn from an aquifer. Clay, although high in porosity, has a low permeability that limits water flow (Table 9.1). Usually, sand, gravel, and fractured rock are good water-bearing deposits that can be developed as a water supply. Table 9.2 summarizes the ranges of hydraulic conductivity for various aquifer materials.

In some geologic formations, groundwater may be confined under pressure between two impervious layers. This "confined" aquifer may create what is termed an artisan condition. For artesian flow to be possible there must be a pervious stratum that is continuous from a region upslope where water can percolate into the aquifer to a downslope region where the aquifer is confined between upper and lower impervious layers. When a well is installed through the upper impervious layer into the confined aquifer, water will rise up the well to a level depending on the hydrostatic pressure on the aquifer at the well location. If the pressure is high enough, water will flow out of the well under this artesian condition. The more normal condition for both confined and unconfined aquifers is that groundwater must be pumped from the well. Figure 9.3 illustrates the geological conditions that foster these various sources of groundwater. For a more detailed presentation of groundwater, the reader is referred to Freeze and Cherry (1979), Todd (1980), and Sterrett (2007).

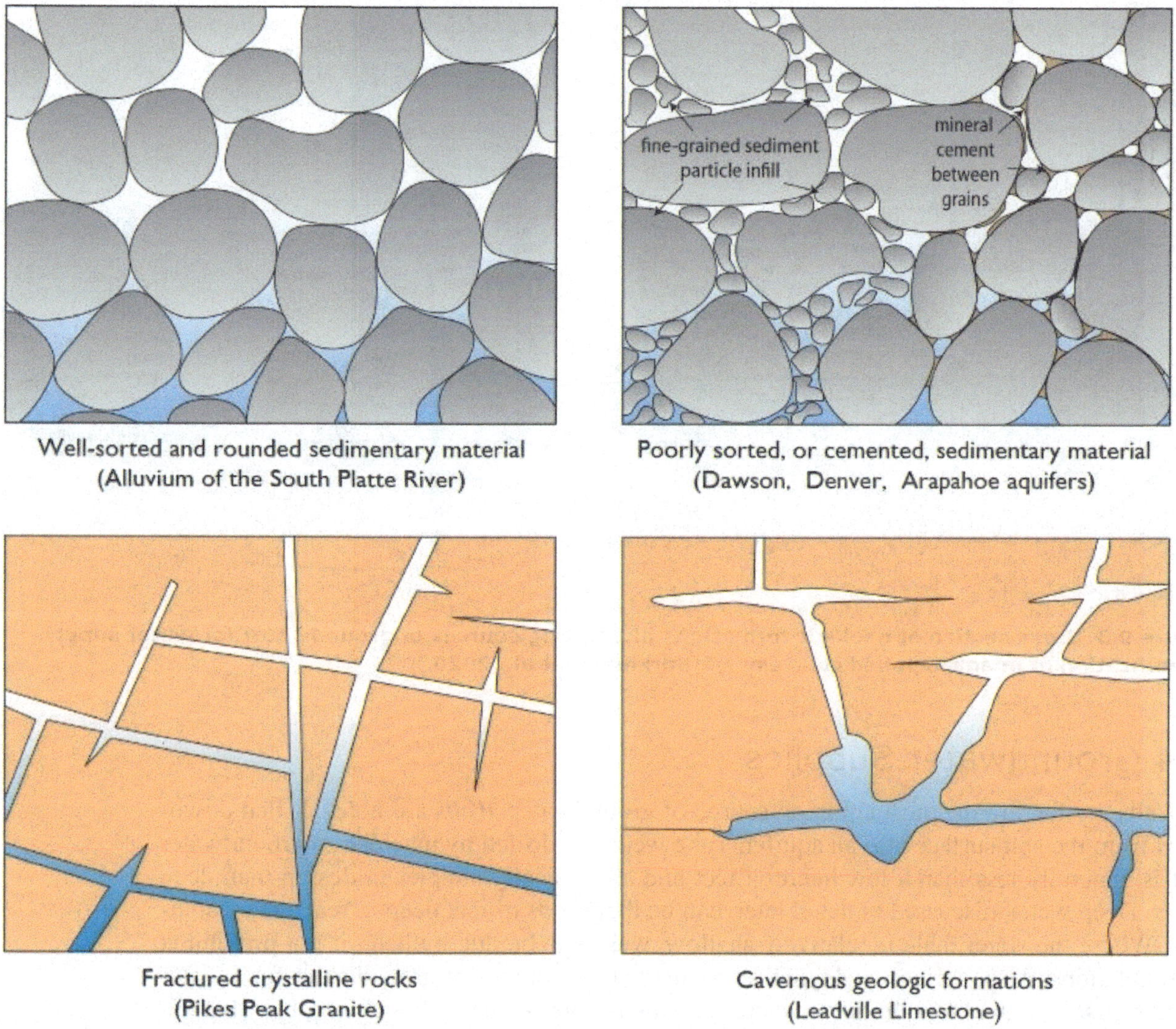

Figure 9.2. Several types of interstices found in substrata that can store groundwater: well-sorted sedimentary deposit having high porosity (top left), poorly-sorted sedimentary deposit having low porosity (top right), rock rendered porous by fracturing (bottom left), and rock rendered porous by dissolution (bottom right). (Image courtesy of Barkmann et al., 2020.)

Table 9.1. Approximate characteristics of groundwater aquifers (adapted from Schwab et al., 1992).

Aquifer Material	Total Porosity (%)	Specific Yield (%)	Relative Permeability
Dense limestone	5	2	1
Dense shale	5	2	1
Sandstone	15	8	700
Gravel	25	22	5,000
Sand	35	25	800
Clay	45	3	1

Table 9.2. Range of hydraulic conductivity values for various types of aquifers (adapted from Driscoll, 1986).

Aquifer Formation or Material	Range of Hydraulic Conductivity (ft/d)	
	Low	High
Fine to coarse gravel	10^{1}	10^{4}
Fine to coarse sand	10^{-3}	10^{2}
Silt and loess	10^{-4}	10^{0}
Glacial till	10^{-8}	10^{0}
Karst limestone	10^{-2}	10^{2}
Shale	10^{-9}	10^{-5}
Sandstone, well cemented, unjointed	10^{-6}	10^{-4}
Sandstone, friable	10^{-4}	10^{-1}
Unfractured igneous and metamorphic rocks	10^{-9}	10^{-6}

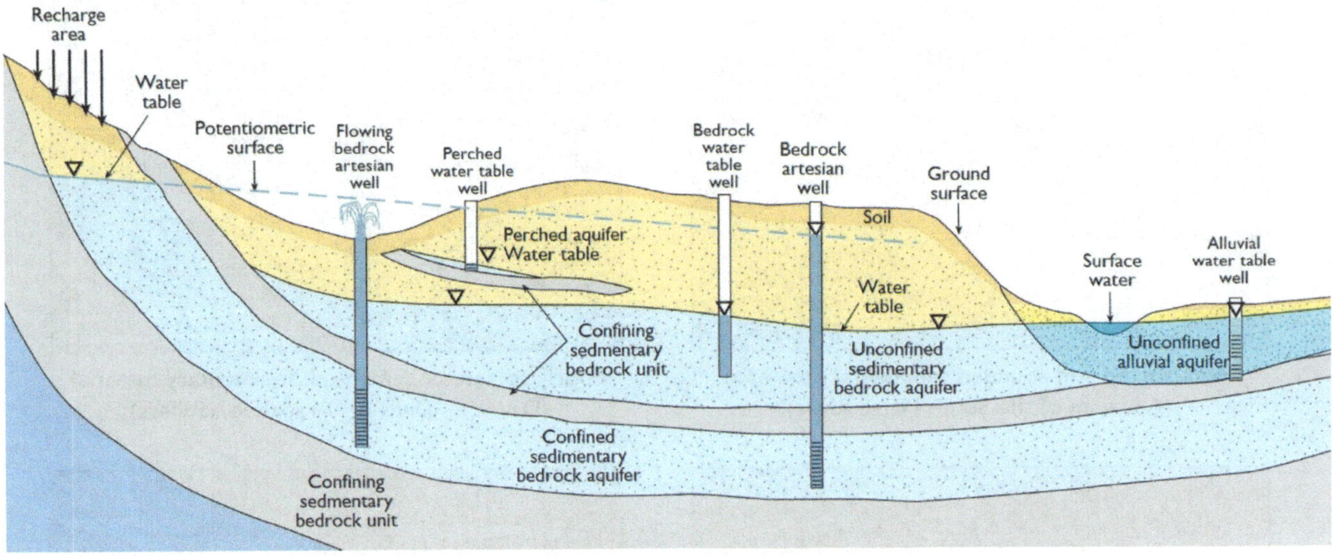

Figure 9.3. Cross section of geologic formations illustrating sources of groundwater for water supply. (Modification of image supplied courtesy of Barkmann et al., 2020.)

9.4 Groundwater Supplies

Wells, by far, are the most common source of groundwater. ***Wells*** are holes drilled downward from the soil surface into an aquifer. Tube wells are drilled by machine to groundwater that is generally less than a few hundred feet and are typically simpler in design than deep wells. Deep wells, discussed in detail later, can be thousands of feet deep to reach deep aquifers. Where the water table is relatively shallow, wells can be dug by hand if not prohibited by regulations. Also, springs or dugout reservoirs can be sources of groundwater for a water supply in some situations. In areas where the water table is only a few feet below the soil surface, dugout reservoirs or open pits can provide access to groundwater. Occasionally, these water sources can be developed into a suitable water supply, if allowed by national and local regulations.

9.4.1 Shallow Wells

Globally, many rivers and streams have flood plains that are irrigated. In many instances, the water table is only a few feet below the level of the water course. In these areas, ***shallow wells*** are dug by hand or machine into the groundwater. These wells are typically less than 50 feet deep and less than 10 feet in diameter. Stone, brick, or other materials may be used to stabilize the walls.

Since power requirements for pumping water are a function of total dynamic head and flow rate (Chapter 8), low-flow wells in shallow aquifers have a much lower power requirement than for wells with high pumping lifts and wells with high flow rates. For small-scale irrigation systems using water from a shallow aquifer, a greater variety of options are available for lifting the water, including treadle pumps, pedal pumps, and water wheels (Figure 9.4). Along with small fuel-powered pumps, small electric pumps powered by a solar panel (solar pumps) are used more often for pumping systems with low power requirements than high-power pumping systems (Figure 9.5).

Figure 9.4. Systems for accessing water from shallow aquifers: (a) treadle pump (photo courtesy of iDE, International Development Enterprises); (b) pedal pump (photo courtesy of Maya Pedal Guatemala), and (c) a water wheel in Rajasthan, India (photo courtesy of Carl Anders).

Figure 9.5. Water delivery system for groundwater from the shallow alluvial aquifer of the Shashe River, Zimbabwe. A solar pump is used to deliver water from the concrete-lined storage pond to a nearby field, and the net provides shade for fish. (Photo courtesy of Annelieke Duker, IHE Delft Institute for Water Education.)

9.4.2 Tube or Cased Wells

When the water table is relatively shallow, on the order of a few hundred feet, ***tube wells*** are used frequently. Tube wells consist of a casing with screen or perforations near the bottom of the well. The casing, normally made of steel, PVC plastic, concrete, or fiberglass, is installed during or after the drilling process to stabilize the hole and allow water, but not subterranean particles, to move into the well. The lower portion of the casing is perforated, slotted, or screened with openings sized to minimize the entry of subterranean particles from the aquifer medium into the well. Refer to section 9.4.4 for a discussion on screen type and size of openings in the screen or casing.

9.4.3 Deep Wells and Well Hydraulics

For ***deep wells***, the casing and screen diameter can range from a few inches to a few feet and can range in depth from less than 50 feet to more than several thousand feet.A cross section of a well installed in homogeneous material overlying an impervious rock formation is shown in Figure 9.6. Under static conditions when the well is not being pumped, the water level in the well will rise to the static water table position (Figure 9.7). When pumping begins, the water level in the well is lowered and water from the surrounding material flows into the well. The water table around the well is lowered to the general form of an inverted cone. The vertical distance from the static water table to the water level at the well is known as the drawdown. If pumping continues at a constant rate, the shape of the water table surrounding a well will become nearly stable. The horizontal distance from the well to where the water table is not noticeably lowered by drawdown is known as the radius of influence.

There is a definite relationship between drawdown and discharge from a well. Typical relationships are shown in Figure 9.8. For thick aquifers or artesian formations, the relationship is nearly a straight line. As the aquifer becomes thinner, less discharge occurs for the same drawdown as in a thick aquifer.

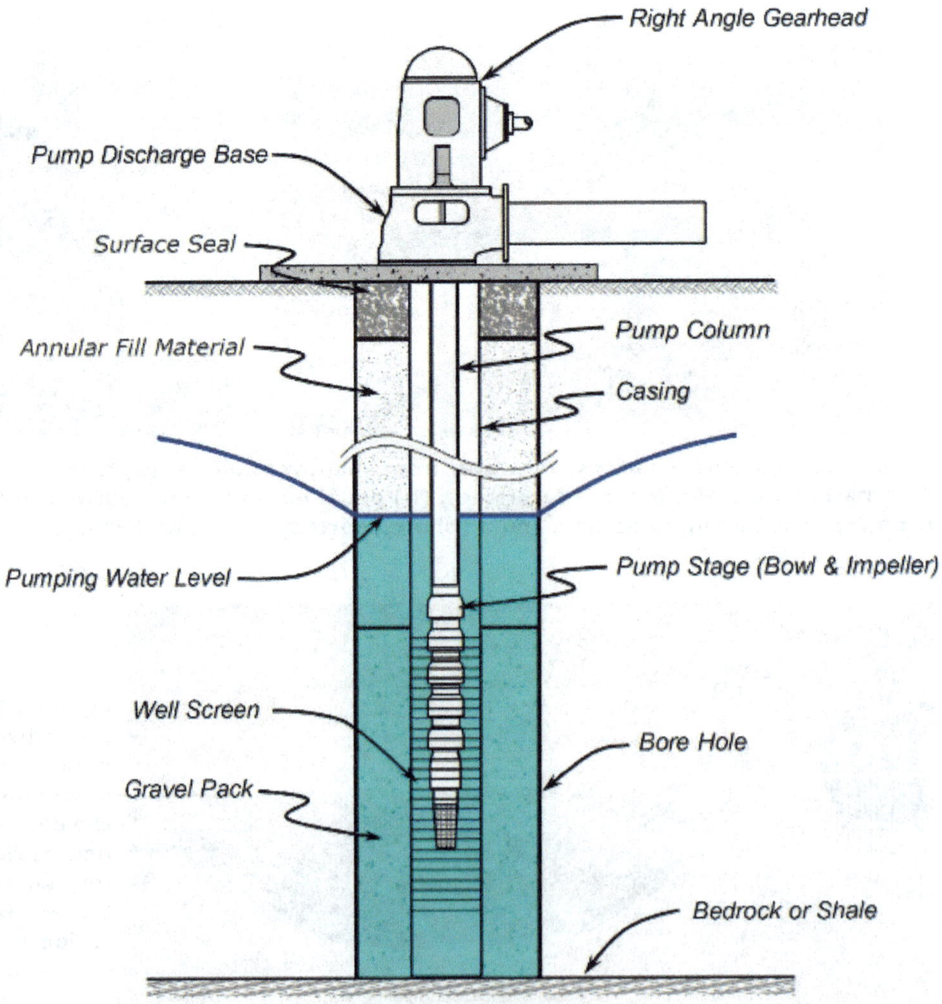

Figure 9.6. Well constructed in a sand and gravel formation. A casing and screen are always used. A gravel pack is optional.

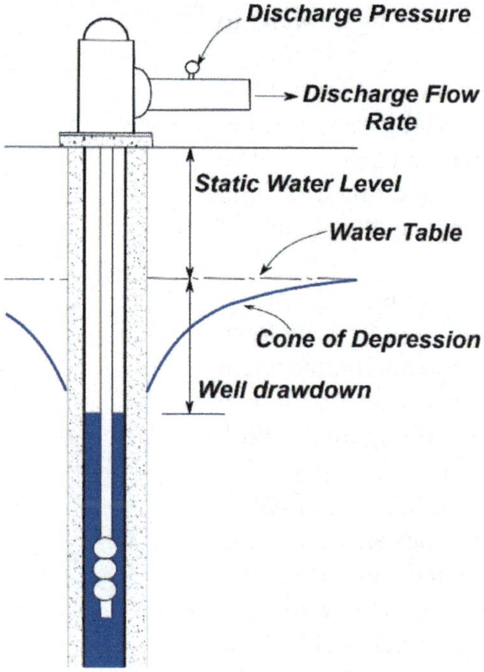

Figure 9.7. Well hydraulics including static water level and drawdown.

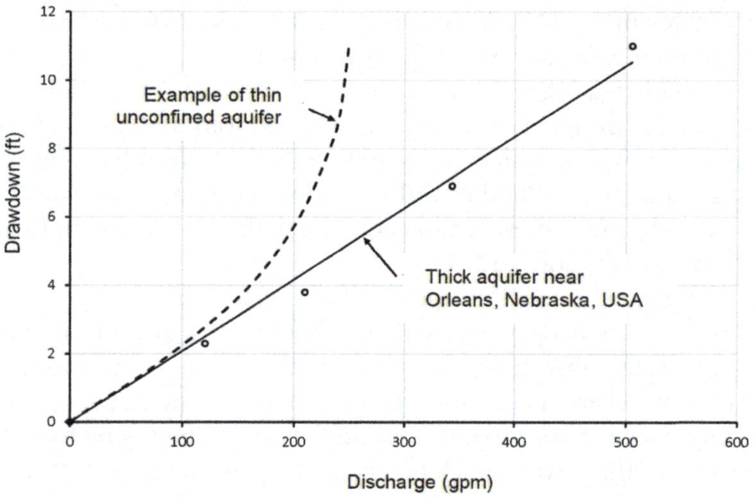

Figure 9.8. Typical relationships between drawdown and discharge of wells.

Drawdown (s) is the difference between static water level (SWL) and the pumping water level in the well (PWL) and is calculated as:

$$s = PWL - SWL \tag{9.1}$$

When a well functions like the straight line in Figure 9.8, the specific capacity, SC, is constant and is calculated as:

$$SC = \frac{Q}{s} \tag{9.2}$$

where Q is discharge in gallons per minute (gpm) and s is drawdown in feet. Specific capacity is a useful term when predicting drawdown in a well for a given discharge because:

$$s = \frac{Q}{SC} \tag{9.3}$$

See Example 9.1 for application of Equations 9.1, 9.2, and 9.3.

Example 9.1

Given: Well and water table illustrated in Figure 9.7

SWL = 100 ft

PWL = 120 ft

Q = 800 gpm

Find: At 600 and 900 gallons per minute:

s

SC

PWL

Solution:

$$SC = \frac{Q}{s} = \frac{800\text{gpm}}{20\text{ ft}} = 40\text{ gpm/ft}$$

For 600 gpm:

$$s = \frac{Q}{SC} = \frac{600\text{ gpm}}{40\text{ gpm/ft}} = 15\text{ ft}$$

$$PWL = 100 + 15 = 115\text{ ft}$$

For 900 gpm:

$$s = \frac{Q}{SC} = \frac{900\text{ gpm}}{40\text{ gpm/ft}} = 22.5\text{ ft}$$

$$PWL = 100 + 22.5 = 122.5\text{ ft}$$

9.4.4 Well Construction

In constructing and developing a successful well, several features require careful consideration. These features include the method of drilling, well alignment, depth of well, casing material, casing perforations, gravel packing, well development, and well testing. Driscoll (1986) presents significant detail on well design and specifications, as well as well drilling methods. The specifications for wells and the certification of well drillers is often regulated by states in the U.S. Also, it is not uncommon for states to regulate the spacing of wells to prevent interference with neighboring wells.

Most irrigation wells are drilled with cable or rotary tools. With cable tools, a heavy bit is repeatedly dropped onto material at the bottom of the well. Crushed material is removed periodically with a bailer. Wells up to 5,000 feet deep have been drilled with a cable tool. The most common method of drilling, however, is by rotary tools. A bit is rotated by a drilling pipe and a mud slurry is pumped through the pipe to the bit to carry cuttings up the outside of the pipe to the soil surface.

An irrigation well should normally penetrate the water-bearing formations as deeply as possible. The deeper well will usually provide a greater yield of water per foot of drawdown of the water table. It is imperative that the well is vertical for the installation and replacement of pumps. Particularly for wells that have a gravel pack, misalignment because of gravel wedging somewhere below the surface causes the casing to be pushed out of alignment. With proper drilling, well development, and maintenance, a well should last several decades.

Casing materials for wells include steel, wrought iron, concrete, plastic, and fiberglass. Many states have specifications for casing depending on the type of well. Well casings are perforated, slotted, or screened near the bottom of the well to facilitate entering groundwater. Properly-sized perforations prevent subterranean particles from flowing into the well with the water.

Wells drilled in unconsolidated material with rotary drills are usually gravel packed. The selection of the right gravel material is crucial to prevent particles from moving into the well. The gravel needs to be large enough to permit sufficient water flow but with small size pore spaces to prevent solids from moving. The gravel packing material cannot be more than five times the average size of the substrata material if the well is to be stabilized. With these size restrictions, many times a specially manufactured well screen is used rather than the gravel pack. Screens are constructed of brass, bronze, galvanized steel, stainless steel, plastic, or fiberglass to resist corrosion.

Immediately after a well is constructed, it is normally "developed." The purpose of well development is to make the well sand-free and maximize the flow of water from the aquifer. To prevent pump damage, materials like clay, drilling mud, silt, and sand are removed from the vicinity of the well casing that is screened. There are several methods to develop a well. They are all designed to loosen fine particles so they can be pumped from the well before the permanent pump is installed. The pump used during well development is designed specially to be tolerant of fine particles. The most common method to develop a well is surging. The pump is turned on and then off to allow water to surge back into the well thereby drawing fine materials into the well to be removed by pumping during a repeat of the surge cycle. Another surging technique uses a surge block. The surge block is a tool fastened to the end of the drill. As the drill is moved up and down, it produces a pumping action to draw fine particles into the well.

After the well is developed, a temporary pump is installed for a pumping test. During the pumping test the flow rate (discharge) and drawdown are measured simultaneously. This information is required to select the proper size of pump. Sterrett (2007) provides a practical reference for planning and installation of water wells.

9.5 Surface Water Supplies

To the irrigator, there is great value and need for a dependable water supply that is flexible with respect to the frequency of available water, the rate of water delivery, and how long the water is available. These expectations for a surface water supply are more easily accomplished by pressurized delivery systems than by open channels. Nevertheless, the predominant means of delivering irrigation supplies from irrigation projects is by open channel. Pressurized delivery systems include pipelines which may vary from being underground and permanent to portable, temporary pipe on the soil surface.

9.5.1 Open Canals

Conveyance canals or ditches are frequently used to deliver water from surface storage or wells. A system of open canals often distributes water great distances from its source to the field. Figure 9.9 shows a large open canal and smaller lateral canals.

Figure 9.9. Canal delivery system in western Nebraska, which delivers water from Seminoe Dam (inset).

Losses of water by seepage from canals can be a major concern. Water seeping out the bottom of the canal is especially high in earthen canals at the beginning of an irrigation season when soil intake rates are high. Figure 5.8 gives approximations of canal seepage losses depending upon soil texture for unlined canals. Proper soil compaction at optimum moisture content can almost eliminate seepage in some soils.

Irrigation canals are sometimes lined to minimize seepage losses. In addition to reducing seepage, canals are lined to ensure against interrupted operation resulting from channel failure; to provide a more efficient cross section by increasing sideslopes, by reducing the roughness coefficient, by eliminating vegetative growth, and by reducing maintenance. Canals can be lined with a variety of materials. The most common lining material by far is concrete, but other materials include brick, rock masonry, asphalt, soil cement, rubber, colloid clay, and plastic. Concrete meets all the requirements for a lining better than any other material. Its principal disadvantages are high initial cost and possible damage from soil swelling and shrinking, soil chemicals, and freezing and thawing. Concrete can be applied in a variety of ways but continuous pouring with slip-form equipment is the most common.

The purpose of irrigation delivery systems is to provide water to the field in a timely and reliable manner (Figure 9.10). To improve reliability and increase flexibility frequently requires some type of automation of the delivery system. Water is delivered by one of three possible scheduling techniques: demand, arranged, and rotation. A "demand" schedule allows for complete flexibility on the frequency, rate, and duration of water delivery. A common example is a municipal water system; the user can open the faucet at any time (flexibility in frequency), receive a low or high flow rate (flexibility in flow rate), and take the water as long as desired (flexibility in duration). An "arranged" schedule requires the user to request the rate and duration of a water delivery in advance. The advance notice required to receive and to turn the water off is typically one to two days. Arranged schedules often require that the water be turned on or off at a specific time of the day. In a "rotation" schedule, all flow entering a small canal is delivered to only one field. The length of time water is delivered to a field depends upon its size. After delivering water for the prescribed period to one field the flow is shifted to the next.

In addition to the reliability that water is delivered when and as promised, there are two other aspects. One aspect is that the flow remain at the prescribed rate; the second is that flows and water levels

Figure 9.10. On-farm ditch providing water to small fields near Delhi, India (top), lateral canal with a weir to provide sufficient head (water surface elevation) for siphon tubes in Nebraska. (Bottom photo courtesy of Steve Melvin, Nebraska Extension.)

in the canal are controlled so that canal structures and soil banks are not damaged.

Many water delivery systems are now automated and there are many types of automated systems. Methods of automatic control differ based upon the control of flow rate or water level in the canal, the control based on measures at the upper end or the lower reaches of the delivery system, and the control being local or remote. More information on developing surface water supply systems for irrigation, including small earth dams, is presented in Huffman et al. (Huffman et al., 2013).

9.5.2 Pressurized Delivery Systems

Pipelines are used extensively to deliver water, especially when the capacity required is low enough for standard pipe sizes or the advantage of a closed delivery system outweigh those of a canal system (Figure 9.11). There are pipeline delivery systems where the pipe is 10 feet or more in diameter. Some of the advantages of buried pipelines include: few problems from damage caused by animals; no vegetative problems; land over the pipeline can be utilized; buried pipelines do not obstruct cross traffic; pipelines do not have to follow elevation contour lines; lower maintenance costs; less hardware required for controlling flows; and less threat of drownings.

Figure 9.11. Installation of a buried pipeline for irrigation water delivery.

Disadvantages of pipelines compared to canals include: initial cost may be higher than canals; and pipelines may plug from sediment or debris more easily.

Pipelines for water delivery systems are increasing in popularity. The conversion is especially rapid in expanding urban areas. Some irrigation districts use monolithic (cast-in-place) concrete pipe for low-pressure conditions. Reinforced concrete pipe are being used uphill and downhill from a supply canal. The uphill pipelines are supplied by pumps while the downhill laterals are normally gravity fed.

In some locations, the downhill laterals have sufficient slope and length to develop the pressure required to operate sprinkler systems without booster pumps. Many pipelines operate with a pressure head of 2 feet or less and lead directly to surface irrigation systems or booster pumps to provide the head for sprinkler or microirrigation systems.

9.6 Surface Water-Groundwater Interaction

It is recognized that surface water and groundwater are connected, to the point that they have been referred to as a single resource (Winter et al., 1998). For example, many streams are gaining streams (gaining water from the adjacent aquifer), while some streams are losing streams. Water resources managers need to account for these interactions when planning at a watershed or basin scale. A specific application for irrigation is the impact that a groundwater well can have on a nearby stream (Figure 9.12), which is called stream depletion (Barlow and Leake, 2012). In this case, the well is pumping water stored in the aquifer which would

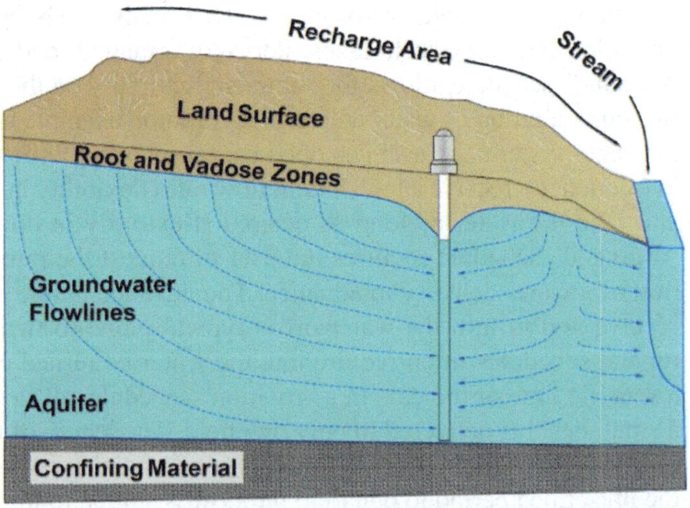

Figure 9.12. Illustration of stream depletion, when a pumping well intercepts groundwater that would have flowed into the stream as baseflow (modified from Winter et al., 1998).

have flowed to the stream; however, in some cases, a portion of the pumped water can actually come from the stream water, depending on the pumping rate, the length of time of pumping, and how close the well is to the stream.

9.7 Reclaimed Water Supplies

Reclaimed water includes raw and treated sewage water from industries and municipalities (wastewater), runoff from the low end of surface irrigated fields (generally called tail water), and water from subsurface drainage systems. Concerns from tail water are fertilizers, pesticides, and suspended soil particles. From drainage water, the concerns are nutrients, chemicals, and salts. The highest concerns, however, are use of sewage water. The concerns focus on the potential risks of disease from bacteria, virus, and pathogens (Pachepsky et al., 2011). Apart from being a health risk, it is also an environmental issue. If sewage water used for irrigation enters surface waters, such as lakes and rivers, it can contaminate these waters and harm ecosystems. Soils do filter a large amount of pollutants from wastewater. Studies indicate up to 90% of pollutants can be removed but the filtered water may still contain bacteria and viruses.

Developing countries report much higher levels of pathogens in irrigation waters than developed countries (Thurston-Enriquez et al., 2002). For example, wastewater irrigation provides a quarter of all vegetables produced in Pakistan. Globally, an area the size of Germany is primarily irrigated with human sewage water. This translates into a health risk for all people who consume the foods grown on these irrigated lands.

Although standards for the use of reclaimed wastewater exist for food crops eaten raw in the United States, using reclaimed wastewater to irrigate food crops is seldom practiced. In developing countries, raw or partially treated wastewater is often used to grow food crops. Throughout the world wastewater use has become significant and this has encouraged many countries to develop regulations to control water quality to reduce health and environmental risks.

Wastewater use will become more and more attractive for irrigation, given the current and future problems of water scarcity for irrigation. The amount of collected and treated wastewater is sure to increase significantly with population growth, rapid urbanization, and improvement in sanitation service. More information on using reclaimed water for irrigation can be found in Waller and Yitayew (2016).

9.8 Summary

Water is supplied for irrigation from both surface and groundwater. The right to use water for irrigation varies among states and countries, and irrigators should check on which laws apply in their area. Groundwater is extracted by wells that vary from a few feet to thousands of feet. The complexity of the well design depends upon its depth. Surface waters are conveyed to fields by a series of open canals or buried pipelines. Reclaimed water will become a larger source of irrigation water in the future as the demand for fresh water increases.

Questions

1. Prepare a table that summarizes the attributes of the two common doctrines of surface water rights with respect to:

   a. How a water right is acquired,

   b. Quantity of water that can be used,

   c. Types of water use allowed,

   d. How or if a water right can be lost, and

e. Where water acquired through these two doctrines may be used.

2. Prepare a diagram of the regolith. Show the vadose zone, phreatic zone, crop root zone, capillary fringe, and the water table.
3. What method of drilling wells is most common in your area?
4. What discharge rates for irrigation wells are typical in your area? What discharge rates are common for domestic wells?
5. If unlined canals can have significant losses of water because of seepage, why are earthen dams effective?
6. In the city nearest to where you live, is water delivered by canal or pipeline? Are farms near you irrigated from canal or pipeline systems or are on-site wells used?
7. A new irrigation well was drilled to a depth of 300 feet. The static water level was 80 feet. The well was test pumped at 1200 gpm and the pumping water level was 140 feet. The planned irrigation system will have a flow rate of 900 gpm. Determine the expected draw-down and pumping water level at 900 gpm.

References

Aiken, J. D. (1980). Nebraska ground water law and administration. *Nebraska Law Review, 59*, 917-1000.

Barkmann, P. E., Broes, L. D., Palkovic, M. J., Hopkins, J. C., Bird, K. S., Sebol, L. A., & Fitzgerald, F. S. (2020). ON-010 Colorado Groundwater Atlas. Geohydrology. Golden, CO: Colorado Geological Survey. Retrieved from https://coloradogeologicalsurvey.org/water/colorado-groundwater-atlas/

Barlow, P. M., & Leake, S. A. (2012). Streamflow depletion by wells-understanding and managing the effects of groundwater pumping on streamflow. Circular 1376. Reston, VA: USGS.

Dieter, C. A., Maupin, M. A., Caldwell, R. R., Harris, M. A., Ivahnenko, T. I., Lovelace, J. K., Barber, N. L., & Linsey, K. S. (2018). Estimated use of water in the United States in 2015. Circular 1441. Reston, VA: USGS.

Driscoll, F. G. (1986). *Groundwater and wells* (2nd ed.). St. Paul, MN: Johnson Division.

Freeze, R. A., & Cherry, J. A. (1979). *Groundwater.* Englewood Cliffs, NJ: Prentice-Hall.

Huffman, R. L., Fangmeier, D. D., Elliot, W. J., & Workman, S. R. (2013). *Soil and water conservation engineering* (7th ed.). St. Joseph, MI: ASABE.

Meinzer, O. E. (1923). The occurrence of ground water in the United States. Water Supply Paper 489. Reston, VA: USGS.

National Agricultural Law Center. (2020). Water Law: An overview. University of Arkansas Division of Agriculture. Retrieved from https://nationalaglawcenter.org/overview/water-law/

Pachepsky, Y., Shelton, D. R., McLain, J. E., Patel, J., & Mandrell, R. E. (2011). Irrigation waters as a source of pathogenic microorganisms in produce: A review. In D. L. Sparks (Ed.), *Advances in agronomy* (Vol. 113).

Schwab, G. O., Fangmeier, D. D., Elliot, W. J., & Frevert, R. K. (1992). *Soil and water conservation engineering* (4th ed.). New York, NY: John Wiley & Sons.

Sterrett, R. J. (2007). *Groundwater and wells* (3rd ed.). St. Paul, MN: Johnson Division.

Suarez, D. L. (2012). Irrigation water quality assessments. In W. W. Wallender & K. K. Tanji (Eds.), *Agricultural salinity assessment and management. ASCE manuals and reports on engineering practice No. 71.* Reston, VA: ASCE.

Thurston-Enriquez, J. A., Watt, P., Dowd, S. E., Enriquez, R., Pepper, I. L., & Gerba, C. P. (2002). Detection of protozoan parasites and microsporidia in irrigation waters used for crop production. *J. Food Prot., 65*(2), 378-382.

Todd, D. K. (1980). *Groundwater hydrology* (2nd ed.). New York, NY: John Wiley & Sons.

Waller, P., & Yitayew, M. (2016). *Irrigation and drainage engineering.* Springer.

Winter, T. C., Harvey, J. W., Franke, O. L., & Alley, W. M. (1998). Ground water and surface water: A single resource. Circular 1139. Reston, VA: USGS.

Chapter 10
Surface Irrigation

10.1 Introduction

Surface irrigation is the oldest irrigation application method in the world. In fact, according to Price and Purcell (2011), the practice was used as early as 6000 years ago in Mesopotamia. Approximately 84% of the world's irrigation (FAO, 2021) and 35–45% of the U.S. irrigation uses the surface method (FAO, 2021 and USDA, 2019). Surface irrigation includes border, furrow, and basin irrigation (Figure 10.1). Surface irrigation requires less pressure than does sprinkler or microsystems. In addition, worldwide, water is commonly furnished to the surface irrigated field using only gravity to deliver and distribute the water; pumping is not required. In the U.S. Midwest, much of the water for surface irrigation is pumped from groundwater and the primary energy cost using surface irrigation is due to lifting the water to the soil surface. If the topography of the land is such that surface irrigation is possible with only moderate leveling, surface irrigation may be less expensive than other methods.

(a) (b) (c) (d)

Figure 10.1. (a) Gated pipe furrow irrigation, (b) large-scale basin irrigation, (c) small-scale basin irrigation, and (d) border irrigation (photo d courtesy of Jan Feyen, KU Leuven, Belgium).

Figure 10.2. Water application in surface irrigation: (a) gated pipe, (b) siphons.

Figure 10.3. (a) Land grading and (b) planing in preparation for surface irrigation.

In all surface irrigation systems, water is applied at the inlet end and the water then flows to the downstream end. A portion of the water infiltrates as it advances across the field. Water is usually applied through gated pipes, siphons, or gates as shown in Figure 10.2.

Surface irrigation can be an efficient application method if the soils and fields are well suited to this method. But, it can be very inefficient if the soils and other factors are not properly considered when developing and managing the system. The soil infiltration rate is especially critical in the efficient operation of surface irrigation systems. If the infiltration rate of the soil is too high, the depth of water that infiltrates near the inlet will be much larger than that at the last point to receive water, the downstream end. The land slope and its uniformity also have a major impact on surface irrigation. Slopes that are too steep cause excess runoff and erosion. Acceptable slopes are usually less than 2%. The uniformity of the slope is also critical so that water does not accumulate in depressions on the surface.

Figure 10.4. Forming furrows for furrow irrigation.

Surface irrigation requires land preparation such as grading and leveling (Figure 10.3). With furrow irrigation, furrow forming or bedding is also required (Figure 10.4).

If a surface-irrigated field is too long, or the inlet flow is too small, a long period of time may be required for water to reach the downstream end of the field. This usually results in nonuniform distribution of water and excessive deep percolation (Figure 10.5). Runoff of water from the downstream end of a field can be one of the largest losses of water for surface systems. Relatively uniform distribution of water in furrow irrigation may require that 20 to 30% of the applied water runs off the field. If this water is not captured in a runoff recovery or reuse system, the application efficiency (E_{LQ}) is usually less than 60 to 70%.

In this chapter, the fundamentals of surface irrigation will be presented and discussed to illustrate the importance of proper application and management. Guidelines for good management will be presented for surface irrigation.

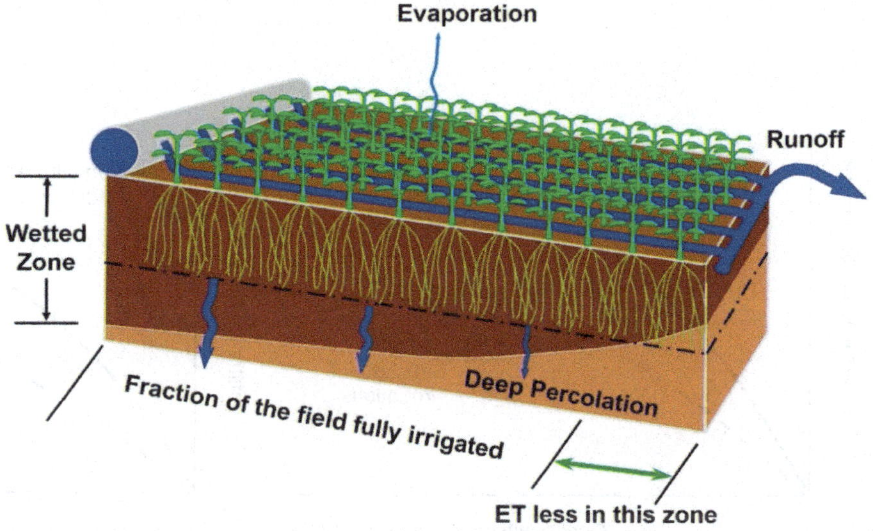

Figure 10.5. Illustration of surface irrigation showing deep percolation, runoff, and evaporation.

10.2 Advance, Recession, and Infiltration

To the casual observer, surface irrigation looks like a very simple concept. The water is applied at the inlet end and the irrigator allows gravity to move the water across the field. As the water moves across the field, part of it infiltrates and part of it is stored on the soil surface. After the water reaches the end of the field, runoff occurs unless the flow is blocked by an earthen dike. Water is usually not applied to the entire field simultaneously but rather is applied to only a portion of the field at one time. These portions of the field are referred to as sets. A set may be an individual border strip, a single basin, or a group of furrows. The water is applied for a fixed time period called set time.

Even though the concept of surface irrigation appears simple, the science of surface irrigation can be very complicated. This is largely because of the many interactions that occur between the rate of inflow, land slope, roughness of the land slope, uniformity of the land slope, and most importantly, the infiltration rate of the soil during irrigation.

In surface irrigation the soil infiltration rate has a large impact on the ultimate distribution of water and the ultimate amount of water that runs off the edge of the field. This is in contrast to sprinkler and microirrigation where the hardware of the system has more control on how the water is distributed and whether or not the water infiltrates at the desired location. The hardware can be designed so that the application rate is less than the infiltration capacity of the soil allowing all of the water to infiltrate at the point of application. This is not true with surface irrigation. Once the water leaves the inlet end of the field, the manager no longer has control of the water; the soil now has control. Infiltration during surface irrigation can vary significantly on land that is cultivated annually. It depends upon whether it is the first irrigation of the season or whether it is a subsequent irrigation, and where tractor tires have traveled and compacted the soil. Some of the variations in infiltration are illustrated in Figure 10.6. There can be many other sources of infiltration variability within the field.

In practice, many surface irrigators have developed an art of irrigating, rather than applying science to irrigation management. What we hope to do in this chapter is to balance the two: the art and the science. It is unlikely that we will ever get to the point where we can completely

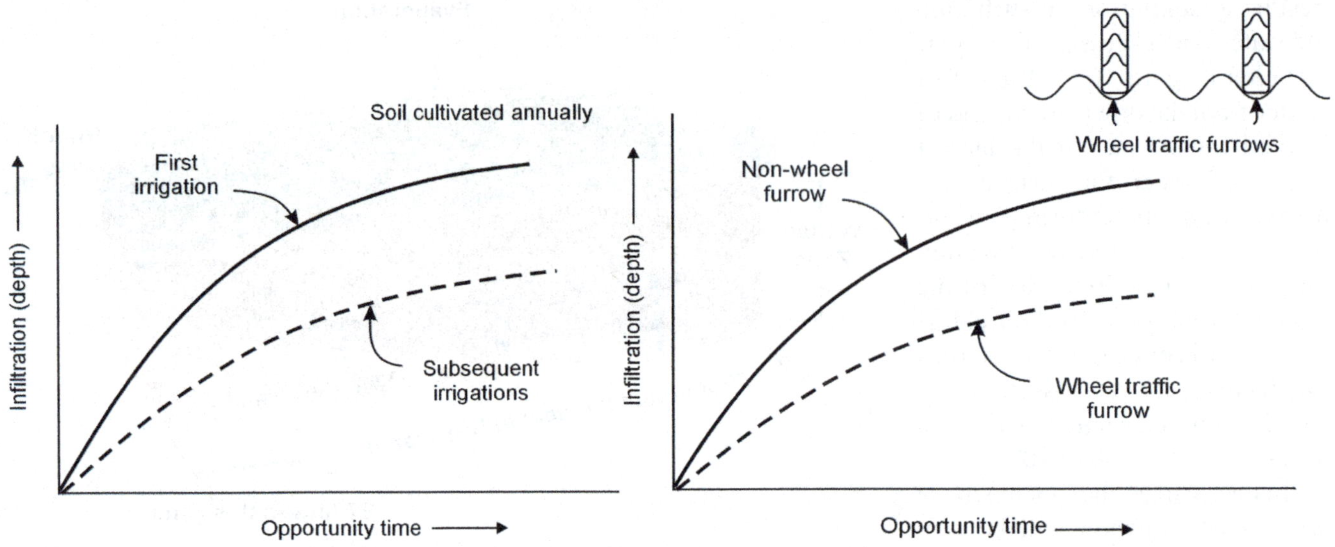

Figure 10.6. Trends in cumulative infiltration as influenced by irrigation sequence and wheel traffic.

manage based on theory alone because there are so many variables that are out of the manager's control.

Let us take a look at the fundamentals that apply to surface irrigation. The first concept is ***advance*** and ***recession*** of water. In Figure 10.7, two curves are shown: the advance curve and the recession curve. The advance curve is a graphical picture of how rapidly water moves from the inlet end to the downstream end of the field, which can be measured directly in the field (Figure 10.8). The curve is not linear. As water moves further and further from the inlet end, the rate at which the wetting front moves decreases. It is typical that it takes about one-third as much time to get halfway across the field as it does to get from the starting point to the downstream end of the field. For example, if it took 3 hours to get to the midpoint of the field, we would estimate approximately 9 hours total to reach the downstream end.

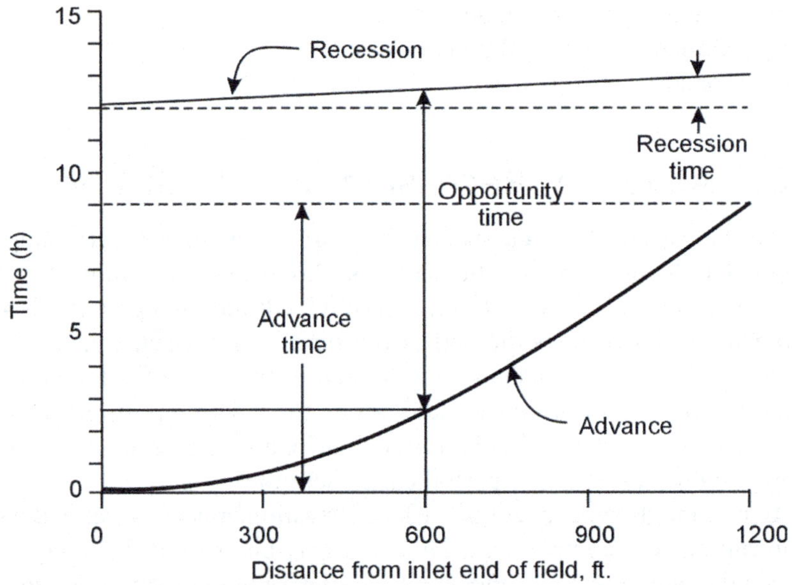

Figure 10.7. Advance and recession curves for surface irrigation.

The recession curve is a plot of how the furrow drains after irrigation has been stopped, and can also be measured directly. Usually, the surface begins to drain from the upstream end. For the example illustrated in Figure 10.7, drainage occurs in approximately 1 hour. This is in contrast to the advance time, which was 9 hours before water reached the downstream end.

Figure 10.8. Students measuring stream size, advance, recession, and runoff in a furrow irrigation system. (Photo courtesy of Laszlo Hayde, IHE Delft Institute for Water Education.)

Why are advance and recession important? The amount of water that infiltrates at any point in the field depends upon how long water was at that point. In our example in Figure 10.7, water was at the inlet end for 12 hours because irrigation was continued for 3 hours after water reached the downstream end of the field. At the downstream end, water arrived after 9 hours of application. Further, the recession took approximately 1 hour at the downstream end. That is, recession stopped at hour 13. So, how long was water present at the downstream end? In this case, 4 hours (13 - 9). At the upstream end water infiltrated for 12 hours, while at the downstream end, water had the opportunity to infiltrate for only 4 hours. You can now see why the amount of infiltrated water would not be uniformly distributed.

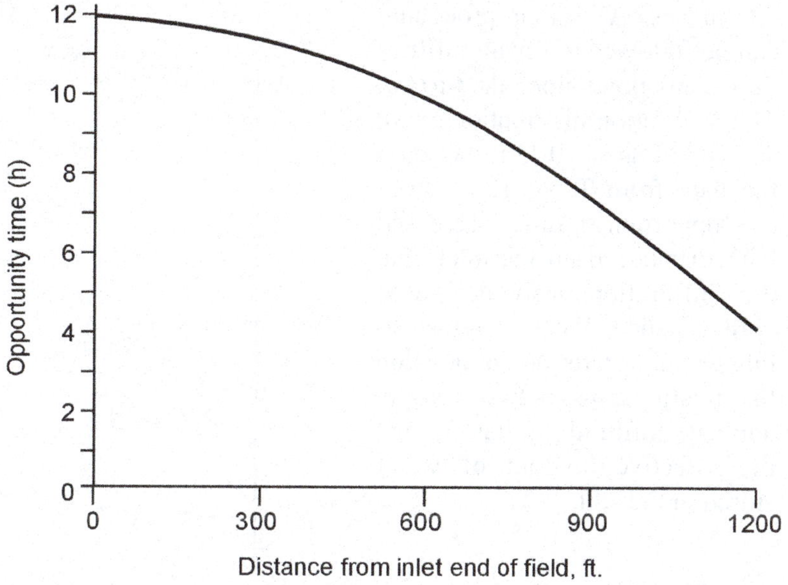

Figure 10.9. Opportunity time for surface irrigation.

The time difference between the recession curve and advance curve is called opportunity time. The opportunity time curve shown in Figure 10.9 is the time difference between the advance and recession curves in Figure 10.7. In this example, opportunity time decreased as you move from the inlet end to the downstream end of the field. If the infiltration characteristics of the soil are uniform throughout the field, we would expect more infiltration at the inlet end compared to the downstream end.

What is necessary to achieve good uniformity? For perfect uniformity the opportunity time curve would have to be horizontal, i.e., equal at all locations within the field. This can only happen if the advance curve and recession curve are parallel to one another. In other words, advance time would have to equal recession time at all points in the field. Even though we commonly picture more opportunity time at the inlet end than at the downstream end, it sometimes happens that recession is slower than advance. In this case, opportunity time would increase with distance from the inlet end.

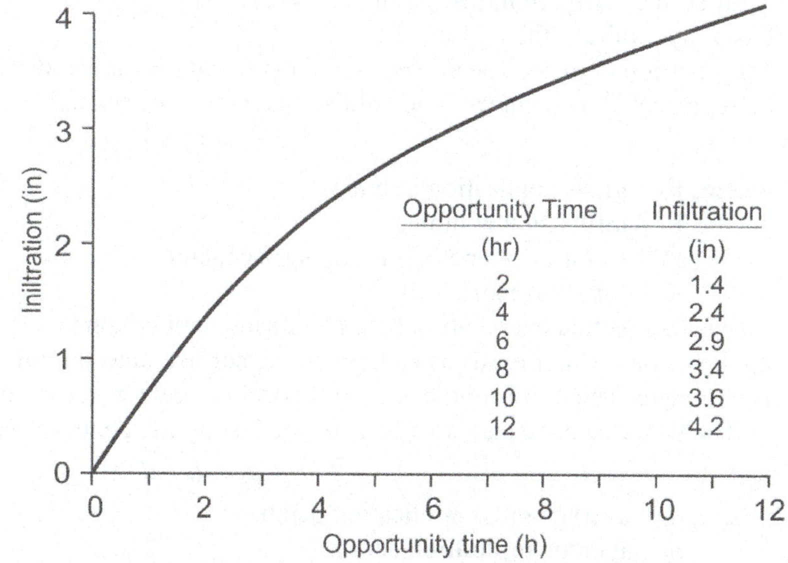

Figure 10.10. Example infiltration vs. opportunity time.

Now, let us look at the development of the infiltration distribution profile. In Figure 10.10, we illustrate an example relationship between the cumulative infiltration and opportunity time. The data are listed in graphical as well as tabular form.

Table 10.1. Data for Figures 10.7, 10.9 and 10.10.

Distance (ft)	Advance Time (h)	Recession Time (h)	Opportunity Time (h)	Infiltration (in)
0	0.0	12.0	12.0	4.2
300	0.8	12.2	11.4	4.0
600	2.7	12.5	9.8	3.6
900	5.5	12.8	7.3	3.2
1200	9.0	13.0	4.0	2.4

In Table 10.1, the advance time, the recession time, and the opportunity time have been tabulated. By combining the opportunity time information with the infiltration characteristics of the soil you can determine the infiltration at any position. Use 600 feet as an example distance. Here the advance time was 2.7 hours and recession occurred at 12.5 hours. Thus, the opportunity time was 9.8 hours. From Figure 10.10 we find that the infiltration would be approximately

3.6 inches. A similar procedure can be followed to obtain infiltration at any point along the furrow. The infiltration distribution curve shown in Figure 10.11 is based on the data from Table 10.1. Since the opportunity time decreased with distance from the inlet end, the infiltration also decreased with distance. We will return to this example later as we develop the relationships between water applied, infiltration, runoff, and the effective amount of water stored in the soil.

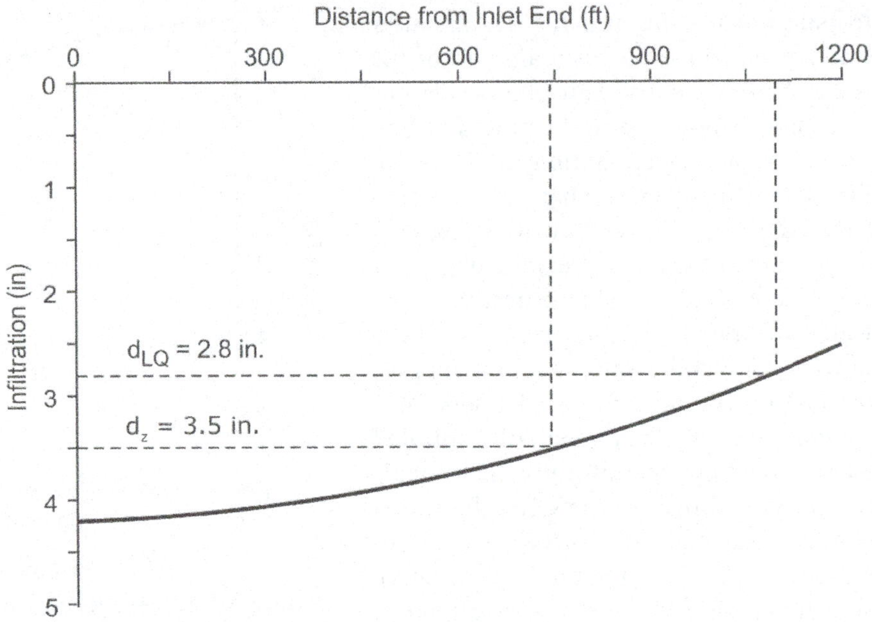

Figure 10.11. Infiltration profile.

10.3 Water Balance

In surface irrigation, just as in basic hydraulics, there must be conservation of mass. The primary components of the mass balance for surface irrigation may be represented as volumes. The volume balance is written as:

$$V_g = V_z + V_s + V_r \tag{10.1}$$

where: V_g = gross application volume,
V_z = infiltration volume,
V_s = storage volume on the soil surface, and
V_r = runoff volume.

We assume that evaporation of water during application is negligible. While water is being applied, some water exists as storage on the surface until the inflow is stopped and recession is complete. Thus, V_s is transient; it only occurs while water is on the surface.

The water balance may also be described using the depth of water:

$$d_g = d_z + d_s + d_r \tag{10.2}$$

where: d_g = average gross application depth,
d_z = average infiltration depth,
d_s = surface storage depth, and
d_r = runoff depth.

As usual, depths represent the volumes divided by the irrigated area.

The gross application depth in ***furrow irrigation*** is calculated as:

$$d_g = 1155\left(\frac{q_s \times t_{co}}{W \times L}\right) \tag{10.3}$$

where: d_g = average gross application depth (in),
q_s = furrow stream size (gpm/furrow),
t_{co} = cutoff time, i.e. set time (hr),
W = spacing of ***watered*** furrows (in), and
L = length of furrow (ft).

or for an entire set:

$$d_g = 1155\left(\frac{Q_t \times t_{co}}{N \times W \times L}\right) \tag{10.4}$$

where: N = number of furrows watered per set, and

Q_t = total inflow rate to the field.

The total inflow rate is equal to the sum of the inflow from the water supply and, when a closed runoff recovery system is used, water reused on the same field. Thus, $Q_t = Q_w + Q_p$ where Q_w = flow rate of the original supply and Q_p = flow rate of the recovery system. The W equals the spacing of the furrows if every furrow is irrigated. If every other furrow is irrigated, then W equals twice the furrow spacing. Often the furrow stream size (q_s) is constant for the duration of the irrigation. When the labor supply is available, efficiency can be improved by reducing furrow stream size after water advance across the field is complete. This is called cutback irrigation.

The gross application depth for ***basins and border irrigation*** is:

$$d_g = 96.3\left(\frac{Q_t \times t_{co}}{W_b \times L_b}\right) \tag{10.5}$$

where: W_b = the width of the border or basin (ft), and

L_b = length of border or basin (ft).

The average infiltration depth (d_z) can be determined from the infiltration profile such as Figure 10.11. It occurs at about 60% of the field's length from the inlet for open-ended systems. In our example, it occurs at 720 feet and equals 3.5 inches. After the irrigation and recession has stopped, the water stored on the surface (d_s) has either infiltrated or has run off; therefore, the depth stored is zero. In Equation 10.2, the only remaining variable is the runoff depth (d_r). The depth of runoff is the total volume of runoff water divided by the area of the irrigation set, basin, or border, or the area irrigated by an individual furrow. By rearranging terms, Equation 10.2 can be used to determine the amount of runoff from a surface irrigated field.

10.4 Efficiency

10.4.1 Calculation of Irrigation Efficiency

The concepts of in-field efficiency and water distribution uniformity as presented in Chapter 5 can be applied to surface irrigation by using the mass balance equations and water distribution graph. This is illustrated in Example 10.1.

The efficiency calculated in Example 10.1 was for a system where runoff is not reused. Later in this chapter we will discuss the use of runoff recovery systems as one method for improving irrigation efficiency. The effects of runoff recovery on efficiency can be determined when two things are known: the amount of runoff and the effectiveness of the runoff recovery system itself, that is, how much of the runoff water is actually captured and applied. The efficiency depends upon whether the recovery system is a closed system in which the runoff water is captured and returned to the field of origin, or whether it is an open system where the runoff is captured from one field and applied to another field with runoff being allowed to leave the second field. These systems are illustrated in Figure 10.12. The equations that apply for calculating efficiency are shown below.

Closed system:
$$E_{LQ} = \left[\frac{d_e}{d_g - d_g R_r R_t}\right]100\% \tag{10.6}$$

Open system:
$$E_{LQ} = \left[\frac{d_e\left(1 + R_r R_t\right)}{d_g}\right]100\% \tag{10.7}$$

where: d_e = effective depth stored,

$d_e = d_{LQ}$ if $d_{LQ} \le$ SWD,

d_e = SWD if d_{LQ} > SWD;
d_g = gross application;
R_r = runoff ratio; and
R_t = return ratio (efficiency of recovery system)
= volume applied from the recovery system divided by volume of runoff.

The runoff ratio is:

$$R_r = \frac{d_r}{d_g} \quad (10.8)$$

Example 10.1

For the furrow-irrigated field with a loam soil described in Figures 10.7 and 10.9 to 10.11 and in Table 10.1, determine the gross application depth, runoff depth, percentage of runoff, and E_{LQ}. Q_t = 760 gallons per minute (gpm), L = 1200 feet, t_{co} = 12 hours, 70 furrows are watered per set, row spacing is 30 inches, and every furrow is watered. Assume the SWD = 3.4 inches. The field slope is 0.3%.

Given: Q_t = 760 gpm
t_{co} = 12 hr
W = 30 in
d_z = 3.6 in (Figure 10.11)
d_s = 0 (since recession is complete)
N = 70

Find: d_g
d_r
Percent runoff
DU
E_{LQ}

Solution:

$$d_g = 1155\left(\frac{Q_t \times t_{co}}{N \times W \times L}\right) \quad \text{(Eq. 10.4)}$$

$$d_g = 1155\frac{(760\text{ gpm})(12\text{ hr})}{(70)(30\text{ in})(1200\text{ ft})} = 4.2\text{ in}$$

$d_r = d_g - d_z - d_s$ (Eq. 10.2 rearranged)

$d_r = 4.2 - 3.5 - 0 = 0.7$ in

$$\text{Percent runoff} = \left(\frac{0.7}{4.2}\right)(100\%) = 17\%$$

The average depth in the low quarter, 2.8 in, is from Figure 10.11.

$$DU = \frac{d_{LQ}}{d_z} = \left(\frac{2.8}{3.6}\right) = 0.78 \quad \text{(Eq. 5.2)}$$

If $d_{LQ} < SWD$, $d_e = d_{LQ}$

$$\text{Thus, } E_{LQ} = \left(\frac{2.8}{4.2}\right)(100\%) = 67\% \quad \text{(Eq. 5.11)}$$

Example 10.2

What is the efficiency of the system given in Example 10.1 if a closed runoff recovery system were used? Assume R_t = 0.85

Given: d_g = 4.2 in
d_z = 3.5 in
R_t = 0.85

Find: R_r
E_{LQ}

Solution:

$$R_r = \frac{d_r}{d_g} \quad \text{(Eq. 10.8)}$$

$$R_r = \frac{(4.2\text{ in} - 3.5\text{ in})}{4.2\text{ in}} = 0.17$$

$$E_{LQ} = \left[\frac{d_e}{d_g - d_g R_r R_t}\right]100\% \quad \text{(Eq. 10.6)}$$

$$E_{LQ} = \left[\frac{2.8}{4.2 - 4.2(0.17)(0.85)}\right]100\% = 78\%$$

Thus, efficiency increased from 67% to 78% by the addition of runoff recovery.

Example 10.3

Repeat Example 10.2 for an open-ended runoff recovery system.

$$E_{LQ} = \left[\frac{d_e(1 + R_r R_t)}{d_g}\right]100\% \quad \text{(Eq. 10.7)}$$

$$E_{LQ} = \left[\frac{2.8(1 + 0.17 \times 0.85)}{4.2}\right]100\% = 76\%$$

The efficiency of this system is slightly lower than the closed system because some runoff is escaping the field irrigated with the runoff water.

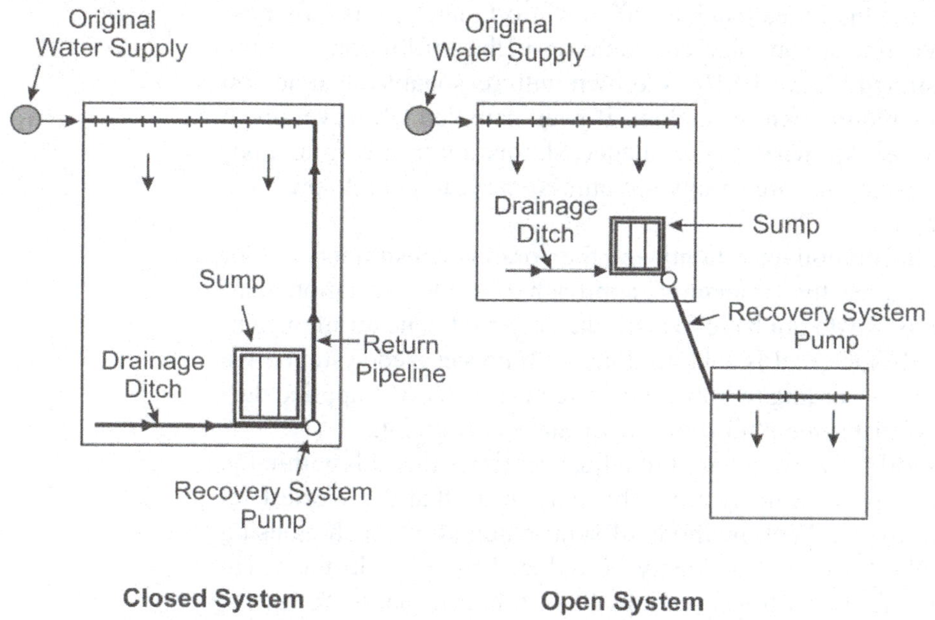

Figure 10.12. Closed and open runoff recovery systems.

10.4.2 Improvement of Surface Irrigation Systems

The application efficiencies of field-scale systems are often reported to be quite low, in the range of 40-50%. Much work has been done to develop methods for improving the efficiencies including the following:

- Converting earthen field ditches to lined-ditches or gated pipe delivery systems to reduce seepage and/or evaporation.
- Recovery or reuse of tailwater to reduce runoff losses.
- Improved land forming methods especially with the use of laser and GPS controlled land grading equipment (Dedrick et al., 2007) for improved application uniformity.
- Cutback irrigation and blocked-end systems to reduce runoff losses (USDA, 2012).
- Surge flow irrigation (Walker and Skogerboe, 1987) to control infiltration and runoff, and to improve uniformity of infiltration.
- Automation of water delivery systems (Humphreys, 1986; Koech et al., 2010; Koech et al., 2014) and semi-automation to better match set times to optimal set times.
- Development of computer-based models for improvement in design and selection of more efficient management options such as set-time and stream size (Bautista et al., 2009).

In this book we concentrate mainly on the set-time and stream size management options as well as tailwater recovery and the management of surge flow systems.

10.5 Management of Sloping Furrow Irrigation Systems

Good management of surface irrigation systems is extremely important. The manager must respond to the effect of infiltration variability on the performance of the system during each irrigation. In addition to satisfying the water needs of the crop, the goals of management might include low runoff, low deep percolation, or that the sum of these two losses be minimized. We'll discuss management practices to *minimize the sum of runoff and deep percolation*. In management of surface irrigation, the irrigator has control of three things: set time (i.e. cutoff time), stream size, and the soil water deficit before water is applied. All three can be changed without changing the system characteristics.

Many, if not all, textbooks, management guides, and computer software establish set time and stream size recommendations so that a required or preplanned desirable irrigation depth infiltrates in a large proportion of the field area (such as 90%). Unfortunately, it is only possible to compute the optimum set time-stream size combination if the infiltration vs. time relationship, such as the one illustrated Figure 10.10, is known with reasonable accuracy on the planned day of irrigation. This requirement is seldom, if ever, satisfied. However, even when the infiltration characteristics are known with confidence, simulation results from models can result in set times or stream sizes that are simply too unreasonable to put into practice. Labor constraints are often a problem.

Given these two problems, the infiltration uncertainty and the possible constraints of labor availability, we have chosen to take a reactive or adaptive approach to surface irrigation management. As explained in Chapter 6, we do not have to refill the crop root zone during irrigation to meet the ET requirements. In fact, that is seldom done with pressurized systems. We simply adjust the irrigation schedule according to how much effective water was applied during each irrigation. Here we follow that same philosophy with surface irrigation.

To overcome the labor-set time dilemma we attempt to adjust set times that fall within the constraints of the irrigator's labor supply. In many areas this may mean that the shortest set time possible is 12 hours or even longer. With methods of semi-automation, such as using surge irrigation valves for example, set times can easily be reduced by 50%. In the Great Plains of the U.S., we commonly refer to set times in intervals of 6 hours, that is, 6, 12, 18, and 24 hours. For example, using a surge irrigation system which irrigates 2 sets simultaneously, 6-hour set times require that the irrigator return to the field only once every 12 hours.

The irrigator can also change stream size. If a water supply rate to a field is constant, then furrow stream size can be changed by changing the number of furrows per set. This is illustrated in the following equation:

$$q_s = \frac{Q_t}{N} \tag{10.9}$$

where: q_s = furrow stream size,
Q_t = total inflow rate to the field, and
N = number of furrows irrigated per set.

Maximum furrow stream size must be kept below the flow that will cause erosion and be low enough so that the furrow has adequate capacity to prevent overflowing. The maximum nonerosive stream size is *approximated* by:

$$q_{max} = \frac{10}{S} \tag{10.10}$$

where: q_{max} = maximum nonerosive stream size (gpm) and
S = field slope (%).

For example, if the field slope is 0.4%, then the maximum nonerosive stream size would be 25 gpm. As stated, Equation 10.10 is an approximation. The NRCS (USDA, 2012) provides more specific guidelines for permissible maximum water velocities to prevent soil erosion in furrows. The important point is that stream size can also be a constraint to the management of furrow irrigation systems.

Another factor that the irrigator can change is the soil water deficit by controlling the frequency of irrigation. The maximum soil water deficit allowable is equal to the management allowed deficit (AD). If an irrigator is having difficulty attaining a high efficiency because of excessive irrigation, the soil water deficit can be increased, up to AD, by irrigating less frequently.

How do stream size, set time, and AD interact? In Section 10.2 we indicated that to obtain a perfectly uniform distribution of water, the advance curve and recession curve have to be parallel. Unfortunately, the tradeoff for uniform distribution is excessive runoff. On the other extreme, if the irrigator's goal is to reduce runoff, then it might be desirable just to get the water to the end of the field and then shut it off or even shut it off before advance is complete.

Obviously, this will result in low runoff, but will also result in poor distribution of infiltration and high deep percolation. So, what is the optimum compromise between runoff and deep percolation that results in the highest system efficiency? A distance-based management parameter that is useful for this determination is called "cutoff ratio." It is defined as:

$$CR = \frac{t_L}{t_{co}} \quad (10.11)$$

where: CR = cutoff ratio,

t_L = advance time to the end of the field, and

t_{co} = cutoff time (set time).

A rapid water advance (low t_L) results in a low cutoff ratio. Conversely, a slow water advance (high t_L) will yield a high cutoff ratio. Low cutoff ratios result in large amounts of runoff and good uniformity. High cutoff ratios result in poor distribution of water, high deep percolation, and low runoff. This concept is illustrated in Figure 10.13.

The cutoff ratio that provides maximum efficiency, where the sum of runoff losses and deep percolation are minimized, is dependent upon the soil characteristics and whether or not the system has runoff recovery. In Figures 10.14 to 10.16 you see that efficiency varies with cutoff ratio and by soil texture for sloping furrow irrigation systems. Here the fine textured soils include clays, silty clays, silty clay loams, and clay loams. Silts, silt loams, loams and sandy clays are considered medium textured soils, and sandy clay loams, sandy loams, loamy fine sand and fine sand are course textured soils. The efficiency term in the graph is the application efficiency of the low quarter (E_{LQ}). Figures 10.14 to 10.16 are based on the assumption that water advance time to the downstream end of the field exceeds water recession time following cutoff. This condition will be met in most cases for long, sloping furrows. This condition might not be met on fields with inadequate slope or small fields with short furrow length (such as smallholder farms). In that case, the principles for border or basin irrigation systems (Section 10.6) may be applicable.

Based on Figure 10.15, if the soil is of medium texture and a closed runoff recovery system is used, the maximum efficiency occurs at a cutoff ratio of about 0.40. Without runoff recovery, the maximum efficiency occurs at about 0.70. How can a manager use these curves? Suppose, because of time constraints, the irrigator can only change sets every 12 hours. If the medium textured soil is considered and the system has runoff recovery, the peak efficiency would occur with a cutoff ratio of 0.40. The desired

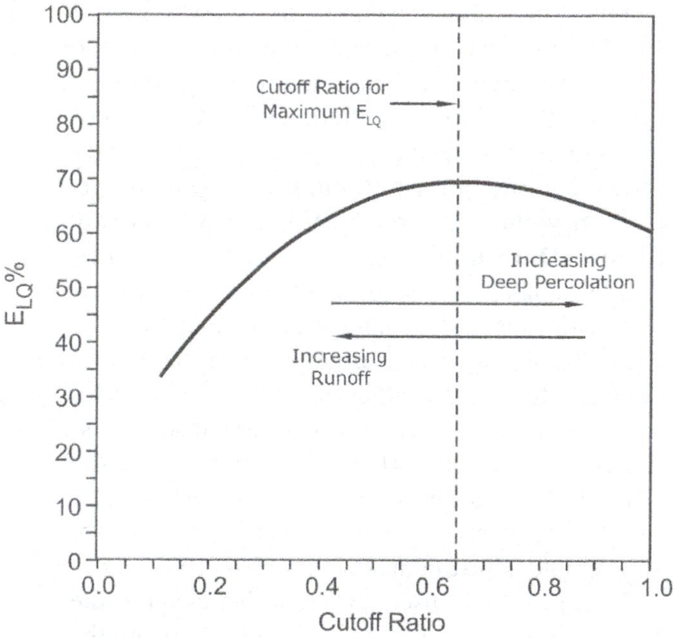

Figure 10.13. Conceptual graph illustrating E_{LQ} vs. cutoff ratio for sloping furrows.

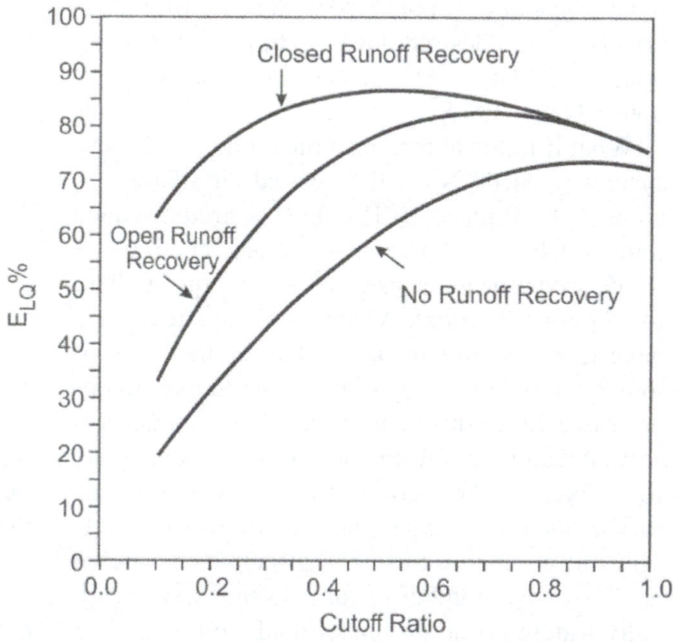

Figure 10.14. E_{LQ} vs. cutoff ratio for sloping furrows with fine textured soils (clays, silty clays, silty clay loams, and clay loams). Assumes advance time exceeds recession time.

advance time is then 0.40 × 12 hours or 4.8 hours. Hence, the irrigator would adjust the furrow stream to achieve the 4.8 hours advance time. Of course, the stream size that has been determined may not be feasible if it exceeds the maximum nonerosive stream size for that slope condition.

The expected maximum efficiency for the system described above would be about 85% (Figure 10.15). For these efficiencies to be attainable, the depth infiltrated at the low quarter, d_{LQ}, must be less than the soil water deficit, SWD. If this is not true, the figures are not applicable and the manager should consider allowing a higher SWD before irrigation without exceeding AD. If SWD already equals AD, then other practices that reduce infiltration depths, such as every other furrow irrigation or shorter set times, must be considered.

As discussed above, usually the irrigator does not know the soil's infiltration characteristics prior to irrigation. The irrigator learns these characteristics by irrigating a portion of the field. Once the advance time is known for a given furrow flow rate, or stream size, then the irrigator can make the appropriate adjustments to maximize efficiency.

In the example used so far in this chapter, the furrow stream size was 11 gpm (760 ÷ 70) and the advance time was 9 hours. According to Figure 10.15, the cutoff ratio that would result in maximum efficiency is about 0.70 with no runoff recovery. Thus, the desired advance time is 0.70 × 12 hours or 8.4 hours. This is close to the measured 9-hour advance time.

What if in the above example a runoff recovery system is used? Now, the desired cutoff ratio is about 0.40 (Figure 10.15). The desired advance time is 0.40 × 12 hours or 4.8 hours. The ratio of the desired time to the original time is equal to 0.53 (4.8 hours ÷ 9 hours). What would the stream size have to be for this to occur? Or, another way of looking at it, how many furrows would have to operate to achieve this goal? Table 10.2 contains correction factors for the number of furrows to irrigate for a fixed Q_t. The ratio of the new advance time to the old one is 0.53. Interpolating from Table 10.2, the number of furrows that should be watered is 63% of the number of furrows that were originally watered (find this under medium textured soil, $N_2/N_1 = 0.63$). Thus, the irrigator should irrigate 44 furrows (0.63 × 70) instead of the original 70. The furrow stream size would now be 760 gpm ÷ 44 = 17 gpm per furrow. If the furrow slope in this example is 0.3%, the maximum nonerosive stream size is 33 gpm. Thus, the 17 gpm flow rate is acceptable.

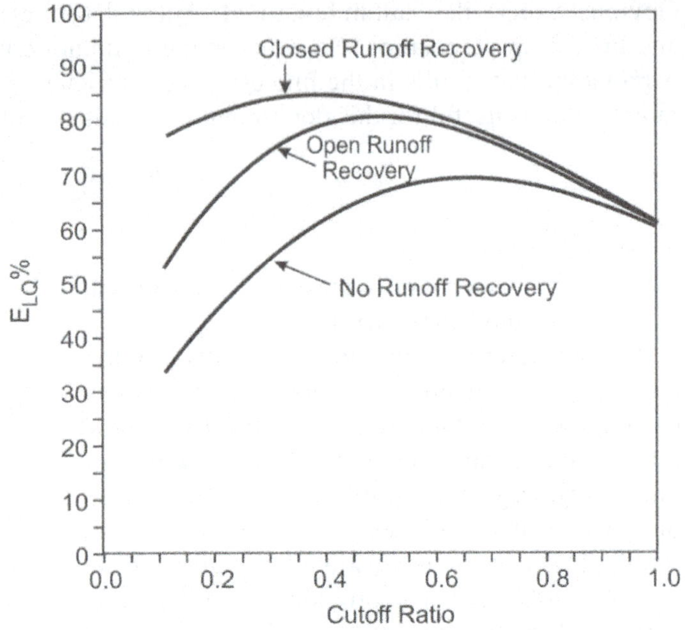

Figure 10.15. E_{LQ} vs. cutoff ratio for sloping furrows with medium textured soils (silts, silt loams, loams and sandy clays). Assumes advance time exceeds recession time.

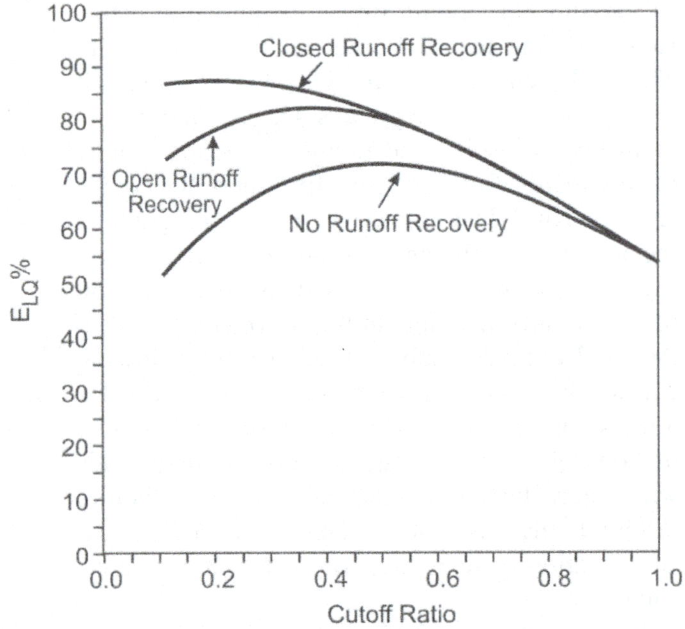

Figure 10.16. E_{LQ} vs. cutoff ratio for sloping furrows with coarse textured soils (sandy clay loams, sandy loams, loamy fine sand and fine sand). Assumes advance time exceeds recession time.

Table 10.2. Correction factor (N_2/N_1) for predicting how many furrows to irrigate per set with a constant water supply. (Table based on equation in Cahoon et al., 1995.)

T_{L2}/T_{L1}	Fine	Medium	Coarse	T_{L2}/T_{L1}	Fine	Medium	Coarse	T_{L2}/T_{L1}	Fine	Medium	Coarse
0.1	0.1	0.2	0.4	1.3	1.3	1.2	1.1	2.5	2.6	2.0	1.5
0.2	0.2	0.3	0.5	1.4	1.4	1.3	1.2	2.6	2.7	2.0	1.5
0.3	0.3	0.4	0.6	1.5	1.5	1.4	1.2	2.7	2.8	2.1	1.6
0.4	0.4	0.5	0.7	1.6	1.6	1.4	1.2	2.8	2.9	2.2	1.6
0.5	0.5	0.6	0.7	1.7	1.7	1.5	1.3	2.9	3.1	2.2	1.6
0.6	0.6	0.7	0.8	1.8	1.9	1.6	1.3	3.0	3.2	2.3	1.6
0.7	0.7	0.8	0.9	1.9	2.0	1.6	1.3	4.0	4.3	2.8	1.9
0.8	0.8	0.8	0.9	2.0	2.1	1.7	1.4	5.0	5.4	3.3	2.1
0.9	0.9	0.9	1.0	2.1	2.2	1.7	1.4	6.0	6.6	3.8	2.2
1.0	1.0	1.0	1.0	2.2	2.3	1.8	1.4	7.0	7.7	4.3	2.4
1.1	1.1	1.1	1.0	2.3	2.4	1.9	1.5	8.0	8.9	4.8	2.5
1.2	1.2	1.1	1.1	2.4	2.5	1.9	1.5	9.0	10.0	5.2	2.7
								10.0	11.2	5.6	2.8

N_2 = Correct number of furrows to water per set.
N_1 = Original number of furrows watered per set.
T_{L2} = Desired advance time.
T_{L1} = Original advance time.

What is the depth infiltrated at the low quarter for the new condition? Table 10.3 relates the depth of low quarter to the gross depth applied and the cutoff ratio. The infiltration factor given in Table 10.3 is defined as:

$$\text{Infiltration factor} = \frac{d_{LQ}}{d_g} \tag{10.12}$$

In Example 10.1, the cutoff ratio was 0.75 and the gross depth applied was 4.2 inches. The depth of low quarter would be equal to 4.2 inches times the factor from Table 10.3 (0.70) or 2.9 inches. This closely agrees with our original graphical analysis (Figure 10.11), 2.8 inches. This number can now be compared with the SWD before irrigation. If it exceeds the SWD, then the irrigator has two choices. The first, and easiest to implement, is to change the irrigation frequency so that SWD is higher during irrigation. Again, the constraint is that AD is the upper limit. The second approach is to change the irrigation cutoff time so that less water infiltrates during the irrigation and thus the depth of low quarter might be maintained less than the soil water deficit.

The irrigation frequency now will be dependent upon the effective water applied. As we discussed in Chapter 5, the effective water applied will be equal to the d_{LQ} if it is less than or equal to SWD. The effective water applied is equal to SWD if d_{LQ} is greater than SWD. In our example, where the effective water applied was 2.9 inches, the irrigation interval should be about 10 days if ET is equal to 0.3 inches per day.

The runoff ratio (R_r, Equation 10.8) must be known to calculate E_{LQ} when a runoff recovery system is used. Table 10.4 gives runoff ratios for various conditions.

Table 10.3. Infiltration factors (ratio of d_{LQ} to d_g) for sloping furrows. (Assumes advance time exceeds recession time.)

Cutoff Ratio	Infiltration Factors		
	Soil Texture		
	Fine	Medium	Coarse
0.1	0.19	0.32	0.50
0.2	0.32	0.45	0.61
0.3	0.42	0.55	0.68
0.4	0.51	0.62	0.71
0.5	0.59	0.66	0.72
0.6	0.65	0.69	0.71
0.7	0.70	0.70	0.69
0.8	0.73	0.69	0.66
0.9	0.74	0.66	0.61

Table 10.4. Runoff ratios (R_r) for sloping furrows. (Assumes advance time exceeds recession time.)

Cutoff Ratio	Runoff Ratios		
	Soil Texture		
	Fine	Medium	Coarse
0.1	0.81	0.68	0.50
0.2	0.68	0.53	0.36
0.3	0.56	0.42	0.26
0.4	0.46	0.32	0.19
0.5	0.37	0.24	0.14
0.6	0.28	0.18	0.09
0.7	0.21	0.12	0.06
0.8	0.14	0.08	0.03
0.9	0.08	0.04	0.02

Notice that the efficiency is lower in Example 10.4 than in Examples 10.2 and 10.3, even though the lower cutoff ratio was supposed to increase efficiency. What went wrong? In Example 10.4, the d_{LQ} > SWD. To improve efficiency to its potential, the d_{LQ} must either be reduced or SWD must be increased. Suppose AD = 3.4. The SWD cannot be increased without yield reduction, so a reduction of d_{LQ} must be attempted. Let us try a 6-hour set time.

An alternative to free flow at the furrow outlet is to block the downstream ends with a dike to prevent runoff. This is usually practical when field slopes are low. While blocking the ends prevents runoff, poor distribution of water can occur because of the ponded water behind the dike (Figure 10.17). Cutoff ratio guidelines that result in maximum efficiency have been established for all the cases discussed so far in this chapter. They are presented in Table 10.5. In general, when the ends are blocked, recommended cutoff ratios are higher than for the nonblocked case. This minimizes the size of the pond and the quantity of deep percolation beneath the pond.

Example 10.4

If runoff recovery is included in the system described in Examples 10.1 to 10.3 and a cutoff ratio of 0.40 is achieved, determine the system's E_{LQ}. Assume that SWD = 3.4 inches and the recovery system returns the water to the same field (closed system).

Given: CR = 0.40 N = 44 furrows
SWD = 3.4 in W = 30 in
Q_t = 760 gpm

Find: d_g d_{LQ}
d_e E_{LQ}

Solution:

$$d_g = 1155\left(\frac{Q_t \times t_{co}}{N \times W \times L}\right) \quad \text{(Eq. 10.4)}$$

$$d_g = 1155\left(\frac{(760\text{ gpm})(12\text{ hr})}{(44)(30\text{ in})(1200\text{ ft})}\right) = 6.65\text{ in}$$

$$d_{LQ} = d_g(\text{Infiltration factor}) \quad \text{(Eq. 10.12)}$$

$$d_{LQ} = (6.65)(0.62) = 4.1\text{ in} \quad \text{(Table 10.3)}$$

Since d_{LQ} > SWD, d_e = SWD = 3.4 in

For a medium textured soil,
CR = 0.40 and R_r = 0.32

$$E_{LQ} = \left(\frac{d_e}{d_g - d_g R_r R_t}\right)100\%$$

$$E_{LQ} = \left(\frac{3.4}{6.65 - 6.65(0.32)(0.85)}\right)100\% = 70\%$$

Example 10.5

Determine the number of furrows to irrigate in a set and the E_{LQ} for the conditions in Example 10.4 if a 6-hour set time is used.

Given: t_{co} = 6 hr
CR = 0.40
When N = 44, t_L = 4.8 hr
SWD = 3.4 in

Find: t_L d_g
N d_{LQ}
E_{LQ}

Solution:

$$t_L = (6\text{ hr})(0.4) = 2.4\text{ hr}$$

$$\frac{t_{L1}}{t_{L2}} = \frac{2.4}{4.8} = 0.5$$

$$\frac{N_2}{N_1} = 0.6$$

$$N_2 = (44)(0.6) = 26 \quad \text{(Table 10.2)}$$

$$q_s = \frac{Q_t}{N} = \frac{760\text{ gpm}}{26} = 29\text{ gpm}$$

Since q_{max} is 33 gpm for the 0.3% slope, 29 gpm is okay.

$$d_g = 1155\left[\frac{Q_t t_{co}}{NWL}\right] \quad \text{(Eq. 10.4)}$$

$$d_g = 1155\left[\frac{(760\text{ gpm})(6\text{ hr})}{(26)(30\text{ in})(1200\text{ ft})}\right] = 5.0\text{ in}$$

$$d_{LQ} = 5.0\,(\text{Infiltration Factor})$$

$$d_{LQ} = (5.0)(0.62) = 3.1\text{ in} \quad \text{(Table 10.3)}$$

d_e = 3.1 in since d_{LQ} is less than SWD

R_r = 0 (Table 10.4)

$$E_{LQ} = \left(\frac{d_e}{d_g - d_g R_r R_t}\right)100\% \quad \text{(Eq. 10.6)}$$

$$E_{LQ} = \left(\frac{3.1}{5.0 - 5.0(0.32)(0.85)}\right)100\% = 85\%$$

This is close to the maximum achievable efficiency. By changing the management and adding runoff recovery, the efficiency was improved from 67% (Example 10.1) to 85% (Example 10.5).

If ET = 0.3 inches per day, the field should be irrigated every 10 days since d_e = 3.1 inches.

In Examples 10.1 to 10.5, we reacted to what occurred in the field, i.e., we reacted to how fast the water advanced across the field. We refer to this as ***reactive management of surface irrigation***. For someone irrigating a field for the first time, the data shown in Table 10.6 can help keep the flow rates, advance times, and field lengths within reasonable range.

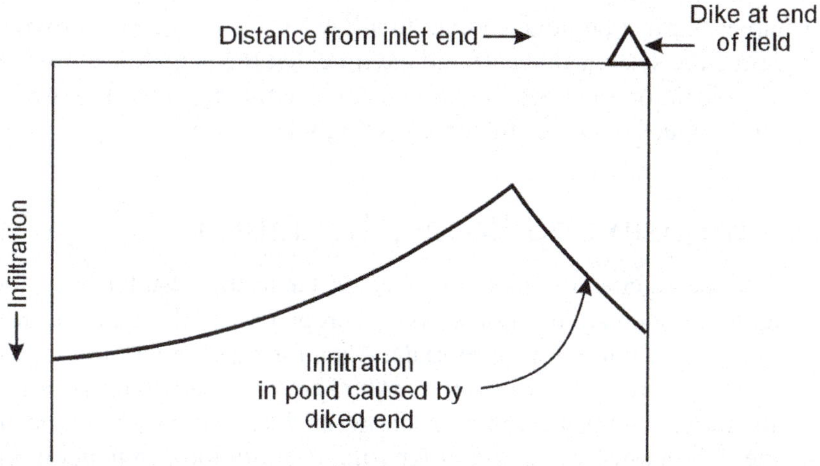

Figure 10.17. Infiltration profile for blocked or diked end sloping furrows.

Table 10.5. Recommended cutoff ratios to achieve maximum efficiency for sloping furrows.

Type of System	Soil Texture		
	Fine	Medium	Coarse
No reuse	0.90	0.70	0.50
Closed reuse system	0.50	0.40	0.20
Open reuse system	0.70	0.50	0.35
Blocked ends (low slope, 0.1%)[a]	0.95	0.85	0.70
Blocked ends (moderate slope, 0.5%)[a]	0.90	0.80	0.65

[a] Based on data from Cahoon et al., 1995.

Are there alternatives to changing set time, stream size, and soil water deficit? As we have illustrated, one option for improving irrigation efficiency is to recover and reuse runoff water. The facilities necessary for recovering runoff are discussed in Section 10.7. Another option is to consider the furrow spacing. Alternate furrow or irrigating every other furrow should be considered if application depths are too large. In general, this practice will reduce infiltration by about 25 to 30%. For the same stream size, changing to every other furrow will increase advance time by about 30 to 40%, because of the longer time it takes for the wetting fronts between the irrigated furrows to meet. Watering every other furrow is usually a practice that can be used to reduce infiltration because even though the advance time is longer, the set size is twice as large as for every furrow irrigation.

Another field factor that can be changed, although not often desirable, is to reduce the furrow length. If the maximum nonerosive stream size is the limiting factor in achieving high efficiency, then furrow length should be reduced so that optimum advance times can be achieved.

Surface irrigation efficiency can sometimes be improved by land smoothing. Land smoothing and laser grading will remove low and high spots and pot holes and provide uniform surface slopes. This will increase the advance rate of the water and uniformity of application.

Other options that can be used to overcome some of the constraints in surface irrigation are automation and semi-automation. This would eliminate the constraint of set time. Semi-automation of surface irrigation can be easily accomplished using surge flow irrigation valves, which will be discussed later. Another option is to use timers to terminate the inflow at the desired time in the absence of the irrigator. For example, if the irrigator can only return to the field every 12 hours but an 8-hour set time is desired, the timer could be set to shut off the water at 8 hours. The limitation of this procedure is that the system capacity, as discussed in

Table 10.6. Furrow irrigation management recommendations for various soil types.

Soil Texture	Basic Infiltration Rate (in/hr)	Basic Infiltration Rate (gpm/100 ft)	Maximum Furrow Length (ft)	Recommended Stream Size[a] (gpm/100 ft)
Loamy sand	2.0–5.0	2.4	600	4.8
Sandy loam	0.5–4.0	1.9	800	3.8
Fine sandy loam	0.2–2.0	1.7	1000	3.4
Silt loam	0.2–1.5	1.1	1100	2.2
Silty clay loam	0.05–0.25	0.1	1300	1.4

[a] Actual stream size must be less than maximum nonerosive stream size.

Chapter 5, must be large enough to allow for the off time or down time that occurs between the time the system shuts off and the time that the irrigator returns to restart it.

If infiltration rates are too high to achieve the desired efficiency, then furrow packing and smoothing using special tillage tools might be helpful.

10.6 Basin and Border Irrigation

The management guidelines given so far in this chapter have focused on sloping furrow irrigation systems. For closely-spaced crops like alfalfa and orchard crops, basin and border systems are often more appropriate. Also, furrows sometimes are used in level basins which contain row crops. For ***basin irrigation systems***, since the bottoms of basins are level and all of the water is retained within the basins by dikes, no runoff will occur. Thus, deep percolation is the only loss of water (again ignoring evaporation). In general, high stream sizes and low set times are appropriate since runoff is not an issue with these closed-level systems. Kay (1986) provides management guidelines for these systems. Readers can use computer simulation models such as WinSRFR (Bautista et al., 2012) to develop optimal management strategies for their individual site.

Border irrigation has many similarities to furrow irrigation in that the borders or bays have slope in the direction of flow and that they can be open-ended at the downstream end. Therefore, in some cases the management guidelines given for sloping furrow irrigation can apply to borders as well. However, sometimes the flow resistance of the closely spaced vegetation in borders can result in a significant amount of water stored on the soil surface before cutoff which can lead to long recession times and even to recession time exceeding advance time. When this occurs, the assumptions that we made in Section 10.5 would be invalid. In fact in some cases, optimal management is to stop inflow before advance is complete (Kay, 1986).

10.7 Runoff Recovery

10.7.1 Options for Managing Runoff

One of the challenges with surface irrigation is to achieve uniformity of infiltration while minimizing runoff from the field. Water must be present at the downstream end of the field long for uniform infiltration. This creates a potential for runoff. Runoff is an inherent problem with border and furrow irrigation systems.

As discussed above, blocking the downstream end of the field is one method for retaining runoff. When the slope is low enough, the retained water will spread back over a relatively large portion of the field. However, if the slope is too large, the ponded water infiltrates into only a small area. The result is poor water distribution. Blocking can also reduce the yield of crops that are sensitive to prolonged submergence.

Another option for minimizing runoff water is cutback irrigation. The concept here is to use a large inflow rate during the advance phase. Following advance, the inflow is reduced to a rate that approximately equals the steady-state infiltration rate of the wetted area. Without automation, this practice is labor intensive and requires good management. The correct cutback flow rate is difficult to estimate without considerable experience.

Recovering or reusing runoff water is another option. With a runoff recovery system, the runoff water is captured and returned to the field of origin or is delivered to another field. With runoff recovery, either less water from the original source is required to irrigate the same land area or more land can be irrigated with an equal volume. In either case, irrigation efficiency is increased. Runoff recovery has many other advantages including reduced nuisance problems associated with runoff, reduced energy requirements for irrigation, reduced labor, increased crop yields, and easier compliance with local regulations.

Often runoff causes nuisance problems downstream of the irrigated field. This can cause conflicts between neighboring farmers because surface drainage problems may occur on the downstream land. Capturing runoff reduces these problems.

If the original supply water is pumped, runoff recovery saves energy when the total dynamic head required to pump the runoff water is less than that required for the original supply. Usually less labor is required for irrigating if runoff recovery is employed. With less worry about the fate of runoff, irrigators do not monitor the water as closely or change sets as often. Crop yields sometimes are improved with runoff recovery if it results in more completely irrigating the downstream end of the field.

An important advantage of runoff recovery can be the ability to comply with water laws and regulations. In some regions, especially where groundwater is being depleted by irrigation, regulations limit the total volume of water that can be pumped from the aquifer. Sometimes the regulations specifically state that runoff cannot leave the irrigated farm. Runoff recovery systems facilitate compliance with these types of regulations.

10.7.2 Description of Runoff Recovery Systems

A runoff recovery system (Figure 10.18) has the following components:

- Drainage ditches for collecting and conveying runoff from the downstream edge of the field to the storage facility.
- A sump or reservoir for storing the runoff water.
- Inlet facilities to the sump or reservoir. These include a desilting basin for settling sediment from the runoff water, screens for removing trash from the water, and a chute, drop, or pipe inlet to deliver the water to the sump without causing serious erosion.
- A pump and power unit for withdrawing the water from the sump and, if necessary, pressurizing it for conveyance.
- A conveyance system, pipelines or open channels, for transporting the water from the storage facility to the field of use. Runoff water can either be returned to the field of origin or be delivered to another field for application. Often, using the water on a different field reduces initial costs because the runoff water is conveyed a shorter distance and normally down slope. If runoff is the only source of water for the receiving field, a very accurate estimate of the volume of runoff from the field or origin is necessary.

(a)

(b)

Figure 10.18. (a) Runoff recovery reservoir, and (b) sump and pump for runoff recovery.

10.7.3 Design of Runoff Recovery Systems

The design of only the reservoir and pumping facilities will be discussed here. Two alternative designs, a continuous pump and an intermittent pump, will be considered.

For the continuous-pump system, the reservoir is designed to store the runoff from one irrigation set (plus allow for any necessary freeboard and unusable or dead storage). The capacity of the pump should equal the time averaged rate of runoff or, stated another way, equal to the volume divided by the time of cutoff (set time). The volume of storage that is required depends on the field and management conditions, but typically is from 20 to 55% of the volume applied to one irrigation set. If the runoff ratio, the ratio of the volume of runoff to the total volume applied, is known *and the runoff recovery reservoir is full at the start of the irrigation,* the following equations apply:

$$V_r = \frac{R_r\, t_{co}\, Q_w}{1 - R_r F} \tag{10.13}$$

$$Q_r = \frac{V_r}{t_{co}} = \frac{R_r\, Q_w}{1 - R_r F} \tag{10.14}$$

where: V_r = runoff volume from one set (active volume of return reservoir),
Q_r = capacity of the runoff recovery pump,
R_r = runoff ratio,
t_{co} = cutoff time,
Q_w = inflow rate to the field from the original source,
F = design factor,
$F = 1$, if the runoff is returned to the field of origin, and
$F = 0$, if the runoff is delivered to another field.

Since some of the runoff water will not be recovered, due to seepage and evaporation, these design equations contain a margin of safety. However, it still is important to include an additional margin of safety by using a high estimate of R_r. Suggested values for R_r are from 0.30 to 0.40.

As the name implies, the continuous-pump system operates continuously, or nearly so. There is very little flexibility in the management of these systems. The intermittent system allows for more flexibility. In this case, the reservoir is designed to store the runoff from two or more irrigation sets and the pump only operates on an intermittent basis. This makes management easier. The return pump must have more capacity than that for the continuously-operating system. Usually, the recovery pump will have a capacity in direct proportion to the reservoir volume. That is, if the reservoir can store the runoff from two sets, the pump would have twice the capacity as the pump for a continuously-operating system. The irrigator can then operate this system when adequate water is present in the reservoir. This system is particularly useful where the water is used to irrigate another field.

Rainfall runoff should be diverted away from the storage reservoir to minimize the accumulation of sediment in the reservoir. A gate on the reservoir inlet can be used to prevent the undesired inflow. If the runoff water is being returned to the field of origin, and if the original supply is groundwater, a check valve should be installed on the water supply pump to prevent the backflow of contaminated water to the groundwater reservoir in the event that the supply pump fails. If the recovered water is used to irrigate a different field, be aware of the potential of unwanted pesticides that may accumulate in the runoff from the field of origin.

Example 10.6

A continuous pump recovery system that returns the runoff to the field of origin is to be designed.

Given: Q_w = 500 gpm
t_{co} = 360 min
R_r = 0.30
F = 1

Find: V_r
Q_r

Solution:

$$V_r = \frac{R_r t_{co} Q_w}{1 - R_r F} \quad \text{(Equation 10.13)}$$

$$V_r = \frac{(0.30)(360\text{ min})(500\text{ gpm})}{[1 - (0.30)(1.0)]} = 77{,}143\text{ gal} = 2.84\text{ ac-in}$$

$$Q_r = \frac{V_r}{t_{co}} = \frac{R_r Q_w}{1 - R_r F} \quad \text{(Equation 10.14)}$$

$$Q_r = \frac{(0.30)(500\text{ gpm})}{[1 - (0.30)(1.0)]} = 214\text{ gpm}$$

10.8 Surge Flow Irrigation

10.8.1 The Surge Flow Process

Surge irrigation or surge flow is the process of intermittently applying water in surface irrigation (Yonts et al., 1996) as compared to continuous flow where water is applied for the entire irrigation set time. Surge irrigation was first studied as a method of reducing the amount of runoff that occurred during irrigation (Stringham and Keller, 1979). It was discovered that the time required for water to move to the end of the field was reduced by applying water intermittently rather than continuously.

Water can be applied intermittently by cycling irrigation water between two irrigation sets. In years past, irrigation water was cycled when it was not getting to the end of a field. The irrigator would move on to subsequent sets and return in 1 or 2 days to finish irrigating the partially watered sets. The second time, the irrigation water could be moved all the way to the end of the field because the soil surface had sealed where previously wetted by irrigation and thus more water was available at the point where flow had stopped. This same process is used with surge irrigation, except 3 to 6 cycles are used and the cycling is done automatically for short durations of 20 minutes to 2 hours.

Figure 10.19. Tee-type surge irrigation valve.

When water first contacts the soil in an irrigation furrow, the infiltration rate is high. As the water continues to run, the infiltration rate at that point in the furrow is reduced to a near constant rate. If water is shut off and the furrow is briefly allowed to dry, the surface soil particles consolidate and form a surface seal in the furrow. When water is reintroduced to the furrow, the infiltration rate is low due to this sealing action. The result is more water moving down the furrow rather than infiltrating into the soil in the initial reach of the furrow. Surge flow can increase irrigation performance by providing a more uniform application.

Rather than turning the water on and off to achieve an on-off cycle, an irrigation surge valve (Figure 10.19) is used to alternate flow between two irrigation sets. Figure 10.20 shows one method of using a surge valve. Cycle times used with surge irrigation vary with soil texture and slope. Fine-textured soils respond less to surge irrigation than do coarse- textured soils that have higher initial infiltration rates. If field slope is so steep that it causes a rapid rate of advance, the effects of surge irrigation will be reduced. If the infiltration rate of a soil is low due to soil texture or compacted layers, surge irrigation is likely to be ineffective in reducing the irrigation advance times below those for continuous flow.

Surge flow has been used to reduce irrigation runoff in some cases by using short duration cycles after the water has reached the end of a field. This helps maintain high uniformity of water application and improve overall irrigation performance. Another application of a surge valve is to use it for semi-automatic operation.

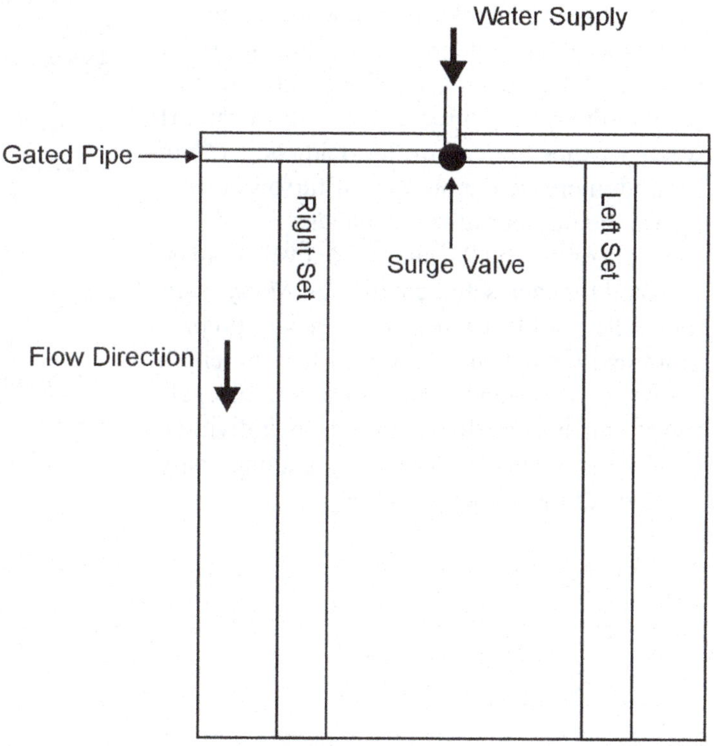

Figure 10.20. Field installation of a surge valve (Yonts et al., 1991).

The surge controller provides a 2-set semi-automated furrow irrigation system. For example, if an irrigator is limited to returning to the field to every 12 hours, two 6-hour sets can be accomplished in that time frame.

Surge flow may not always reduce the advance time of water down the furrow. If it does not, there may still be benefits of labor savings and runoff reduction.

10.8.2 Management of Surge Flow Irrigation

Normally intermittent application is accomplished by using surge valves to alternate the water between a left and right irrigation set (Figure 10.20). The irrigation on times, during which water is applied to one side of the surge valve, are normally between 20 minutes and 2 hours. For each irrigation, an equal amount of off time occurs during each cycle. This will not be the case when different cycles times are used to compensate for an irregular shaped field. A cycle time—the time it takes to complete a full on time and off time cycle—is based on furrow length, soil texture, and field slope. The number of surge cycles used should be based on field length and field condition. Long fields and fields with high intake soils will require more cycles (5 to 6); shorter fields with low intake soil will need fewer cycles (3 to 4).

It is common to advance water during each surge cycle a distance that is equal to that fraction of the number of surge cycles used. For example, if using 4 surge cycles during advance, divide the field into 4 parts and advance the water one-fourth of the field distance during the first surge cycle. The time required to move the water that distance is the Cycle 1 on time. For the second and subsequent on times, multiply the factors given in Table 10.7 by the Cycle 1 on time. Table 10.7 provides the on-time factors for four and six cycles during advance.

Following the final advance cycle, set the valve for the cutback or post-advance phase. During cutback, the valve cycles the water at a shorter frequency between the two irrigation sets until irrigation is complete. Table 10.7 gives the on time factors for the post advance or cutback cycles.

If water does not reach the end of the field by the last surge cycle, adjustments are necessary. Options include increasing the number of surge cycles or decreasing the number of furrows in the set to increase furrow flow rate. If water reaches the end of the field sooner than desired, increase the number of furrows or decrease the number of surge cycles.

Cycle times and the number of cycles can be adjusted for each set of conditions. Many commercially sold valves will have preprogrammed cycle times based on furrow length or expected advance time. In addition, cycle times can be developed based on individual conditions. The valve will automatically change at those times selected.

Table 10.7. Surge irrigation on time factors for four and six surge cycles during advance. (Table based on Yonts et al., 1996, and Fekersillassie and Eisenhauer, 2000.)

Cycle No.	Four Cycles		Six Cycles	
	Fraction of Field	On Time Factor	Fraction of Field	On Time Factor
1	0.25	1.0	0.17	1.0
2	0.50	1.9	0.34	1.9
3	0.75	2.4	0.51	2.4
4	1.00	2.9	0.68	2.9
5	-	-	0.85	3.3
6	-	-	1.00	3.7
Post advance or cutback	-	0.8–1.6	-	1.5–3.0

Example 10.7

Determine the number of cycles and the on time for each cycle for a 1,200-ft long field. Assume that 4 cycles are desired during advance. Water reaches 300 ft (1/4 the length) in 30 min. The irrigator plans to return to the field every 12 hr to change sets.

Given: On time for Cycle 1 = 30 min

4 cycles during advance are desired

360 min total on time for each side of valve (based on 720 min return interval to field)

Solution:

Surge irrigation schedule using a 12-hr set time.

Cycle Advance	Surge on Time (min)	Left Set	Right Set	Total Time 1 Cycle (min)	Cumulative Time (min)
1	30	7:00 a.m.	7:30	60	60
2	57	8:00	8:57	114	174
3	72	9:54	11:06	144	318
4	87	12:18	1:45	174	492
Cutback					
5	38	3:12	3:50	76	568
6	38	4:28	5:06	76	644
7	38	5:44	6:22	76	720
New Set		7:00 p.m.			

10.9 Summary

Surface irrigation includes the methods of furrow, border, and basin irrigation. It is the oldest form and most commonly used method of irrigation in the world. Water is usually delivered to surface irrigation sets through gated pipe, siphons, or gated inlets. Water flows over the land by gravitational force. The land must be graded to a uniform surface with slopes from 0 to 2 %.

With surface irrigation it is important to properly proportion any water losses so that the total of runoff and deep percolation is minimized. This is accomplished by choosing the appropriate irrigation frequency, inflow rate or stream size, and set time.

To control or reduce runoff losses, the irrigator can choose to either block the downstream end of the field, use cutback irrigation, or use runoff recovery. All methods have advantages and disadvantages.

Surge flow irrigation can be used to help reduce set times and reduce infiltration rates of the soil so that infiltration is more uniform within the field. Surge flow is accomplished with special valves that are equipped with a programmable controller to cycle the water as desired.

Questions

1. What is the major factor that determines the effective depth of water applied in a surface irrigation system without runoff recovery? How can this factor be modified?
2. Graph an advance-recession curve that characterizes uniform water distribution. Explain each component and the required conditions.
3. Without making considerable alterations to a furrow irrigation system, what changes can a manager make to minimize runoff and deep percolation? How do these adjustments

minimize water losses?

4. Graph and compare advance-recession curves for surface irrigation on a fine-textured soil and a coarse-textured soil. Explain the difference expected.

5. Given a furrow irrigated field with a medium textured soil and data below, determine: d_g, d_z, d_{LQ}, d_r, percent runoff, DU, and E_{LQ}. Assume that the advance and recession curves given in Figure 10.7 and the data in Table 10.1 apply to this problem.

   $Q = 900$ gpm — $N = 40$ watered furrows, every other furrow watered
   $L = 1{,}200$ ft — Furrow spacing = 30 in
   $T_{co} = 12$ h — $SWD = 4.0$ in

   You'll need to use Tables 10.3 and 10.4 for this problem.

6. If the slope in Question 5 is 0.3%, is maximum non-erosive stream-size being exceeded?

7. What would the E_{LQ} be for the conditions of Question 5 if a closed reuse system were installed and $R_t = 0.85$?

8. a. For the medium textured soil used in Question 5, determine the desired cutoff ratio to achieve maximum efficiency with a closed runoff recovery system (use Figure 10.15).

   b. How many gates should be opened to achieve the cutoff ratio given in 8.a.?

   c. Is the maximum non-erosive stream-size being exceeded?

   d. What is the theoretical maximum efficiency for the advance time in Question 8.a. without a 1-hour recession? Assume $R_t = 0.85$ and loam soil (use Figure 10.15).

9. Discuss the differences in efficiency between Questions 7 & 8.

10. Does the d_{LQ} exceed SWD in any cases given in Questions 7 or 8?

11. A farmer is making a conversion from continuous flow furrow irrigation to surge flow. Your job is to determine how to set the controller and estimate the savings in water due to surging. The following conditions apply:

   $Q = 800$ gpm
   Row length = 1,200 ft
   Row spacing = 30 in
   Every other furrow is watered
   $ET = 0.25$ in/d
   Irrigation frequency = 6 d (assume $SWD = ET \times$ time interval between irrigations)
   Net (effective) irrigation required per year = 10 in
   Field slope = 0.3%

   Continuous flow:
   Furrows per set = 45
   Set time (cutoff time) = 12 hr
   Average depth infiltrated (d_z) = 2.7 in
   $DU = 70\%$
   Advance time = 9 hr

   Surge flow:
   Furrows watered per side of valve = 30
   Set time = 6 hr on each side of valve (12 hr total for both sides)
   Average depth infiltrated = 2.0 in
   $DU = 80\%$
   Inflow time for advance = 4.5 hr

   a. Determine the on-times for each surge cycle using 4 advance cycles and 2 cycles after advance is complete (post-advance).

   b. Determine the gross depth applied and effective depth applied for each irrigation and for the year for both surge flow and continuous flow.

   c. How much less water was applied by surging? Express your answer in inches/year and percent savings.

   d. Determine the E_{LQ} for the continuous flow and surge systems.

References

Bautista, E., J. L Schlegel, and T. S. Strelkoff. (2012). *WinSRFR 4.1 software and user manual.* Maricopa, AZ: USDA-ARS Arid Land Agricultural Research Center.

Bautista, E., Clemmens, A. J., Strelkoff, T. S., & Schlegel, J. (2009). Modern analysis of surface irrigation systems with WinSRFR. *Agric. Water Manag., 96*(7), 1146-1154.

Cahoon, J. E., Mandel, P., & Eisenhauer, D. E. (1995). Management recommendations for sloping blocked-end furrow irrigation. *Appl. Eng. Agric., 11*(4), 527-533.

Dedrick, A. R., Gaddis, R. J., Clark, A. W., & Moore, A. W. (2007). Land forming for irrigation (Ch. 10). In G. J. Hoffman, R. G. Evans, M. E. Jensen, D. L. Martin, & R. L. Elliott (Eds.), *Design and operation of farm irrigation systems* (2nd ed.). St. Joseph, MI: ASABE.

FAO. (2021). AQUASTAT Main database. Rome, Italy: United Nations FAO.

Fekersillassie, D., & Eisenhauer, D. E. (2000). Feedback-controlled surge irrigation: II. Operating criteria. *Trans. ASAE, 43*(6), 1630-1641.

Humphreys, A. S. (1986). *Automated farm surface irrigation systems worldwide. International Commission on Irrigation and Drainage.* New Delhi: United India Press.

Kay, M. (1986). *Surface Irrigation: Systems and practice.* England: Cranfield Press.

Koech, R. K., Smith, R. J., & Gillies, M. H. (2014). A real-time optimisation system for automation of furrow irrigation. *Irrig. Sci., 32*(4), 319-327.

Koech, R., Smith, R., & Gillies, M. (2010). Automation and control in surface irrigation systems: Current status and expected future trends. Southern Region Engineering Conference. November 11-12. Toowoomba, Australia

Price, P. R., & Purcell, J. D. (2011). Surface irrigation (Ch.19). In L. E. Stetson, & B. Q. Mecham (Eds.), *Irrigation* (6th ed.). Falls Church, VA: Irrigation Association.

Stringham, G. E., & Keller, J. (1979). Surge flow for automatic irrigation. *Proc. ASCE Irrigation and Drainage Specialty Conference.* Reston, VA: ASCE.

USDA. (2012). Surface irrigation (Ch. 4) Part 623. In *Irrigation. National engineering handbook.* Natural Resources Conservation Service. Washington, DC: USDA.

USDA. (2019). 2018 Irrigation and Water Management Survey. Vol. 3. Special Studies, Part 1. AC-17-SS-1. Washington, DC: USDA.

Walker, W. R., & Skogerboe, G. V. (1987). *Surface irrigation: Theory and practice.* Prentice-Hall.

Yonts, C. D., Eisenhauer, D. E., & Cahoon, J. E. (1991). Fundamentals of surge irrigation. NebGuide G91-1018. Lincoln, NE: University of Nebraska Cooperative Extension.

Yonts, C. D., Eisenhauer, D. E., & Fekersillassie, D. (1996). Impact of surge irrigation on furrow water advance. *Trans. ASAE, 39*(3), 973-979.

Chapter 11 Sprinklers

11.1 Introduction

At this point we know how crops use water and have determined the amount of irrigation needed in the near term. We understand how much water can be supplied from ground and surface water sources. We need to know how to apply that water to our field. Many producers have chosen to utilize sprinkler irrigation. So now we need to determine how sprinkler systems operate and think about how to manage them for efficient water use. Questions to answer include how large of sprinkler nozzle is required, how rapidly does the system apply water, and therefore, how many hours should I operate the system? How much pressure does our pump need to provide? What do we need to do to achieve uniform distribution of water across the field?

Irrigation water is often applied using sprinkler systems. In fact, much of the growth in irrigated area in the U.S. during the past two decades has been due to development of sprinkler irrigation. Sprinkler irrigation has expanded because of the ability to irrigate many crops and landscapes efficiently that could not be irrigated effectively with surface irrigation. Advancements in controller technology have allowed automation of some sprinkler systems to further enhance efficiency and minimize labor inputs. This chapter focuses on the fundamentals of sprinkler irrigation while the following chapters focus on specific types of sprinkler systems. The National Engineering Handbook from the USDA-NRCS (2016) and the chapter by Martin et al. (2007) are references that provide detailed analysis, description of standard practices, and equipment specifications for sprinkler systems.

11.2 System Components

Sprinkler devices, frequently called heads (Figure 11.1), are the nucleus of sprinkler systems. Sprinkler devices consist of a sprinkler body that may be stationary or rotate due to water pressure. The water exits through a ***nozzle*** installed in the sprinkler body. The nozzle is smaller than the pipe leading to the sprinkler or the sprinkler body itself. The small diameter causes pressure to build in the pipe and sprinkler body. The ***discharge***, flow rate, of water through the nozzle is related to the amount of pressure and the diameter or size of the nozzle.

One sprinkler shown in Figure 11.1 is an ***impact sprinkler*** because the water from the nozzle sprays onto the spoon of the sprinkler arm. The impact of the jet on the spoon causes the arm to rotate away from the jet. The arm is connected to a spring that stores energy as the arm rotates. The spring decelerates the arm rotation, eventually causing the arm to stop. The energy stored in the spring is then released to accelerate in the opposite direction. As the arm returns to its original position it strikes the sprinkler body. This impact causes the sprinkler body to rotate through a small angle.

Many impact sprinklers use two nozzles. The nozzle that causes the sprinkler arm to rotate is the ***drive*** or ***range nozzle*** (Figure 11.1). The second nozzle often has a slit to enhance breakup of the water jet. This is called a ***spreader nozzle*** and it increases the amount of water applied close the sprinkler while the range nozzle throws water further.

Figure 11.1. Examples of sprinkler irrigation devices. (Upper left photo is a modification of a photo provided by jjharrison.com.au/ CC-BY-SA-3.0 through Wikimedia Commons; upper right photo is courtesy of Senniger Irrigation; bottom two photos are courtesy of Nelson.)

Impact sprinklers were invented in the 1930s and became the standard type of sprinkler device. Spray head devices were later developed where the jet impinges onto a plate or pad (Figure 11.1). The jet is divided into either several streams or a smooth surface of water. Some devices spray onto pads that rotate or vibrate due to the impact of the water jet. Spray pad devices are used extensively on center pivot and linear-move irrigation systems.

Low-angle impact sprinklers and spray heads were developed for systems that position the device above the crop. The low angle reduces drift and evaporation. Large or high-volume gun sprinklers are operated at a high pressure and are designed to throw water hundreds of feet. Only one gun is generally used at a time. The gun often travels across the field in a continuous motion. Some sprinklers are also made to rotate throughout part of a circle. The ***part-circle*** sprinkler has a special mechanism so that a latch engages when the sprinkler has rotated to the desired angle. It then rotates in the opposite direction to the original position. The part-circle sprinkler has been very useful at the edges of fields, on guns, and in landscape and turf applications where irregularly shaped areas are irrigated.

The nozzle is used to build pressure causing a water jet to discharge from the device. Several properties of that jet are important for successful operation of a sprinkler system. It is desirable to have nozzles that throw water as far as possible using as little pressure as possible. In addition, it is desirable for the droplet to breakup so that the application is uniform. Large drops are often desirable because they have smaller drift or evaporation issues; however, large droplets can pack unprotected soil surfaces. Trade-offs often occur as some criteria are contradictory.

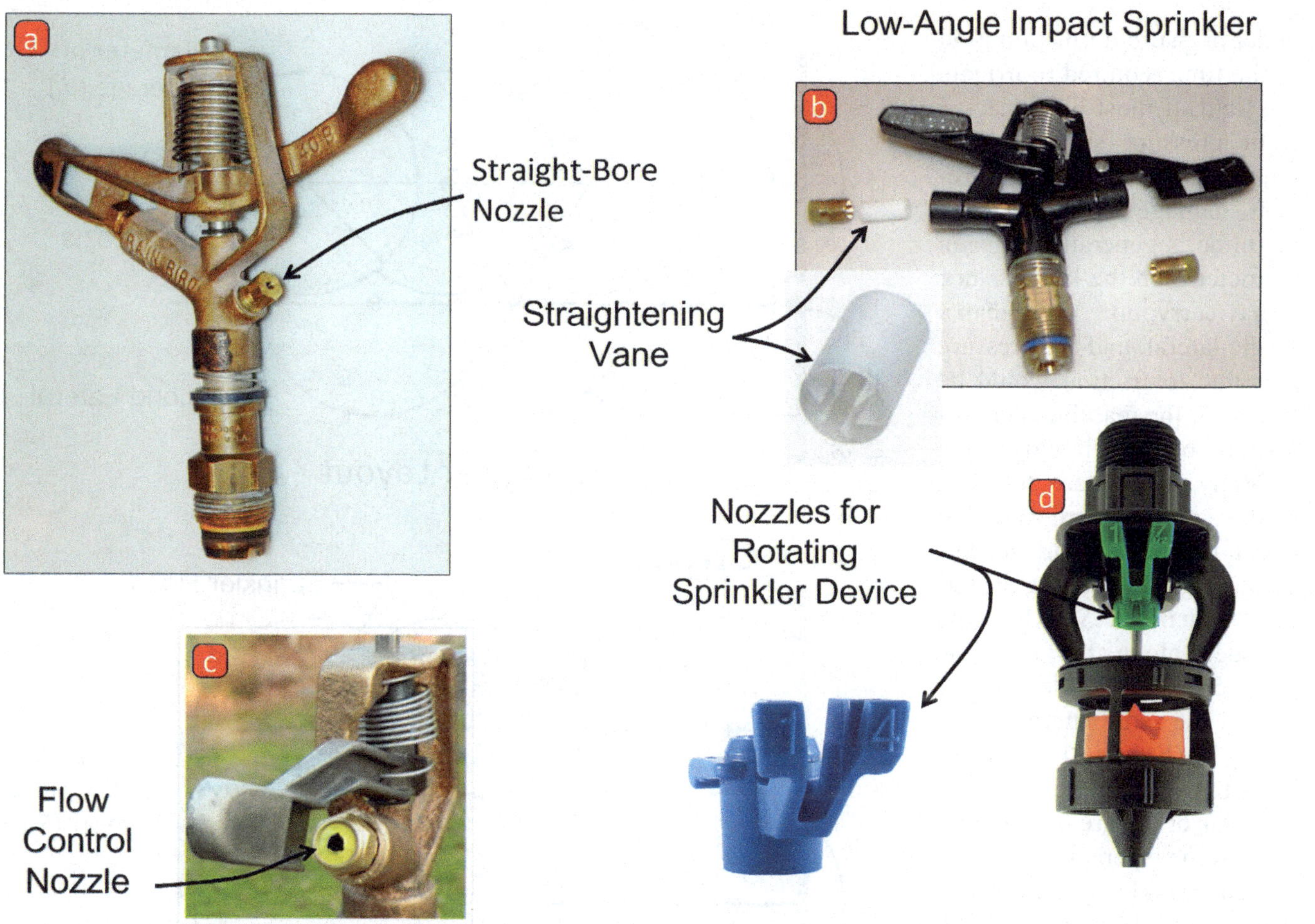

Figure 11.2. Examples of nozzles used in sprinkler irrigation. (Lower right, photo courtesy of Senninger Irrigation.)

Many types of nozzles have been developed to accommodate these objectives (Figure 11.2). The original, and still popular, nozzle was the straight bore nozzle. It is usually made of brass and machined to be very smooth to reduce turbulence in the nozzle. The hole in the center of the nozzle matches commercially available drill bit sizes. Small plastic inserts called straightening vanes are sometimes used upstream of the nozzle to reduce turbulence and increase the distance of throw. Vanes are frequently built into the body of spray devices.

In the 1970s an energy crisis occurred causing the cost of sprinkler irrigation to increase dramatically. As illustrated in Chapter 8, decreasing pump pressure can reduce operating costs. When straight bore nozzles are operated at low pressure, the jet does not breakup very well leading to poor uniformity and soil compaction. As a result, low-pressure nozzles were developed. Some are shown in Figure 11.2.

Water is provided to the sprinkler device from a pipeline called a lateral (Figure 11.3). The sprinkler lateral may be located below the ground surface as with solid-set and turf irrigation systems. For some systems, the lateral lays on the soil surface, while for continuously moving systems the lateral is carried above the soil surface by a series of towers. A smaller diameter pipe—the riser—is used to conduct water from the lateral to the sprinkler device for some systems (Figure 11.3). Risers are primarily used to position the sprinkler above the crop and/or structural elements of the irrigation system to prevent canopy or structural interference with the jet.

Water is supplied to the lateral with the ***mainline*** (Figure 11.3) which is an enclosed pipeline conveying water from the source at the inlet of the mainline to outlet of the mainline at the lateral. The mainline may serve several laterals simultaneously. The mainline is under pressure for the duration of the time required to irrigate the field. It must be protected from pressure surges, vacuums, and other factors to prevent damage or leaks. The mainline is generally larger in diameter than the lateral since it may carry more water than a single lateral, and the pressure loss due to friction would be larger for the mainline than for a lateral of equal length.

Many systems are designed so that sprinklers are spaced at equal intervals along the lateral. The spacing along the lateral is denoted S_L, while the distance along the mainline between successive laterals, or sets of the same lateral, is denoted S_m (Figure 11.3). Sprinklers may be laid out in a rectangular or square orientation. For some permanently installed sprinkler systems heads may be placed in a triangular orientation as shown in Figure 11.3. In this orientation the sprinklers are placed an equal distance S from adjacent sprinklers in an equilateral orientation.

The diameter of coverage of the individual sprinkler is a critical property for the system (Figure 11.3). Sprinklers and laterals need to be placed close enough to overlap providing uniformity; therefore, sprinklers and laterals are spaced closer than the diameter of coverage. The crosshatched areas in Figure 11.3 are representative areas for computing the rate and uniformity of water application. The area for the equilateral triangular orientation is $A = 0.433\ S^2$ while the area for rectangular or square orientations is $A = S_m \times S_L$.

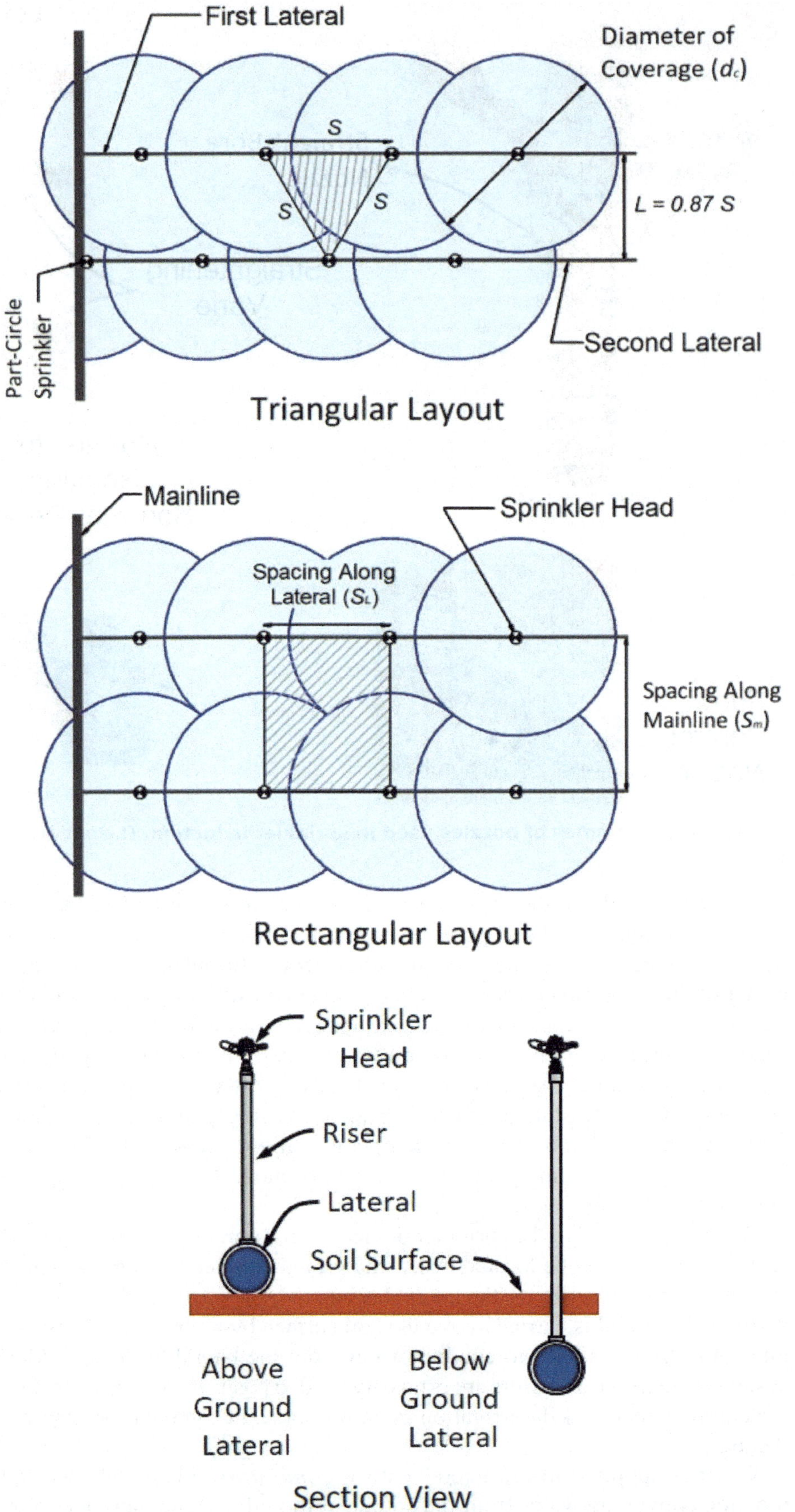

Figure 11.3. Components and layout of typical sprinkler systems.

11.3 Sprinkler Performance

The performance of sprinkler systems depends on the operation of individual sprinkler heads. The goal of sprinkler irrigation is to apply water uniformly at a rate that does not cause runoff or erosion. The system should meet crop water requirements and attain the highest practical efficiency—and of course must be cost-effective.

The ***discharge***, or volume flow rate of water, leaving the nozzle is important and can be described by:

$$q_s = 29.82\, C_d\, D^2 \sqrt{P} \qquad (11.1)$$

where: q_s = discharge through the nozzle (gallons per minute, gpm),
C_d = discharge coefficient for the sprinkler head,
D = inside diameter of the nozzle orifice (inches),
P = pressure of the water at the inlet to the sprinkler device (pounds per square inch, psi); and
29.82 is a unit conversion and geometric constant.

Example 11.1

The discharge from a sprinkler depends on the pressure at the nozzle and the diameter of the nozzle orifice. Would a 20% increase in nozzle diameter produce more flow than a 20% increase in pressure?

Given: A straight bore nozzle is used in a sprinkler. The discharge is 10 gpm.

Solution:

Let q_{s1} = 10 gpm be the initial flow rate.

Use Equation 11.1 to develop a term called the discharge ratio where 2 denotes the new condition and 1 the original condition.

$$\frac{q_{s2}}{q_{s1}} = \frac{D_2^2 \sqrt{P_2}}{D_1^2 \sqrt{P_1}}$$

For a 20% increase in diameter $D_2 = 1.2 \times D_1$ and $P_2 = P_1$

$$q_{s2} = q_{s1} \left(\frac{1.2 \times D_1}{D_1} \right)^2 = 1.44 \times q_{s1} = 14.4 \text{ gpm}$$

So, a 20% increase in diameter provides a 44% $\left(\frac{14.4 - 10}{10} \times 100\% \right)$ increase in flow.

For a 20% increase in pressure $P_2 = 1.2\, P_1$ and $D_2 = D_1$

$$q_{s2} = q_{s1} \sqrt{\frac{1.2\, P_1}{P_1}} = 1.1\, q_{s1} = 11 \text{ gpm}$$

So, a 20% increase in pressure only changes the discharge by 10%.

Changing the nozzle size increases flow more than an equal percentage change of pressure.

The value of the discharge coefficient is about 0.96 but depends on the design of the nozzle and sprinkler head. The inside diameter of nozzles is customarily referred to as the ***nozzle size***. Performance for a range of nozzle sizes and pressures is summarized in Table 11.1 for straight bore nozzles. Some manufacturers produce nozzles sized by the diameter in 64ths of an inch; others may use 128ths of an inch. For example, a nozzle diameter of one-quarter inch is referred to as a size of 16 (Table 11.1) or 32 depending on which system is used. Table 11.1 represents the typical discharge for a broad range of sprinkler nozzles. The discharge from a specific design of nozzle and sprinkler may vary from data presented in Table 11.1. Data from the relevant manufacturer should be used for specific systems. The total discharge from a sprinkler head with two nozzle outlets is the sum of the discharge from each nozzle for that pressure. The discharge for pressures between those listed in Table 11.1 can be determined by interpolation.

The second important characteristic of sprinkler performance is the ***diameter of coverage*** (also referred to as wetted diameter) as illustrated in Figures 11.3 and 11.4. The diameter of coverage is the maximum diameter wetted by the sprinkler at a rate that is significant for the intended use of the sprinkler. For example, the diameter of coverage for agricultural sprinklers is usually determined to be the maximum radial distance where the water application rate equals 0.01 inches per hour. Usually, the wetted diameter is measured in an indoor laboratory with no wind. The diameter of coverage is affected by the design of the sprinkler body and nozzle. Representative diameters of coverage are given in Table 11.2 for impact sprinklers with straight bore nozzles. The data are for sprinklers where the water jet exits from the sprinkler head at an angle of 23° above the horizon for the range nozzle. The diameter may vary for other designs and should be determined from data by the manufacturer. The diameter of coverage also depends on the height of the device above the crop or the ground surface; therefore, the intended usage of the device is important.

The straightening vanes shown in Figure 11.2 are used to reduce turbulence in the sprinkler barrel; thus, producing a larger diameter of coverage. Straightening vanes increase the diameter of coverage from 5% to as much as 20% depending on the design of the sprinkler head and the specific nozzle.

Nozzles are designed to operate within a specified pressure range. When used outside that range the performance changes in an undesirable way. The patterns shown in Figure 11.5 illustrate the impact of pressure on the distribution of water. When the pressure is within the proper range, the pattern is nearly elliptical with distance from the sprinkler. When the pressure is too high the water jet breaks up into a high percentage of small drops. In some cases, the jet atomizes into very small drops.

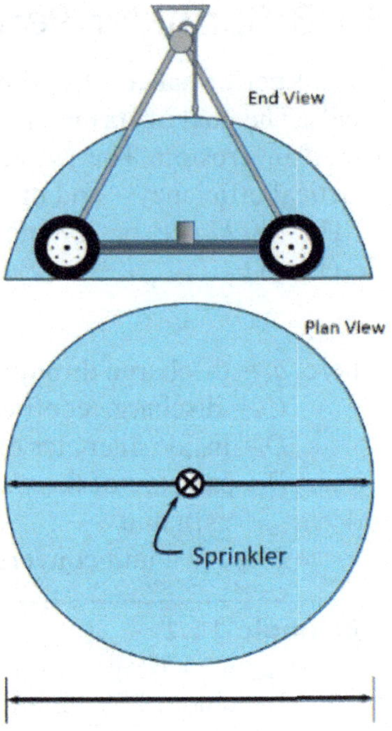

Figure 11.4. Diameter of coverage for a sprinkler.

Table 11.1. Discharge (gpm) for straight bore nozzles of various sizes operating at a range of pressures.

Nozzle Size		Nozzle Pressure (psi)															
in	64th in	25	30	35	40	45	50	55	60	65	70	75	80	85	90	95	100
3/32	6	1.2	1.4	1.5	1.6	1.7	1.8	1.9									
7/64	7	1.7	1.9	2.0	2.2	2.3	2.4	2.6									
1/8	8	2.2	2.5	2.7	2.9	3.0	3.2	3.4	3.5	3.6	3.8	3.9	4.0				
9/64	9	2.8	3.1	3.4	3.6	3.8	4.0	4.2	4.4	4.6	4.8	5.0	5.1				
5/32	10	3.5	3.9	4.2	4.5	4.7	5.0	5.2	5.5	5.7	5.9	6.1	6.3				
11/64	11	4.3	4.7	5.1	5.4	5.7	6.0	6.3	6.6	6.9	7.2	7.4	7.6				
3/16	12	5.1	5.6	6.0	6.4	6.8	7.2	7.5	7.9	8.2	8.5	8.8	9.1				
13/64	13	6.0	6.5	7.1	7.6	8.0	8.4	8.9	9.2	9.6	10.0	10.3	10.7				
7/32	14	6.9	7.6	8.2	8.8	9.3	9.8	10.3	10.7	11.2	11.6	12.0	12.4				
15/64	15	7.9	8.7	9.4	10.1	10.7	11.2	11.8	12.3	12.8	13.3	13.8	14.2				
1/4	16	9.0	9.9	10.7	11.4	12.1	12.8	13.4	14.0	14.6	15.1	15.7	16.2				
17/64	17	10.0	11.2	12.1	12.9	13.7	14.4	15.1	15.8	16.5	17.1	17.7	18.3				
9/32	18	11.0	12.5	13.5	14.5	15.4	16.2	17.0	17.7	18.5	19.2	19.8	20.5				
5/16	20	14.0	15.5	16.7	17.9	19.0	20.0	21.0	21.9	22.8	23.6	24.5	25.3				
11/32	22	17.0	19	20	22	23	24	25	26	28	29	30	31	32	32	33	34
3/8	24	20.0	22	24	26	27	29	30	32	33	34	35	36	38	39	40	41
13/32	26	23.0	26	28	30	32	34	35	37	38	40	41	43	44	45	47	48
7/16	28	27.0	30	33	35	37	39	41	43	45	46	48	50	51	53	54	55
15/32	30	31.0	35	38	40	43	45	47	49	51	53	55	57	59	60	62	64
½	32	33.0	37	40	43	45	48	50	52	54	57	58	60	62	64	66	68
17/32	34	38.0	42	45	48	51	54	57	59	62	64	66	68	70	72	74	76
9/16	36	42.0	47	51	54	57	60	64	66	69	72	74	76	79	81	83	86
5/8	40	52.0	58	62	67	71	75	78	82	85	88	92	94	97	100	103	105
11/16	44	63.0	70	76	81	84	90	95	99	103	106	110	114	117	121	124	127

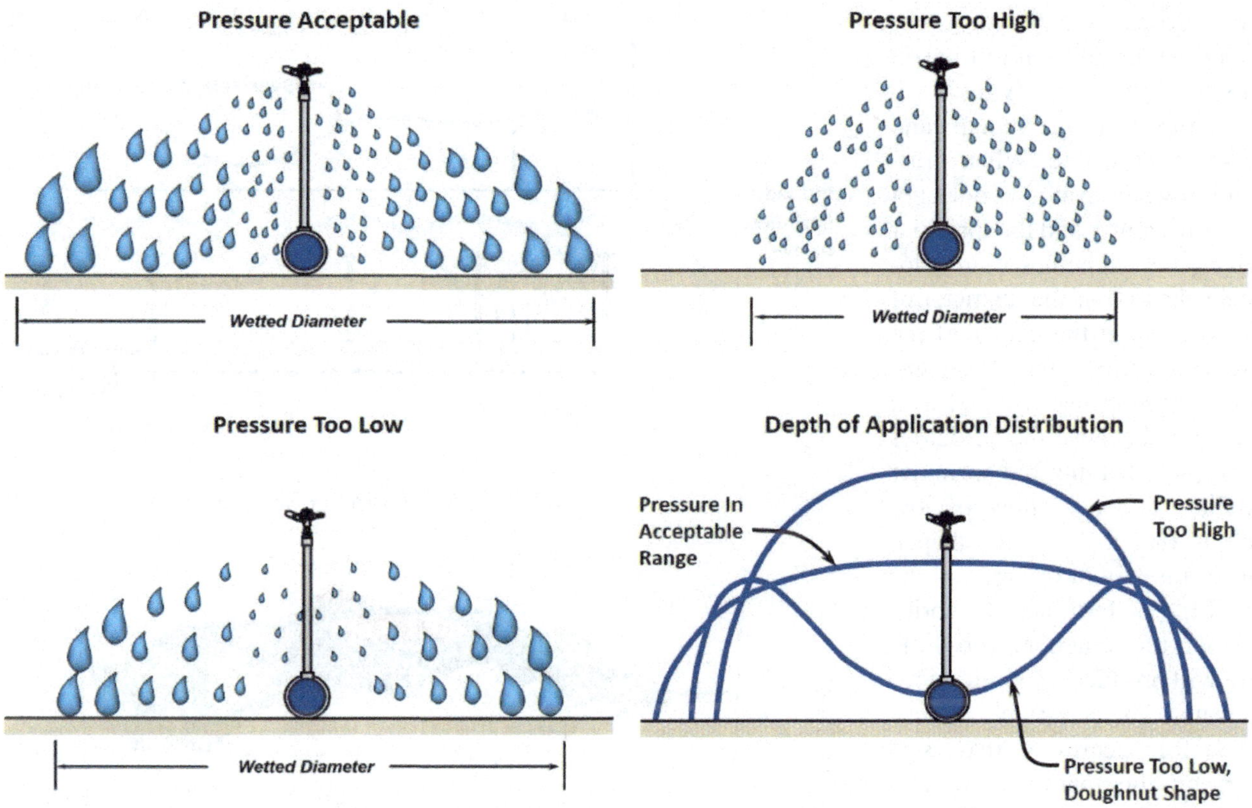

Figure 11.5. Sprinkler distribution as affected by operating pressure.

Table 11.2. Diameter of coverage (ft) for impact sprinklers with straight bore nozzles.[a]

Nozzle Size		Nozzle Pressure (psi)															
in	64th in	25	30	35	40	45	50	55	60	65	70	75	80	85	90	95	100
3/32	6	64	66	68	69	70	71	72									
7/64	7	65	67	69	70	71	72	73									
1/8	8	78	79	80	81	82	83	84	85	86	86	87	87				
9/64	9	80	81	82	83	84	85	86	87	88	89	90	91				
5/32	10	82	85	87	88	89	90	91	92	93	94	95	96				
11/64	11	83	88	90	92	93	95	96	97	98	99	100	101				
3/16	12	85	91	94	96	98	100	101	102	103	104	105	106				
13/64	13	91	97	100	103	105	107	109	111	113	114	116	117				
7/32	14	92	99	102	105	108	110	113	115	117	118	120	122				
15/64	15	93	100	104	107	110	112	115	117	119	121	123	125				
1/4	16	94	102	105	109	112	115	118	120	122	124	127	129				
17/64	17	95	103	107	110	114	117	119	122	125	127	129	131				
9/32	18	96	104	108	112	116	119	122	125	127	130	132	134				
5/16	20	121	124	127	130	133	136	140	143	145	147	149	151				
11/32	22	122	128	134	138	142	146	150	154	158	162	164	166	170	172	174	176
3/8	24	124	130	136	142	146	150	154	158	162	166	168	172	174	178	180	182
13/32	26	128	136	144	150	154	158	162	166	168	172	174	178	180	184	186	188
7/16	28	132	138	158	154	158	162	166	172	174	178	180	184	186	190	192	194
15/32	30	132	144	154	160	164	168	172	176	180	182	186	188	192	194	196	198
1/2	32	132	146	156	166	170	174	178	182	186	188	192	194	198	200	202	204
17/32	34	132	146	158	166	176	180	184	188	192	196	198	202	204	208	210	212
9/16	36	132	146	158	172	180	188	192	194	198	202	204	208	210	212	216	218
5/8	40	132	146	158	172	184	190	198	202	204	208	210	214	216	220	222	224
11/16	44	132	146	158	172	184	194	200	208	212	216	218	220	224	226	230	232

[a] For a brass impact sprinkler where the exit angle of the range nozzle is 23° above the horizontal.

Small drops decelerate very quickly in the air and fall to the soil close to the sprinkler, giving a reduced diameter of coverage and higher application rate. When pressure is too low, the water jet does not breakup sufficiently and the sprinkler primarily wets an annular area located near the end of the diameter of coverage. Areas at the center of the circle receive little water. The diameter of coverage is also reduced with low pressures because the velocities of the droplets leaving the sprinkler are smaller. The net effect of low pressure is that a doughnut shaped pattern results with a dry area in the middle of the pattern near the sprinkler. Either too much or too little pressure will produce a poor distribution of water. The acceptable operating range for specific sprinklers is provided by the manufacturer and should be followed. Straightening vanes reduce droplet breakup, which can lead to a doughnut shaped pattern. Usually, the minimum operating pressure of sprinklers with vanes is higher than those without vanes to prevent the doughnut shaped pattern.

Sprinkler systems require that the water pattern from one sprinkler overlap with adjacent sprinklers. When the sprinklers are properly designed and located, the overlap pattern will be like that shown in Figures 11.3 and 11.6. The depth of water applied to a point is the sum of water from all sprinklers reaching that point. In Figure 11.6 the total depth would be $d_1 + d_2$ for the point shown. Some irrigators attempt to reduce costs by extending the spacing between laterals or between sprinklers on a lateral. When that is done, the overlap is inadequate and poor uniformity results. The upper portion of Figure 11.6 shows the water pattern where sprinklers are spaced a distance S_L between sprinklers. The

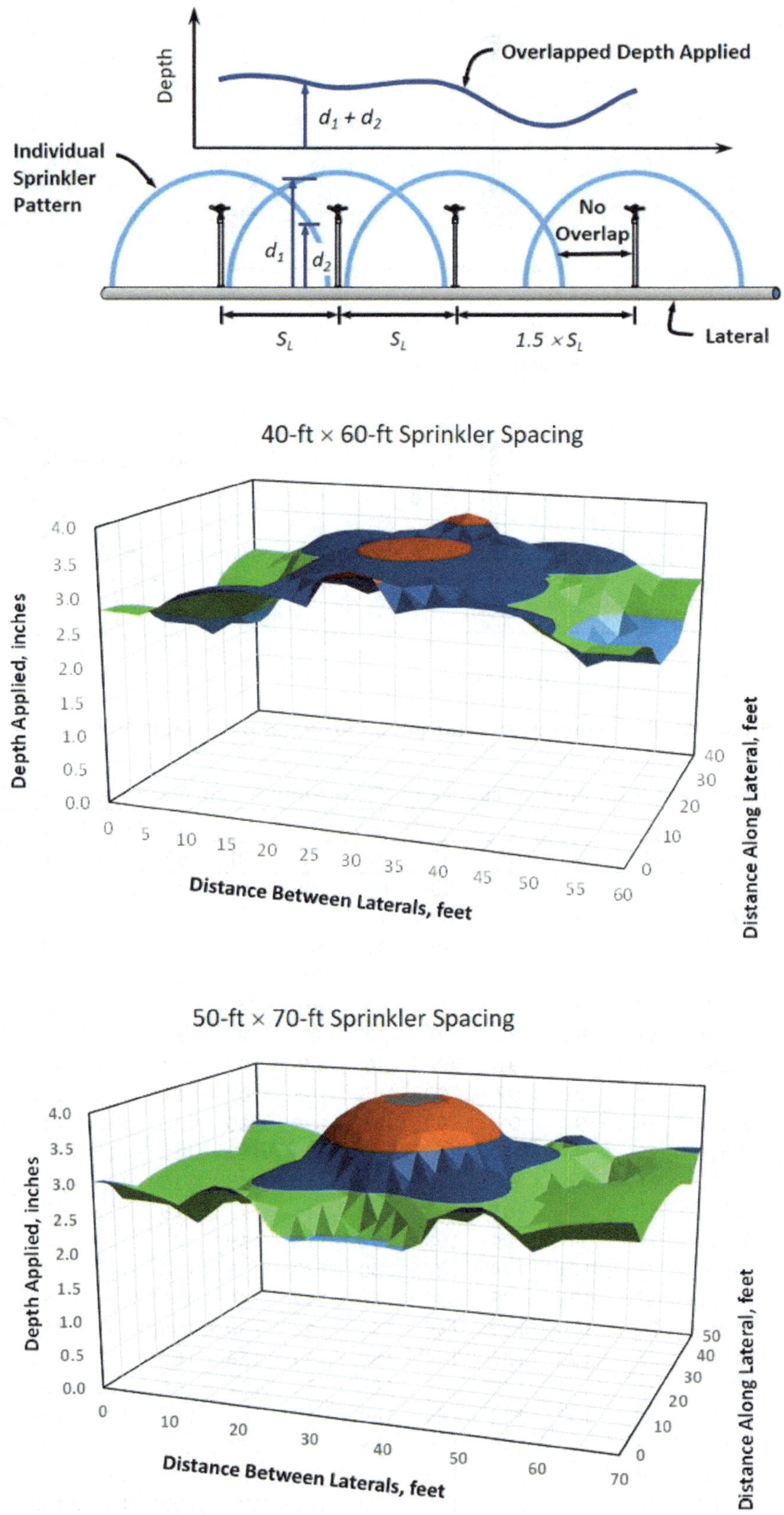

Figure 11.6. Illustration of the effect of sprinkler spacing on uniformity of water application.

depth of application is relatively uniform. When the spacing is increased to 1.5 × S_L, the depth of application between the sprinklers decreases.

The lower and middle portions of Figure 11.6 show the three-dimensional distribution of water between sprinklers along and between laterals. The middle figure shows the distribution when sprinklers are spaced 40 ft apart along the lateral and 60 ft between laterals. The bottom figure shows the distribution when the spacing is increased to 50 ft along and 70 ft between laterals. The system was designed to apply 3 inches of water during a 10-hour application time with an operating pressure of 50 psi. The required sprinkler discharge for the 40 ft × 60 ft spacing is 7.48 gpm and 10.9 gpm for the 50 ft× 70 ft spacing. The diameter of coverage for the two spacings was 106 and 112 ft, respectively. The wider spacing leads to a poorer distribution with a peak depth centered in the representative area.

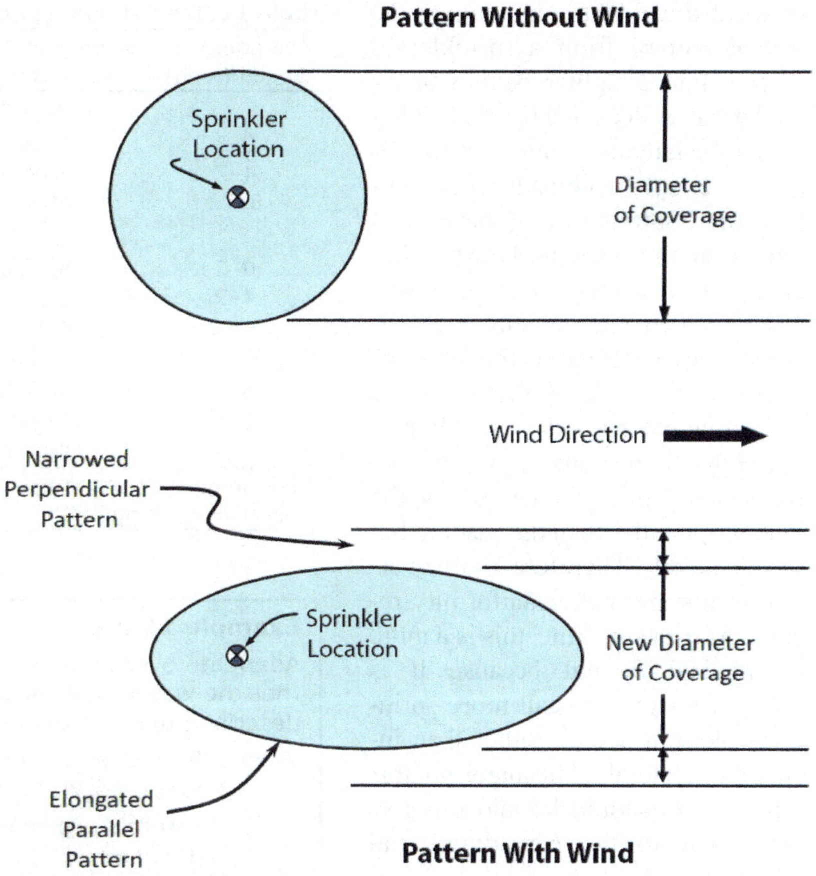

Figure 11.7. Plan view of the effect of wind on the distribution of water from a sprinkler.

Overlap requirements have been developed for sprinkler systems. The recommendations depend on the wind speed and direction. Consider a plan view of the wetted pattern of a single sprinkler as shown in Figure 11.7. The pattern for calm conditions will be circular. As the wind speed increases the pattern is displaced downwind giving the elongated pattern. Note that the wetted pattern is not only displaced downwind, but also is narrower perpendicular to the direction of wind travel. This occurs because the wind blowing perpendicular to the water jet causes droplets to travel a curved path, originally perpendicular to the wind but later in a more downwind direction. The perpendicular wind may also cause the jet to breakup into a distribution with a higher percentage of smaller drops which do not travel as far.

The narrowing of the wetted pattern perpendicular to the wind direction has an important impact on sprinkler spacings and on the orientation of the lateral relative to the predominant wind during the irrigation season. The diagram in Figure 11.8 shows the effect of wind on two orientations of laterals relative to the wind direction. With no wind the individual sprinkler pattern is circular, and the laterals appear to have adequate overlap. When the laterals are oriented parallel

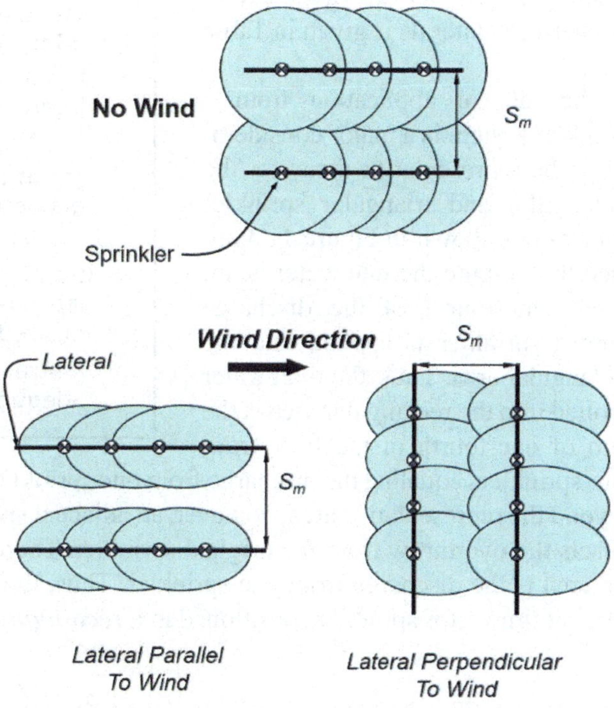

Figure 11.8. Effect of wind direction on overlap of sprinkler distribution pattern.

to wind travel, the wind causes the wetted pattern from a sprinkler to narrow into a tighter pattern along the lateral. A dry zone may result between the laterals because of insufficient overlap. To adjust for this problem, more laterals would be needed with a smaller spacing between laterals. This leads to a more expensive system. When the laterals are oriented perpendicular to the wind as shown in Figure 11.8, the pattern of an individual sprinkler is still narrower due to the wind; however, now the spacing of sprinklers along the lateral is smaller than the spacing between laterals. Therefore, more overlap occurs and better uniformity results. At the same time, this is a more economical system because it is more efficient to install more sprinklers along a lateral, rather than install more laterals. The major point is that laterals should be laid out perpendicular to the pervading wind when possible.

The spacing of sprinklers and laterals depends upon wind conditions. Recommendations for maximum spacing of sprinklers along the lateral and between laterals is given in Table 11.3.

Table 11.3. Maximum spacing of sprinklers.

Average Wind Speed (mph)	Maximum Spacing Between Sprinklers on the Lateral	Maximum Spacing Between Laterals Along the Mainline
	Rectangular Spacing	
0–3	50% of diameter	60% of diameter
4–7	45% of diameter	60% of diameter
8–12	40% of diameter	60% of diameter
	Square Spacing	
0–3	55% of diameter	-
4–7	50% of diameter	-
8–12	45% of diameter	-
	Equilateral Triangle Spacing	
0–3	60% of diameter	[a]
4–7	55% of diameter	[a]
8–12	50% of diameter	[a]

[a] For an equilateral triangle pattern, the spacing between laterals is 0.866 × sprinkler spacing.

Example 11.2

Adequate overlap is necessary with sprinkler systems to ensure that the water application is reasonably uniform. Will the layout described below provide acceptable uniformity?

Given: Impact sprinklers with two nozzles (23° exit angle for range nozzle) are spaced 40 ft apart along a lateral.

Laterals are spaced at intervals of 60 ft along the mainline.

The nozzle sizes are 11/64 in × 3/32 in and the operating pressure is 50 psi.

Wind in the area usually averages 5 mph.

Solution:

The diameter of coverage for the range nozzle (11/64 in) is 95 ft, from Table 11.2.

From Table 11.3 the maximum sprinkler spacing along the lateral is 45% of the diameter of coverage.

Maximum spacing along lateral = 0.45 × 95 = 43 ft

The maximum spacing between laterals is 60% of the diameter of coverage.

Maximum spacing between laterals = 0.60 × 95 = 57 ft

So, this layout just fails the criteria for the spacing between laterals in Table 11.3 and uniformity may be less than desired.

A straightening vane for the nozzle would probably provide adequate coverage.

The rate of application from a sprinkler system is a major consideration. The representative areas for the rectangular and triangular sprinkler orientations shown in Figure 11.3 are used to compute the rate water is applied. One-fourth of the discharge from a sprinkler is applied into the rectangular area. Thus, the total water applied into the rectangular area is the sum of one-fourth of the flow from four sprinklers equaling the discharge from one sprinkler. Water from a sprinkler may be applied beyond the representative area; however, an adjacent sprinkler applies water into the area which offsets the overthrow from the original sprinkler. Therefore, the effective water discharge into the area is the discharge from one sprinkler. Thus, the application rate—volume per unit area per unit time—for sprinklers positioned in a ***rectangular spacing*** is given by:

$$A_r = \frac{96.3\ q_s}{S_L\ S_m} \tag{11.2}$$

where: A_r = the rate of water application (in/hr),
q_s = the sprinkler discharge rate (gpm),
S_L = spacing of sprinklers along the lateral (ft),
S_m = spacing of laterals along the mainline (ft), and
96.3 is for unit conversion.

For a square spacing S_L and S_m are equal so that the denominator becomes S_L^2.

The effective water supply into a triangular space is half of the discharge from a sprinkler, i.e., a sprinkler applies water into six triangles surrounding the sprinkler. This yields the application rate of a system with sprinklers oriented in an ***equilateral triangle*** spacing as:

$$A_r = \frac{111.2\, q_s}{S^2} \tag{11.3}$$

where S = the spacing of sprinklers in the triangular orientation and all other parameters are as previously defined.

The application rate of the sprinkler system is important for two reasons. First, the depth of water applied for a given time is proportional to the application rate:

$$d_g = A_r\, T_o \tag{11.4}$$

where: d_g = the gross depth of water applied per irrigation (in) and
T_o = the actual time of operation (hr).

For example, if the application rate was 0.4 in/hr, then an irrigation that lasted 10 hours would apply 4 in of water. The time of operation (T_o) is the time that water is applied. The quantity of water determined from Equation 11.4 is a gross application and must be reduced by the application efficiency to determine the amount of water provided to the crop.

Second, when the application rate of the sprinkler system exceeds the infiltration rate of the soil, water will accumulate on the soil surface. If enough water accumulates, runoff will begin. The maximum application rate that is acceptable for different soils and slopes is summarized in Table 11.4. These are general recommendations and should be adjusted upward for production practices that enhance infiltration, especially where adequate crop residue protects the soil, and downward for practices that reduce infiltration.

Table 11.4. Maximum recommended water application rates for soils (inches/hr).[a]

Slope (%)	Coarse Textured Soils (sands, fine sands, and loamy fine sands)	Medium Textured Soils (sandy loams, fine sandy loams, and silt loam soils)	Fine Textured Soils (silty clay loams, clay loams, and clayey soils)
Soil Surface Not Protected			
0–5	0.50–0.75	0.25–0.50	0.10–0.25
6–8	0.40–0.60	0.20–0.40	0.08–0.20
9–12	0.30–0.45	0.15–0.30	0.06–0.15
13–20	0.20–0.30	0.10–0.20	0.04–0.10
> 20	0.10–0.20	0.05–0.10	0.02–0.05
Turfgrass or Heavy Residue Cover			
0–5	0.85–1.30	0.50–0.95	0.15–0.35
6–8	0.70–1.00	0.40–0.75	0.10–0.25
9–12	0.50–0.75	0.30–0.55	0.10–0.20
13–20	0.35–0.50	0.20–0.40	0.05–0.15
> 20	0.15–0.35	0.10–0.20	0.03–0.05

[a] Based on recommendations of the Rain Bird Corporation and Pair et al., 1983.

Example 11.3

A sprinkler irrigation system is used to irrigate a young row crop with an unprotected soil surface. The sprinkler spacing is 40 ft between sprinklers and 60 ft between the laterals.A pressure of 50 psi is available at the design location along the lateral. Wind in the region averages 5 mph during the irrigation season.The soil texture is silt loam.The sprinklers are brass impact sprinklers with straight bore nozzles and the range nozzle has an exit angle of 23° above horizontal.

Determine the smallest nozzle size that is acceptable and the application rate of the system.

Is this system acceptable for a silt loam soils with a slope of 2%?

Given: Sprinkler spacing 40 ft x 60 ft — Soil is silt loam

P_a = 50 psi — Wind = 5 mph

Brass impact sprinklers with straight bore and range nozzles at 23° angle

Solution:

a. The maximum spacing of sprinklers along the lateral is 45% of the diameter of coverage for a wind speed of 5 mph in Table 11.3. The maximum spacing between laterals is 60% of the diameter of coverage for that wind speed.

Since we know the actual spacing of the sprinklers and the lateral spacing, we need to determine the diameter of coverage (d_c) needed for this system:

d_c needed = maximum of $S_L/0.45$ and $S_m/0.60$ or

d_c needed = maximum of 40/0.45 = 89 ft and 60/0.60 = 100 ft.

The spacing of laterals along the mainline is the most limiting based on the criteria in Table 11.3.

Thus, a diameter of coverage of 100 feet is needed for sprinklers in this system.

From Table 11.2, a nozzle size of 3/16 inch will provide a diameter of coverage of 100 feet when operated at a pressure of 50 psi, so the nozzle should provide adequate overlap.

b. From Table 11.1, a 3/16-inch nozzle operated at 50 psi produces a discharge of 7.2 gpm.

Using Equation 11.2 the application rate would be:

$$A_r = \frac{96.3\ q_s}{S_L\ S_m} = \frac{96.3 \times 7.2\,\text{gpm}}{40\,\text{ft} \times 60\,\text{ft}} = 0.29\ \text{in/hr}$$

c. The maximum recommended application rate for silt loam soil with little cover is between 0.25 and 0.5 in/hr (Table 11.4).

Therefore, the 3/16-inch nozzle meets the overlap and application rate limitations.

11.4 Lateral Design

Sizing laterals is fundamental to sprinkler irrigation. Laterals must be large enough to carry the needed flow without excessive pressure loss. The general criteria constraints the variation of sprinkler discharge along the lateral. The difference between discharge from the sprinkler with the largest flow to the sprinkler with the smallest flow should be less than 10% of the average discharge. Since discharge from a sprinkler is related to the square root of pressure, 10% discharge variation is equivalent to a maximum permissible pressure variation of 20% (i.e., since $q_s \propto \sqrt{P}$ if the maximum discharge ratio is 1.1, then the maximum ratio for pressure variation would be $1.1^2 = 1.21$ or about 1.2).

Pressure varies along a lateral due to elevation changes and friction loss in the pipe and fittings. The pressure distribution along a lateral placed on level ground is illustrated in Figure 11.9. The pressure at the inlet of the lateral is determined by the pressure available from the mainline. The pressure loss in the first several lengths of the lateral is nearly the same as for a conveyance pipe without outlets along the pipe. However, as water is discharged from sprinklers the flow in the lateral decreases with distance. Ultimately, the flow in the last section of the lateral is that discharged from the last sprinkler. Of course, there is very little loss in the lateral for such a small flow.

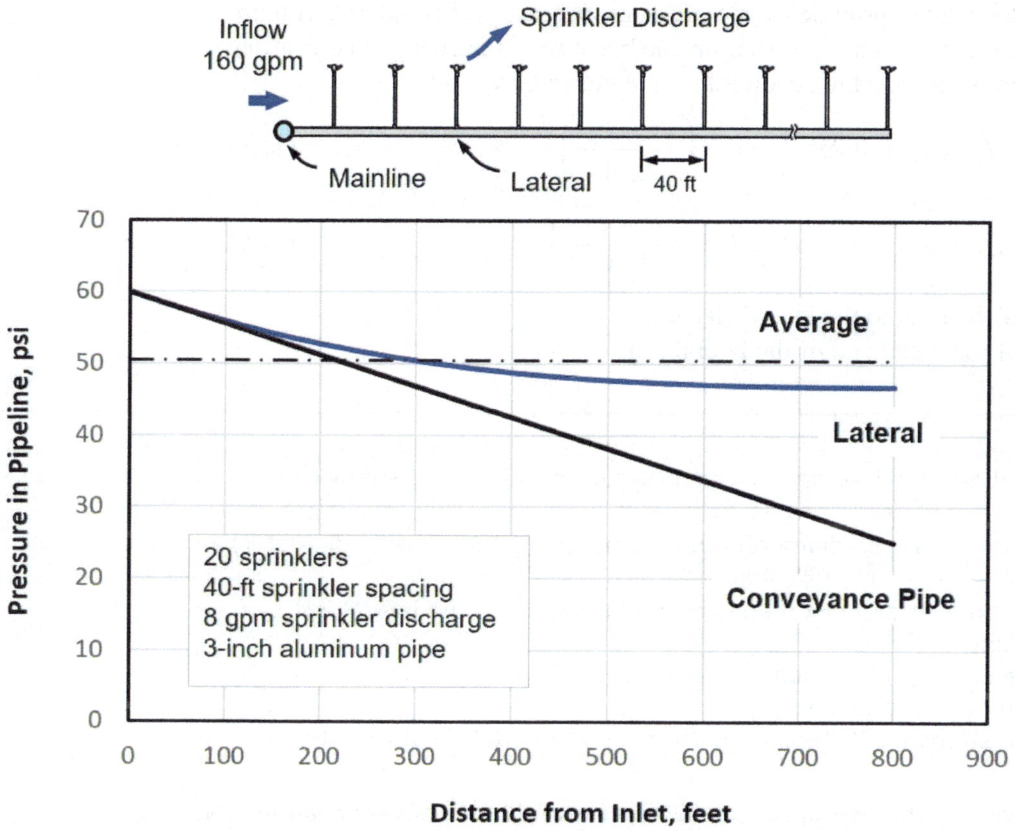

Figure 11.9. Pressure distribution along a lateral placed on a level surface.

The example in Figure 11.9 represents a 3-inch aluminum pipe lateral with sprinklers spaced 40 feet apart. The lateral is 800 feet long with 20 sprinklers averaging 8 gpm discharge per sprinkler. The friction loss in a 3-inch conveyance pipe 800-ft long with an inflow of 160 gpm is approximately 35 psi. Since the inlet pressure is 60 psi, the pressure at the end of the pipe is about 25 psi. The pressure loss in a lateral with the same pipe is only about 13 psi. The friction loss for this lateral is about 37% of the loss encountered in a conveyance pipe of the same diameter and inflow (the same as listed in Table 8.3 for 20 sprinklers). The average pressure along the lateral is about 50 psi and the average pressure occurs about 38% of the way along the lateral from the inlet. About 75% of the total head loss along the lateral occurs betweenthe inlet and the point where the average pressure occurs.

Sprinkler systems are usually designed by selecting the nozzle size for the average pressure along the lateral. Then the pressures at the ends of the lateral are computed. The pressures for a lateral on level ground can be computed by:

$$P_i = P_a + 0.75\,P_l \tag{11.5}$$

$$P_d = P_a - 0.25\,P_l \tag{11.6}$$

where: P_i = pressure at the inlet into the lateral (psi),

P_a = average pressure along the lateral (psi),

P_d = pressure at the distal end of the lateral (psi), and

P_l = pressure loss along the lateral (psi).

The maximum pressure loss along a lateral on the level is 20% of the average, or design, pressure of the lateral:

$$\text{Max } P_l = 0.20\,P_a \tag{11.7}$$

When a lateral runs up or down hill, the change in elevation causes changes in pressure. An elevation change of 10 ft is equal to a pressure change of 4.3 psi. Thus, when laterals run downhill there is less pressure variation from the inlet to the distal end than for laterals on level ground because the downslope provides some pressure increase. When laterals run uphill, the pressure in the lateral drops because of friction and because of the change in elevation. Equations 11.5 and 11.6 can be adjusted to account for changes in elevation:

$$P_i = P_a + 0.75\ P_l\ - 0.5\left(\frac{E_i - E_d}{2.31}\right) \quad (11.8)$$

$$P_d = P_a - 0.25\ P_l + 0.5\left(\frac{E_i - E_d}{2.31}\right) \quad (11.9)$$

where: E_i = the elevation of the inlet to the lateral (ft) and
E_d = the elevation of the distal end of the lateral (ft).

Example 11.4

Given: A sprinkler lateral designed for an average pressure of 50 psi has sprinkler heads with one 5/32-inch nozzle.

The sprinkler lateral is 4-inch diameter aluminum pipe (3.90 in inside diameter) with sections 30 feet long. The lateral is 1,320 feet long.

Find: The pressure at the inlet and distal ends of the lateral when the lateral is:

On level ground

Runs down a uniform 2% grade

Runs up a uniform 2% grade

Which of these systems meet the criteria for pressure variation along laterals?

Solution:

There are 44 sprinklers on the lateral (i.e., 1,320 feet with 30 feet between sprinklers).

The average sprinkler discharge is 5 gpm for 5/32-inch nozzles at 50 psi (Table 11.1)

The inflow to the lateral is 220 gpm (5 gpm/sprinkler × 44 sprinklers).

The friction loss in 4-inch diameter aluminum pipe for a flow of 220 gpm is 1.87 psi/100 ft from Table 8.2a. The loss for a conveyance pipe is then 1.87 × 1320/100 = 24.7 psi.

The multiple outlet friction factor (F) for a lateral with 44 sprinklers is about 0.36 (Table 8.3) so the friction loss for the lateral is:

$$P_1 = F\,P_m = 0.36 \times 24.7 \text{ psi} = 8.9 \text{ psi}$$

The pressure at the inlet to the lateral for level ground is:

$$P_i = P_a + 0.75\,P_l = 50 + 0.75 \times 8.9 = 56.7 \text{ psi}$$

The pressure at the distal end of the lateral for level ground is:

$$P_d = P_a - 0.25\,P_l = 50 - 0.25 \times 8.9 = 47.8 \text{ psi}$$

a. The pressure variation along the lateral is 8.9 psi compared to the average pressure of 50 psi. The variation is 17.8% of the average pressure and is less than the permissible variation so the lateral meets the standard.

b. When the lateral runs down a 2% grade, the elevation change along the lateral is:

$$E_i - E_d = 0.02 \times 1320 = 26.4 \text{ ft}$$

The inlet is 26.4 feet above the distal end. The pressures at the inlet and distal ends are:

$$P_i = P_a + 0.75\ P_l - 0.5\left(\frac{E_i - E_d}{2.31}\right) = 50 + 0.75 \times 8.9 - 0.5\left(\frac{26.4}{2.31}\right) = 51 \text{ psi} \quad \text{(Eq. 11.8)}$$

$$P_d = P_a - 0.25\ P_l + 0.5\left(\frac{E_i - E_d}{2.31}\right) = 50 - 0.25\,(7.7) + 0.5\,(26.4/2.31) = 53.8 \text{ psi} \quad \text{(Eq. 11.9)}$$

Here the pressure variation is only 2.8 psi, well within the allowable variation.

c. When the lateral runs uphill the elevation of the inlet is below the distal end so the value of ($E_i - E_d$) = −26.4 feet. Using this value the pressures at the ends of the lateral are P_i = 62.4 psi and P_d = 42.1 psi.

The pressure variation is 20.3 psi or 41% of the average pressure, which exceeds the criteria.

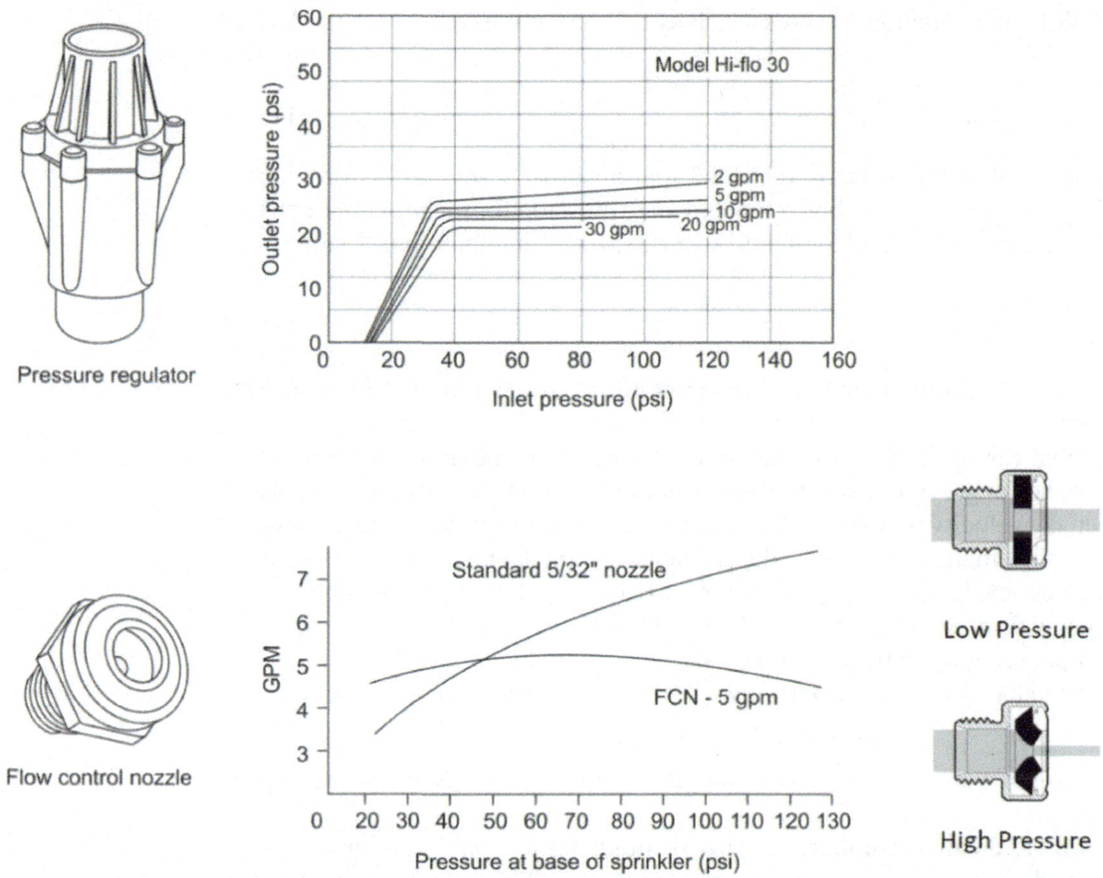

Figure 11.10. Performance of a pressure regulator at several flow rates, and a flow control nozzle (courtesy of Nelson Irrigation Corp.).

It is not always possible to satisfy the allowable pressure variation limitation for laterals. In such cases ***pressure regulators*** or ***pressure compensating nozzles*** (flow control) can be used to control sprinkler discharge. A regulator can provide nearly constant outlet pressure for a range of inlet pressures (Figure 11.10). This provides the same pressure to sprinklers along the lateral and produces high uniformity; however, a higher inlet pressure is required so that all sprinklers receive the design pressure. Regulators may not be required along the entire lateral if the same end of the lateral is always next to the mainline. Some pressure is lost as water flows through the regulator so inlet pressure must be increased to overcome this loss. The regulators also increase the operating and the initial cost of the system. Pressure compensating nozzles serve the same purpose and may reduce both the operating and installation costs compared to pressure regulators. Some designs of pressure compensating nozzles include flexible orifices that contract at high pressure and expand under low pressure. The change of the orifice size regulates the flow as shown in Figure 11.10. Compensating nozzles generally have a smaller operating range than regulators.

11.5 Maximum Lateral Inflow

The maximum inflow to a sprinkler lateral is limited by two conditions: the maximum permissible pressure variation and the maximum acceptable water velocity in the lateral pipe. The maximum permissible pressure variation along the lateral limits the maximum inflow as described in the previous section (Section 11.4). The maximum pressure variation along the lateral is 20% of the average operating pressure, therefore:

$$\text{Max } P_l = 0.20 \times P_a \qquad (11.10)$$

Using the Hazen-Williams equation for pressure loss the maximum inflow for the lateral can be determined:

$$\text{Max } P_l = 4.56\ F \left(\frac{Q_{max}}{C}\right)^{1.852} \frac{L}{D^{4.866}} \tag{11.11}$$

where F is the multiple outlet factor from Table 8.3 and all other terms are as previously defined. The above equations can be combined to yield an expression for the maximum inflow that is permissible for a lateral of given length and size (i.e., fixed diameter and C value):

$$Q_{max} = \left(\frac{0.2\ P_a\ D^{4.866}\ C^{1.852}}{4.56\ F\ L}\right)^{1/1.852} \tag{11.12}$$

where Q_{max} is the maximum inflow to the lateral to maintain pressure variation less than 20% of the average pressure.

The second factor that can limit the inflow rate to the lateral is the maximum allowable velocity in the pipeline. The danger of damage to the pipeline and its components due to pressure surges increases when the velocity of water in the pipeline increases. Sprinkler laterals can withstand higher velocities than mainlines because the sprinklers on the lateral allow water and air under high pressure to escape before damaging the pipe. However, there is still an upper limit to the velocity of water flow in sprinkler laterals. Commonly the upper limit is 7 feet per second, while 10 feet per second can be used if the valve closes gradually and the pipe is filled slowly. The maximum velocity can also determine the maximum inflow for the pipeline:

$$Q_{max} = 2.445\ v_{max}\ D^2 \tag{11.13}$$

where v_{max} = the maximum water velocity in the lateral and all other terms are as previously defined.

Equations 11.12 and 11.13 can be combined to provide limits for the maximum inflow to the sprinkler lateral so that the velocity is below the maximum permissible and the pressure variation along the lateral is less than 20% of the average pressure. The smallest value from the two equations defines the maximum inflow. These equations were used to develop charts for the maximum inflow for aluminum pipe with couplers forty feet apart as shown in Figure 11.11. Similar relationships can be developed for other types of pipe material to use as general guidelines for laterals.

Solution of equation 11.13 for the maximum inflow at a velocity of seven feet per second gives flows of 410, 260, and 145 gpm for 5-inch, 4-inch, and 3-inch pipe respectively as shown in Figure 11.11.

Results in Figure 11.11 show that the velocity limit (7 ft/sec) determines the maximum inflow for the initial lengths of the lateral. As the lateral length increases, the friction loss limitation determines the maximum inflow rate. Results in Figure 11.11 can be used for laterals with different sprinkler outlet spacings because the velocity limits the inflow for short laterals. By the time friction loss becomes determinate a significant number of sprinklers will be included and the friction factor (F) for laterals will be nearly the same for either sprinkler spacing. Thus, the friction loss will be comparable since all other factors are the same for friction loss calculations.

Example 11.5

Determine the maximum sprinkler discharge for a 5-inch aluminum pipe lateral that is 1,000 feet long where the average pressure is 50 psi. Sprinklers are spaced 40 feet apart along the lateral.

Given: P_a = 50 psi $\quad$ D = 5 in

L = 1,000 ft $\quad$ S_L = 40 ft

C = 120 (aluminum pipe with couplers, Table 8.1)

q_s = discharge of individual sprinklers

Solution:

From Figure 11.11 the maximum lateral inflow is about 410 gpm.

For a lateral 1,000 ft long, 25 sprinklers would be needed if spaced at 40 ft.

Thus, each sprinkler could average up to 16.4 gpm, which is a high flow for most applications.

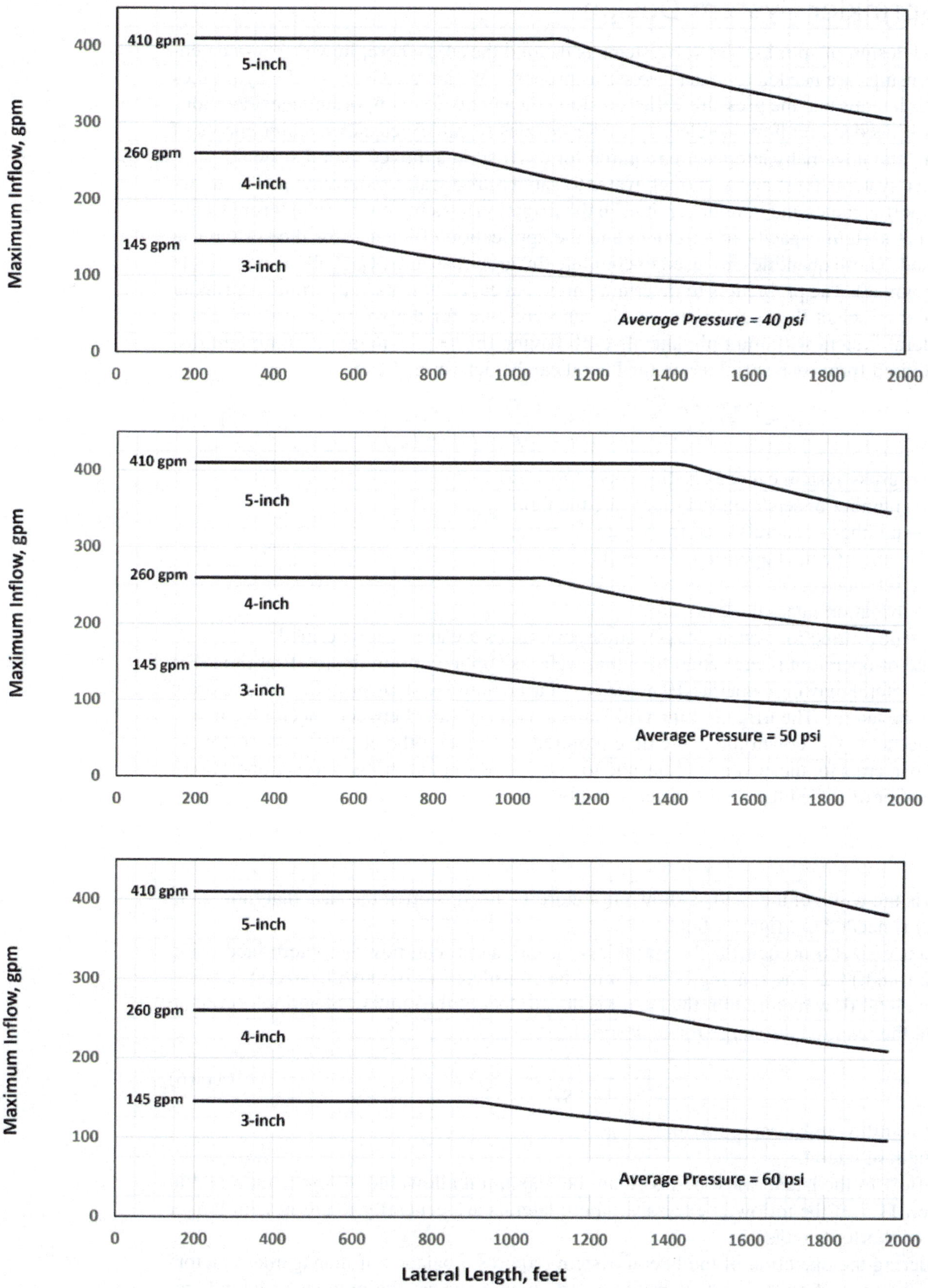

Figure 11.11. Maximum inflow for three diameters of aluminum sprinkler lateral pipe with outlets 40 feet apart (*C* value = 120), and three average pressures along the lateral. A maximum velocity of 7 feet per second is used for this chart.

11.6 Sprinkler System Design

Detailed design of sprinkler lateral systems is beyond the scope here; however, some general relationships are needed to manage systems properly. We have considered the hydraulics of sprinkler laterals and the pressure variation along the lateral. Two important considerations that are still needed are: how to select sprinkler nozzles to satisfy capacity requirements of the system, and how many laterals are required for the field in a moved lateral system.

Sprinkler systems must apply enough water to satisfy crop water requirements and to account for inefficiencies and nonuniformities in the irrigation system and the field. From Chapter 5, the net system capacity requirement and the application efficiency for the system can be estimated. These quantities are used to compute the gross capacity, Q_c (gallons per minute per acre, gpm/ac). The problem is to determine how that capacity is used to arrange sprinkler laterals, and to select the appropriate nozzles and pressure for the sprinkler system. For a moved lateral system with multiple laterals each having the same length, the minimum discharge required from each sprinkler on the lateral can be determined from:

$$q_s = \left(\frac{Q_c\ S_L\ S_m}{43560}\right)\left(\frac{N_s}{N_l}\right)\left(\frac{T_s}{T_o}\right)\left(\frac{I_i}{I_i - T_d}\right) \tag{11.14}$$

where: Q_c = gross system capacity requirement (gpm/ac),
N_s = number of sets required to irrigate the field,
N_l = number of laterals used to irrigate the field,
T_o = time of actual operation per set (hr),
T_s = set time (hr),
I_i = irrigation interval (days), and
T_d = downtime for system (days); other parameters are as already defined.

The time of operation is the actual time that water is applied during the total set time. For example, a lateral may only operate 10 hours out of a 12-hour set. This provides time to drain and move the lateral. The irrigation interval is the amount of time between successive irrigations of the field. The downtime is the time required to maintain the engine, system, etc., to prepare the laterals for the next irrigation, and for any harvesting, farming, or other operations.

The number of sets in the field is determined by:

$$N_s = \frac{W_f}{S_m} \tag{11.15}$$

where W_f is the width of the field as shown in Figure 11.12. It is often the case that more than one lateral is needed to irrigate a field.

The nozzle size(s) needed for the sprinklers on the lateral can now be determined using Tables 11.1 and 11.2. The spacing criteria must be considered as the nozzle size(s) is determined. The total flow required for the lateral is the product of the number of sprinklers on the lateral and the required discharge for each sprinkler:

$$Q_i = \frac{q_s\ L}{S_L} \tag{11.16}$$

where: Q_i = inflow to lateral (gpm) and
L = length of lateral.

The inflow to the lateral must be less than the maximum allowable inflow determined in the Section 11.5. If the inflow is excessive, more laterals are generally required with longer set times or shorter lengths.

Considering the operation of the lateral system requires a balance of management factors including the time of operation, the sprinkler and lateral spacing, the number of laterals required, and the application efficiency. Pressure and flow limitations must also be considered for proper operation. Often, a trial-and-error procedure is needed to balance all factors, and

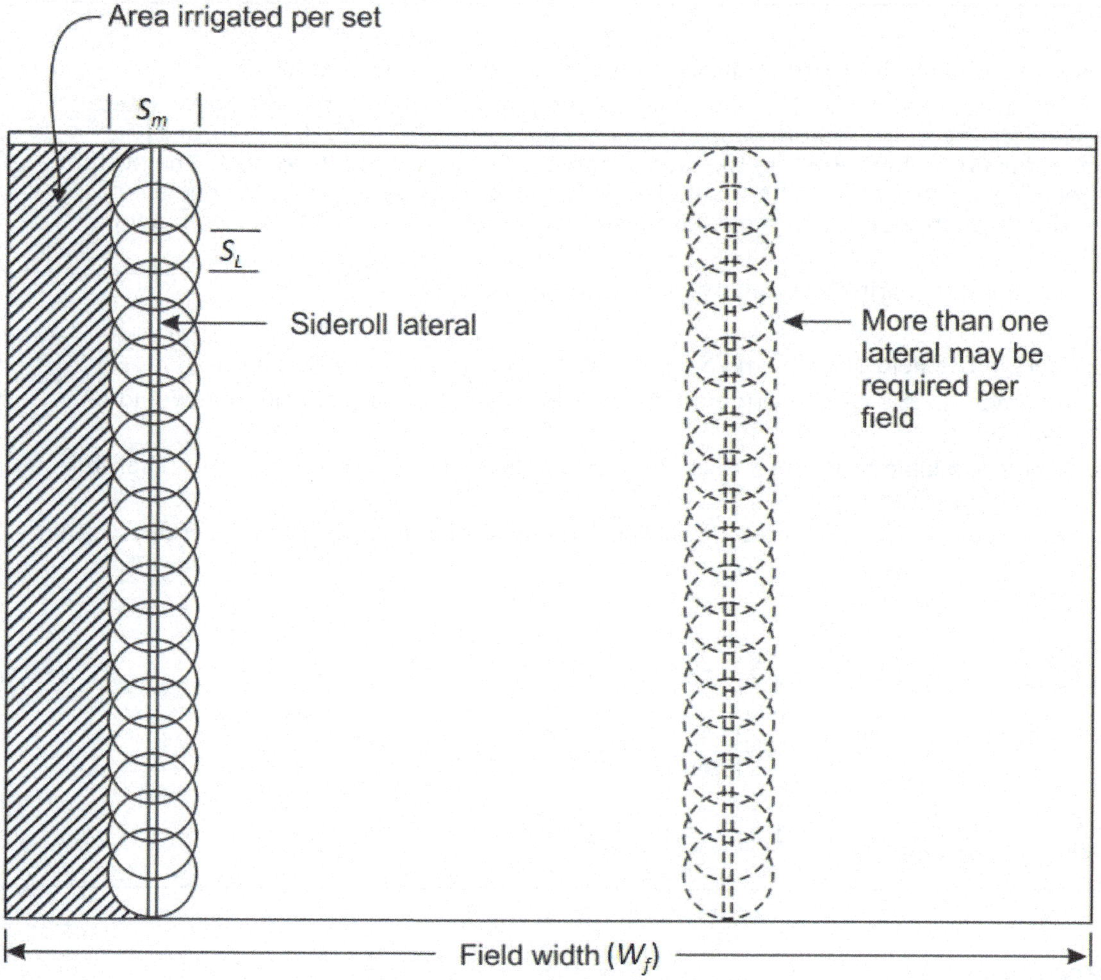

Figure 11.12. Field layout for a moved-lateral sprinkler system.

tradeoffs frequently are required. The landowner's and/or irrigation manager's preferences for operation should be incorporated into a management plan for the system.

The layout of laterals on sloping fields can be crucial. It is generally best to run the mainline up and down the hillslope while positioning the lateral so that it is relatively level. If the lateral must run up and down the hill, it is best to run the lateral downslope if possible. The prevailing wind direction and speed during the irrigation season should also be considered.

Example 11.6

Compute the minimum sprinkler discharge required for the system described below.

Given: A square field (1,200 feet × 1,200 feet) is irrigated with a portable set-move (moved lateral) sprinkler system. The gross system capacity has been determined to be 6.0 gpm/ac. The spacing of sprinklers is 40 feet along the lateral and 50 feet between lateral sets. The system operates for 10 hours out of a 12-hour set. The field must be irrigated at least once every 10 days and 2 days are needed to move laterals to the beginning side and for equipment maintenance.

Find: Compute the minimum sprinkler discharge required for the system.

Solution:

The number of sets in the field are $N_s = W_f/S_m$ = 1,200 ft/50 ft = 24 sets. (Eq. 11.15)

With 12-hour set times, 2 sets can be irrigated daily so 12 days of continual irrigation would be required with one lateral.

We only have 8 days available to irrigate since 2 out of 10 days are used for downtime. Therefore, 2 laterals will be needed ($N_l = 2$).

Each lateral must irrigate 12 sets taking 6 days. Thus, the irrigation interval could be 8 days rather than 10.

Then using Equation 11.14:

$$q_s = \left(\frac{Q_c\ S_L\ S_m}{43{,}560}\right) \times \left(\frac{N_s}{N_l}\right) \times \left(\frac{T_s}{T_o}\right) \times \left(\frac{I_i}{I_i - T_d}\right)$$

$$q_s = \left(\frac{(60\ \text{gpm/ac})\ (40\ \text{ft})\ (50\ \text{ft})}{43{,}560\ \text{ft}^2/\text{ac}}\right) \left(\frac{24\ \text{sets}}{2\ \text{laterals}}\right) \left(\frac{12\ \text{hr}}{10\ \text{hr}}\right) \left(\frac{8\ \text{days}}{8-2\ \text{days}}\right)$$

$$q_s = \frac{6.0\ \text{gpm}}{\text{ac}} \times \frac{0.55\ \text{ac}}{\text{sprinkler}} \times 1.2 \times 1.33 = 5.3\ \text{gpm/sprinkler}$$

11.7 Frost Protection

Agricultural and horticultural plants are produced in regions where cold temperatures may damage crops. If the plant temperature drops below a critical value, production may be lost on annual crops and perennial species may be damaged. Damage can result from two types of cooling. An ***advective freeze*** occurs when the ambient air temperature drops below a critical level and wind increases the convective heat transfer from the cold air to plants. There is little that irrigation can do to protect plants from an advective freeze. In fact, wetting the foliage can cool plants substantially causing increased damage. In addition, the buildup of ice on plants and irrigation systems can cause structural damage.

Radiant frost occurs in a clear, calm, and dry environment where energy is radiated from plants into the atmosphere. The ambient air temperature is generally above critical temperatures that causes damage, but outgoing radiation cools plants 1° to 4°F below the air temperature. In addition, crops draw energy from the air immediately surrounding the plants, thus, air in contact with plants is cooler than above the canopy. Light winds reduce the turbulence above plants allowing the plant surfaces to cool further. Frost begins to form on plants when the canopy temperature drops below the dew point temperature of the air. The dew point may be lower than critical temperature in dry environments.

Leaves, blossoms, and young fruit are usually the most sensitive to frost damage and are frequently killed at temperatures between 26° to 30°F. Lethal temperatures for more hardy plant parts are related to the stage of development; thus, protection may be more important at one time than another.

Managing for frost protection requires an understanding of the processes involved when water changes phases. Water can exist as a vapor, liquid, or solid. Changing phases involves energy exchange. Evaporation requires about 585 calories of energy per gram of water. The reverse process, condensation, releases energy. Melting ice requires energy, and freezing water releases an equal amount of energy. Sublimation is where ice is transformed directly into water vapor without going through the liquid state. Sublimation requires a great deal of energy.

What happens during sprinkling to provide frost protection? Consider an irrigation sprinkler operating while the air temperature is 33°F. Irrigation water is usually much warmer than critical temperatures, for example groundwater in northern climates where frost protection is needed averages about 50 to 55°F. After water leaves the nozzle, the droplets begin to cool and evaporate. Cooling the droplets adds energy to the air providing some frost protection. However, large amounts of water are needed since only 1 calorie is released per gram for each °C of temperature change of the water. With time, the droplets cool to the wet bulb temperature of the air, which can be below 33°F. If droplets reach plants before reaching the wet bulb temperature, water will evaporate from the plant surface, drawing energy from plants—further cooling plants. If sprinkling only wets the crop canopy so that evaporation occurs, the plants will be cooled below the ambient air temperature, and sprinkling could damage the crop rather than protect it.

What must happen to provide protection? The processes that release energy, thereby warming plants and the air, include condensation and freezing. These processes must occur at a faster rate than the inverse processes of evaporation, melting, and sublimation. The irrigation system should be operated to provide that environment.

Coating plants with a water film can maintain temperatures above the critical damage temperature. Energy is lost from the outer surface of the water film by radiation, convection, and evaporation. The heat of fusion is released from the thin film as the water freezes. If the film is maintained, the temperature will remain near 32°F as freezing supplies the energy lost from the outer surface of the water film. The ice coating on the plant must be continually in contact with unfrozen water until the air warms so that the wet bulb temperature of the air is above the critical temperature. Usually sprinkling is required until the ice formed on the plants completely melts the next morning. Sprinkling above the crop has provided frost protection; however, results have been mixed and protection is not a certainty.

The appropriate application rate for frost protection depends on several factors and general recommendations are risky as evidenced by the failures of overcrop sprinkling. Yet, results from Gerber and Harrison (1964) provide an initial estimate of the required application rate for frost protection (Table 11.5). The most practical rates range from 0.1 to 0.3 inches per hour. Repeat frequency of leaf or foliage wetting must be once each minute. Sprinkling must begin by the time the wet bulb temperature reaches 4°F above the lethal temperature of the plant parts to be protected. Sprinkling must continue until the wet bulb temperature is back above the lethal temperature by about 4°F. Systems are usually operated until the plant is free of ice, due to rising air temperature. Recommended minimum temperature for turning the irrigation system on or off for frost control of apple trees in Washington is given in Table 11.6.

Research has shown that overcrop sprinklers can be operated intermittently to provide frost protection while minimizing the amount of water applied. The cycling frequency affects the water application rate and frost protection. The foliage configuration of the plants, especially the

Table 11.5. Sprinkling rate (in/hr) necessary for frost protection (adapted from Gerber and Harrison, 1964).

Temperature of a Dry Leaf (°F)[a]	Wind Speed (mph)[b]				
	0–1	2–4	5–8	10–14	18–22
27	0.1	0.1	0.1	0.1	0.2
26	0.1	0.1	0.1	0.2	0.4
24	0.1	0.2	0.3	0.4	0.8
22	0.1	0.2	0.5	0.6	-
20	0.2	0.3	0.6	0.8	-
18	0.2	0.4	0.7	-	-
15	0.3	0.5	0.9	-	-
11	0.3	0.7	-	-	-

[a] The temperature of a dry leaf is the expected minimum leaf temperature on an unprotected leaf. This will range from 1°F below air temperature on nights with light wind to 3° to 4°F below air temperature on very calm nights.

[b] Note: These rates assume that relative humidity does not affect frost protection. Thus, the rates should be used as a first approximation in determining the application rate for design and planning. The rates should not be used to manage an actual sprinkler irrigation system.

amount of foliage overlap, has a significant effect on success. The portion of the wetted area that receives water is also important for selecting an application rate and cycle frequency.

Undertree sprinkling can provide frost protection. Undertree sprinklers often produce small water droplets below the canopy, an area Barfield et al. (1990) termed the misting zone. Water droplets cool and evaporate, transferring energy from the water into the air surrounding the plants. If the humidification of the air causes ice formation on the plants, energy will be released that can increase frost protection. Evaporation from the soil increases the humidity, increasing the efficiency of undertree sprinkling. As the relative humidity increases the emissivity of the air decreases, reducing the outgoing long-wave radiation, and the degree of frost damage. The level of protection is dependent on the amount of water applied and the aerial extent of the freezing surface. Part of the heat from freezing and cooling of water is carried into the ground by infiltrating water, part goes into warming the air, and part into evaporation. Heat is transferred from the frosty buds by radiation, convection, and by condensation which occurs on the coldest plant tissues. Ambient air temperature increases of about 2°F are common, although increases up to 4°F have been found. Most of the systems use small (5/64 to 3/32 in), low-trajectory (< 7°) sprinkler heads at 40 to 50 psi. Application rates range from 0.08 to 0.12 inches per hour or slightly more than half of typical overtree requirements.

Table 11.6. Temperatures to start and stop overtree frost protection (adapted from Washington State Irrigation Guide, 1985).

Critical Temperature (°F)	Dewpoint Temperature Range (°F)	Minimum Turn-On or Turn-Off Air Temperature[a] (°F)
32	3–10	45
	10–16	43
	16–21	41
	21–24	39
	24–28	37
	28–31	35
	31–32	33
30	0–9	42
	9–15	41
	15–20	38
	20–24	36
	24–30	32
28	0–8	39
	8–14	37
	14–19	35
	19–23	33
	23–27	31
	27–28	29
26	0–10	35
	10–16	33
	16–20	31
	20–24	29
	24–25	27

[a] Absolute minimum temperature for turning the irrigation system on or off. It is recommended that the system be turned on or off 2°F or 3°F higher than the indicated minimum.

Undertree sprinkling appears to be promising; however, the process is not fully understood, and has not been tested as extensively as overtree sprinkling. Additional testing is needed before recommendations can be developed.

Sprinkler systems can provide frost protection in addition to evapotranspiration and salinity management requirements. Frost protection can pay high dividends during short periods. The rate and timing of water application is often more important than the volume of water applied. In some cases, one sprinkler system can accommodate both the primary uses and frost protection. In other cases, frost protection requires performance that the system cannot satisfy, and a second irrigation system may be required. The design of the secondary system is much different than for the primary system and additional information from specific references must be consulted. In any case, careful management is required for frost protection. The air temperature and ice formation should be carefully monitored.

This discussion on frost protection highlights the processes and provides some very general management practices. However, the process is sensitive to local meteorological conditions that change rapidly. Success requires monitoring of ambient conditions and reliable information for crop susceptibility during sensitive grow stages. Local management guides must be used for each plant species. Care must be taken to minimize runoff, deep percolation, and depletion of scarce water supplies. Snyder and Melo-Abreu (2005) provide a thorough treatise of frost protection and this or similar references as well as local information should be consulted for successful frost protection.

11.8 Summary

Sprinkler irrigation is the most broadly used method of irrigation in the United States. Sprinkler systems consist of the sprinkler device, the lateral pipe that delivers water to a series of sprinklers along the pipe, a mainline that provides water to the inlet of each lateral, and a source of water provided under pressure—usually by pumping. Efficient irrigation depends on understanding the performance characteristics of sprinkler systems. Key parameters are the water flow rate from the nozzle and the diameter wetted by the sprinkler. These factors depend on the pressure provided to the sprinkler and the diameter of the sprinkler nozzle. Uniformity of application is achieved by limiting the variation of pressure along laterals and mainlines. Pressure variation is controlled by selecting pipe sizes that limit pressure loss and maintain flow velocities below limits that may cause pressure surges or water hammer. Simultaneously, the application rate of water depends on the discharge from nozzle(s) and the representative area for an individual sprinkler. The representative area is determined by the spacing between sprinklers on the lateral and the distance between laterals. The uniformity of application is also controlled by the spacing of sprinklers along laterals and between laterals, relative to the diameter of coverage of a sprinkler. Finally, the depth of water applied depends on how long water is applied and the rate of application. The depth of water applied must be adequate to meet crop needs over the time between irrigations. These relationships are described in this chapter, which lays the foundation for subsequent chapters on specific types of sprinkler irrigation systems.

Questions

1. What wetting pattern and distribution would be expected if the operating pressures is higher than recommended for a sprinkler? What role does water drop size play?
2. Should laterals be positioned perpendicular or parallel to the predominant wind? Explain.
3. How is sprinkler spacing important to uniformity, investment cost, and application rate?
4. If the limits of pressure variation along the lateral cannot be met, what options are available to resolve the problem?
5. Define and explain the two conditions limiting the maximum inflow rate into a sprinkler lateral.
6. Determine the discharge from a 1/4-inch nozzle operated at 30 psi. What size nozzle (in 128ths of an inch) would you select to increase the flow (gpm) by 25% at the same pressure?
7. What is the application rate for an irrigation system that has a triangular spacing of 30 ft and the discharge from each sprinkler is 4.2 gpm? Is this acceptable for a silt loam soil with 8% slope?
8. How long would a sprinkler system need to operate to apply 3 in of water if the sprinkler discharge was 10 gpm and the sprinkler spacing was 40 ft × 60 ft? If the application efficiency was 75%, how long would it take to apply a net depth of 3 in?
9. A 4-inch sprinkler lateral of aluminum pipe is 1,320 ft long with sprinklers spaced at 40 ft along the lateral. The average pressure along the lateral is 45 psi and the average discharge is 8 gpm per sprinkler. The lateral lays on level land. Does this lateral conform to the criteria for pressure variation?
10. Sprinklers are placed in a rectangular orientation with a spacing of 30 ft × 40 ft. The gross system capacity is 5 gpm/ac. There are 14 laterals in the field. The field is irrigated every

other day and can only be irrigated at night from 11 p.m. till 6 a.m. The laterals are connected to a controller that automatically cycles from one lateral to another. What discharge is needed for each sprinkler?

11. A grower wants to protect an orchard from frost by sprinkling. The solid-set system was designed to meet ET requirements. The following information is available:

    Sprinkler discharge = 12 gpm
    Sprinkler spacing = 40 ft × 60 ft
    An automatic controller is available to control set times
    Number of laterals in the field = 25

    How could this system be used for frost protection if the dry leaf temperature is 24°F and the wind speed is 6 mph? What dangers might exist?

References

Barfield, B. J., Perry, K. B., Martsolf, J. D., & Morrow, C. T. (1990). Modifying the aerial environment. In G. J. Hoffman, T. A. Howell, & K. H. Solomon (Eds.), *Management of farm irrigation systems* (pp. 827-869). St. Joseph, MI: ASAE.

Gerber, J. F., & Harrison, D. S. (1964). Sprinkler irrigation for cold protection of citrus. *Trans. ASAE, 7*(4), 404-407.

Martin, D. L., Kincaid, D. C., & Lyle, W. M. (2007). Design and operation of sprinkler systems. In *Design and operation of farm irrigation systems* (2nd ed., pp. 557-631). St. Joseph, MI: ASABE.

Pair, C. H. (Ed.). (1983). *Irrigation: Formerly Sprinkler Irrigation.* Irrigation Association.

Snyder, R. L., & Melo-Abreu, J. D. (2005). Frost protection: fundamentals, practice, and economics. In *Frost protection: Fundamentals, practice, and economics* (Vol. 1, pp. 1-240).

USDA-NRCS. (2016). Sprinkler irrigation. Chapter 11 in *Part 623 National engineering handbook.* Washington, DC: USDA-NRCS. Retrieved from https://www.wcc.nrcs.usda.gov/ftpref/wntsc/waterMgt/irrigation/NEH1

WAIG. (1985). *Washington state irrigation guide.* Part WA685. Irrigation method and design criteria with reference subpart L: Frost protection. USDA-SCS.

Chapter 12
Moved-Lateral, Gun, and Traveler Sprinkler Systems

12.1 Introduction

Sprinkler devices were invented at the end of the nineteenth century with over seventeen patents issued before 1890 (Morgan, 1993). Since then, sprinkler irrigation has become widespread. It is used around the world on many types of crops and soils. Water is delivered through pipes under pressure directly to the application location, thereby minimizing field conveyance losses while supplying crops on undulating terrain and/or highly permeable soils. Sprinkler systems can be efficient when properly designed and managed. Success depends on understanding characteristics and capabilities while operating within resource and management limitations. What questions should be asked to determine operator goals and restrictions? How should the irrigation system be configured to efficiently meet crop needs while satisfying constraints? What management plan would be most effective? How should you monitor the system to evaluate performance? Concepts presented in this chapter will allow you to address these issues.

The USDA-NASS (2018) lists the seven types of sprinkler irrigation systems shown in Table 12.1. Survey results show that center-pivot irrigation systems represent approximately 85% of the sprinkler irrigated land in the U.S. in 2017. Linear-move irrigation systems are mechanized systems with characteristics much like center pivots, yet only represent approximately 1% of the irrigated land. The remaining five types of irrigation systems constitute approximately 14% of the sprinkler irrigated land in the United States. While that area is much smaller than for center pivots, it still is significant. The USDA-NASS database includes the number of farms that employed the types of systems. The acres irrigated per farm for center pivots is much larger than other types of sprinkler irrigation. The extent of periodically moved systems for the ten states with the most area is listed in Table 12.2. Most of the area for side-roll and hand-move systems is in the states in or west of the Rocky Mountains. Solid-set systems are used in some states east of the Rocky Mountains, yet California, Washington and Oregon dominate. Big-gun systems are more uniformly distributed across the country.

These data represent irrigation development in the United States. Periodically moved systems are significant internationally, especially in areas with small landholdings or developing areas that lack financial resources needed for drip or mechanized systems. The characteristics and management practices for systems except center pivots and linear-move systems are

Table 12.1. Data on sprinkler irrigation from USDA (2019).

Type of Sprinkler System	Number of Farms	Total Area (acres)	Acres per Farm	Percent of Sprinkler Area
Center Pivot	49,923	26,800,613	537	85%
Side Roll	16,130	1,788,443	109	6%
Solid Set	18,216	1,206,860	66	4%
Traveler	7,518	596,059	79	2%
Linear Move	3,669	469,408	122	1%
Other	8,673	401,318	46	1%
Hand Move	22,266	394,194	17	1%
Total	126,395	31,656,895		100%

Table 12.2. Rank of top ten states for each type of periodically moved system (data from USDA, 2019).

Rank	Side Roll		Hand Move		Solid Set		Big Gun	
	State	Acres	State	Acres	State	Acres	State	Acres
1	Idaho	406,429	Oregon	115,405	California	454,924	Oregon	112,182
2	Utah	258,816	Idaho	73,733	Washington	253,939	Michigan	61,636
3	Montana	198,603	California	50,262	Oregon	88,463	Washington	48,632
4	Oregon	196,155	Washington	44,429	Arizona	58,051	Georgia	43,631
5	Texas	86,163	Montana	33,825	Idaho	46,463	Texas	36,217
6	California	83,396	Utah	13,745	Florida	42,564	Florida	31,256
7	Washington	82,373	Colorado	6,973	Texas	38,368	Missouri	24,766
8	Colorado	79,972	Texas	6,714	Wisconsin	34,903	California	24,433
9	Illinois	42,338	Wyoming	6,548	Georgia	29,471	New Jersey	22,336
10	Wyoming	36,645	New Jersey	5,106	Michigan	18,995	N. Carolina	21,305

examined in this chapter. Center pivots and linear-move systems are considered in Chapter 13. Detailed design of moved-lateral, solid-set and gun-based systems involves matching all components and ensuring that hydraulic principles are satisfied—see Keller and Bliesner (1990) and/or Stetson and Mecham (2011) for design procedures. Most management situations depend on systems already in place, so design considerations are only discussed to help evaluate alternatives when system changes are needed or when monitoring system performance.

12.2 Periodically Moved Laterals

The layout and design of sprinkler laterals in general were presented in Chapter 11. The typical layout of moved-lateral systems is illustrated in Figure 12.1. Water is supplied from a pump into the mainline which conveys water to the lateral. The mainline may be lengths of aluminum pipe aboveground. If multiple laterals operate simultaneously, valves may be

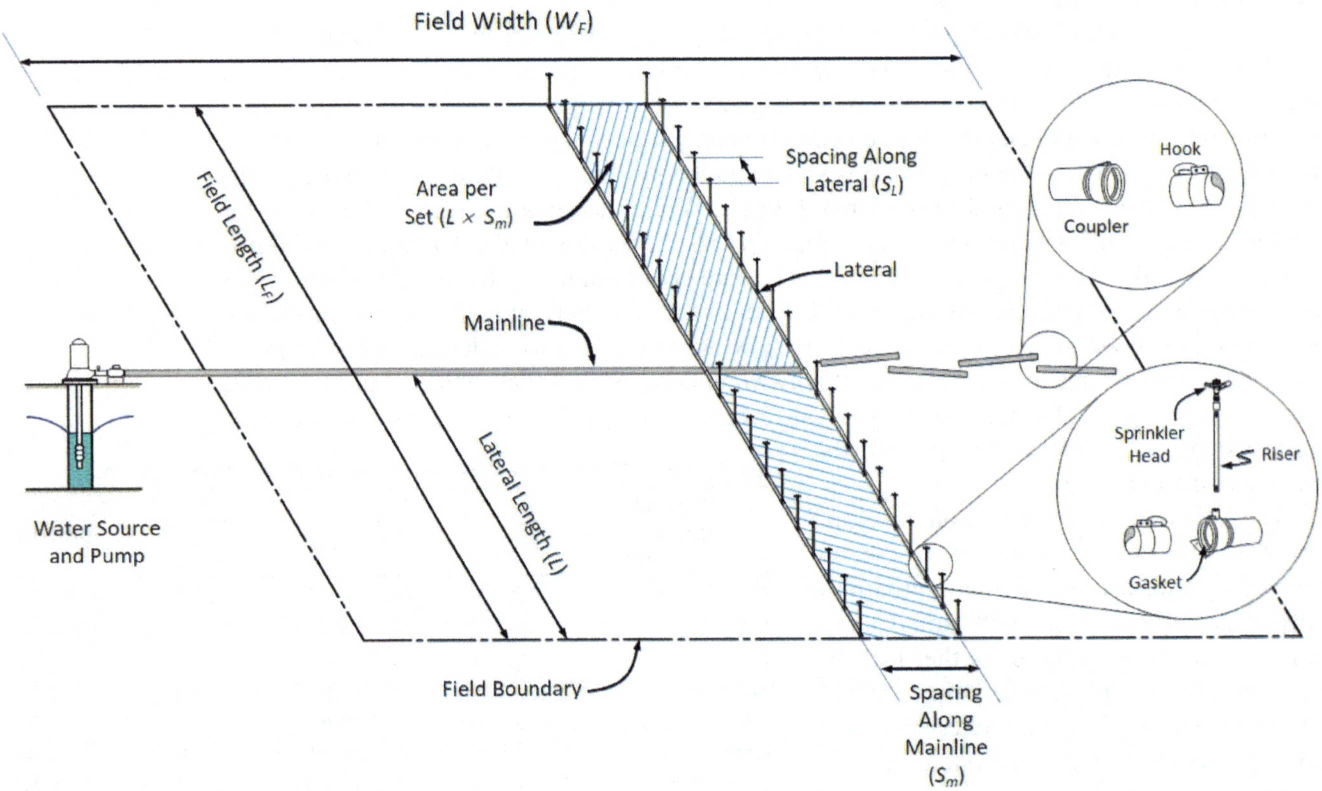

Figure 12.1. Typical layout for a moved-lateral sprinkler system for two successive sets.

placed along the mainline to adjust the pressure into a lateral and to shutoff a lateral for moving without shutting down the pumping plant. Buried PVC pipe may also be used for mainlines with riser pipes and valves to connect to laterals. Hand-move and tow-line systems often require sprinkler risers several feet tall to position the sprinkler above the crop so that the canopy does not interfere with the water jet. Moved-lateral systems use impact or rotating sprinklers that provide a large diameter of coverage allowing for wider spacings and fewer sets. Effective operation depends on selecting the proper sprinkler and nozzle for the chosen spacing of sprinklers along the lateral and the width between sets as discussed in Chapter 11. Laterals should be large enough to limit pressure variations along the lateral to less than 20% of the average operating pressure. Pressure regulators and flow control nozzles may be necessary in some instances to meet uniformity goals. Ultimately, the system should be designed to provide the water requirements of the crop while maintaining efficiency by minimizing deep percolation, evaporation, and runoff. Operator preferences must be incorporated into the management plan.

12.2.1 Types of Systems

Moved-lateral sprinkler systems are composed of a lateral that is periodically repositioned across the field. The lateral consists of individual pieces of pipe connected with a coupler and latching system. Individual pieces of pipe are often referred to as a joint, a length, or a section of pipe. The simplest sprinkler system is a ***hand-move*** system where the lateral joints are carried from set to set by hand and the lateral is reassembled at the new set (Figure 12.2a and 12.2b). In some cases, aluminum pipe is used for the mainline (Figure 12.2c). In other cases, the mainline is buried and risers with hydrants are connected to the lateral (Figure 12.2d). While these systems are versatile, they require considerable labor, especially if the soil surface

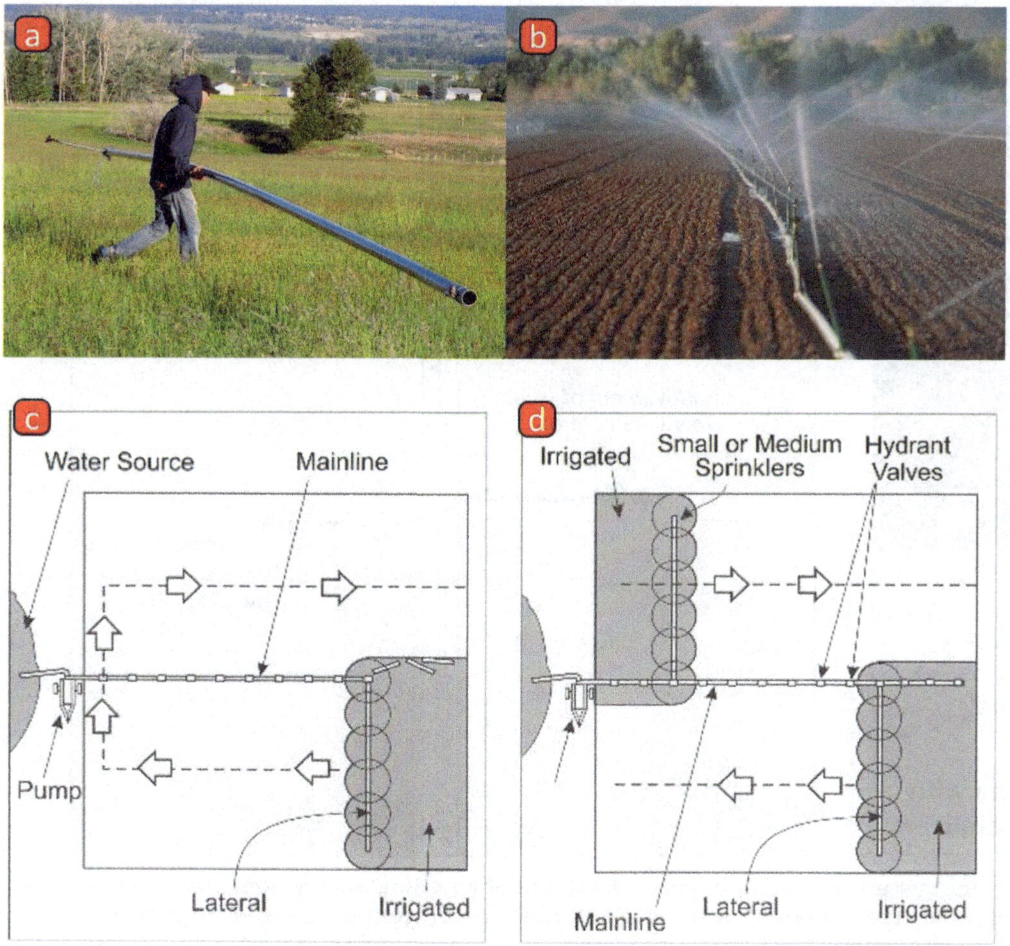

Figure 12.2. Hand-moved sprinkler system. (Photos a and b are courtesy of USDA-NRCS. Figures c and d were adapted from Turner and Anderson, 1980)

remains muddy after irrigation or the soil surface is not protected by the crop. Moving laterals is also difficult when crops are tall. Moving the lateral is much easier when the soil is covered such as with grass or alfalfa. Once the lateral reaches a field boundary it must be disassembled and transported to the next location to be irrigated. The lengths of pipe, sprinkler types and nozzle sizes are usually the same for all joints of the lateral to avoid confusion when repositioning the lateral. The substantial effort required to move the lateral promotes large application depths per irrigation to minimize the number of moves.

Hand-move irrigation systems are the cheapest to buy and maintain. Maintenance involves replacing gaskets used to seal adjoining lengths of pipe—usually replaced biennially or triennially. Sprinklers and nozzles should also be replaced periodically. Sprinkler replacement depends on the amount of annual use; however, sprinklers often last more than five years. The pipe has a long life; thus, investment and maintenance costs are small while labor requirements are quite high. In some cases, hand-move systems are used for special purposes such as leaching salts during the off-season where surface methods are used during the irrigation season. Hand-move systems are extremely portable so they can be used on fields where supplemental irrigation is not required every season.

To alleviate the labor of carrying laterals, several mechanical adaptations were developed to reposition the lateral to the next set. One method uses a tractor, or other power source, to pull the lateral across the field from one set to another. This type of system is called a ***tow-line, skid-tow*** or ***drag-line*** system. The components and plan view of the system as shown in Figure 12.3. As shown in the plan view, the pipeline is towed in a zigzag fashion across the

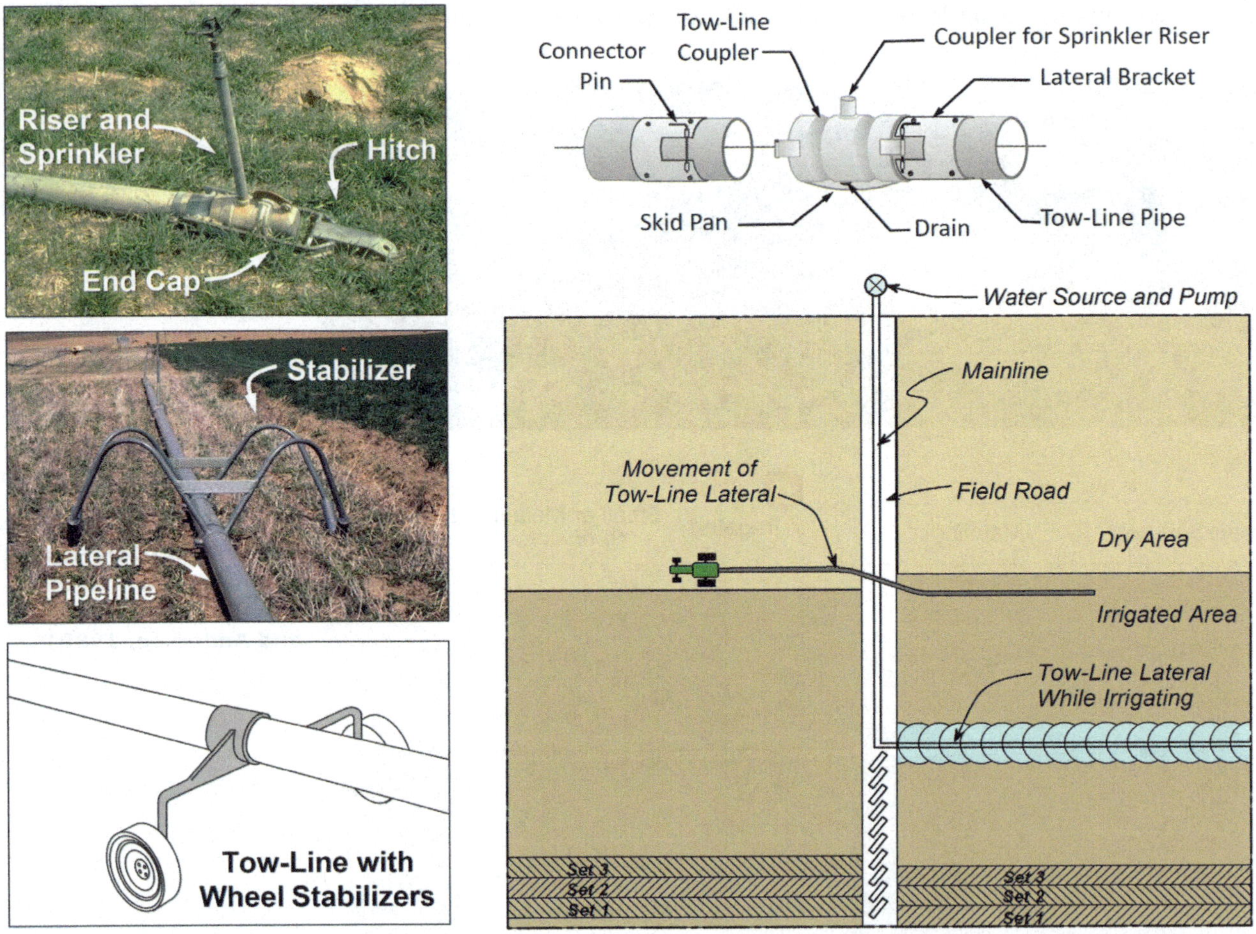

Figure 12.3. Tow-line sprinkler system and components. (Diagram of tow-line with wheel stabilizers is courtesy of Turner and Anderson, 1980. Plan view of field and tow-line system is adapted from Turner and Anderson, 1980.)

field. The lengths of pipe are held together with tow-line couplers that connects two lengths of pipes. Connecting pins were used within the coupler to allow quick disassembly when relocating laterals to the starting set. A skid pan held the pipe above the soil and protected the drain that was included in the coupler. The pipeline is supported by devices, called outriggers or stabilizers, that are clamped onto the pipeline to prevent the lateral from twisting during movement which prevents sprinkler risers from tipping over and breaking. Wheels were also used as shown in Figure 12.3 to provide stability; thus, some systems are called ***wheel-tow*** systems. To reduce abrasion on the aluminum pipe—especially for rough terrain—steel skid pans can be clamped at the midpoint of the joints to carry the pipe above the soil.

Tow-line systems work well on low-growing crops where the lateral pipe can slide freely across the soil and crop surface. The pipe must be pulled between the rows when the system is used for row crops. This is easily done for low-growing crops such as soybeans or grain sorghum. For tall crops, such as corn, the tractor will flatten one or two rows of corn when the pipe is pulled. Sometimes producers plant a few rows of a low-growing crops in the alley where the pipe is pulled. Others plant the tow alley to a permanent grass. In any case there will be a loss of production area for tall crops. The end cap and hitch shown in Figure 12.3 are installed on both ends of the lateral. The lateral is connected to the mainline using a flexible hose.

Tow-line systems are more expensive to purchase and maintain than hand-move systems. The same pipe and sprinklers are needed as for hand-move systems, but stabilizers and couplers increase investment cost. Friction between the soil and the pipe often causes wear that shortens the life of tow-line systems. Less labor is needed to move laterals from one set to the next than for hand-move systems; however, more time is needed to disassemble and reposition the pipeline once the lateral reaches the field boundary.

The third type of moved-lateral systems is the ***side-roll, wheel-move, wheel-line,*** or ***hand-roll*** system. With this system wheels are clamped directly onto the lateral pipeline (Figure 12.4c). Pipe used on side-roll systems is usually thicker walled than for hand-move or tow-line systems. The joints of the lateral are rigidly fastened—sometimes using gears between joints—to remain connected while applying torque when moving the system. The pipeline is moved by rotating the pipeline directly or in some cases to a drive shaft that runs parallel to the lateral. For mechanically powered systems torque is applied by an engine located on a

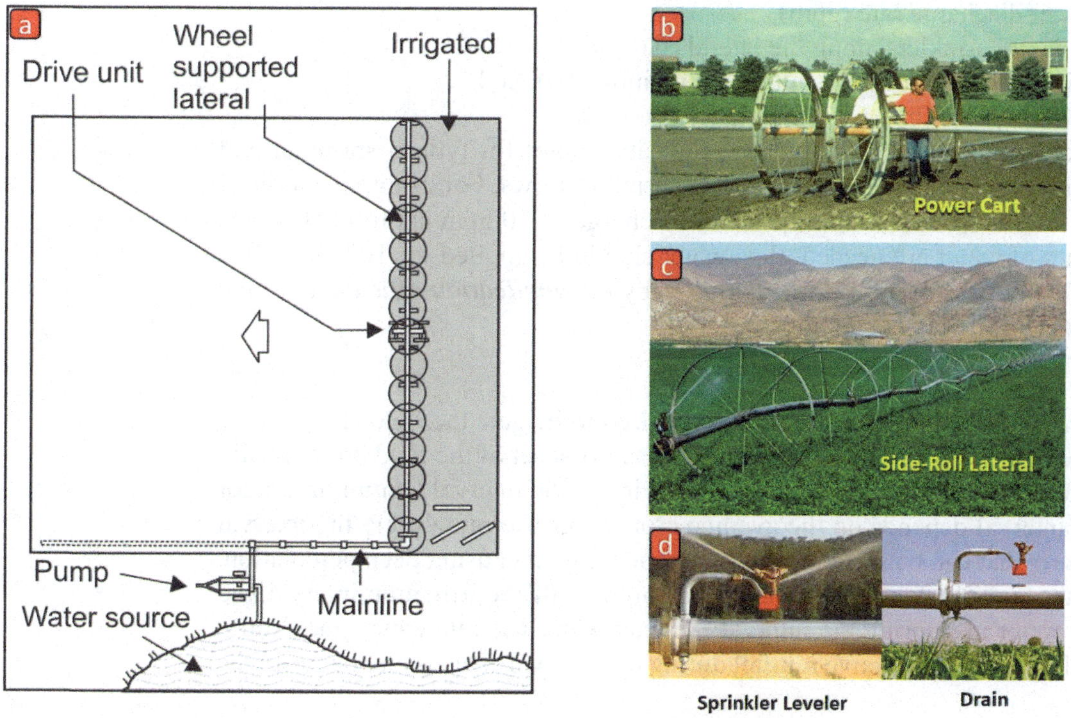

Figure 12.4. Side-roll sprinkler system and components. (The plan view shown in drawing a is adapted from Turner and Anderson, 1980. Picture c is courtesy of Rain for Rent and picture d is courtesy of Wade Rain, Inc.).

chassis (Figure 12.4b) located either at one end of the lateral or along the center of the lateral where access is convenient. The engine slowly turns the pipeline or drive shaft, and the wheels rotate across the field. Hand-roll systems do not have an engine and the wheels are rotated by hand. The clearance below side-roll systems is typically about 3 feet but can be as much as 6 or 8 feet for special large diameter wheels. Sometimes braces are needed to keep the lateral in place after moving. Some alignment by hand to straighten laterals may be needed after moving. Sprinkler levelers (Figure 12.4d) include a swivel that kept sprinklers vertical if the pipeline did not rotate to the perfect angle. All moved-lateral systems should be drained prior to moving as demonstrated with the drain shown in Figure 12.4d for side-roll systems.

12.2.2 Operational Characteristics

Managing sprinkler irrigation systems involves adjusting several variables to meet crop water needs, avoid deep percolation and align with management goals and constraints. Moved-lateral systems are repositioned from set to set across the field. The time that the lateral is in one location is called the ***set time***. These systems require significant effort and labor to move each set; therefore, the minimum set time acceptable to operators is about 8 hours with a maximum of 3 sets per day. The most common set time is 12 hours, but 24-hour sets may be used for high clay content soils or for salinity management—low application rates over long periods enhance leaching and minimize runoff. Laterals must be drained before moving which can take up to 1 hour. Therefore, the time that water is applied, the application time, will be less than the set time.

The depth of water applied is often large for periodically moved systems. The depth is determined by the average ***application rate*** (A_r) and the time of application as described from Equations 11.3 and 11.4. The application rate is determined by the flow rate from the sprinkler (q_s), the spacing of sprinklers along the lateral (S_L) and the spacing between lateral sets along the mainline (S_m):

$$d_g = A_r \times T_o = \frac{96.3\ q_s}{S_L\ S_m} \times T_o \tag{12.1}$$

where: d_g = the gross depth of application (in),

A_r = the average application rate, inches/hour,

q_s = the sprinkler discharge rate (gpm),

S_L = the spacing of sprinklers along the lateral (ft), and

S_m = the spacing of lateral locations along the mainline (ft), and

T_o = the application time (hr).

The depth of water applied per hour (i.e., the application rate) for typical sprinkler and lateral spacings are listed in Table 12.3 for a range of sprinkler flows. For example, a typical system using a 40 ft × 60 ft spacing with a sprinkler discharge of 10 gpm applies 0.4 inches per hour. If four inches of water are needed, then water should be applied for 10 hours. This is the gross depth of application and must be multiplied by the ***application efficiency*** for the low quarter (E_{LQ}) to determine the net depth (d_n):

$$d_n = d_g \times E_{LQ} \tag{12.2}$$

The ***irrigation interval*** (I_i) is the amount of time required to irrigate the field. This can be thought of as the time between consecutive irrigations of the first set of the field. Periodically moved laterals are operated to utilize the longest possible irrigation interval to minimize labor input. The irrigation interval depends on the average crop water use rate during the interval and the amount of water that can be stored in the root zone without causing deep percolation. The net depth of water required for an irrigation is the product of the irrigation interval and the average net crop water use during the interval. The net water use rate equals the evapotranspiration minus the expected effective rainfall during the interval. This is analogous to the

Table 12.3. Depth of water applied per hour (i.e., the application rate), (in/hr).

Lateral Spacing, S_L (ft)	Spacing Along Mainline, S_m (ft)	Representative Area, $S_L \times S_m$ (ft²)	Sprinkler Discharge, q_s (gpm)							
			4	6	8	10	12	15	20	25
20	40	800	0.48	0.72	0.96	1.20	1.44	1.81		
30	40	1200	0.32	0.48	0.64	0.80	0.96	1.20	1.61	
40	40	1600	0.24	0.36	0.48	0.60	0.72	0.90	1.20	1.50
20	50	1000	0.39	0.58	0.77	0.96	1.16	1.44	1.93	
30	50	1500	0.26	0.39	0.51	0.64	0.77	0.96	1.28	1.61
40	50	2000	0.19	0.29	0.39	0.48	0.58	0.72	0.96	1.20
50	50	2500	0.15	0.23	0.31	0.39	0.46	0.58	0.77	0.96
20	60	1200	0.32	0.48	0.64	0.80	0.96	1.20	1.61	
30	60	1800	0.21	0.32	0.43	0.54	0.64	0.80	1.07	1.34
40	60	2400	0.16	0.24	0.32	0.40	0.48	0.60	0.80	1.00
50	60	3000	0.13	0.19	0.26	0.32	0.39	0.48	0.64	0.80
60	60	3600	0.11	0.16	0.21	0.27	0.32	0.40	0.54	0.67
30	70	2100	0.18	0.28	0.37	0.46	0.55	0.69	0.92	1.15
40	70	2800	0.14	0.21	0.28	0.34	0.41	0.52	0.69	0.86
50	70	3500	0.11	0.17	0.22	0.28	0.33	0.41	0.55	0.69
60	70	4200	0.09	0.14	0.18	0.23	0.28	0.34	0.46	0.57
70	70	4900		0.12	0.16	0.20	0.24	0.29	0.39	0.49
40	80	3200	0.12	0.18	0.24	0.30	0.36	0.45	0.60	0.75
50	80	4000	0.10	0.14	0.19	0.24	0.29	0.36	0.48	0.60
60	80	4800		0.12	0.16	0.20	0.24	0.30	0.40	0.50
70	80	5600		0.10	0.14	0.17	0.21	0.26	0.34	0.43
80	80	6400		0.09	0.12	0.15	0.18	0.23	0.30	0.38

net system capacity (C_n, in/d) discussed in Chapter 4. Thus, the required net depth of application is given by:

$$d_r = I_i \times C_n \tag{12.3}$$

The net application depth must be less than or equal to the allowable depletion (AD) determined from scheduling:

$$AD = R_d \times AWC \times f_{dc} \tag{12.4}$$

where: AD = allowable depletion before irrigating, in,
R_d = root depth for scheduling, ft
AWC = available water capacity, in/ft, and
f_{dc} = allowable depletion, fraction.

The relationship between the allowable depletion and the net crop water use is shown in Figure 12.5. The solid blue lines represent the cumulative net crop water use during an irrigation interval. The horizontal dashed lines represent the allowable depletion for six soils using a critical allowable depletion of 50%, a management root zone depth of 4 ft, and the available water capacities consistent with Table 2.3. For example, the allowable depletion for a silt loam soil for these conditions is 4.3 inches. Applying more water would cause deep percolation. If the average crop water use rate was 0.30 inches/day, then the longest acceptable irrigation interval would be (4.3 in ÷ 0.3 in/d = 14.3 days) or rounding down to 14 days. Sandy loam soils only hold about 2.9 inches for these conditions. Also, since the irrigation interval will be shorter than for the silt loam you would expect that the net water use rate would be higher for the shorter period. So, for a water use rate of 0.35 inches/day the maximum irrigation interval for sandy loam would be about eight days. Sandy soils have small water holding capacities which leads to short irrigation intervals, requiring short set times or more laterals in an equally sized field. The irrigation intervals shown in Figure 12.5 represent the maximum acceptable

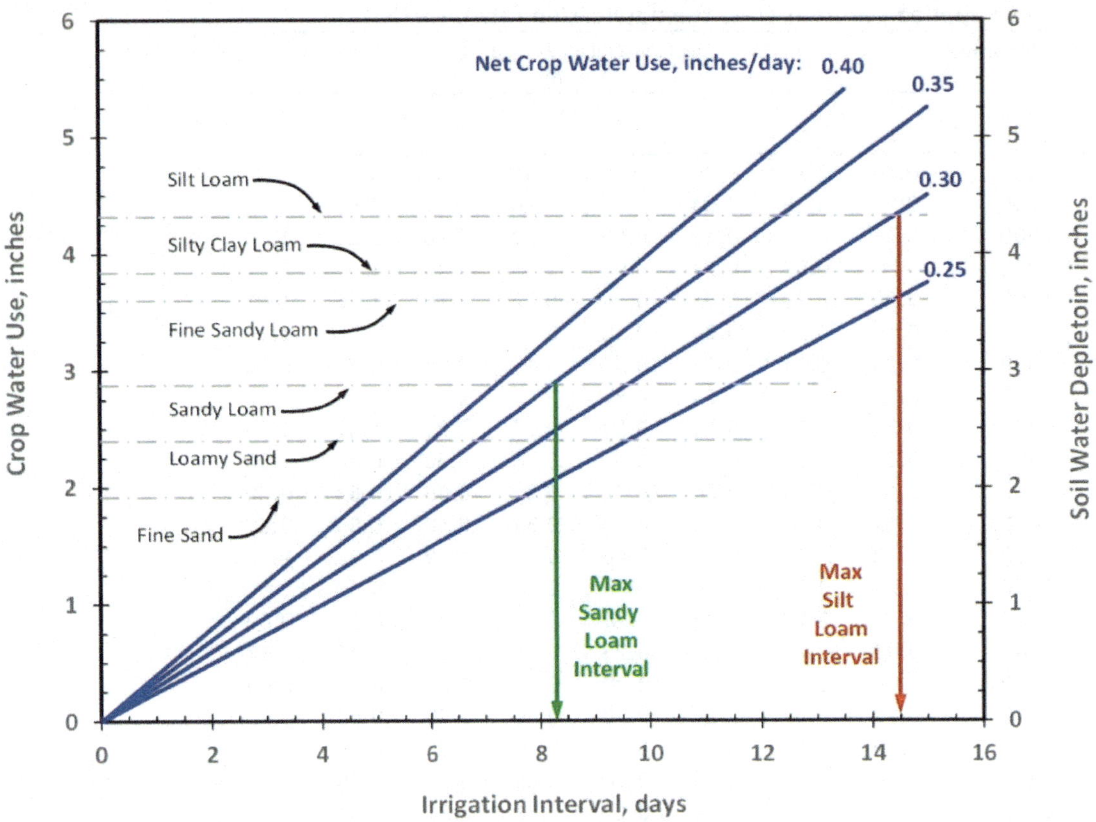

Figure 12.5. Maximum allowable irrigation interval for 4-ft root zone depth and 50% allowed depletion.

values. Shorter intervals could be used such as seven days for the sandy loam soil so that scheduling activities are more tractable. The irrigation interval is based on the more extreme water use periods during the season. The actual irrigation interval will depend on irrigation scheduling during the season.

Water quantities in Figure 12.5 represent net irrigation depths and must be increased to determine the gross depth to apply. The application efficiency for well managed systems could be about 75% (see Table 5.4). Therefore, the gross depth for the silt loam soil is 5.6 inches (i.e., 0.3 in/d × 14 d ÷ 0.75) and 3.7 inches (i.e., 0.35 in/d × 8 d ÷ 0.75) for the sandy loam soil. If using the seven-day interval for the sandy loam soil, the gross depth would be 3.3 inches.

The irrigation interval accounts for time that water is applied, time for draining and moving laterals, and time to relocate the lateral to the starting set. If relocating the lateral takes one day, then water would only be applied for 13 days for the silt loam and 7 days for the sandy loam soil—assuming an eight-day interval. If an irrigator selected a 12-hour set time, then two sets can occur per day and the total sets possible for one lateral for the silt loam would be 26 (i.e., 13 days × 2 sets/day). One lateral would only allow 14 sets per lateral for the sandy loam soil.

The *number of laterals* required for the field depends on the irrigation interval and the field dimensions. The number of laterals also depends on the number of sets in the field and the number of sets possible during the irrigation interval for one lateral. Since fields for periodically moved systems are usually rectangular, the amount of land irrigated per set is usually constant. An example of a moved-lateral system is shown in Figure 12.6. This is an eighty-acre field with laterals on the left and right halves with the mainline running down the middle of the field. The left side of the sketch shows the layout of the sets, and the right side shows the lateral position and field information.

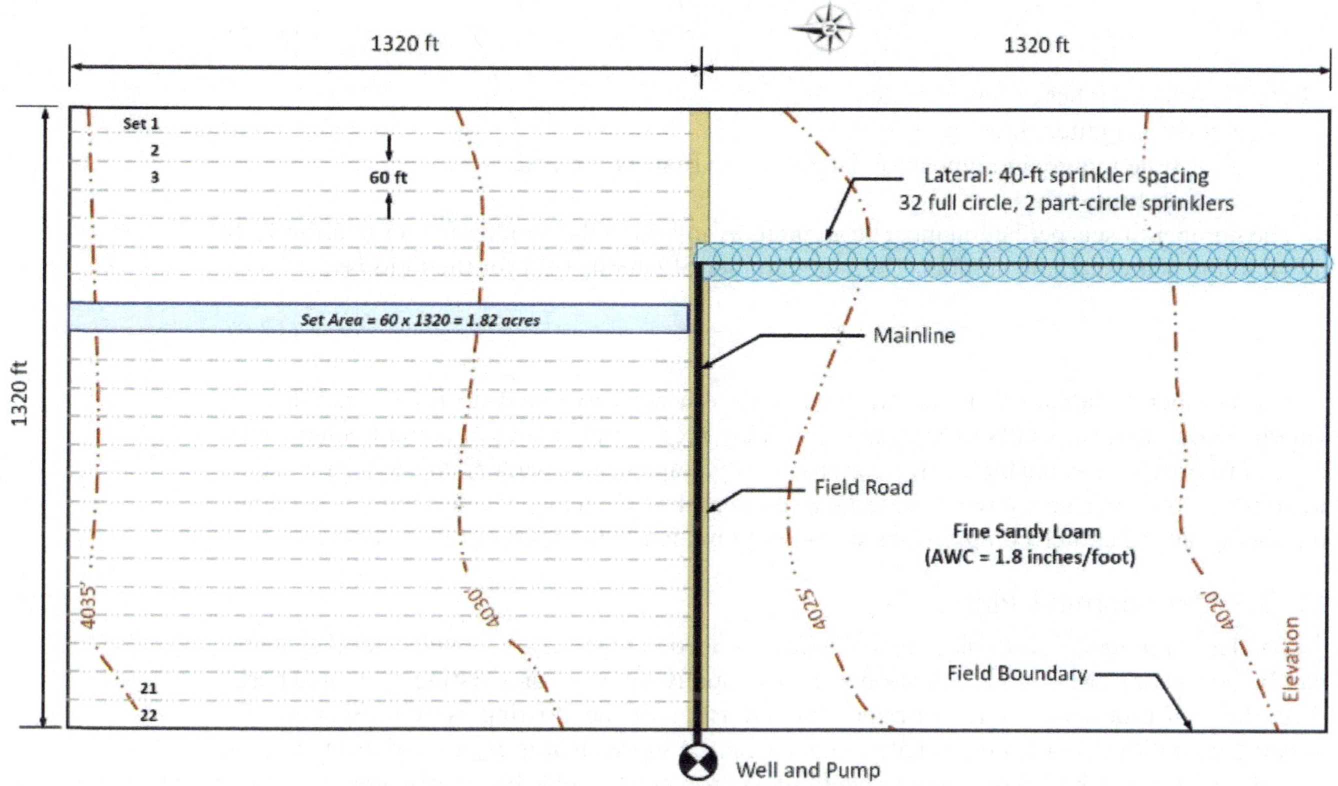

Figure 12.6. Plan view of field layout for moved lateral examples. Note that both sides of the field are irrigated.

The amount of area per set (A_s) is shown in Figures 12.1 and 12.6 and is given by:

$$A_s = \frac{L \times S_m}{43560} \tag{12.5}$$

where: A_s = the area irrigated per set, acres
L = the length of the lateral, ft, and
S_m = the spacing of laterals along the mainline, ft.

Then the number of sets (N_s) in the field is the area of the field (A_f) divided by the area per set:

$$N_s = \frac{A_f}{A_s} \tag{12.6}$$

The number of sets must be an integer, so the value from Equation 12.6 should be rounded up to the nearest integer.

For example, the area of the field in Figure 12.6 is 80 acres (2640 ft × 1320 ft ÷ 43560 ft^2/acre) and the area per set is 1.82 acres; thus, 80 acres ÷ 1.82 acres/set gives 44 sets for the field. Equation 11.15 can be expanded for when the lateral length is less than the field length to also give the number of sets:

$$N_s = \frac{W_f \times L_f}{S_m \times L} \tag{12.7}$$

where the length and width of the field are illustrated in Figure 12.6. Equation 12.7 and the layout in Figure 12.6 show that 44 sets are required for this field. The spacing of 60 feet between laterals is common for moved lateral systems as that is the length of two pipe sections for portable mainline pipe.

The number of laterals depends on the number of sets that can be irrigated with one lateral during the interval:

$$N_p = \frac{24\, I_i - T_m}{T_s} \tag{12.8}$$

where: N_p = the number of sets per lateral
I_i = the irrigation interval, d
T_m = time required to reposition lateral to the first set, hr, and
T_s = set time, hr/set.

The number of sets per lateral must be an integer; so, round the number from Equation 12.8 down to the nearest whole number. Then the number of laterals (N_L) for the field is:

$$N_L = \frac{N_s}{N_p} \tag{12.9}$$

The number of laterals must be an integer, so you need to round up for the number of laterals. Rounding of results to integer values may require relaxation of some management criteria to provide reasonable configurations. For example, increasing f_{dc} to 55% on a sandy loam soil allows one more day for the irrigation interval which may provide a more acceptable number of laterals. Such compromises would not threaten crop yields in most cases.

12.2.3 Management Plan

The large number of variables and calculations needed to describe moved-lateral systems can be perplexing. Managing irrigation systems usually involves an existing system. Therefore, the first process is to describe the characteristics of the existing system. Second, the existing conditions should be evaluated to determine if the system is capable of efficient irrigation. The third step is to develop a plan to meet crop needs and achieve efficiency.

A management spreadsheet such as in Figure 12.7 can help alleviate confusion and facilitate development of a management plan. Users enter parameter values into the shaded boxes and then compute results for the unshaded cells. These data are critical, but they are characteristics of the system and only need be considered once unless major changes are made which would require redesign and new investment.

The Moved-Lateral Management Spreadsheet is divided into four portions. The first portion is an inventory of field characteristics. The values entered for this example are from the field shown in Figure 12.6—an actual field from western Nebraska. The second portion includes design variables which represent the considerations made when designing the system. These are choices made during design; however, these values are generally constant once a system has been installed. Designs can be modified which would result in changing parameters, but once changes were made the variables would be constant again. Thus, design variables are only defined once and are not modified routinely by managers. You can think of field characteristics and design variables as a description of the system you are called upon to manage.

The third portion represents the parameters that describe system performance based on field characteristics and design parameters. These boxes contain either calculation results or parameter values derived from earlier sections. Values in these boxes do not represent choices. The fourth portion includes management variables which can be changed annually or within the season. These values are adjusted to provide the desired performance. Conflicts can arise in selecting management values and issues must be resolved when compromise is needed.

Values for the mainline portion of the operation results section include the total inflow which equals the flow per lateral times the number of laterals. The remaining values are based on application of procedures developed in Chapter 8 for friction loss and flow velocity. One should check values against guidelines for friction loss and velocity.

The lateral information represents calculations described in Chapters 8 and 11. Note that the friction loss calculations are for the most critical lateral that runs uphill—the lateral north of the mainline. This example is for 5-inch aluminum pipe for the side roll. The Hazen-Williams C value is taken as 120 and the wall thickness for the side-roll pipe is 0.072 inches.

Moved-Lateral Management Spreadsheet

Field Properties:

Field:	Example			Type of System	Sideroll
Soil Texture	Fine Sandy Loam			Water Holding Capacity, in/ft	1.8
Field Length, feet	2,640	Field Width, feet	1,320	Field Size, acres	80

Water Supply:

Total Inflow, gpm	700	System Capacity, gpm/acre	8.75

Design Variables:

Mainline:

Diameter, inches	8	Length, feet	1,320	Elevation Change, feet	0

Lateral Information:

Lateral Length, feet	1,320	Number of Laterals	2	Average Wind Speed, mph	7
Nominal Pipe Size, inches	5	Inflow Goal, gpm	350	Elevation Change, feet	8
Spacing Along Mainline, ft	60	Spacing Along Lateral, feet	40		

Sprinkler Information:

Range Nozzle Size, 128th of inch	28	Spreader Nozzle Size, 128th of inch	0		
Design Pressure, psi	55	Design Discharge, gpm	10.3	Wetted Diameter, ft	113
Number Full Circle Spr.	32	Number of Half Circle Spr.	2	Lateral Flow, gpm	340

Operation Results:

Mainline:

Mainline Flow, gpm	680	Inside Diameter, inches	7.872	C Value	120
Friction Loss Factor, psi/100 ft	0.49	Friction Loss in Mainline, psi	6.5	Total Pressure Loss, psi	6.5
Flow Velocity, feet/sec	4.5				

Lateral Information:

Inside Diameter, inches	4.856	C Value	120	Friction Loss Factor, psi/100 ft	1.44
Enclosed Pipe Friction Loss, psi	19.0	Multi-Sprinkler Factor(F)	0.36	Lateral Friction Loss, psi	6.8
Elevation Change, psi	3.5	Pressure Loss Along Lateral, psi	10.3	Pressure Loss / Des. Pressure, %	19%
Space Limit - Sprinklers, %	45%	Space Limit - Laterals, %	60%	Application Rate, inch/hour	0.41
Max Sprinkler Spacing, ft	51	Max Lateral Spacing, ft	68	Flow Velocity, ft/sec	5.9
Acres per Set	1.82	Number of Sets in Field	44	Number of Sets per Lateral	22

Management Variables/Results:

Soils/Plant Info:

Crop	Sugar Beets	Root Depth, feet	4	Critical Depletion, %	50%
Allowable Depletion, inches	3.6	Application Effic. Low Qtr., %	75%		

Times:

Set Time, hours	12	Drain/Move Time, hours	1.0	Application Time, hours	11.0
Reposition Time, hours	12	Irrigation Interval, days	11.5	Down Time, %	12%
Gross Depth Applied, inches	4.5	Net Depth Applied, inches	3.4	Net Crop Water Use, in/day	0.30

Comments/Issues:

Figure 12.7. Example of the Moved-Lateral Management Spreadsheet.

Four-inch pipe can also be used for the side-roll lateral and the wall thickness is the same as for the 5-inch pipe. The spacing limits between sprinklers and laterals are derived from Table 11.3. The application rate is also computed. The number of sets in the field and per lateral are also shown. Completion of these sections and validation that characteristics meet established guidelines establishes the foundation for managing the system.

The fourth section of the sheet is for management choices. These parameters are frequently changed or adjusted to achieve short-term management goals for crop and system performance. The management decisions must be entered into the shaded cells. Management choices vary throughout the season and annually. The following examples illustrate the computation process for the field in Figure 12.6. The second example illustrates computation of the parameter values for the lateral information portion of the operation results in Figure 12.7. The last section in the Moved-Lateral Management Spreadsheet involves the core of managing the system. Example 12.3 illustrates the decision-making process for the management section.

The side-roll system is designed to meet crop water use during the middle of the season when water use rates are at the peak value. During other times of the year the system will have excess capacity. Early in the growing season the root depth will be less than 4 feet, so the root zone would not store a net application of 3.38 inches. When the system has excess capacity the irrigation interval can be extended through accurate scheduling. Timers can also be used to shutoff the pump for a shorter application time for the same set time. This will reduce the net depth when the root zone is swallower or when water demands are less. Once a management plan has been developed it is important to ensure that the system is operating properly—as designed.

Example 12.1

You were retained to evaluate the side-roll system shown in Figures 12.6. Measurements for the system are shown in Figure 12.7. Verify the calculations for the mailine portion of the system to determine if the system meets management guidelines.

Solution:

1. Data from Figure 12.6 shows that the field is 2640 ft long and 1320 ft wide giving an area of 80 acres: Area = (2640 × 1320)/43560 = 80 acres .
2. The available water supply is about 700 gpm and the fine sandy loam soil holds about 1.8 inches of water per foot of soil.
3. Designers choose to orient the pipeline so that the mainline runs through the middle of the field which requires the least amount of mainline pipe—1320 ft. This results in little elevation change along the mainline.
4. The design included sprinklers with a 28/128 or 7/32 inch nozzle operating at 55 psi which produces a sprinkler discharge of 10.3 gpm (see Table 11.1). The lateral incorporates 32 full-circle sprinklers and two half-circle spirnklers. This results in (10.3 gpm/sprinkler × the equivalent of 33 sprikilers) = 340 gpm per lateral.
5. With two laterals the flow in the mainline is 680 gpm.
6. Pressure loss due to friction is computed using Equation 8.11b. The inside diameter of 8-inch aluminum pipe is 7.898 inches and the roughness coefficient is C = 120. So the loss is:

$$P_f = 456\left(\frac{Q}{C}\right)^{1.852}\left(\frac{1}{d^{4.866}}\right) = 456\left(\frac{680}{120}\right)^{1.852}\frac{1}{7.898^{4.866}} = 0.49 \text{ psi}/100\,\text{ft}$$

7. The maximum length of the mainline is 1320 ft so the maximum friction loss is $P_L = 0.49 \times 1320/100 = 6.5$ psi. Since there is negligible elevation change the pressure loss is 6.5 psi and inlet pressure to the mainline will need to be about 61.5 psi for the highest case.
8. The velocity of flow in the mainline is:

$$v = 0.409\,Q/D^2 = 0.409 \times 680 / 7.898^2 = 4.45 \text{ ft/sec}$$

The velocity is less than the 5 ft/sec limit for mainline, thus the mainline is appropriate and the calculations are correct.

Example 12.2

The next phase in the assessment of the system in Figures 12.6 is to compute values for the lateral portion of the side-roll system using the data in Figure 12.7. The assessment should determine if calculations are correct and if the lateral satisfies management guidelines.

Solution:

1. The side-roll system uses 5-inch aluminum pipe that has a wall thickness of 0.072 inches; hence, the inside diameter is 4.856 inches.
2. Pressure loss due to friction is computed using equation 8.11b. with a roughness coefficient of C = 120. So the loss is:

$$P_f = 456\left(\frac{Q}{C}\right)^{1.852}\left(\frac{1}{d^{\,4.866}}\right) = 456\left(\frac{340}{120}\right)^{1.852}\frac{1}{4.856^{\,4.866}} = 1.44\,\text{psi}/100\,\text{ft}$$

The friction loss for an enclosed pipe would be $1.44 \times 1320 / 100 = 19.0$ psi

3. The multiple outlet factor (F) for the sprinkler lateral from Table 8.3 is about 0.36; therefore, the friction loss in the lateral is $0.36 \times 19 = 6.84$ psi.
4. The lateral north of the field runs uphill with the change in elevation of about 8 ft which is equivalent to 8 ft / 2.31 ft/psi = 3.5 psi.
5. Then, the total pressure loss along the critical lateral is 6.84 + 3.5 = 10.3 psi.
6. The average sprinkler pressure is 55 psi from Figure 12.7, so the ratio of pressure variation to the average sprinkler pressure is 10.3/55 = 19% which is smaller than the 20% limit.
7. The average wind speed for the middle of the season is listed as 7 miles/hr. From Table 11.3, the maximum spacing of sprinklers for a retangular sprinkler orientation is 45% of the sprinkler spacing and 60% of the lateral spacing.
8. The wetted diameter for the 7/32-inch nozzle at 55 psi is 113 ft from Table 11.2.
9. The maximum sprinkler spacing along the lateral is then $0.45 \times 113 = 50.9$ feet and the maximum lateral spacing is $0.6 \times 113 = 67.8$ feet. Both actual spacings are less than the maximum, so the spacings are acceptable.
10. The application rate for the side roll is: $A_r = \dfrac{96.3\ q_s}{S_L\ S_m} = \dfrac{96.3 \times 10.3}{40 \times 60} = 0.41\ \text{in/hr}$
11. The irrigated area per lateral is: $L \times S_m / 43560 = 1320 \times 60 / 43560 = 1.82$ acres/set.
12. The number of sets is: $N_s = \dfrac{W_f \times L_f}{S_m \times L} = \dfrac{1320 \times 2640}{60 \times 1320} = 44$ sets and 22 sets/lateral.
13. The flow velocity in the lateral is $v = 0.409\ Q/D^2 = 0.409 \times 340/4.856^2 = 5.9$ ft/sec which is less than the 7 ft/sec limitation. Note that this is the velocity of inflow to the lateral.

Results show that the calculations are accurate and the lateral conforms with management guidelines.

Example 12.3

The final assessment of the system in Figure 12.6 is to ascertain if the management parameters will achieve operational objectives.

Solution:

1. The soil and crop information must be used to compute the allowable depletion. The root depth and critical depletion for the sugar beet crop are estimated from Chapter 4 and local sources to be 4 feet and 50% respectively. This gives an allowable depletion for the fine sandy loam soil of:
$$AD = R_d \times AWC \times f_{dc} = 4 \text{ ft} \times 1.8 \text{ in/ft} \times 0.5 = 3.6 \text{ in}$$
2. The times are dependent on operator preferences. The choice was for 12-hour sets with one-hour downtime to move the lateral. It also requires one-half day to reposition the lateral back to the first set after an irrigation.
3. The irrigation interval depends on the number of sets per lateral and the number of sets per day:
$$I_I = \frac{12 \text{ sets/lateral}}{2 \text{ sets/d}} + 0.5 \text{ d} = 11.5 \text{ d}$$
4. The gross depth of application is the product of the application rate times the application time:
$$d_g = A_R \times T_a = 0.41 \text{ in/hr} \times 11 \text{ hr} = 4.5 \text{ in}$$
5. Since the application efficiency is 75% the net depth is:
$$d_n = AE_{LQ} \times d_g = 0.75 \times 4.5 = 3.38 \text{ in}$$
6. The net depth of application is less than the allowable depletion of 3.6 inches so deep percolation should not be a problem.
7. With an irrigation interval of 11.5 days and a net depth of 3.38 inches, the net crop water use that the system can satisfy is:
$$C_n = \frac{3.38 \text{ in}}{11.5 \text{ d}} = 0.29 \text{ in/d}$$
8. The net crop water use rate should be compared to local conditions to decide if the crop water needs will be met. The value of 0.29 in/day is barely acceptable for this location but should be adequate most years.

The calculations in Figure 12.7 appear to be correct. The system and management decisions should meet crop water requirements efficiently.

12.2.4 Pressure Distribution

You will recall from Chapter 11 that the pressure variation along the lateral should not vary more than 20% of the average pressure. When a larger pressure range occurs, irrigation is not applied uniformly; therefore, excess water must be applied to adequately irrigate the drier portions of the field. The best way to evaluate the pressure variation along the lateral is to measure the pressure at critical points. The sketch in Figure 12.8 shows locations along a lateral that should be monitored. Usually, the highest pressure will occur near the inlet to the lateral. It could occur directly at the inlet or at a low-lying sprinkler close to the inlet. The lowest pressure locations will generally be near the distal end of the lateral. This could be at the end or at a high elevation near the end. These are good locations to measure the pressure.

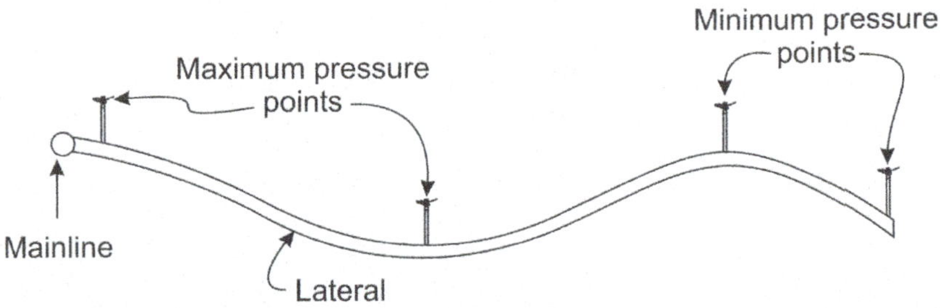

Figure 12.8. Locations along a lateral where pressure measurements should be made.

The pressure can be measured in several ways. First a good quality pressure gauge that is accurate to within 1 or 2 psi should be used. The gauge can be attached to a pitot tube as shown in Figure 12.9 to measure the pressure. Care must be taken to place the pitot tube directly in the center of the jet from the sprinkler. The tube can be moved around in the jet to determine the maximum pressure reading. The maximum pressure read on the pitot is generally the true pressure. The pitot tube can be made from ¹/₈-inch flexible copper tubing attached to the gauge with an appropriate tube fitting. The pressure can also be measured by removing the sprinkler from the riser and directly attaching the pressure gauge. This, however, will require more time to conduct the test.

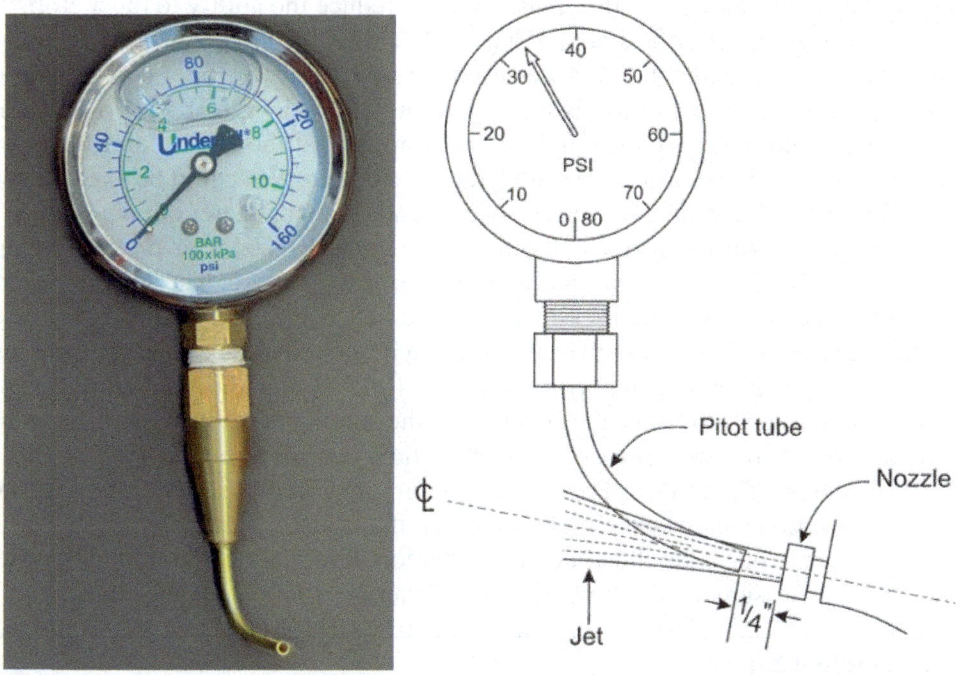

Figure 12.9. Pressure gauge connected to a pitot tube to measure nozzle pressure.

Instead of measuring pressure you can also measure the discharge from the sprinkler at selected locations along the lateral. This can be done by placing a soft, flexible hose over the nozzle and measuring the time required to fill a container to a specified volume. Several measurements should be made on each nozzle to determine the mean flow rate. While pressure can vary 20% of the mean along the lateral, discharge is only allowed to vary 10%.

After performing the pressure tests, you should compute the average pressure. The average pressure (P_a)for the lateral will be approximately:

$$P_a = P_{\min} + 0.25\,(P_{\max} - P_{\min}) \tag{12.10}$$

As previously stated, the maximum acceptable range of pressure for the four points shown in Figure 12.8 is 20% of the average pressure. If the ***flow rate*** was measured the variation should be less than 10% of the average discharge. The average discharge equals the minimum discharge plus about 40% of the variation in flow that you measured.

Suppose that the pressure or discharge variation is too large, what could the problem be and how can you correct the situation? First compare the pressure or discharge at the lateral inlet to the design value. If the pressure or discharge is too small the problem may be with the pump or mainline system and the entire lateral is simply under pressurized.

Excessive variation in pressure may be due to a lateral that is too long, or the pipe diameter may be too small. The lateral could also run up a hill that was not considered in design. Correcting problems can be difficult. It is probably infeasible to replace the lateral unless the variation is bad, or the lateral is worn and will be replaced soon anyway. Recall from Chapter 11 that about half of the pressure loss occurs in the first third of the lateral. Thus, the initial section of the lateral could be replaced with larger diameter pipe to reduce the pressure loss. This is practical for side-roll and hand-move systems where the larger pipe will always be located near the mainline. This solution will not work for tow-line systems since the larger pipe would be at the distal end of the lateral half of the time.

The nozzles could be replaced with a smaller size to reduce the average flow rate of the lateral and to increase pressure at the downstream end. This will reduce the average discharge along the lateral. Some will think that this will reduce the ability to meet crop water requirements. However, the critical area along the lateral is the area receiving the smallest amount of water. Assuming that this area was used for scheduling, the poor uniformity contributed to deep percolation or runoff in the early portions of the lateral. If the flow rate at the critical distal end is not reduced, the depth of water applied at the critical area will be the same or more than when the nozzles were too large. Smaller nozzles could be used on sprinklers near the mainline if the smaller nozzles provide an adequate diameter of coverage.

Another alternative is to install either flow control nozzles or pressure regulators in the sections of the lateral that are likely to have excess pressure. These devices reduce the discharge in the high-pressure areas and produce better uniformity. It may not be necessary to install regulators along the entire lateral. Keep in mind that there is a pressure loss of about 5 psi across regulators, so you may not want to install them in the areas already low in pressure. You may also need to change the nozzle(s) in the sprinklers equipped with pressure regulators. Pressure regulators are more expensive than flow control nozzles, but they also operate over a wider range of pressures. Pressure regulators may be needed all along a tow-line lateral since the sprinklers are changing locations on the landscape every set.

An example may help illustrate the evaluation of lateral distribution and some alternatives for solving pressure distribution problems (Example 12.4).

Example 12.4

You evaluated the pressure distribution along a lateral and need to determine if the lateral conforms to pressure variation guidelines. Recommend changes if the lateral is inadequate.

Given: The pressure at the second sprinkler nozzle is 45 psi and the nozzle pressure at the next to last sprinkler on the lateral is 32 psi.

The lateral is a 4-inch aluminum pipe with sprinklers spaced every 40 ft along the lateral.

The first sprinkler is 40 feet from the mainline.

Risers 5 ft tall are used to elevate sprinklers above the crop.

The sprinklers have one 15/64 nozzle per sprinkler except the first and last sprinklers which are 5/32 nozzles.

The land is generally flat.

Solution:

The average pressure will be approximately:

$$P_a = P_{min} + 0.25\,(P_{max} - P_{min}) = 32 + 0.25\,(42 - 32) = 35 \text{ psi} \qquad \text{(Eq. 12.3)}$$

20% variation of the average is 7 psi (35 × 0.2)

The pressure variation along the lateral is 13 psi (45 - 32).

Since the pressure variation of 13 psi is larger than 20% of the average pressure, the lateral does not conform to the pressure variation guideline.

The cheapest way to bring the pressure variation into an acceptable range would be to use pressure regulators. Pressure regulators cause a pressure loss of approximately 5 psi so the nozzle size may need to be adjusted to provide the needed flow.

Analyzing solutions for existing laterals is complex, so a spreadsheet program was developed to assist evaluation. The program is called ***Lateral Analysis***. Performance for an existing lateral is shown in Figure 12.10. The shaded cells are where operators input data about the lateral. The unshaded areas cannot be changed. The variation of nozzle pressure for the existing lateral is 14.7 psi which represents about 40% of the average pressure—double the guideline.

Suppose pressure regulators are used to minimize variation. Examining the data for the sprinklers along the lateral in Figure 12.10, the pressure at the distal end of the lateral is about 33 psi. A 35-psi pressure regulator would give about the same nozzle pressure. Additionally,

Sprinkler Lateral Analysis

Pipe Information:

Portion of Lateral	Nominal Diameter, inches	Length of Portion, feet	Hazen-Williams C Value	Wall Thickness, inches	Inside Diameter, inches
First	4	1320	120	0.050	3.900
Second	4	0	120	0.050	3.900
Total Length. Feet		1320			

Spacing of Sprinklers Along Lateral, feet	40
Lateral Spacing, feet	60
Height of Sprinkler Riser, feet	5.0
Pressure at the Distal End of Lateral, psi	35
Lateral Slope, % (+ is Uphill & - is downhill)	0.00
Sprinkler Discharge Coefficient	28.93
Number of Sprinklers from Inlet with Regulators	10
Pressure Rating for Regulators	35

Nozzle Sizes	Range	Spreader
Size of Principle Sprs. 128ths	30	0
Size of Nozzle End Sprs. in 128ths	20	0

Part Circle Sprinkler @ Inlet	Y	1
Part Circle Sprinkler @ Distal End	Y	1
Number of Full Sprinklers		32
Number of Pipe Joints		33

Performance Analysis

Values	Pressure in Lateral, psi	Nozzle Pressure, psi	Sprikler Discharge, gpm
Maximum	49.4	37.7	9.76
Minimum	35.0	32.8	9.11
Average	38.8	34.3	9.32
Variation	14.4	4.9	0.65
Percent Variation	37%	14%	7%

Lateral Inflow, gpm	Inflow Velocity, ft/sec	Average Application Rate,in/hr
307	8.3	0.37

Distribution Analysis

Sprinkler #	Distance from Lateral Inlet, feet	Relative Elevation, feet	Regulator Used (Y/N)	Pipe Inside Diameter, inches	Pressure in Lateral, psi	Nozzle Pressure, psi	Sprinkler Discharge, gpm	Flow in Lateral, gpm
1	40	0	Y	3.900	49.4	35.0	9.40	302
2	80	0	Y	3.900	48.2	35.0	9.40	293
3	120	0	Y	3.900	47.0	35.0	9.40	283
4	160	0	Y	3.900	45.9	35.0	9.40	274
5	200	0	Y	3.900	44.8	35.0	9.40	265
6	240	0	Y	3.900	43.8	35.0	9.40	255
7	280	0	Y	3.900	42.9	35.0	9.40	246
8	320	0	Y	3.900	42.1	34.9	9.39	236
9	360	0	Y	3.900	41.3	34.1	9.28	227
10	400	0	Y	3.900	40.6	33.4	9.18	218
11	440	0	N	3.900	39.9	37.7	9.76	208
12	480	0	N	3.900	39.3	37.1	9.68	199
13	520	0	N	3.900	38.7	36.5	9.61	189
14	560	0	N	3.900	38.2	36.0	9.54	179
15	600	0	N	3.900	37.7	35.6	9.48	170
16	640	0	N	3.900	37.3	35.1	9.42	160
17	680	0	N	3.900	36.9	34.8	9.37	151
18	720	0	N	3.900	36.6	34.4	9.33	142
19	760	0	N	3.900	36.3	34.2	9.29	132
20	800	0	N	3.900	36.1	33.9	9.25	123
21	840	0	N	3.900	35.8	33.7	9.22	114
22	880	0	N	3.900	35.7	33.5	9.20	105
23	920	0	N	3.900	35.5	33.3	9.18	95
24	960	0	N	3.900	35.4	33.2	9.16	86
25	1000	0	N	3.900	35.3	33.1	9.14	77
26	1040	0	N	3.900	35.2	33.0	9.13	68
27	1080	0	N	3.900	35.1	33.0	9.12	59
28	1120	0	N	3.900	35.1	32.9	9.12	50
29	1160	0	N	3.900	35.0	32.9	9.11	40
30	1200	0	N	3.900	35.0	32.9	9.11	31
31	1240	0	N	3.900	35.0	32.8	9.11	22
32	1280	0	N	3.900	35.0	32.8	9.11	13
33	1320	0	N	3.900	35.0	32.8	4.05	4

Figure 12.10. Existing conditions in the Lateral Analysis program for the example lateral.

regulators cause about 5 psi loss when regulation is not active. Thus, a nozzle pressure of 38 psi without regulation will give an outlet pressure of about 33 psi. The first ten sprinklers have a nozzle pressure above 38 psi when regulators were not used. So, regulators are installed on the first ten sprinklers.

The lateral analysis program was used to evaluate the results when using the ten regulators. Table 12.4 shows a comparison of the performance analysis when no regulators were used on the lateral and when 10 regulators were used at the inlet of the lateral. The results show that using regulators reduced the nozzle pressure variation to 14% and the discharge variation to about 7%. Both quantities are within the acceptable guidelines for uniformity. Ten regulators represent an investment of approximately $100 which would work for a long time, so pressure regulation is a relatively inexpensive and efficient way to achieve the uniformity goals. Of course, it is essential that the regulators are always used at the inlet to the lateral which would require some organization for hand-move systems. The spreadsheet can be used to analyze laterals and refining designs for special needs. Laterals with two pipe diameters can also be evaluated.

Table 12.4. Comparison of sprinkler lateral performance with and without regulators.

Values	Lateral Pressure (psi)	Nozzle Pressure (psi)	Sprinkler Discharge (gpm)
Performance—No Regulators			
Maximum	49.7	47.6	10.96
Minimum	35.0	32.8	9.11
Average	38.8	36.7	9.62
Variation	14.7	14.7	1.85
Percent variation	38%	40%	19%
Performance—With Regulators for First 10 Sprinklers			
Maximum	49.4	37.7	9.76
Minimum	35.0	32.8	9.11
Average	38.8	34.3	9.32
Variation	14.4	4.9	0.65
Percent variation	37%	14%	7%

12.2.5 Uniformity Issues

Poor uniformity is evidenced by plant water stress in areas receiving less water. The problem can be due to pressure distribution, but other factors are possible. One issue is stretching the spacing between laterals or sprinklers along the lateral. When the spacing is excessive for prevailing wind conditions the overlap is inadequate to provide uniformity. Poor uniformity may also arise from worn sprinklers and nozzles. The sprinkler bearing may be worn causing the sprinkler to rotate slowly or stick in locations during rotation. Bearings can be replaced but it is often best to replace the entire sprinkler with this amount of wear. Wear of brass straight-bore nozzles can be evaluated by matching the diameter to a drill bit of that size. If the nozzles are worn significantly, they can be replaced very economically; however, sprinklers should be checked to ensure that they should not also be replaced.

The diameter of coverage of some sprinklers can be increased by inserting straightening vanes. The straightening vane shown in Figure 11.2 decreases turbulence and increases the diameter of coverage which may provide the coverage needed for acceptable uniformity. The sprinkler jet with a vane does not breakup as quickly as sprinklers without vanes. This also provides more throw and helps fight wind effects; however, vanes may lead to poorer distributions about an individual sprinkler resulting from a doughnut pattern. Straightening vanes are inexpensive and easily installed, so vanes can be evaluated for a few sprinklers. If vanes do not improve performance, then other alternatives should be considered.

The spacing between lateral sets is often limited by the length of mainline pipe. Aluminum pipe is commonly available in either 20-, 30- or 40-ft lengths. This dictates the width of sets and ultimately the uniformity. Operators can adapt to this problem by ***offsetting*** the lateral each set. Suppose that mainline joints are 30 ft long and two joints are used for a set width of 60 feet. For odd numbered irrigations, the lateral could be placed at locations of 30, 90, 150, etc., feet from the field boundary. The lateral is then placed between these setting for even numbered irrigations or at locations of 0, 60, 120, 180, etc. feet from the field boundary. Offsets place laterals halfway between the previous set and the ***cumulative uniformity*** of water application generally improves. Offsetting may cause issues when the first and/or last set is along the field boundary or where lanes are required for tow-line systems in tall crops.

12.2.6 Uniformity Evaluation

The ultimate evaluation of uniformity is to measure the distribution using an array of collector cans to compute the coefficient of uniformity as described in Chapter 5. It is impractical to measure the distribution along the entire lateral; thus, a representative area should be selected near the downstream end of the lateral where uniformity will be lowest. The configuration of catch cans is illustrated in Figure 12.11 for a lateral with sprinklers 40 ft apart along the lateral and a set width of 60 ft. The spacing of cans should be selected so that each container represents the same area. A common denominator should be determined that is convenient—either 5, 10 or 20 ft for Figure 12.11. In this case a ten-foot spacing was selected for collector spacing. Initially, the column of cans is placed one-half of the nominal can spacing from the lateral and the first row is one-half the can spacing from the sprinkler— i.e., first can is placed 5 ft from the lateral and 5 ft from the sprinkler. The remaining rows and columns of cans are space the full distance (10 ft) apart. This orientation ensures that each container represents the same area (10 ft × 10 ft) which simplifies computation of uniformity. Cans are placed on both sides of the lateral to evaluate the effect of wind. Tests should be conducted when wind, temperature, and humidity conditions are representative of the area.

It is impractical to measure the depth of water applied by all laterals for moved-lateral systems, so it is necessary to numerically overlap the catch data from one lateral. The lateral for the second set in Figure 12.11 should be operated for the test. Water is measured on both sides of the lateral. An adequate distance along the lateral should be tested to avoid bias from one or two sprinklers. Sprinklers should be evaluated to ensure they represent the system. However, the number of containers increases quickly. For example, the layout in Figure 12.11 requires 112 catch cans. Cans should be placed at least one row beyond adjoining laterals if wind is expected during the test. The system should be operated long enough to provide adequate water to accurately measured the depth in the cans. The water caught in cans is measured with a graduated cylinder. The diameter of the top of several catch containers should also be measured. The volume caught is converted to a depth by dividing the volume by the area of the top of the container.

The depth of water from the second lateral must be *overlapped* to determine the depth applied to the area by adjoining laterals. The depth of water applied during successive sets is computed based on the distance of the lateral from the point of interest. The following example shows how to overlap depths to evaluate the uniformity.

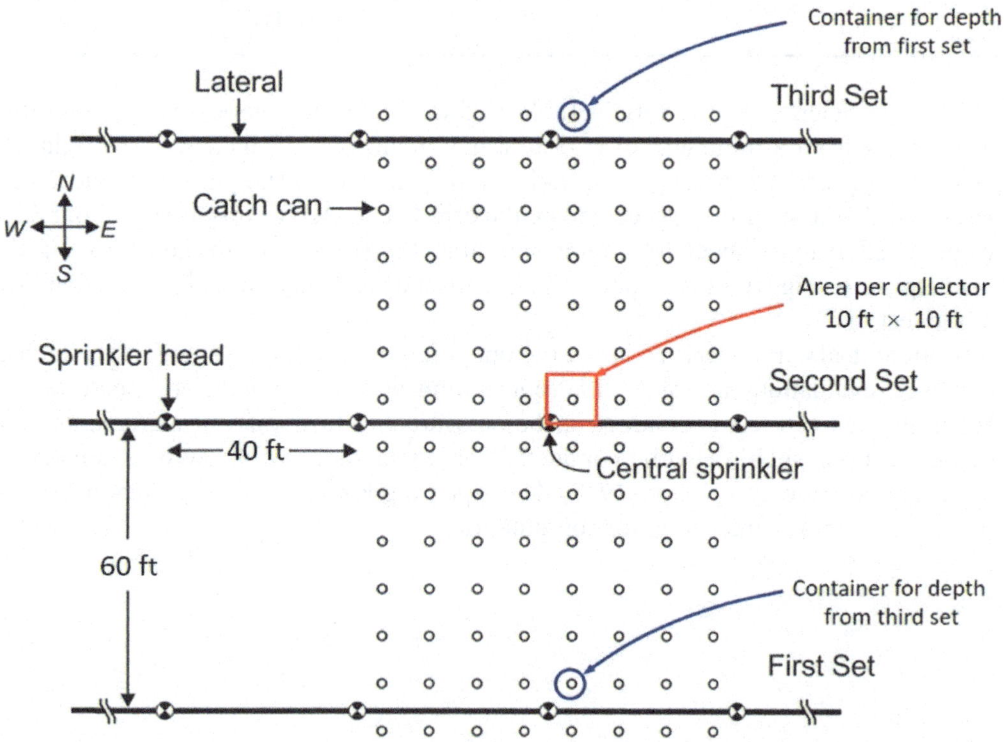

Figure 12.11. Layout for testing the uniformity of application using catch can collectors.

Example 12.5

The uniformity of a moved-lateral system was measured by placing catch cans around the central lateral (second set) as shown in Figure 12.11. The central lateral was operated for 2 hours to provide measurable quantities of water. Results of the test are summarized in Table 12.5.

Find: The depth of water applied between laterals and the coefficient of uniformity.

Solution:

1. The volume of water caught in the cans and the depth of water are given in Table 12.5.
2. Consider the can located 5 feet north and 5 feet east of the central sprinkler (delineated by red box). A volume of 235 cm^3 of water was caught during the test.
3. This is equivalent to: 235 cm^3 / 81.1 cm^2 = 2.90 cm.
4. Dividing by 2.54 cm/in gives a depth applied of 1.14 inches (cell with red border).
5. However, this is only the water applied when the lateral is located at the second set.
6. The depth applied by the first and third sets of the lateral must be determined.
   The distance from the first set is 65 ft north of lateral one (5 ft east of the sprinkler).
   The depth in the container was 24 cm^3 which is equivalent to 0.12 in (blue cells).
7. The depth when the lateral is at set 3 is equivalent to the depth caught in the container 55 feet south of the lateral from set 2.
   The depth caught was 19 cm^3 which is equivalent to 0.09 in (blue cell).
8. The overlapped depth for the three sets is 1.14 + 0.12 + 0.09 = 1.35 in (red cell).
9. This procedure was used to compute the depth for all container locations in Table 12.5.
   The average depth of water caught in the containers was 1.26 in.
   The deviation from the mean depth is 1.35 – 1.26 = 0.09 in for the red cell.
10. The average deviation for all containers is 0.13 in.
11. Then, the coefficient of uniformity is then given by:

$$CU = 100\left(1 - \frac{\sum |d_i - d_a|}{n\ d_a}\right) = 100\left(1 - \frac{0.13}{1.26}\right) = 90\%$$

The coefficient of uniformity in Table 12.5 is 90, which is good, even though the application ranges from a minimum of 0.98 to a maximum of 1.57 inches over a relatively small distance. The areas between the sprinklers along the lateral (i.e., those 15 and 25 ft east and west of the central sprinkler) seem to be the driest. The DU for this area is about 84%, so you would need to apply about 20% more water than the average depth (i.e., 1.26 ÷ 0.84 = 1.5 in) to adequately irrigate the dry spots. The CU for the entire lateral will be less than for the area of the test.

A great deal can be learned about the operation of sprinkler systems with catch-can tests; however, evaluations are quite time consuming and wind conditions make tests difficult. When a catch can test is conducted, the pressure and flow rate measurements described in earlier sections should also be conducted. This is a short overview of evaluating sprinkler systems. Merriam and Keller (1978) developed a good reference on system evaluation that provides examples and charts for computation.

Table 12.5. Results of catch can evaluation for an example system. Red and blue cells are used in example.

Diameter of Top of Can (in) = 4				Area of Container (cm^2) = (6.452 π D^2/4) = 81.08				
Distance North of Central Sprinkler (ft)	Distance East of Central Sprinkler (ft)							
	-35	-25	-15	-5	5	15	25	35
	Volume of Water Caught (cm^3)							
65	23	20	19	23	24	19	18	22
55	76	63	60	73	76	61	58	70
45	118	98	94	114	119	94	90	109
35	160	134	127	154	161	128	123	147
25	191	160	153	185	193	153	147	176
15	212	177	169	205	214	170	163	196
5	233	195	186	225	235	187	179	215
-5	210	176	167	203	212	168	161	194
-15	191	160	152	184	192	154	147	176
-25	163	136	129	156	163	132	125	149
-35	112	93	87	105	111	93	87	100
-45	50	40	35	43	47	45	39	41
-55	22	17	12	15	19	23	18	14
-65	0	0	0	0	0	0	0	0
	Depth of Water Applied (in)							
65	0.11	0.10	0.09	0.11	0.12	0.09	0.09	0.11
55	0.37	0.31	0.29	0.35	0.37	0.30	0.28	0.34
45	0.57	0.48	0.46	0.55	0.58	0.46	0.44	0.53
35	0.78	0.65	0.62	0.75	0.78	0.62	0.60	0.71
25	0.93	0.78	0.74	0.90	0.94	0.74	0.71	0.85
15	1.03	0.86	0.82	1.00	1.04	0.83	0.79	0.95
5	1.13	0.95	0.90	1.09	1.14	0.91	0.87	1.04
-5	1.02	0.85	0.81	0.99	1.03	0.82	0.78	0.94
-15	0.93	0.78	0.74	0.89	0.93	0.75	0.71	0.85
-25	0.79	0.66	0.63	0.76	0.79	0.64	0.61	0.72
-35	0.54	0.45	0.42	0.51	0.54	0.45	0.42	0.49
-45	0.24	0.19	0.17	0.21	0.23	0.22	0.19	0.20
-55	0.11	0.08	0.06	0.07	0.09	0.11	0.09	0.07
-65	0.00	0.00	0.00	0.00	0.00	0.00	0.00	0.00
	Overlapped Depth for the Three Sets (in)							
55	1.39	1.16	1.10	1.34	1.40	1.11	1.06	1.28
45	1.50	1.25	1.19	1.45	1.51	1.20	1.15	1.38
35	1.57	1.31	1.24	1.51	1.57	1.26	1.20	1.44
25	1.47	1.23	1.17	1.41	1.48	1.19	1.14	1.34
15	1.27	1.05	0.99	1.20	1.27	1.04	0.98	1.15
5	1.35	1.13	1.05	1.28	1.35	1.11	1.04	1.22
	Absolute Deviation from Mean Depth (in)							
55	0.13	0.10	0.16	0.08	0.14	0.15	0.20	0.02
45	0.24	0.01	0.07	0.19	0.25	0.06	0.11	0.12
35	0.31	0.05	0.02	0.24	0.31	0.00	0.06	0.18
25	0.21	0.03	0.10	0.15	0.22	0.07	0.12	0.08
15	0.01	0.21	0.27	0.06	0.01	0.22	0.28	0.11
5	0.09	0.13	0.21	0.02	0.09	0.15	0.22	0.04
					Coefficient of Uniformity = 90			

12.3 Solid-Set Systems

Two forms of solid-set systems are available. One is a permanently installed system as illustrated in Figure 12.12. A typical design includes buried mainline and laterals, often with PVC pipe. Special risers are used to bring sprinklers to the required height for the crops irrigated. Risers usually include some type of flexible connection near the soil surface to prevent rupturing the lateral if sprinkler risers are damaged during farming operations. Solenoid valves are used at the inlet to allow irrigation of a set as frequently as desired and with a variable irrigation interval. Solenoid valves may be above ground, as shown in Figure 12.12, or can be buried in irrigation valve boxes to provide access for repair. The solenoid valves are connected to an electronic controller that can be programmed to open and shut valves for desired frequency and duration. The solenoid valves may be connected to the controller with direct wiring or by wireless control. Controllers with web access can communicate to office computers or portable devices for real-time control. Some controllers now allow integration of irrigation scheduling into the controller programming as described by Davis and Dukes (2016) and Haghverdi et al. (2021). Controllers can also interface with soil water monitoring to provide information on the crop water status.

Portable solid-set systems are also available as shown in Figure 12.13. These systems are essentially a series of hand-move laterals connected to a mainline. Some systems such as shown in Figure 12.13c can be move mechanically to allow field operations and to reposition laterals. Manual or automatic values can be used to turn on and off laterals which allows for varying set times and irrigation intervals as needed. These systems are cheaper and more flexible than permanently installed systems. Many characteristics of portable solid-set systems are the same as periodically moved laterals.

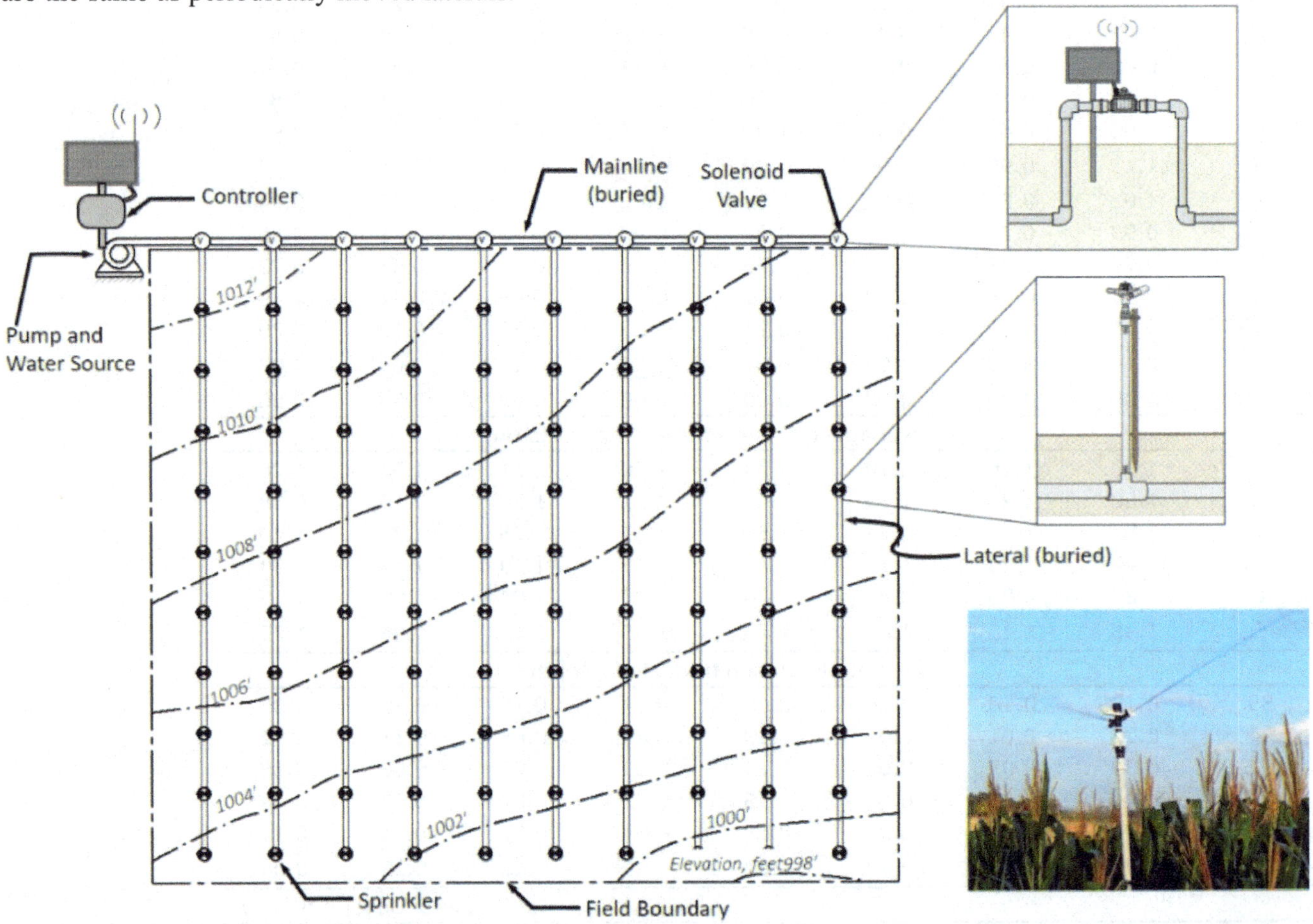

Figure 12.12. Plan view and some components of a permanent solid-set system. (Lower right photo is courtesy of Senninger Irrigation.)

Figure 12.13. Examples of portable solid-set systems (photos a and b are courtesy of Hunter Industries; photo c is courtesy of Westlake Pipe & Fittings).

12.3.1 System Design

Solid-set systems, especially permanent systems, are expensive to install and, therefore, should be carefully designed. Permanent solid-set systems can be tailored to specific fields conditions to minimize installation and operation costs. The size of mainlines, manifolds and laterals can be reduced in an incremental fashion to achieve pressure loss and flow guidelines while saving investment costs. For example, the pipe for the distal portion of the lateral may be smaller than at the inlet since flows decrease along the lateral. One lateral in the system will ultimately determine the maximum pressure required from the pump. Mainline, submain and lateral sizes for other portions of the system may be smaller to reduce investment cost. The size of nozzles along the lateral can be varied for solid-set systems which allows for enhanced uniformity with little investment.

Each lateral can be specifically designed for local conditions. Thus, some laterals may operate at different average pressures depending on the location in the network and elevation of the lateral. The discharge and application rate can be designed to apply the desired depth at the appropriate application rate to avoid runoff and erosion. The set time can be short to apply small depths each irrigation. The cost of solid-set systems depends on the number of laterals that are needed. Therefore, a common problem is that the distance between laterals is extended to reduce investment costs. This is critical because once the system is installed it is expensive to retrofit the system to operate appropriately.

The piping network in buried solid-set systems will probably be PVC. This has proven to be an economical pipe for construction and operation. However, the pipe cannot take large pressure surges. Therefore, special precautions should be taken to prevent pipeline damage due to water hammer or vacuum. Vacuum relief valves must be installed at the high locations in the field to allow air to enter when the system is shutdown. High-pressure surges can be dealt with in several ways. A high-pressure relief valve can be installed in areas where pressure reached peak values. Surge tanks can be installed, especially at the pump, to absorb some of the pressure surge ahead of the PVC pipeline. Special valves can also be used to throttle the flow at the pump until pressure develops in the mainline. This prevents the pressure surge that occurs when flowing water reaches the end of an enclosed pipeline. These valves can also be adjusted to maintain a constant downstream pressure. This is useful when a reduced number of laterals are operated. The pressure control of the valve keeps the operating pressure of the pipeline within an acceptable range. Depending on the characteristics of the pump, the pressure ahead of control valves may rise to high levels when a small flow rate is pumped; therefore, variable speed pumps or other controls may be needed.

Lateral spacing is not contingent on the length of mainline pipes for permanent solid-set systems; therefore, the lateral spacing should align with farming equipment operation. One of the major inconveniences with solid-set systems is that you must farm around the risers. If the lateral spacing is adjusted to match typical or critical equipment width, then farming practices are easier.

Laterals must be designed to prevent frost damage. The laterals should be drained when cold weather threatens. Drain valves can be used to drain the pipelines every irrigation, but this may not be desirable, especially if the laterals and mainlines are large containing substantial amounts of water. In this case, a long time is required to drain the pipes and drainage accumulates at lower elevations along the lateral or mainline. Thus, a good deal of water drains, and wet areas can develop that are difficult to accommodate. An alternative is to use compressed air to force water out of the system. The end of the mainline and laterals can be equipped with a manual ball valve. The valve is shut when irrigating. The valve is opened when the pipeline is drained, then compressed air is supplied into the mainline. Water in the mainline can be expelled first. Then valves for the laterals are sequentially opened to force water from the laterals. Valves may be needed on some risers to prevent compressed air from bleeding through sprinklers on rolling terrain. Pipelines must also be installed deep enough so that farming operations do not either crush or damage the pipe while tilling. Any control or power cabling should be installed consistent with local codes.

Obviously, solid-set systems must be carefully designed and installed. An experienced designer should be employed for sophisticated systems. Intricate software programs are available to customize designs to local needs.

12.3.2 Management Problems

Management problems with solid-set systems are the same as for moved laterals. The spacing of sprinklers is too large for the selected sprinkler, the lateral is too long or small to meet pressure guidelines, etc. These are more difficult to correct for solid-set systems since the system is often installed below ground. However, solid-set systems are usually not limited by the set time or irrigation interval. So, sprinklers and nozzles can be changed to meet performance requirements and then the set time can be adjusted to provide the depth of water needed. Pressure regulators and flow control nozzles are still a viable option if changes are necessary.

Maintenance requirements for solid-set systems are more demanding than periodically moved systems. There are more electromechanical components in the system that periodically fail. The operator should conduct periodic inspections to ensure that the automatic system is working properly. This is critical because solid-set systems are often used on valuable crops where stress is expensive.

Many solid-set systems are connected to automatic controllers that allow producers to start the system anytime they desire. The controller must be programmed so that each lateral is operated at the desired time. Many controllers have more than one program that can be stored so that different depths can be easily applied. These functions work well. The controller can start irrigating any time on any day of the week without operator assistance. Unfortunately, this automatic operation is often not well linked with scheduling so that the system operates when it does not need to or does not operate when it is needed. Thus, systems have the potential for efficient operation if users take the time to learn how to operate the system and train assistants.

12.4 Guns

Large sprinklers called guns have been developed for stationary and moving irrigation systems. A gun has a single large diameter nozzle that discharges large flows and throws water long distances. Stand-alone guns can be portable or installed at a fixed site (Figure 12.14). Water can be supplied from buried pipelines with valves that allow the gun to be moved from riser to riser, or water can be supplied from portable pipes. Guns can be operated to provide overlap and uniform irrigation. In other cases, guns may be used for irregular areas where uniformity is a secondary objective.

Big guns are also useful for distribution of wastewater from animal feeding operations or effluent from processing centers. An example of animal wastewater application is illustrated in Figure 12.14d. Guns are well suited to wastewater application because the nozzle are large, high pressures are used and all discharge is through one nozzle, so the flow velocity remains high minimizing clogging or plugging.

The performance of Model 150 and 200 guns from Nelson Irrigation Corporation is listed in Table 12.6. The performance includes the discharge and wetted diameter for some nozzle sizes and a range of pressures. Three types of nozzles are included. Taper nozzles are shaped to minimize turbulence and provide high flows and large throw distances. The ring nozzle creates more turbulence which reduces flow and throw (Figure 12.15). A hybrid nozzle called the tapered-ring nozzle falls between the tapered and ring nozzles. The jet from a tapered nozzle does not break up as completely and provides larger droplets. Large droplets travel further and retain their velocity longer than small drops. This provides increased throw but also causes a donut pattern and compacts bare soils. Conversely, ring nozzles provide more breakup; thus, the wetted diameter is less than for tapered nozzle and the water application from ring nozzles is gentler. The discharge of the ring nozzle varies from 60 to 75% of the flow from the tapered nozzle. The wetted diameter of the ring nozzle is about 85% of that for the tapered nozzle (Figure 12.15).

Data in Table 12.6 show that the discharge ranges from 100 to over 1200 gpm from a single gun. Simultaneously the wetted diameter varies from 250 to over 600 feet depending on the nozzle size and operating pressure. Such a large operating range makes guns flexible and adaptable to many conditions. The large jet can compact bare soil surfaces and reduce infiltration rates. Energy in large drops moving at high velocity may cause leaf damage for sensitive plants. Operating pressures range from 50 to 130 psi making gun-based systems expensive to operate. These systems also experience significant pressure loss in conveyance systems because large flows must be delivered to the end of the supply pipeline. These systems also require pipe with higher pressure ratings which increases investment costs. Guns may apply water at high application rates that cause runoff and erosion, especially on steep slopes and clayey soils.

Figure 12.14. Illustration of gun sprinklers (photos a and b are courtesy of Nelson; photo c is courtesy of Wade Rain, Inc.).

Table 12.6. Performance of 150 and 200 guns from Nelson Irrigation Corp.

150 Series Big Guns—24° Trajectory[a]								150 T Taper Bore Nozzles						
Size:	0.7 in		0.8 in		0.9 in		1.0 in		1.1 in		1.2 in		1.3 in	
Press. (psi)	Flow (gpm)	Diam (ft.)	Flow (gpm)	Diam (ft.)	Flow (gpm)	Diam (ft.)	Flow (gpm)	Diam (ft.)	Flow (gpm)	Diam (ft.)	Flow (gpm)	Diam (ft.)	Flow (gpm)	Diam (ft.)
50	100	250	130	270	165	290	205	310	255	330	300	345	350	360
60	110	265	143	285	182	305	225	325	275	345	330	365	385	380
70	120	280	155	300	197	320	245	340	295	360	355	380	415	395
80	128	290	165	310	210	335	260	355	315	375	380	395	445	410
90	135	300	175	320	223	345	275	365	335	390	405	410	475	425
100	143	310	185	330	235	355	290	375	355	400	425	420	500	440
110	150	320	195	340	247	365	305	385	370	410	445	430	525	450
120	157	330	204	350	258	375	320	395	385	420	465	440	545	460

150 Series Big Guns—24° Trajectory[a]								150 R Ring Nozzles						
Size:	0.86 in		0.97 in		1.08 in		1.18 in		1.26 in		1.34 in		1.41 in	
Press. (psi)	Flow (gpm)	Diam (ft.)	Flow (gpm)	Diam (ft.)	Flow (gpm)	Diam (ft.)	Flow (gpm)	Diam (ft.)	Flow (gpm)	Diam (ft.)	Flow (gpm)	Diam (ft.)	Flow (gpm)	Diam (ft.)
50	100	245	130	265	165	285	205	300	255	320	300	335	350	350
60	110	260	143	280	182	300	225	315	275	335	330	350	385	365
70	120	270	155	290	197	310	245	330	295	350	355	365	415	380
80	128	280	165	300	210	320	260	340	315	360	380	380	445	395
90	135	290	175	310	223	330	275	350	335	370	405	390	475	405
100	143	300	185	320	235	340	290	360	355	380	425	400	500	415
110	150	310	195	330	247	350	305	370	370	390	445	410	525	425
120	157	315	204	335	258	360	320	380	385	400	465	420	545	435

200 Series Big Guns—27° Trajectory[a]								200 T Taper Bore Nozzles								
Size:	1.05 in		1.2 in		1.3 in		1.4 in		1.5 in		1.6 in		1.75 in		1.9 in	
Press. (psi)	Flow (gpm)	Diam (ft.)	Flow (gpm)	Diam (ft.)	Flow (gpm)	Diam (ft.)	Flow (gpm)	Diam (ft.)	Flow (gpm)	Diam (ft.)	Flow (gpm)	Diam (ft.)	Flow (gpm)	Diam (ft.)	Flow (gpm)	Diam (ft.)
60	250	345	330	375	385	390	440	410	515	430	585	445	695	470	825	495
70	270	360	355	395	415	410	480	430	555	450	630	465	755	495	890	515
80	290	375	380	410	445	430	515	450	590	470	675	485	805	515	950	535
90	310	390	405	425	475	445	545	465	625	485	715	505	855	535	1005	555
100	325	400	425	440	500	460	575	480	660	500	755	520	900	550	1060	575
110	340	410	445	450	525	470	605	495	695	515	790	535	945	565	1110	590
120	355	420	465	460	545	480	630	505	725	530	825	550	985	580	1160	605
130	370	425	485	465	565	485	655	515	755	540	860	560	1025	590	1210	620

200 Series Big Guns—27° Trajectory[a]								200 R Ring Nozzles						
Size:	1.29 in		1.46 in		1.56 in		1.66 in		1.74 in		1.83 in		1.93 in	
Press. (psi)	Flow (gpm)	Diam (ft.)	Flow (gpm)	Diam (ft.)	Flow (gpm)	Diam (ft.)	Flow (gpm)	Diam (ft.)	Flow (gpm)	Diam (ft.)	Flow (gpm)	Diam (ft.)	Flow (gpm)	Diam (ft.)
50	230	325	300	355	350	370	410	390	470	405	535	420	640	435
60	250	340	330	370	385	390	445	410	515	425	585	440	695	455
70	270	355	355	385	415	405	480	425	555	440	630	455	755	475
80	290	370	380	400	445	420	515	440	590	455	675	470	805	490
90	310	380	405	415	475	435	545	455	625	470	715	485	855	505
100	325	390	425	425	500	445	575	465	660	480	755	500	900	520
110	340	400	445	435	525	455	605	475	685	490	790	510	945	535
120	355	410	465	445	545	465	630	485	725	500	825	520	985	545
130	370	415	485	450	565	470	655	490	755	505	860	525	1025	550

[a] The diameter of throw is approximately 2% less for the 24° trajectory angle and 5% less for the 21° trajectory angle.

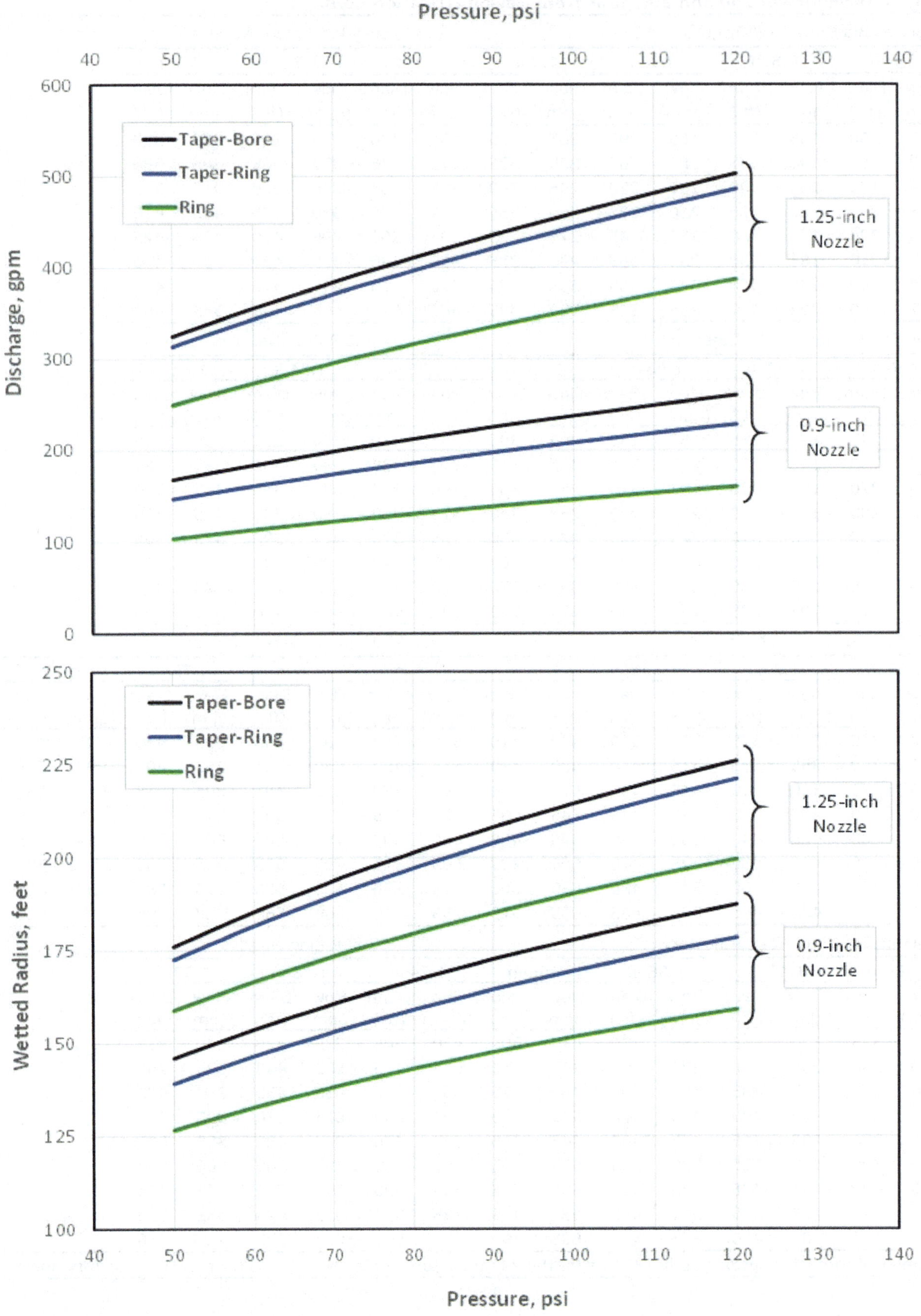

Figure 12.15. Performance of types of nozzles for guns. (Data from Nelson Irrigation Corp.)

12.5 Travelers

A semi-automated sprinkler system was developed in the 1960's to reduce labor and to adjust application depths to match soil and crop requirements. The early versions consisted of rotating booms mounted on a cart that was periodically moved. That design was replaced with a towable cart that could be pulled continuously across the field to provide a moving sprinkler system. Initially the travelers were pulled by winding up cable on a cart. The other end of the cable was anchored at the end of the travel lane. Big guns were developed that operated at high pressures but could throw water hundreds of feet. Utilization of traveler systems has decreased in the United States to about 2% of the sprinkler irrigated land. However, systems are utilized more extensively internationally.

A modern cable-tow, or soft-hose, traveler is shown in Figure 12.16. Water is supplied to the traveler with a flexible hose, called a lay-flat hose. The hose is looped behind and to the side of the traveler when positioning the cart to the edge of the field. This avoids interference with the cart when moving toward the center of the field. This style of traveler is pulled by a winch that rolls cable from the field boundary toward an anchor—often the tractor used to reposition the cart for subsequent irrigations. The water source can also be in the middle of the field. This allows a hose that is half the length of the cable. The effective diameter of the cable reel increases as cable is rewound on the cart. This could cause variable speed of movement as the cart moves along the tow-path. Special controls are used to vary the speed of the cable winch to provide uniform speed of travel. The traveler

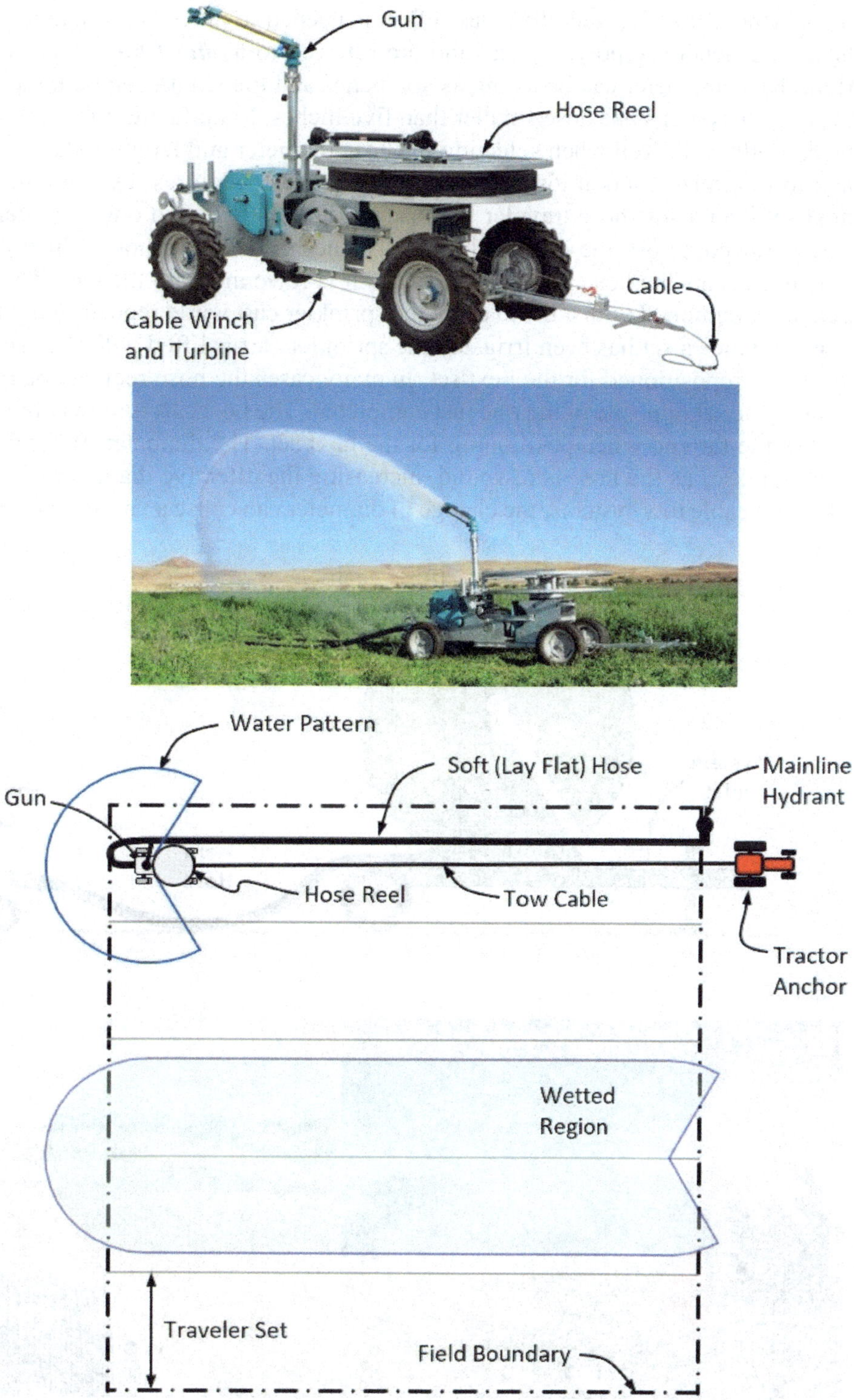

Figure 12.16. Soft-hose traveler irrigation system. (Photos at top are courtesy of Yüzüak Irrigation Sprinklers.)

stops when it reaches the anchor point. The hose is then drained and rewound onto the hose reel. The cart is then located at the edge of the next towpath and the process repeats. Guns apply water beyond the edge of the traveler set as shown in Figure 12.16.

Later designs of travelers eliminated the cable and used a fortified hose to drag the sprinkler cart along the path (Figure 12.17). Hoses were designed to supply water to the cart and with enough tensile strength to pull the sprinkler cart through the field. This eliminates the need for a cable to move the gun through the field. The sprinkler cart is smaller than the cart for the soft-hose traveler; therefore, less effort is needed to move the gun across the field. These hoses are generally polyethylene and are referred to as ***hard hoses*** because they are quite rigid. Hose diameter can be as big as six inches and the length can be up to 2,000 feet; however, most systems use hose smaller than five inches. Manufacturer recommendations should be carefully followed when selecting the hose diameter and length to have adequate strength and to minimize friction loss. The hard-hose traveler requires less time to reposition to the next set than a soft-hose traveler because the hose is rewound onto the reel as the sprinkler cart is towed across the field. Additionally, the hard hose is not drained during rewinding which prevents the hose from collapsing as it is rewound onto the cart. The hose-reel cart is generally equipped with a lift to carry the sprinkler cart while repositioning the system or for storage. Once a set has been irrigated the sprinkler cart is lifted with the primary cart and the system is repositioned for the next set. In many cases the hose reel can be rotated in place to irrigate the set opposite of the one just completed. The host cart is slowly pulled from the hose reel to the far end when positioning for the next set. The diameter of the coiled hose on the reel increases as the hose is rewound, increasing the effective diameter of the hose reel. Just like with cable tow systems, the change of diameter can cause a variation of travel speed along

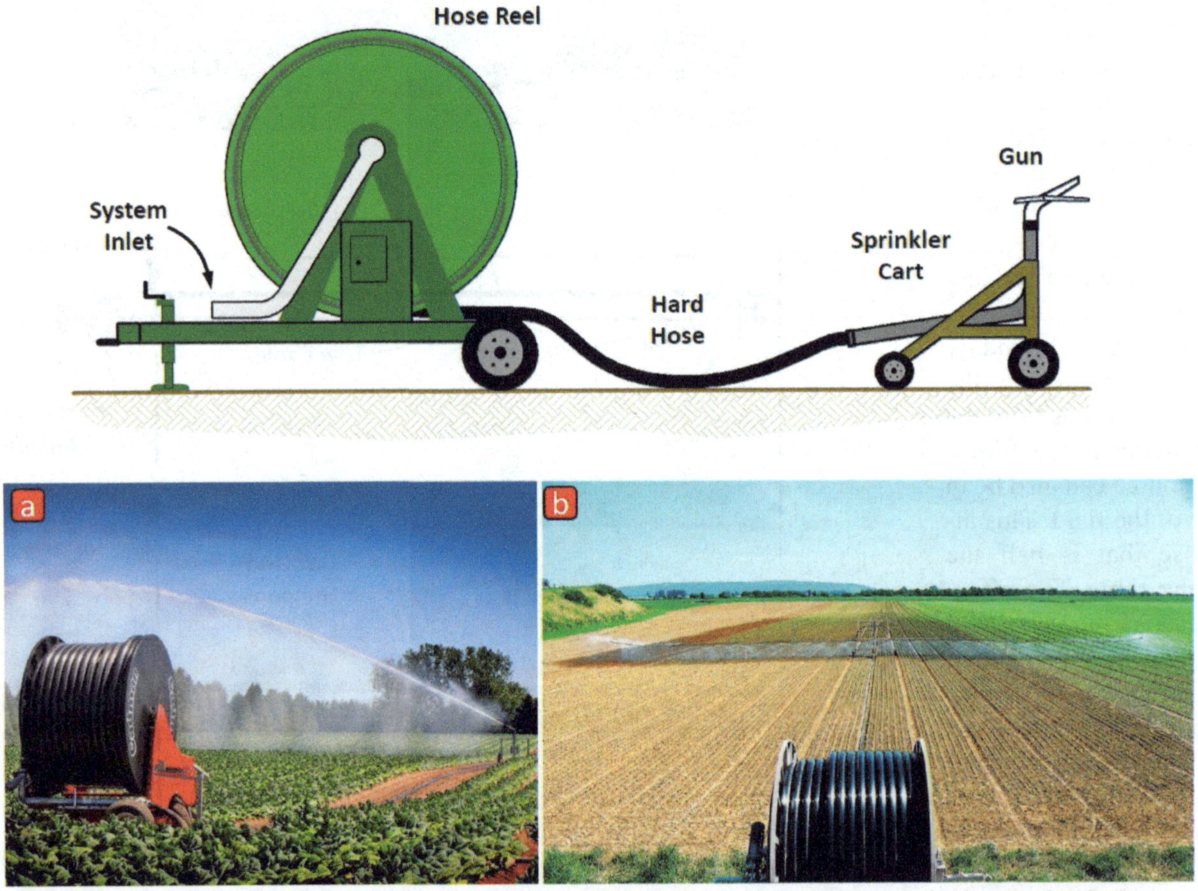

Figure 12.17. Hard-hose traveler diagram with photo of a traveler with a gun (photo a is courtesy of Cadman Power Equipment) and with large boom (photo b is courtesy of Bauer Group).

the towpath; therefore, these travelers must adjust the speed of rotation of the reel to maintain a constant sprinkler cart velocity.

Recently a large boom has become available which replaces the big gun (Figure 12.17b). The advantage of the boom, which does not rotate, is that uniform application of water is more achievable, and less pressure is required to apply water across the set (Peters and McMoran, 2008). Friction loss in the hose remains the same as for a gun with equal flow; however, the operating pressure is less. Wind effects are also diminished with a boom configuration.

All travelers have slope limitations. Slope along the towpath changes the effort required to transport the gun. Slope perpendicular to the towpath may cause the cart to slide downslope. Slope also affects the acceptable application rate to avoid runoff. Manufacture recommendations should be followed regarding slope.

The traveler has little clearance and the hose and/or cable must be pulled across the soil surface. Therefore, the traveler operates along a travel lane. This is often a grass or alfalfa strip for row crops so that the hose can move easily.

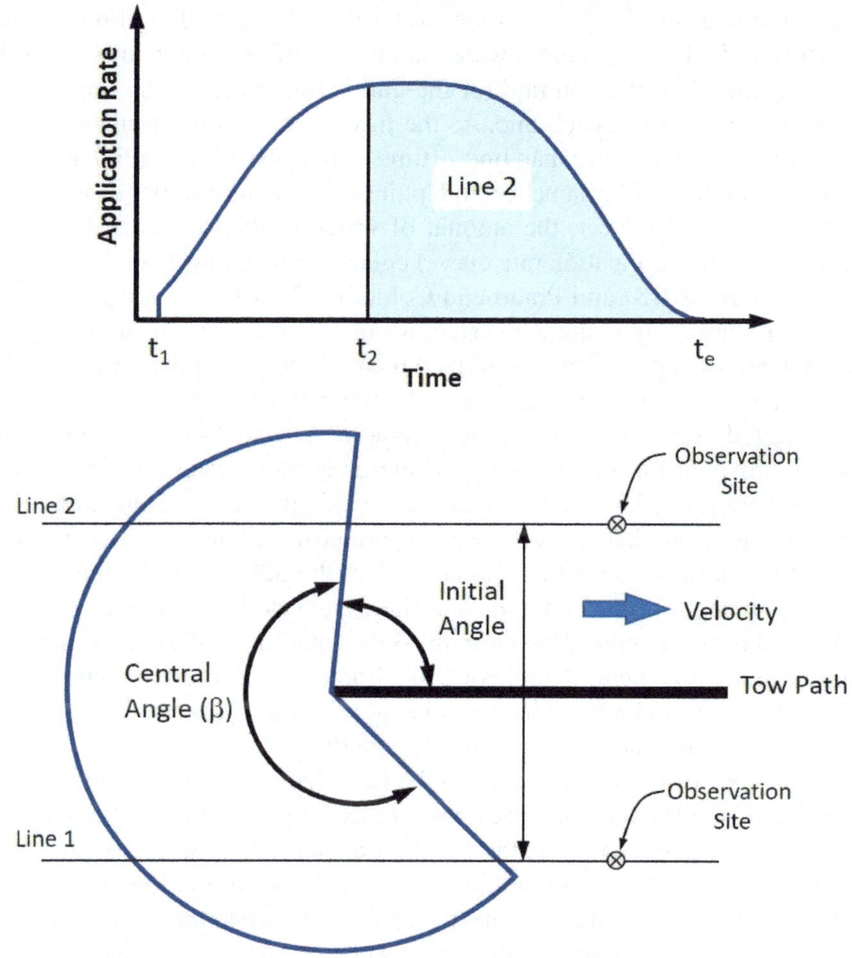

Figure 12.18. Operational characteristics of a big gun traveler.

The big gun does not make a complete circle during operation (Figure 12.18). The gun is designed to operate over an arc and then it automatically reverses to the starting position of the arc. The arc and starting position should be set according to the manufacturer's recommendations to provide uniform irrigation. If the arc is too large, excess water will be applied near the lane where the traveler is towed.

12.5.1 Gun Performance

Efficient irrigation with travelers depends on understanding the characteristics of the moving gun. Table 12.6 lists the discharge and wetted diameter for guns with varying nozzle sizes and pressures. Certainly, these are important characteristics; however, the distribution of water about the gun is also critical. This distribution is affected by settings of the angles of operation for the gun as illustrated in Figure 12.18. The aerial view of the sprinkler pattern shows that two angles are involved. The gun begins rotation at the initial angle and progresses through the central angle. When the gun completes rotation through the central sector the gun reverses rotation. The reversal continues until it reaches the initial angle. The initial angle can be arbitrary relative to the line of travel. The central angle of the sector can be independently adjusted also. The water application process is complicated because the gun is moving at a relatively constant velocity. The plan view of the gun in Figure 12.18 includes two lines equidistant from the towpath. The initial angle was set so that the bottom portion of the circular sector, along line 1, receives more water than the area along line 2. The upper portion of the sector, along line 2, receives water about 60% of the time compared to line 1.

The water application rate for the gun is illustrated in the upper portion of Figure 12.18. Consider a point on each line. As the traveler moves the water pattern reaches the point on line 1 (at time t_1) earlier than line 2 (time t_2) because of the initial angle. After time t_2 the water application rate is the same for both points. The same amount of water is applied at each point after time t_2. However, the amount of water applied between time t_1 and t_2 (the unshaded portion of the application rate curve) enlarges the application at point 1 relative to point 2.

Ge et al. (2018) and Prado and Colombo (2020) analyzed the distribution of water for a pass of a traveling irrigation system using either a small or medium size gun (Figure 12.19). The depth of application was divided by the average depth applied for the ratio on the vertical axis. The distance perpendicular to the towpath was normalized by dividing the distance by the wetted radius of the gun. These authors estimated the distribution of water perpendicular to the towpath for one pass of a traveler irrigation system equipped with a small gun with a central angle of 270° and with a medium gun with a central angle of 270° and 320°. The initial angle was set so that the pattern was symmetrical about the towpath. So, the initial angle was 45° for central angles of 270°, and 20° for the 320° central angle. Results show that the application depth peaks about 45% of the wetted radius away from the gun. The patterns from these guns are similar. However, the 320° rotation applies more water near the gun than the same gun with a central angle or 270°. Some manufacturers recommend the central angle be between 220° and 320°. The central angle should be greater than 180° to maintain gun thrust so that the hose and/or cable rewind properly.

The uniformity of application in the field depends on overlapping the water distribution for two passes of the traveler. The degree of overlap depends on the wetted radius of the gun and the spacing between towpaths for the traveler. An example of the overlap for the medium sized gun with the central angle of 270° is shown in the lower portion of Figure 12.19. The dashed line on the left represents the application when the traveler makes one pass for the gun that has a wetted radius of 150 feet. The mirror image of the application occurs for the second pass as shown by the dash line on the right. In this case the spacing between paths and therefore the distance between guns during each pass is 260 feet. The percent overlap is the ratio of the spacing of the towpath relative to the wetter diameter of the gun. In this case the towpath spacing is 260 feet and the wetted diameter of the gun is 300 feet; therefore, the percent overlap is 87%. The blue dots in the diagram represent the depth of water applied as result of overlapping the distribution from each pass of the traveler. The distribution is reasonably uniform. All water comes from the first pass for the first 110 feet, and all water comes from the second pass from 150 to 260 feet. The patterns overlap from 110 feet to 150 feet, so the depths are added for this region. The average depth of application after overlapping was 0.7 inches and the uniformity coefficient which was 91 which is good.

These results were based upon computer simulation for low wind speeds. The authors simulated windy conditions, but those results are site specific. In lieu of predicting the distribution pattern for each traveler and gun configuration, general recommendations have been made for the maximum spacing between towpaths based on the wetted diameter of the gun and the prevailing wind speed (Table 12.7). The recommended maximum spacing for a gun with a wetted diameter of 300 feet under no wind conditions is 240 feet or 80% of the wetted diameter from Table 12.7. Wind distorts the water distribution pattern for sprinklers and especially for guns which throw water high into the air for hundreds of feet. Thus, as the wind speed increases the amount of overlap must increase to maintain uniformity as illustrated in Table 12.7. For example, if wind speeds are over 10 mph, Table 12.7 recommends that the maximum path spacing would be 50% of the wetted diameter of the gun. Therefore, the maximum spacing in windy conditions would be 150 feet for the gun in Figure 12.19.

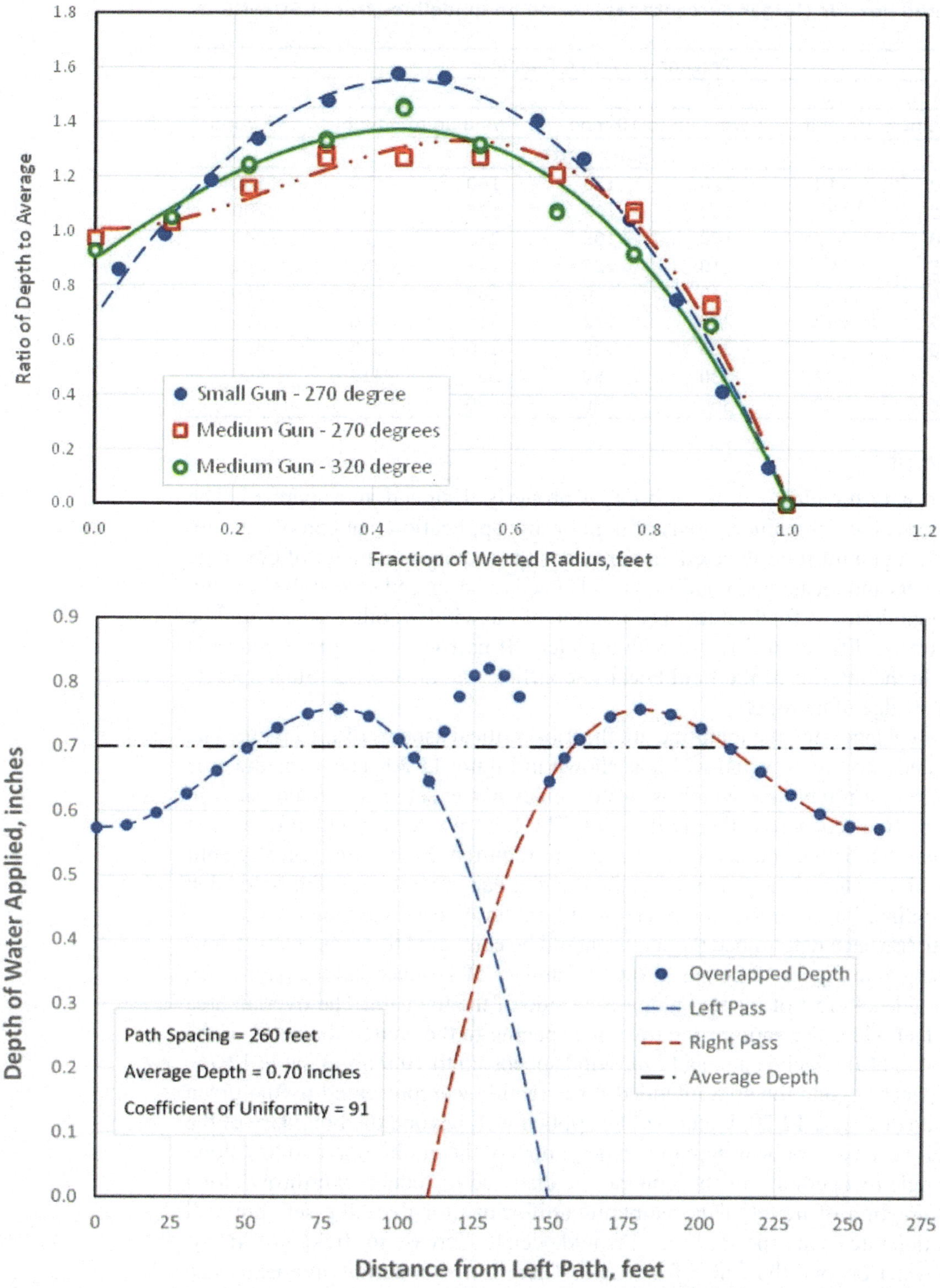

Figure 12.19. Distribution of water from a single pass of a traveler for three types of gun settings, and the overlapped patterns for the medium size gun with a central angle of 270°. Based on data from Ge et al. (2018) and Prado and Colombo (2020).

Table 12.7. Maximum spacing for traveler irrigation systems for ring nozzles (smaller percentages) and tapered nozzles (larger percentages) based on guidelines from USDA-NRCS (2016).

Sprinkler Wetted Diameter (ft)	Percent of Wetted Diameter						
	50	55	60	65	70	75	80
	Wind over 10 mph		Wind up to 10 mph		Wind up to 5 mph		No Wind
	Spacing (ft)						
200	100	110	120	130	140	150	160
250	125	137	150	162	175	187	200
300	150	165	180	195	210	225	240
350	175	192	210	227	245	262	280
400	200	220	240	260	280	300	320
450	225	248	270	292	315	338	360
500	250	275	300	325	350	375	400
550	275	302	330	358	385	412	440
600	300	330	360	390	420	-	-

12.5.2 Field Layout

Water application with travelers can be uniform if properly designed and operated. The traveler constantly moves which reduces areas of high or low application that can occur with stationary sprinklers. A gun must be selected that provides the required diameter of coverage for the layout of the sets and local wind conditions. Sets need to be spaced so that they evenly fit within the field boundaries. A final set that is a fraction of the width of other sets should be avoided since this area is difficult to irrigate with travelers. If narrow sets are required, it is best to locate them in the interior of the field because conflicts can arise when guns operate on a narrow set at the edge of a property.

The location of travel lanes and the mainline are the most critical aspects of the layout. The field should be divided into sets of equal width as shown in Figure 12.20. The set is the area located on either side of a travel lane which is in the center of the set. The 80-acre field in Figure 12.20 requires five sets across the field width. The field was also divided down the middle where the mainline is located, so ten total sets are required. Fields are typically split with the mainline in the center of the field so that the traveler can operate from the field edge back toward the mainline. This minimizes the length of hard hose needed for the traveler.

After the set width has been determined the width should be compared to spacing limitations from Table 12.7. The gun in Figure 12.20 has a wetted diameter of 440 feet and the path width is 264 feet. This provides 88 feet of overlap along each side of the towpath. The overlap area would then be 176 feet wide. The ratio of the tow lane spacing to the wetted diameter of the gun is 60% (i.e., 264 ÷ 440) which is adequate for wind speeds up to 10 mph in Table 12.7.

Travelers apply water beyond the area intended for irrigation as represented by the green areas on the right side of Figure 12.20. Water will be applied well beyond the boundary of the field along all edges. This may not be acceptable to neighbors or the public if a road lies along the property. The angle of operation of the gun can be changed to reduce overthrow along field edges in the direction of travel. This maintains uniformity for the edge sets but will increase the application rate so the speed of travel would need to increase for field uniformity.

Travelers throw water beyond the ends of the field. There is also a deficit area near each end of the towpath because the wetted pattern cannot fully pass over those areas without throwing water long distances beyond the field boundary. Small dry areas also occur if the traveler is stopped exactly at the center of the field. If the hose reel can be moved beyond the centerline the dry areas could be reduced. If water cannot be thrown beyond the ends of the lane, then the traveler should be positioned further into the field which causes larger deficit zones along the ends of the field.

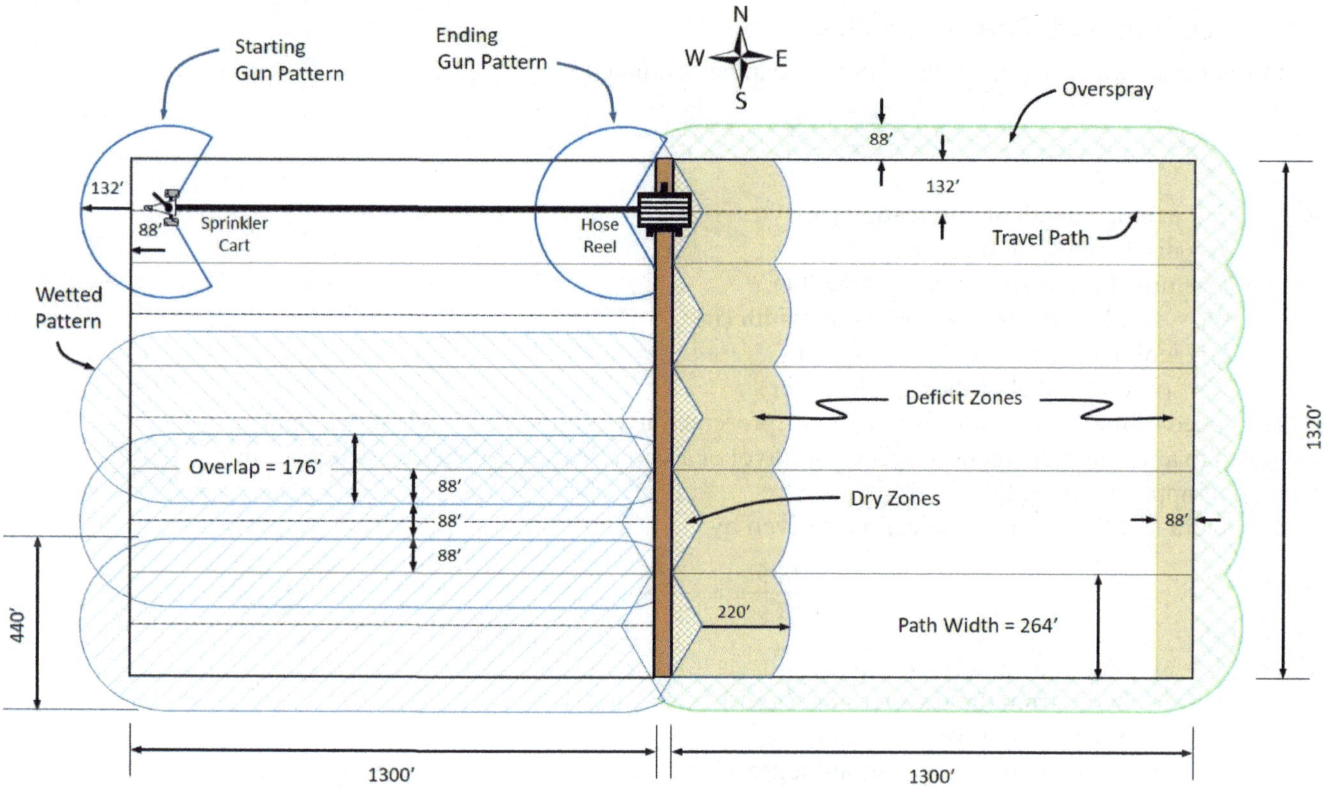

Figure 12.20. Layout for eighty-acre field irrigated with a traveler.

The traveler is flexible as it can irrigate many shapes of fields. The length of the towpath can be adjusted to match fields with variable boundaries. The traveler shuts off automatically when it returns to the hose reel; thus, variable operation times are possible for irregular lengths of lanes.

Traveler systems can only pull a maximum length of hose. The length depends on the model of traveler, the type and size of hose, and the type of movement system (cable or hose reel). Typical characteristics for cable-tow traveler systems for the southeastern area of the United States are listed in Table 12.8. These results assume that the travel lane can be twice the length of lay-flat soft hose. The capacity listed in Table 12.8 is based on approximately five gpm/acre which may be insufficient in more arid regions. Manufacturer specifications should be used for specific conditions and designs.

Table 12.8. Typical characteristics for cable-tow traveler irrigation system from Harrison et al. (2015).

Hose Diameter (in)	Hose Length (ft)	Maximum Travel Distance (ft)	Maximum Capacity (gpm)	Maximum Irrigated Area (acres)	Sprinkler Pressure (psi)	Typical Lane Spacing (ft)	Area Covered per Pass (acres)	Maximum Hose Pull Range (lbs)
2.5	660	1,320	165	33	60–70	180	5.5	1,300–1,900
3	330	660	250	50	70–80	210	3.2	1,000–1,500
3	660	1,320	250	50	70–80	210	6.4	2,000–3,000
3.5	660	1,320	375	75	80–90	240	7.3	3,000–4,000
4	660	1,320	535	107	90–100	300	9.1	3,500–5,000
4	1,320	2640	535	107	90–100	300	18.2	7,000–10,000
4.5	660	1,320	730	145	90–100	300	9.1	4,000–6,000
4.5	990	1,980	730	145	90–100	300	13.6	6,000–9,000
4.5	1,320	2,640	730	145	90–100	300	18.2	8,000–12,000
5	660	1,320	960	192	100–120	330	10	5,000–7,000

12.5.3 Operational Characteristics

The depth of water applied with a traveler can be computed by:

$$d_g = \frac{96.3\ q_s\ T_o}{W_p\ L_l} = \frac{96.3\ q_s}{W_p\ v_t} \qquad (12.11)$$

where: d_g = average depth of application (in),
q_s = discharge from the gun (gpm),
T_o = time of operation for one lane (hr),
W_p = width of the travel lanes = set width (ft),
L_l = length of the travel lane (ft), and
v_t = speed of travel of the traveler (ft/hr).

The speed of travel is $v_t = L_l / T_o$ which is represented in equation 12.11. Most travelers are designed to allow several specific speeds of travel or a variable range of speeds which allows a range of application depths.

The average rate of water application is given by:

$$A_r = \left(\frac{11035}{\beta}\right)\frac{q_s}{W_r^2} \qquad (12.12)$$

where: A_r = average application rate (in/hr),
q_s = discharge from the traveler gun, (gpm),
W_r = wetted radius of gun, (ft), and
β = central angle of gun operation (degrees).

Consider the following example.

Example 12.6

You represent a company that sells traveler irrigation systems. A client recently bought a field and needs to know how much water will be applied per irrigation for an existing traveler system. They also want to know the application rate of the system.

Given: The traveler is depicted in Figure 12.20. The system uses a gun that discharges 585 gpm on a set that is 264 ft wide and 1,300 ft long. The traveler starts 88 ft from the field edge as a setback to balance uniformity against overspray at the end of the path.

The travel speed is adjusted to irrigate the set in 11 hrs of operation and to be moved every 12 hrs.The central angle of rotation for the lateral is 270 degrees and the wetted radius of the gun is 220 ft.

Solution:

1. The velocity of travel is:

$$v_t = \frac{L_l}{T_o} = \frac{1300 - 88\ \text{ft}}{11\ \text{hr}} = 110\ \text{ft/hr}$$

The depth of application is computed using Eq. 12.11:

$$d_g = \frac{96.3\ q_s}{W_p\ v_t} = \frac{96.3 \times 585}{264 \times 110} = 1.94\ \text{in}$$

2. The application rate is given by Eq. 12.12:

$$A_r = \left(\frac{11035}{\beta}\right)\frac{q_s}{W_r^2} = \left(\frac{11035}{270}\right) \times \left(\frac{585}{220^2}\right) = 0.49\ \text{in/hr}$$

The flow rate needed for the traveler (q_s) is determined by revising Equation 11.14 to:

$$q_s = \left(\frac{C_n\ W_p\ L_l}{E_a\ 43560}\right)\left(\frac{N_s}{N_t}\right)\left(\frac{T_s}{To}\right)\left(\frac{I_i}{I_i - T_d}\right) \qquad (12.13)$$

where: C_n = net system capacity requirement (gpm/ac),
E_a = application efficiency (decimal fraction),
W_p = width of the travel path (ft),
L_l = Length of the travel length (ft),
Ns = number of sets in the field,
N_t = number of travelers used,
T_s = set time between moving traveler to next travel lane (hr),
T_o = time of operation, i.e., time water is applied for the travel lane (hr),
I_i = irrigation interval (time between irrigations of the field) (days), and
T_d = down time between irrigations (days).

Equation 12.13 is applied to the system in Figure 12.20 in the following example.

Example 12.7

A field and traveler similar to the system shown in Figure 12.20 has a net system capacity is 4.5 gpm/acre. The irrigation interval is 5.5 days with half a day to reposition.

Find: The area irrigated in one set (i.e., one traveler path) and the flow rate needed for a traveler to service the field.

Solution:

1. The area per set $= \dfrac{W_p \times L_l}{43560} = \dfrac{264 \times 1300}{43560} = 7.88\ \text{acres / set}$
2. The system flow is determined from Eq. 12.13:

$$q_s = \left(\frac{4.5 \times 264 \times 1300}{0.75 \times 43560}\right)\left(\frac{10}{1}\right)\left(\frac{12}{11}\right)\left(\frac{5.5}{5.5-0.5}\right) = 567\ \text{gpm}$$

The pressure loss in the hoses used to supply travelers can be quite high due to the use of hoses that are relatively small for the required flow rates. Small hoses are used because large diameter hoses are harder to pull and much more expensive. The friction loss for a range of diameters of hose and flow rates is given in Table 12.9 and 12.10. Friction loss for the lay-flat used with a cable-tow traveler is shown in Table 12.9. The diameter of lay-flat hose varies depending on the pressure. Values in Table 12.9 are for a tube pressure of about 100 psi. Comparison of losses for hard hoses is slightly higher than for lay-flat

Table 12.9. Pressure loss (psi/100 ft) for lay-flat hose when operated at 100 psi (USDA-NRCS, 2016).

Flow	Nominal Inside Diameter (in)				
(gpm)	2.5	3	4	4.5	5
100	1.6	-	-	-	-
150	3.4	1.4	-	-	-
200	5.6	2.4	-	-	-
250	-	3.6	0.9	-	-
300	-	5.1	1.3	0.6	-
400	-	-	2.3	1.3	-
500	-	-	3.5	2.1	1.1
600	-	-	4.9	2.7	1.6
700	-	-	-	3.6	2.1
800	-	-	-	4.6	2.7
900	-	-	-	-	3.4
1000	-	-	-	-	4.2

Table 12.10. Friction loss in hard hose, psi/100 ft (Hazen-Williams resistance coefficient = 150).

Flow	Hose Inside Diameter (in)							
(gpm)	2.5	2.7	3.0	3.3	3.7	4.0	4.5	5.0
75	1.45	1.00	0.60	-	-	-	-	-
100	2.48	1.70	1.02	0.64	-	-	-	-
125	3.74	2.57	1.54	0.97	0.56	-	-	-
150	5.24	3.61	2.16	1.36	0.78	0.53	-	-
175	6.98	4.80	2.87	1.81	1.04	0.71	0.40	-
200	8.94	6.14	3.68	2.31	1.33	0.91	0.51	0.31
250	-	9.29	5.56	3.50	2.00	1.37	0.77	0.46
300	-	-	7.80	4.90	2.81	1.92	1.08	0.65
350	-	-	10.37	6.52	3.74	2.56	1.44	0.86
400	-	-	-	8.35	4.79	3.28	1.85	1.11
450	-	-	-	-	5.95	4.07	2.30	1.38
500	-	-	-	-	-	4.95	2.79	1.67
550	-	-	-	-	-	5.91	3.33	1.99
600	-	-	-	-	-	-	3.91	2.34
650	-	-	-	-	-	-	4.54	2.72
700	-	-	-	-	-	-	5.21	3.12
750	-	-	-	-	-	-	-	3.54
800	-	-	-	-	-	-	-	3.99

hoses. However, hard hose systems are by far the most common system today.

Pressure losses in the sprinkler cart must be computed. The pressure loss in the traveler depends on the flow rate, speed of travel, type of power unit, and machine design. Performance for a specific machine must be obtained from the manufacturer. Some travelers use water pressure through a turbine to power the hose reel to rewind the hard hose. Other systems use an engine to power the reel. About ten psi is required to power the turbine.

The pressure and discharge relationships for a typical traveler powered with a turbine is shown in Figure 12.21. The upper portion of the figure shows the pressure discharge relationships for the 1.83-inch nozzle used with the gun and the input pressure required for a given discharge from the traveler—developed from manufacturer's data. The difference in the pressure between the nozzle and inlet to the hose reel for the same discharge represents the friction loss in the hose reel, turbine, hard hose, and the sprinkler cart. The pressure loss is quite substantial for travelers. The pressure requirement of the traveler is significant, so operating costs for travelers are high. The lower portion of Figure 12.21 shows the friction loss in the 4.5-inch hard hose and the cart and the reel system with a turbine powered traveler. Most of the loss occurs in the hose, especially at high flow rates.

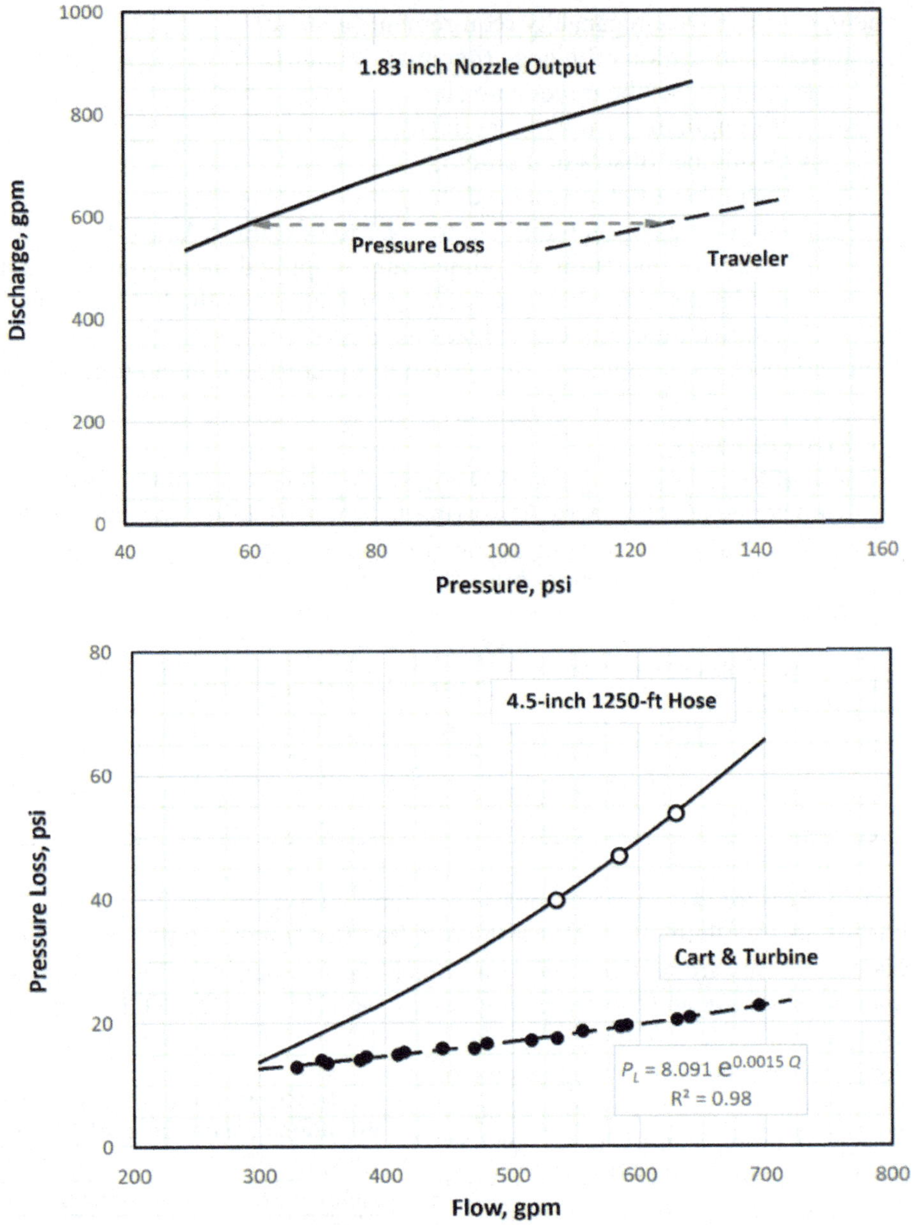

Figure 12.21. Pressure versus discharge and friction loss relationships for a traveler with a 1.83-inch nozzle.

12.5.4 Management

As with other systems, management should start with an assessment of the properties of the existing system and then evaluation of the characteristics of the system compared to crop water needs and guidelines for efficient irrigation with traveler systems. The Traveler Management Spreadsheet is shown on Figure 12.22. This analysis is based on the traveler depicted in Figure 12.20. The 80-acre field was divided into towpaths (sets) that are 1300 feet long and 264 feet wide. This allows 40 ft in the middle of the field to rotate the traveler to irrigate the alternate side of the field and gives an irrigated area of 78.8 acres. The traveler will be posi-

Traveler Management Spreadsheet

Field Properties:

Field:	Example			Type of System	Hard Hose
Soil Texture	Fine Sandy Loam			Water Holding Capacity, in/ft	1.8
Field Length, feet	2,640	Field Width, feet	1,320	Field Size, acres	80
Length of Irrigated Area, feet	2,600	Starting Setback, feet	88	Irrigated Area, acres	78.8
Water Supply:		Available Inflow, gpm	700	System Capacity, gpm/acre	8.88

Design Variables:

Mainline:

Diameter, inches	8	Length, feet	1,320	Elevation Change, feet	0

Traveler Information:

Hose Length, feet	1,250	Number of Travelers	1	Average Wind Speed, mph	7
Hose Size-Diameter, inches	4.5	Height of Nozzle, feet	9	Elevation Change Along Path, feet	8
Length of Set (Tow Lane), feet	1,300	Set (Tow Path) Width, feet	264	Turbine (Y/N)	Y

Sprinkler Information:

Traveler Inlet Pressure, psi	126	Design Nozzle Pressure, psi	60	Nozzle Size, inches	1.83
Discharge Coeff (Cd) =	16.00	Pressure Exponent (a) =	0.50	Nozzle Size Exponent (b) =	2.566
Design Discharge, gpm	584	Wetted Radius of Gun, feet	220	Central Gun Angle, degrees	270

Operation Results:

Mainline:

Mainline Flow, gpm	584	Inside Diameter, inches	7.872	C Value	120
Friction Loss Factor, psi/100 ft	0.373	Friction Loss in Mainline, psi	4.7	Total Pressure Loss, psi	4.7
Flow Velocity, feet/sec	3.9				

Traveler Information:

Inside Diameter, inches	4.50	C Value	150	Hose Friction Loss, psi/100 ft	3.75
Hose Friction Loss, psi	46.9	Nozzle Height, converted to psi	3.9	Elevation Change, converted to psi	3.5
Cart and Reel Pressure Loss, psi	19.4	Total Pressure Loss, psi	73.7	Reqired Inlet Pressure, psi	134
Maximum Lane Width, %	65%	Wetted Diameter of Gun, feet	440	Maximum Set Width, feet	286
Acres per Set	7.88	Number of Sets in Field	10	Number of Sets per Traveler	10

Management Variables/Results:

Soils/Plant Info:

Crop	Sugar Beets	Root Depth, feet	4	Critical Depletion Percentage, %	50%
Allowable Depletion, inches	3.6	Application Effic. Low Qtr., %	75%		

Times:

Set Time, hours	12	Application Time, hours	11.0	Drain/Move Time, hours	1.0
Reposition Time, hours	12	Irrigation Interval, days	5.5	Down Time, %	17%
Velocity of Travel, feet/hour	110	Application Rate, inches/hour	0.49		
Gross Depth Applied, inch	1.9	Net Depth Applied, inches	1.5	Net Crop Water Use, in/day	0.26

Comments/Issues:

Figure 12.22. Traveler Management Spreadsheet for traveler irrigation systems.

tioned 88 feet from the field boundary when starting a set. The layout provides substantial overspray which assumes that transboundary conflicts are immaterial. This layout provides five sets on each side of the mainline that fits the field boundary.

The characteristics of the traveler are based on an actual model available from a manufacturer. This system uses a 4.5-inch inside diameter hard hose that is 1250 feet long. This provides the ability to irrigate a length of up to 1,338 feet (1250 + 88). The traveler used a water turbine to recoil the hose and the gun was set to a central angle of 270 degrees. The nozzle is about 9 ft above the ground. The elevation at the west end of the lane is 8 ft higher than the mainline. The manufacturer shows that the pressure at the inlet to the hose reel should be 126 psi to produce 60 psi to the gun nozzle. The wetted radius of this gun and nozzle configuration is 220 feet (440-ft wetted diameter). The application provides 88 feet of overlap on each side of the set when the set width is 264 feet, and the wetted diameter is 440 feet. The gun discharge computed from equation 12.14 for a nozzle pressure of 60 psi is 584 gpm for this nozzle and gun.

$$q_s = C_d \; P^a \; D_n^{\;b} \tag{12.14}$$

where: q_s = gun discharge, gpm

C_d = discharge coefficient,

a = pressure exponent when pressure is in psi, and

b = the nozzle diameter exponent when the size is in inches.

As listed in Figure 12.22 the discharge coefficient was 16.0, and exponents a and b are 0.50 and 2.566 respectively for this gun and nozzle.

The average wind speed at this location was listed as 7 mph. The maximum set width for this wind speed is given as 65% of the wetted diameter of the gun in Table 12.7. Since the wetted diameter is 440 feet the maximum width is 286 feet. The actual set width of 264 feet is 60% of the wetted diameter which is less than the maximum. Most of the rest of the inputs and operational results have been discussed.

The soil and plant information for the management variables are the same as for the moved lateral systems. The time inputs are as previously discussed. This combination results in a down time of approximately 17%. The irrigation interval is 5.5 days since there are ten sets and two sets are irrigated a day, plus one-half of a day is needed to reposition the traveler. The velocity of travel, application rate and gross depth were computed in previous examples. The net irrigation depth is the product of the gross irrigation depth and the application efficiency giving a net depth of 1.45 inches. The 1.45-inch net depth would support a net crop water use rate of 0.26 in/d over the 5.5-day irrigation interval. This capacity should be compared to regional needs. The data in Figure 12.22 summarizes the capability of the traveler and the outcome of management choices for this field. It also illustrates critical issues for travelers.

Computer simulation programs have been developed to predict the performance of traveler systems. Programs such as that by Rolim and Teixeira (2016) or Smith et al. (2008) can be used to design and evaluate traveler systems and as decision support systems to enhance management. Those resources should be examined for advanced management.

12.5.5 Other Issues

Areas at the ends of the towpaths receive less water than in the middle of the field. These deficits occur because the entire water pattern cannot traverse these areas due to boundary limitation as illustrated in Figure 12.20. Some operators adapt to this issue by leaving the traveler stationary for a period before starting movement of the traveler when irrigating. This will reduce the deficit but results in deep percolation in areas that are watered while the traveler is stationary and receive a full pass of the water pattern.

Uniformity issues due to set widths that exceed recommendations for windy conditions may require system modifications. Towpath spacing does not need to be permanent. If travel lanes are too far apart, they can be changed after harvesting the current crop. Modifications may be needed if mainline risers are at the wrong location, but this is not particularly troublesome. Excessive wind drift may result because towpath spacings are too far apart. Wind drift problems can be partially alleviated by altering the time of day that irrigation is started on the field. Usually, winds are highest during the day. When 12-hour set times are used, the starting time for irrigating the field can be altered by half a day each irrigation so that a set irrigated during the day one irrigation and is irrigated at night the next irrigation.

The speed of travel along the towpath may vary for several reasons. The effective diameter of the reel used to rewind cable or hose increases as more material is pulled in. If the rotation speed of the reel is the same, then the traveler will speed up as more cable or hose is rewound. This use to be a major problem with earlier designs but has been mostly assuaged in modern systems. The amount of drag for soft hoses increases as the traveler moves toward the anchor point. Resistance increases with length so the travel speed may decrease as more power is needed as the length of towed hose increases. The inverse occurs with hard hose systems as the maximum length of towed hose is largest at the start of the set. Rolling terrain also contributes to uneven drag of on the hose. Increased drag exerts more stress on the reel system and can slow rotation due the increased load.

Mainline operation and protection can be troublesome for travelers. Travelers require high operating pressures. Since there is only one sprinkler per lateral pressures increase rapidly when the system is started. This can lead to high pressure surges. The mainline needs to be protected from pressure surges as described for solid-set systems. Valves can be used with electrically powered pumps to prevent the surge. Internal combustion engine powered pumps can be started at a slow speed to minimize the pressure surge.

Safety is more of a concern with travelers than moved-lateral systems. The high pressure required of these systems poses some threat if proper operations are not followed. The large diameter hoses are difficult to move and may require assistance to prevent injury. There are many moving parts on the traveler where operators could become entangled. Safety shields and proper operation and maintenance are required to maintain a safe machine. A unique feature of travelers is that water is propelled great distances from the machine. Research has shown that if the water jet impinges on electrical power lines, some current can be transmitted back to the traveler. This, of course, poses severe safety concerns and should be unconditionally avoided.

12.6 Summary

This chapter describes the characteristics and operational requirements for moved-lateral irrigation systems which includes hand-move, tow-line, and side-roll systems. Solid-set, stationary gun-based systems, and travelers are also discussed. These systems collectively represent about 15% of the irrigated area in the United States in 2018. So, while the extent is not large, they still represented significant areas of irrigated land in the United States and a much more significant footprint around the world. Management of these systems requires a thorough understanding of their attributes, familiarity with the operational characteristics including the friction loss, discharge, overlap, and soil-plant interactions.

The focus of this text is on management of systems therefore little emphasis is placed on designing systems. However, the characteristics of the system related to the design must be understood for proper management. The initial step in developing a management plan is to describe the field layout of the system. Then the current conditions of the irrigation system must be inventoried and analyzed. The performance of the system should then be computed to ensure the system will meet acceptable management and industry guidelines. Finally, a management plan should be developed to operate the system to meet current crop soil and environmental demands. There is also discussion about monitoring of existing systems to ensure that they are performing as expected from the analysis.

Questions

1. Describe the advantages and disadvantages of hand-move, towline, and side-roll irrigation systems. Discuss any issues that would limit the adequacy at these types of systems.
2. Discuss two of the general management problems associated with moved lateral irrigation systems (e.g., side-roll, hand-move, towline).
3. Determine the required set time for a side-roll irrigation system with the following characteristics:
   Spacing along the lateral is 40 ft.
   Spacing along the mainline 60 ft.
   Sprinkler discharge is 8 gallons per minute.
   Application efficiency is 75%.
   Soil water depletion at irrigation is 2 $^1/_2$ in.
4. A tow-line irrigation lateral has a sprinkler spacing of 40 ft. The spacing between adjacent positions for the lateral is 70 ft. The diameter of coverage (d_c) for the sprinklers is 103 ft, and the average wind speed is 7 miles per hour. Is the sprinkler spacing and distance between laterals acceptable for good water distribution?
5. Determine the maximum acceptable irrigation interval for a silty clay loam soil where the root depth is 4 feet deep, f_{dc} is 45% and the anticipated net crop water used rate is 0.3 in/day.
6. Determine how the management plan would change in Figure 12.7 for a crop that had a root depth of 2.5 ft and the soil was a sandy loam soil.
7. How would you change the orientation of the mainline and laterals in Figures 12.6 and 12.7 if three laterals were required for the field? What sizes of mainline would you recommend and what total length of pipe would be needed to minimize investment cost?
8. A lateral in an existing solid-set irrigation system consists of 2½-inch PVC pipe that runs up a hillside with a 2% slope. Sprinklers are 30 feet apart on the 600-ft long lateral. Impact sprinklers with a single $^3/_{16}$-inch nozzle are used. The design called for an average nozzle pressure of 55 psi. Your client has complained about dry areas at the distal end of the lateral.
   a. Do you expect uniformity to be an issue with this system?
   b. How could pressure regulators be used to improve uniformity? Where would you put them and how many would you recommend?
   c. Suppose you decided to change nozzles in the lateral to achieve acceptable uniformity without regulators. What size of nozzle would you recommend at the distal end of the lateral?
9. The gun on a traveler irrigation system discharges 400 gpm and the towpath spacing is 240 feet.
   a. What travel velocity is required to apply a gross depth of 2 inches?
   b. What is the application rate if the wetted radius of the gun is 200 ft?
10. How many gallons of diesel fuel are required to apply an acre-inch of water with the traveler system shown in Figure 12.22?
11. The design net system capacity (C_n) for a moved-lateral irrigation system on an alfalfa field is 0.21 in/d. The system is moved every 12 hr (set time = T_s = 12 hr), with downtime to move the system being 1 hr. Downtime allowed for harvesting alfalfa is 10%. The E_{LQ} is 75% (allows for 10% drift & evaporation; assumes no runoff). Determine the gross

system capacity (C_g) in gpm. If the area of the field is 33 ac, what is the minimum flow rate (Q_{min}) needed for the system?

12. Your client purchased a field identical to the one in Figure 12.6, except the soil is a silt loam and the field has a hard-hose traveler. The traveler has the characteristics listed below. You will need to determine mainline orientation and size, length and width of tow-path, and times for management. Justify any additional assumptions you require.
    a. Develop a traveler management spreadsheet and plan for the field.
    b. Discuss any issues you foresee for this field and propose solutions as needed.

    Traveler and field characteristics are as follows:

    The soil is predominately silt loam.
    A hard hose traveler uses a 1.46-inch diameter nozzle operated at 60 psi.
    Characteristics of the gun are available from Table 12.6.
    Parameters for gun discharge equation are $C_d = 15.97$, $a = 0.50$, and $b = 2.586$.
    The inlet pressure for the hose reel should be 104 psi for 60 psi at the nozzle.
    A 4-inch inside diameter hard hose (Hazen-Williams's coefficient of 150 is appropriate).
    Hard hose is 1320 feet long.
    Cart and turbine losses as a function of flow can de determine from Figure 12.21.
    The irrigator plans to irrigate field beans with a root depth of 3.5 feet and and critical depletion fraction of 45%.

References

Davis, S. L., & D. Dukes, M. D. (2016). Importance of ET controller program settings on water conservation potential. *Appl. Eng. Agric., 32*(2), 251-262.

Ge, M., Wu, P., Zhu, D., & Zhang, L. (2018). Analysis of kinetic energy distribution of big gun sprinkler applied to continuous moving hose-drawn traveler. *Agric. Water Manag., 201*, 118-132.

Haghverdi, A., Singh, A., Sapkota, A., Reiter, M., & Ghodsi, S. (2021). Developing irrigation water conservation strategies for hybrid bermudagrass using an evapotranspiration-based smart irrigation controller in inland southern California. *Agric. Water Manag., 245*, 106586.

Harrison, K., Porter, W. M., & Perry, C. (2015). Factors to consider in selecting a farm irrigation system. University of Georgia Ext. Bull. 882. 1-12.

Keller, J., & Bliesner, R. D. (1990). *Sprinkle and trickle irrigation.* Van Nostrand Reinhold Publ.

Merriam, J. L., & Keller, J. (1978). Farm irrigation systems evaluation. Logan: Utah State University.

Morgan, R. M. (1993). Water and the land: A history of American irrigation. Fairfax, VA: Irrigation Association.

Peters, R. T., & McMoran, D. (2008). Boom-type carts vs. big-guns in northwestern Washington. *Proc. Int. Irrigation Show*, (pp. 27-38).

Prado, G. D., & Colombo, A. (2020). Asymmetric wetted sector angle in water distribution of traveling gun irrigation systems. *Engenharia Agricola, 40*(4), 443-452.

Rolim, J., & Teixeira, J. L. (2016). The design and evaluation of travelling gun irrigation systems: Enrolador software. *Engenharia Agricola, 36*(5), 917-927.

Smith, R. J., Gillies, M. H., Newell, G., & Foley, J. P. (2008). A decision support model for travelling gun irrigation machines. *Biosyst. Eng., 100*(1), 126-136.

Stetson, L. E., & Mecham, B. Q. (2011). *Irrigation* (6th ed.). Falls Church, VA: Irrigation Association.

Turner, J. H., & Anderson, C. L. (1980). *Planning for an irrigation system* (2nd ed.). Athens, GA: American Association for Vocational Instructional Materials.

USDA-NRCS. (2016). Sprinkler irrigation. Chapter 11 in *Part 623 National engineering handbook.* Washington, DC: USDA-NRCS.

USDA (2019). 2018 Irrigation and water management survey. Vol. 3. Special Studies, Part 1. AC-17-SS-1. Washington, DC: USDA.

Chapter 13 Center Pivots and Lateral Moves

13.1 Introduction

In 1948 Frank Zybach invented a device he called the "self-propelled irrigator." This led to the development of ***center pivot*** and ***linear or lateral move*** irrigation systems (Bittinger and Longenbaugh 1962; Heermann and Hein, 1968). A chapter is devoted to these systems because of the unique management needed to capitalize on the capability of these systems. Additionally, the growth of center pivot irrigation during the last three decades far exceeds the growth of any other method of irrigation. In some areas the amount of land irrigated with surface systems has receded. Many of the fields, previously surface irrigated, have been converted to center pivot irrigation. The growth in center pivot irrigation in one year almost equals the total amount of land irrigated with microirrigation in the U.S. The growth has been very substantial and will likely continue. Three principal reasons drive this growth. First, the systems have the ability to be very efficient. They can apply small depths of water at the time that the crop needs irrigation. Second, the systems require less labor than surface or moved lateral systems. In many areas the scarcity of available labor is a limitation to the amount of land that a farmer can irrigate. Third, the systems have the capability to irrigate crops, soils, and terrains that are infeasible with surface, or periodically-moved sprinkler systems.

The basic components of a center pivot (Martin et al., 2007) are illustrated in Figure 13.1. The pivot lateral is a pipeline with sprinkler outlets. The pivot lateral is supported by a tower assembly. The towers include a structure to support the pipeline plus a motor to propel each tower. Today most pivots are powered by electricity. However, some manufacturers use oil hydraulic motors. The pivot base or pivot point is located at the center of the field. The base can be permanently installed or, for smaller systems that are towed from field to field, the base is mobile. Water is supplied to the inlet pipe on the pivot base. The water is pressurized by a pump upstream of the pivot. Water is carried up the pivot base through a rotating elbow to the inlet of the pivot lateral. Power is supplied continuously to motors installed at each tower using a slip-ring assembly. This device contains contacts that allow the pivot to pick up power from the pivot base while the lateral rotates. A control panel is usually located on the pivot base where the operator can adjust the speed of rotation of the pivot and check on other factors. A road is usually necessary so the operator can conveniently reach the pivot base.

The combination of the pivot lateral, the truss support structure, sprinkler devices, and the tower are called a span. The length of a span can vary from 100 to 200 feet. Installation costs are less with longer spans; however, the maximum length of a span is determined by the diameter of the pipe and the slope and undulation of the terrain. The length of a span can vary along the pivot to adjust to the dimensions of the field. Spans of the pivot can be connected together in either a rigid or flexible fashion. For rolling terrain it is necessary to provide a flexible coupler between spans.

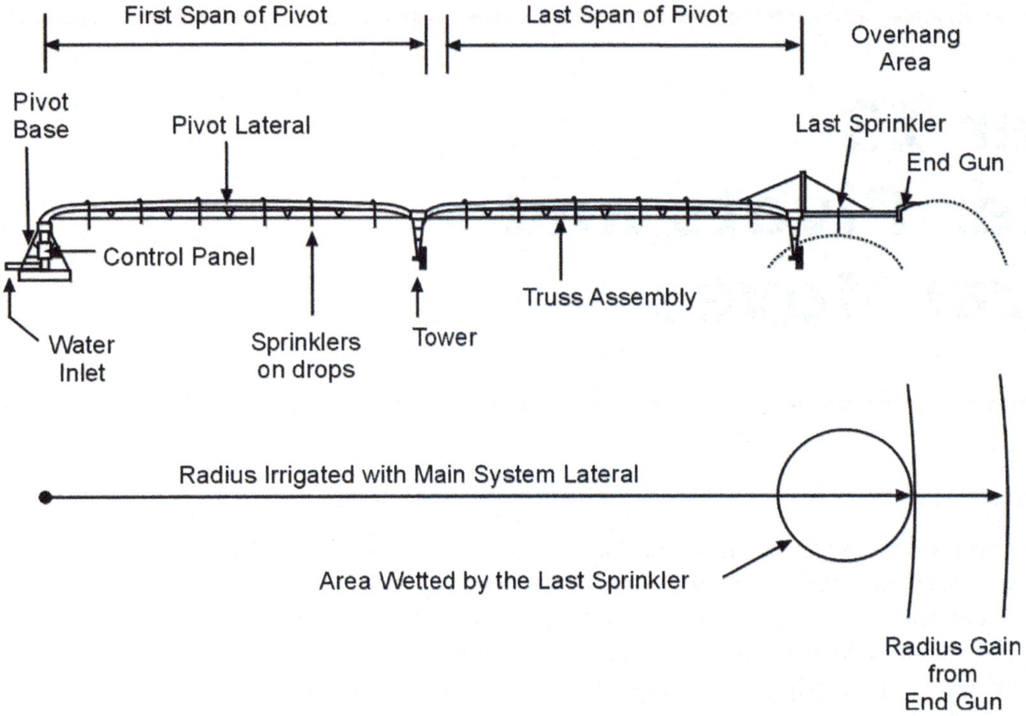

Figure 13.1. Illustration of the components of a center pivot irrigation system.

A pipe called the overhang is often attached beyond the last tower of the pivot (Figure 13.2a). The overhang could be up to 80 feet long. A special sprinkler can be attached to the end of the overhang to increase the amount of area irrigated. This sprinkler is usually called an end gun and is used to water part of the corners not reached by the last sprinkler on the lateral. It only operates when the water from the end gun stays within the field. Since about 1975, there has been a major effort in the center pivot industry to reduce the amount of pressure required to operate center pivots. Systems originally required 75 psi of pressure at the inlet to the pivot. Now, many are designed to operate at 30 psi or less. These pressures are often too low for proper operation of end guns, so a booster pump is installed at the last tower to pressurize the end gun. A valve is used to control when the end gun operates.

(a)

(b)

Figure 13.2. (a) Example of a center pivot irrigation system (Valmont Industries) with seven spans, an overhang, and an end gun. (b) Example of a center pivot with a corner system (photo b courtesy of Lindsay Corporation).

Center pivots can also be equipped with corner watering systems (Figure 13.2b). These systems have a corner lateral that hinges or rotates from the last tower of the main system. The corner system can be guided by GPS or a buried cable that emits a radio signal for the corner tower to follow. Sprinklers on the corner system have special valves that open and close depending on how far the corner lateral has rotated away from the main pivot lateral. Recent developments in center pivot irrigation include remote monitoring and control, high-speed variable frequency drive motors for the towers, low-energy precision application (LEPA) sprinkler packages, variable rate irrigation, and variable frequency drive for the pumping plant (Lamm et al., 2019).

To irrigate rectangular fields, or to irrigate a larger portion of square fields, mechanically moved systems were developed where the lateral moves along a straight line (Martin et al., 2007). These systems are called ***linear*** or ***lateral move*** systems (Figure 13.3). The spans of these systems are nearly identical to those of center pivots. The unique feature of these systems is how the water is supplied to the lateral. Two types of water delivery systems are common:

- systems that drag a supply hose and
- systems that pump water from a canal that runs parallel to the direction of travel.

Systems supplied by either a hose or buried valves are usually pressurized with the main supply pump. Systems that obtain water from a canal carry a pump along with the system to obtain and pressurized the water. The main supply pump, or surface water supply system, must be hydraulically interfaced with the system so that the water supply is continual but does not exceed the canal capacity or the discharge of the system. The water supply features of these systems affect management.

(a)

(b)

Figure 13.3. Hose-fed linear or lateral move irrigation systems. (a) A four-span system driven by electric motors (photo courtesy of Lindsay Corporation). (b) A drive unit for a linear move irrigation system driven by oil hydraulic motors (photo courtesy of T-L Irrigation Company).

13.2 Center Pivot Characteristics

13.2.1 Sprinkler Discharge

Since center pivot laterals rotate around the field, the delivery of water along the lateral is much different than for other lateral-based systems. The area that is irrigated by an individual sprinkler increases with distance from the pivot base (Figure 13.4). The goal in irrigation is to apply the same depth of water to all parts of the field; therefore, the discharge from a sprinkler must be larger near the distal end of the lateral than close to the pivot point. The required discharge is given by:

$$q_s = \frac{2\pi RSC_g}{43560} \qquad (13.1)$$

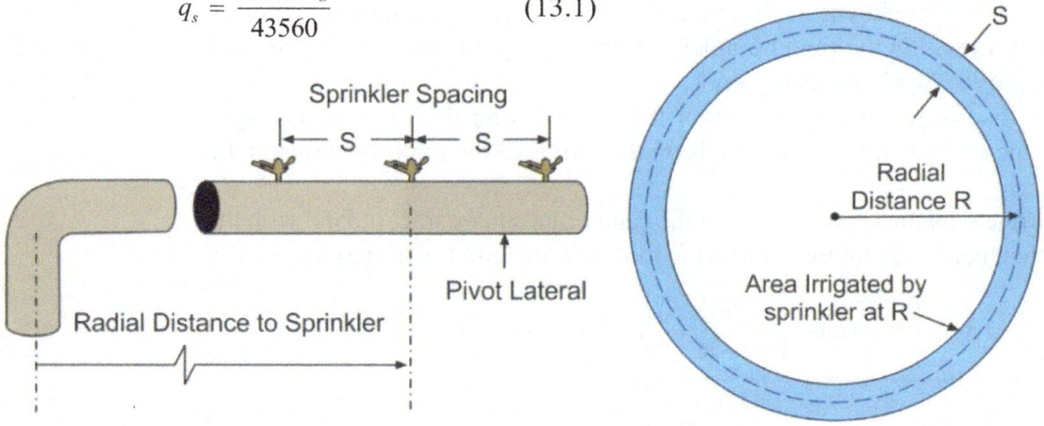

Figure 13.4. Diagram of the area associated with a sprinkler along a center pivot lateral.

where: q_s = the discharge from an individual sprinkler (gpm),
R = the distance of the sprinkler head or spray nozzle from the pivot base (ft),
S = the local spacing between successive sprinklers along the lateral (ft),
C_g = the gross system capacity required for the irrigation system (gpm/ac) = Q/A

The discharge from the sprinkler increases linearly with the distance from the pivot, i.e., a sprinkler 1,000 feet from the pivot will require twice as much discharge as a sprinkler at 500 feet. The discharge also depends on the spacing between sprinklers and the gross system capacity. The system capacity is determined by the crop, climate, and soils as described in Chapters 4 and 5, and does not vary by location along the pivot. The system capacity (C_g) must be determined from the field requirements and should not be determined arbitrarily.

Originally, pivots were manufactured with a constant spacing of about 32 feet between sprinklers. Spacing sprinklers closer together at the distal end allows lower operating pressures to be used while maintaining excellent uniformity. Today pivot laterals are manufactured with sprinkler outlets spaced at 7.5 to 10 feet. Near the pivot base sprinklers are not placed in every available outlet. Somewhere along the lateral the discharge required from a sprinkler becomes too large if outlets are skipped. Then the spacing must be reduced. This generally allows for using the same size of sprinkler device along a major portion of the lateral. Equation 13.1 has been solved in terms of discharge per unit length along the lateral (i.e., q_s/S) for a range of conditions for pivots (Table 13.1).

Table 13.1. Sprinkler discharge requirement per unit length along the lateral (q_s/S), gpm/ft.

Distance from Pivot (ft)	Gross System Capacity (gpm/ac)						
	4	5	6	7	8	9	10
100	0.06	0.07	0.09	0.10	0.12	0.13	0.14
200	0.12	0.14	0.17	0.20	0.23	0.26	0.29
300	0.17	0.22	0.26	0.30	0.35	0.39	0.43
400	0.23	0.29	0.35	0.40	0.46	0.52	0.58
500	0.29	0.36	0.43	0.50	0.58	0.65	0.72
600	0.35	0.43	0.52	0.61	0.69	0.78	0.87
700	0.40	0.50	0.61	0.71	0.81	0.91	1.01
800	0.46	0.58	0.69	0.81	0.92	1.04	1.15
900	0.52	0.65	0.78	0.91	1.04	1.17	1.30
1000	0.58	0.72	0.87	1.01	1.15	1.30	1.44
1100	0.63	0.79	0.95	1.11	1.27	1.43	1.59
1200	0.69	0.87	1.04	1.21	1.38	1.56	1.73
1300	0.75	0.94	1.13	1.31	1.50	1.69	1.88
1400	0.81	1.01	1.21	1.41	1.62	1.82	2.02
1500	0.87	1.08	1.30	1.51	1.73	1.95	2.16
1600	0.92	1.15	1.38	1.62	1.85	2.08	2.31
1700	0.98	1.23	1.47	1.72	1.96	2.21	2.45
1800	1.04	1.30	1.56	1.82	2.08	2.34	2.60
1900	1.10	1.37	1.64	1.92	2.19	2.47	2.74
2000	1.15	1.44	1.73	2.02	2.31	2.60	2.88
2100	1.21	1.51	1.82	2.12	2.42	2.73	3.03
2200	1.27	1.59	1.90	2.22	2.54	2.86	3.17
2300	1.33	1.66	1.99	2.32	2.65	2.99	3.32
2400	1.38	1.73	2.08	2.42	2.77	3.12	3.46
2500	1.44	1.80	2.16	2.52	2.88	3.25	3.61
2600	1.50	1.88	2.25	2.63	3.00	3.38	3.75

13.2.2 Area Irrigated

The area irrigated with a center pivot depends on the radius irrigated with the main lateral and the radius gain when the end gun is turned on. Typically, a center pivot is positioned into a square land area similar to that shown in Figure 13.5. The end gun can only be operated when the spray pattern stays within the field boundary. In the example in Figure 13.5 the end gun operates over an angle (ß) of 42° in each corner.

Usually the end gun discharges water only about half of the time that the main system operates. The time that the end gun operates depends on the radius of the main system and the gain from the end gun.

The amount of area irrigated with a pivot placed in the center of a square tract of land with the end gun operating in all four corners is computed with the following equation (von Bernuth, 1983):

$$A_t = \frac{\pi R_l^2 + \left[\pi - 4\cos^{-1}\left(\frac{R_l}{R_l + R_{eg}}\right)\right]\left(2R_l R_{eg} + R_{eg}^2\right)}{43560} \tag{13.2}$$

where: A_t = the total irrigated area (ac),
R_l = the radius irrigated with the main system lateral (ft),
R_{eg} = the radius gain from using the end gun (ft), and
inverse cosine is evaluated in radians.

Increasing the radius gain from the end gun does not ensure more irrigated area since the angle of the section that can be irrigated with the end gun decreases. The maximum irrigated area will, in fact, occur when the radius gain from the end gun is about 21% of the length of the main pivot lateral. Usually, however, the availability of end gun nozzle sizes, discharge requirement of the end gun, and available system pressure dictate the radius gain from the end gun. Solutions to Equation 13.2 have been developed in Table 13.2. The values in this table apply when all four corners are irrigated. Sometimes a road along the property reduces the angle of operation of the end gun in the corner of the field. Table 13.2 also assumes that the entire area wetted by the end gun is planted to the irrigated crop. This may not be done in some cases if the depth of application tapers off near the end of the radius of coverage of the end gun. This will reduce the values from Table 13.2 slightly, but usually not by a significant amount. The values in Table 13.2 should be adequate for planning and management.

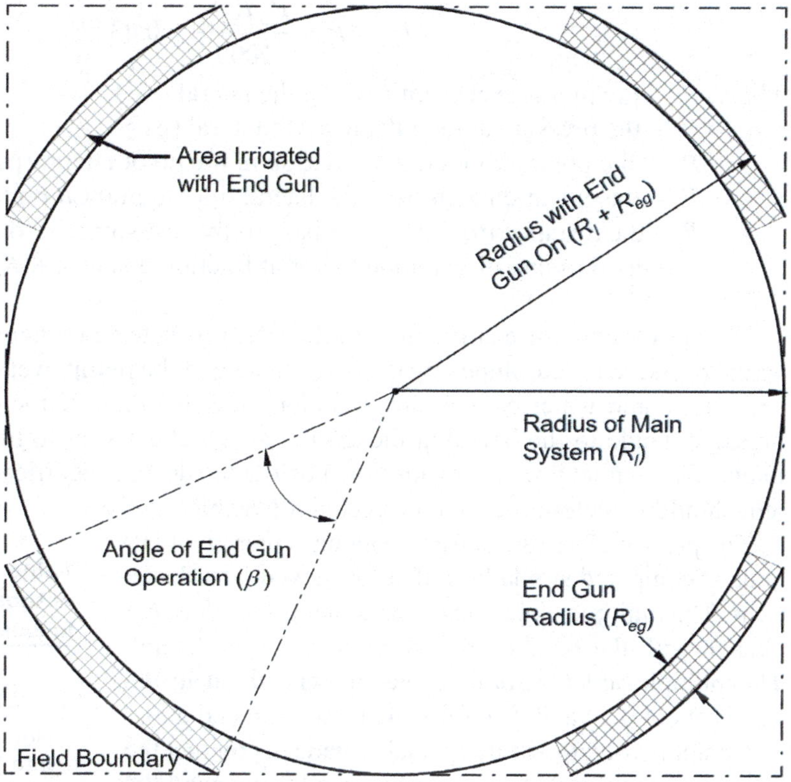

Figure 13.5. Diagram of the effect of end-gun operation on the area irrigated (adapted from Martin et al., 2017).

Table 13.2. Total irrigated area for different lengths of the main system and the end gun coverage.

Radius Irrigated with Main Lateral (ft)	Gain of Wetted Radius from End Gun Operation (ft)							
	0	50	75	100	125	150	200	Maximum Area[a]
800	46	49	50	51	51	51	-	51
900	58	62	63	64	65	65	-	65
1000	72	77	78	79	80	80	80	80
1100	87	92	94	95	96	97	97	97
1200	104	109	111	113	114	115	116	115
1300	122	128	130	132	133	134	135	135
1400	141	148	151	152	154	155	157	157
1500	162	170	172	175	176	178	180	180
1600	185	193	196	198	200	202	204	205
1700	208	217	220	223	225	227	230	231
1800	234	243	246	249	252	254	257	259
1900	260	270	274	277	280	282	286	289
2000	288	299	303	306	309	312	316	320
2100	318	329	333	337	340	343	348	353
2200	349	361	365	369	373	376	381	387
2300	382	394	399	403	407	410	415	423
2400	415	428	434	438	442	445	451	461
2500	451	464	470	475	479	482	489	500
2600	488	502	507	512	517	521	528	541

[a] Maximum area occurs when the radius gain is 21% of the main lateral length.

13.2.3 Pressure Distribution

Nozzle selection and center pivot evaluation require knowledge of the pressure distribution along the pivot lateral. The distribution is unique for pivots since the discharge required of sprinklers increases along the pivot lateral. The pressure at a point along the pivot lateral is given by:

$$P_R = P_0 - \frac{P_{lp} R_l f_p}{1000} - 0.433\,E_g \qquad (13.3)$$

where: P_R = the pressure at point R along the lateral (psi),
P_0 = the pressure at the inlet to pivot lateral (psi),
P_{lp} = the pressure loss due to friction in the pivot lateral (psi/1000 ft),
E_g = the elevation gain from the lateral inlet to point R on the lateral (ft),
R_l = the distance from the pivot base to the last sprinkler on the main lateral (ft), and
f_p = the pressure distribution factor at fraction distance R/R_l (dimensionless) (Figure 13.6).

The desired pressure at the inlet to the lateral is selected when the pivot is designed. The actual pressure is determined by the performance of the pump, wear of sprinklers and pressure regulators, and water or pressure loss along the mainline that supplies the pivot. Adding a pressure gauge to the lateral at the inlet is an excellent way to monitor center pivot performance. If the inlet pressure drops much below the design specification, the cause of the problem should be determined and corrected if feasible.

The pressure loss due to friction along center pivot laterals is computed similarly to that for moved lateral systems. The multiple outlet factor for center pivots does not change with the number of sprinklers along the lateral. The multiple outlet factor for center pivots without an end gun is about 0.54 and 0.56 for systems with an end gun.

Center pivot laterals are specially made to conduct the water and to provide the strength needed to suspend the lateral above the ground. The lateral diameter is also unique for center pivots and moving lateral systems. The typical lateral is made of galvanized steel pipe with a wall thickness of 0.109 inches. A C value in the Hazen-Williams equation (Equation. 8.11) of 140 is typically used to compute friction loss along the pivot. Values for the pressure loss per unit length of center pivot laterals are given in Table 13.3 for typical lateral diameters. Table 13.3 is for laterals that are all one size. As will be illustrated below, 80% of the pressure loss of a pivot lateral occurs in the first half of the lateral. Pressure loss can be reduced

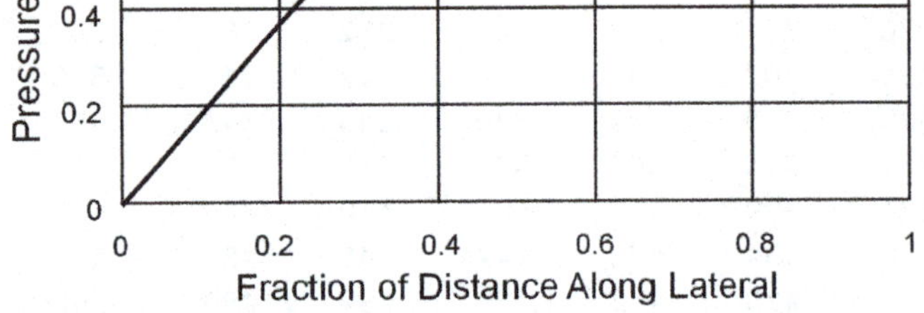

Figure 13.6. Pressure loss distribution factor for center pivot laterals.

Table 13.3. Pressure loss in center pivot laterals without end guns, psi/1,000 feet of pipe. Multiply losses by 1.037 for laterals with end guns.

Hazen Williams Equation C Value = 140; Multiple Outlet Factor for Center Pivots = 0.54				
Flow Rate into Pivot Lateral (gpm)	Outside Diameter of Pipe (in)			
	6	6 5/8	8	10
	Inside Diameter of Pipe (in)			
	5.782	6.407	7.782	9.782
200	0.93	0.57	-	-
300	1.98	1.20	-	-
400	3.38	2.05	-	-
500	5.10	3.10	-	-
600	7.15	4.34	-	-
700	9.51	5.77	3.57	-
800	12.18	7.39	4.43	-
900	15.15	9.20	-	-
1000	18.42	11.18	-	-
1100	21.98	13.34	5.18	-
1200	-	15.67	6.08	2.00
1300	-	18.17	7.06	2.32
1400	-	20.85	8.09	2.66
1500	-	-	9.20	3.02
1600	-	-	10.36	3.41
1700	-	-	11.60	3.81
1800	-	-	12.89	4.24
1900	-	-	14.25	4.68
2000	-	-	15.67	5.15
2100	-	-	17.15	5.63
2200	-	-	18.69	6.14
2300	-	-	20.30	6.67
2400	-	-	-	7.22
2600	-	-	-	8.37
2800	-	-	-	9.60
3000	-	-	-	10.91
3200	-	-	-	12.29
3400	-	-	-	13.75

Table 13.4. Pressure loss in center pivot laterals with two diameters of pipe, psi/1,000 feet. Values are for laterals without end guns. Multiply by 1.037 for systems with an end gun.

Pressure Loss in Laterals Composed of 8- and 6-inch Diameter Pipe

Flow Rate (gpm)	Fraction of Lateral That Is 8-inch O.D. Galvanized Steel Pipe										
	1.0	0.9	0.8	0.7	0.6	0.5	0.4	0.3	0.2	0.1	0.0
900	3.6	3.6	3.7	3.9	4.2	4.7	5.4	6.2	7.1	8.1	9.2
1000	4.3	4.4	4.5	4.7	5.1	5.8	6.6	7.6	8.7	9.9	11.2
1100	5.2	5.2	5.3	5.6	6.1	6.9	7.8	9.0	10.4	11.8	13.3
1200	6.1	6.1	6.2	6.6	7.2	8.1	9.2	10.6	12.2	13.9	15.7
1300	7.1	7.1	7.2	7.6	8.3	9.4	10.7	12.3	14.1	16.1	18.2
1400	8.1	8.1	8.3	8.8	9.6	10.7	12.3	14.1	16.2	18.5	20.8
1500	9.2	9.2	9.4	10.0	10.9	12.2	13.9	16.0	18.4	21.0	23.7
1600	10.4	10.4	10.6	11.2	12.3	13.7	15.7	18.0	20.7	23.7	26.7
1700	11.6	11.6	11.9	12.6	13.7	15.4	17.6	20.2	23.2	26.5	29.9
1800	12.9	12.9	13.2	14.0	15.2	17.1	19.5	22.4	25.8	29.4	33.2

Pressure Loss in Pivot Laterals Composed of 10- and 8-inch Diameter Pipe

Flow Rate (gpm)	Fraction of Lateral That Is 10-inch O.D. Galvanized Steel Pipe										
	1.0	0.9	0.8	0.7	0.6	0.5	0.4	0.3	0.2	0.1	0.0
1900	4.7	4.7	4.8	5.2	5.8	6.7	7.8	9.2	10.8	12.5	14.2
2000	5.1	5.2	5.3	5.7	6.4	7.3	8.6	10.1	11.8	13.7	15.7
2100	5.6	5.7	5.8	6.2	7.0	8.0	9.4	11.1	12.9	15.0	17.1
2200	6.1	6.2	6.4	6.8	7.6	8.7	10.2	12.0	14.1	16.4	18.7
2300	6.7	6.7	6.9	7.4	8.2	9.5	11.1	13.1	15.3	17.8	20.3
2400	7.2	7.3	7.5	8.0	8.9	10.3	12.0	14.2	16.6	19.2	22.0
2500	7.8	7.8	8.1	8.6	9.6	11.1	13.0	15.3	17.9	20.7	23.7
2600	8.4	8.4	8.7	9.3	10.3	11.9	13.9	16.4	19.2	22.3	25.5
2700	9.0	9.0	9.3	10.0	11.1	12.8	15.0	17.6	20.6	23.9	27.3
2800	9.6	9.6	9.9	10.6	11.9	13.7	16.0	18.8	22.1	25.6	29.2
2900	10.2	10.3	10.6	11.4	12.7	14.6	17.1	20.1	23.5	27.3	31.2
3000	10.9	11.0	11.3	12.1	13.5	15.5	18.2	21.4	25.1	29.0	33.2

by using larger diameter pipe for the initial portion of the lateral rather than one diameter for the whole lateral. The pressure loss for systems with multiple pipe diameters are given in Table 13.4. The pressure distribution factor for center pivots is given in Figure 13.6. The elevation gain in Equation 13.3 is elevation of the point of concern minus the elevation at the pivot base. If the pivot base is higher than the point of concern, then $E_g < 0$.

Variation of pressure as the pivot rotates around the field affects the uniformity of water application. It is useful to monitor the pressure of the outer end of the pivot as it rotates around the field. The critical points will be the highest and lowest elevations of the outer half of the pivot lateral. If the pressure varies more than 20% of the design pressure for the end of the lateral, consideration should be given to the use of pressure compensating nozzles or pressure regulators. Lower pressures than expected at the highest elevations may be a sign that the pump is not operating as originally designed or that there are leaks in the system.

13.3 Application Rate

13.3.1 Center Pivots

The rate of water application under a center pivot has unique characteristics that are important in design and management. A tower at the end of a conventional center pivot might move at 2 feet per minute. The first tower on the pivot might move at 1/10 that speed. Since each location should receive the same depth of water each irrigation, water must be applied 10 times as rapidly at the outer tower compared to the inner tower. The very high rate of water application can exceed the soils infiltration rate. If adequate storage is not provided on the soil surface to retain the water while infiltration occurs the water may run downhill. This runoff process can be acute with center pivot irrigation on steep slopes and soils that have low infiltration rates. Yet pivots can work in these conditions if they are properly designed. Therefore, it is important to determine the factors that control the rate of application.

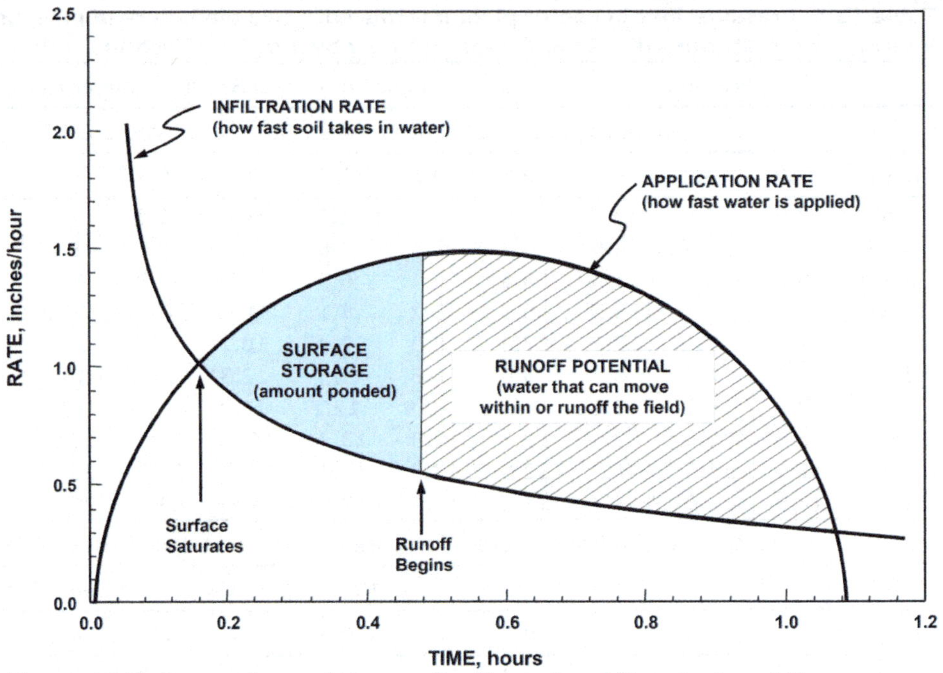

Figure 13.7. Comparison of the application rate of the pivot and the infiltration rate of the soil (adapted from Martin et al., 2017).

The typical application rate is shown in Figure 13.7. The example shows that water will only be applied for about 25 minutes at the distal end of the lateral for this pivot. The rate of water application reaches a peak when the pivot lateral is directly overhead of the point of concern. Then the rate decreases as the pivot continues to move forward. The application pattern is generally described as an elliptical rate.

Two characteristics are important to evaluate the elliptical rate:

- the highest or peak rate of application when the pivot lateral is directly overhead and
- the total length of time that water is applied to the location, called the time of wetting.

The peak application rate is given by:

$$A_p = \frac{0.0177 C_g R}{D_c} \tag{13.4}$$

and the time of wetting is given by:

$$T_w = 72 \frac{D_c d_g}{R C_g} \tag{13.5}$$

where: A_p = the peak application rate (in/hr),
T_w = the time of wetting (hr),
R = the radial distance from the pivot point (ft),
D_c = the diameter of coverage of the sprinkler at position R (ft),
C_g = the gross system capacity (gpm/ac), and
d_g = gross application depth (in).

These relationships show that the peak application rate is totally determined by the design of the center pivot. The length of the pivot lateral, the type of sprinkler used, and the system capacity determine the peak application rate. The peak rate does not change with the speed of rotation of the pivot. The time of wetting at a point is a factor of these variables and the depth of water applied per irrigation. Thus, the time of application can be controlled by management, i.e., by changing d_g.

The rate of infiltration of a hypothetical soil is also shown in Figure 13.7. The diagram shows that the application rate of the pivot exceeds the infiltration rate for most of the time of wetting. During this time water could runoff if it is not stored on the soil surface.

What can be done to reduce runoff? The three design variables, C_g, D_c, and R that affect the peak application rate and the time of wetting could all be changed when the pivot is designed and installed. The system capacity used in selecting an irrigation system is based on the crop needs for the soil and climate at the location. Thus, the system capacity should not be reduced much below the requirement just to prevent runoff. The length of the pivot lateral is determined by the geometry of the field. In some cases, there is a choice between installing one very long system or several shorter systems. The investment cost per acre will be less for the longer system but the potential for runoff is higher.

The primary alternative to reduce runoff problems is to select sprinkler devices that provide the necessary diameter of coverage. This is generally done at the time the system is purchased but can be changed later. Once the pivot is installed the only system management alternative to reducing runoff is to reduce the depth of application. The maximum depth of application and the appropriate types of sprinkler devices are discussed in the next section on sprinkler and nozzle selection.

13.3.2 Linear or Lateral Move

One inherent advantage of linear or lateral move systems over center pivots is that peak application rates are much lower for these systems. This is because with linear move systems, the discharge is distributed uniformly throughout the lateral's length while with center pivots, discharge increases with distance from the pivot point. The peak application intensity of a linear move can be calculated with the following equation:

$$A_p = \frac{122.5Q}{D_c L} \tag{13.6}$$

where: Q = the system's flow rate (gpm),
L = the lateral length, and
D_c = the diameter of coverage of the sprinkler heads on the system.

13.4 Sprinkler and Nozzle Selection

The center pivot operator should be concerned with the following questions regarding the sprinkler and nozzle package installed on a center pivot.

- What type of sprinklers and nozzles to install on the pivot?
- Are the proper sprinklers and nozzles installed at the correct location along the pivot lateral?
- Are the sprinklers working properly?

Determining the proper nozzle size for each sprinkler along the center pivot lateral is complex. The number of nozzles needed, the size of the nozzles, spacing of sprinklers at the point of concern, the diameter of coverage, pressure loss along the lateral, the use of pressure regulators, and the elevation gain around the field are all issues. In addition, every sprinkler along the pivot lateral is considered individually. The details of this design process will not be considered here. Generally, the nozzle sizes along the lateral will be determined by the center pivot or sprinkler manufacturer. The center pivot owner and operator should obtain a copy of the sprinkler package chart. This chart specifies the type of sprinkler and nozzle sizes to be used at a particular location along the lateral. The operator can use the chart to check the final installation to determine if errors were made in shipping or construction.

An important decision for the center pivot manager is the type of sprinkler device to use. Many choices of sprinklers are available. Early center pivots used only impact sprinklers. These have the same performance characteristics as presented in Chapter 11. Impact sprinklers are appealing because they provide a large diameter of coverage which produces lower application rates and less runoff potential. Recently these sprinklers have been made of plastic leading to sprinkler packages that are price competitive because fewer sprinklers are required with the larger diameter of coverage. The primary disadvantages of impact sprinklers are the higher operating pressures required and the higher potential for evaporation and drift losses. To reduce the evaporation potential low angle impact sprinklers have been developed. The range nozzle on conventional impact sprinklers emits water at an angle 23° above the horizon. The low angle sprinklers discharge water at an angle of about 7°. Low-angle sprinklers can be used with special nozzles for operation at lower pressures than conventional sprinklers.

Special spray head devices that discharge water onto pads have been developed for use on center pivots. The pad could either be stationary or moving. The devices can generally be installed in an upright position or inverted. When the pad devices are inverted they are attached to drop tubes that allow the devices to be positioned below the truss assembly, or even lower to apply water just above the crop canopy. Dropping the devices closer to the crop canopy reduces the potential for evaporation or drift but increases the runoff potential. The advantage of the rotating and wobbling pad devices is the increased diameter of coverage requiring fewer devices while providing lower application rates and better uniformity.

Various types of pads can be used with both the stationary and moving spray heads. The face of the pad can be smooth or grooved. The smooth pad produces smaller droplets. Grooved pads produce small streams of water off of the pad leading to larger drops than the smooth pad. The depth of the groove and the number of grooves on the pad determine the size of the droplets. Pads are also designed for use when the device is upright or inverted. If the device is placed on top of the pivot lateral, a concave pad is used to direct the spray toward the soil. When the device is inverted and dropped below the pivot lateral, a flat or convex pad is used to direct the water horizontally to maximize the diameter of coverage. Two issues are important when selecting the type of pad: drift and energy of impact of the droplet. Small droplets contain less kinetic energy when they reach the soil surface than large droplets. If the soil at the site has low aggregate stability, the large droplets (from pads with large grooves) can cause a seal to form at the soil surface leading to lower infiltration rates. If you irrigate when there is little cover on the soil, then smooth or shallow grooved pads would be desirable. Small drops are affected by wind much more than larger drops. If you irrigate in a windy area and infiltration rates are good, you may choose a deeper grooved pad. In windy areas mounting the spray pad devices below the pivot lateral closer to the crop may be a good idea. In areas with low infiltration rates and/or steep slopes, impact sprinklers may still be preferable because of the smaller runoff potential.

13.5 Depth of Water Applied

The depth of water applied each irrigation greatly affects the amount of potential runoff. As indicated earlier, the maximum application rate does not change with the depth of water applied. However, the time required to apply the water is directly proportional to the depth applied. Since the infiltration rate of the soil decreases with time, the longer it takes to apply water the greater the chance of runoff. The example in Figure 13.8 shows that there would be little runoff for an application of 0.8 inches each irrigation. Contrast that to the potential runoff for an application of 2.4 inches. There would certainly be substantially more runoff for the larger irrigation. It is desirable to apply smaller depths of water each irrigation with center pivots than to apply one large irrigation. It is common to apply from 0.70 to 1.25 inches per irrigation with pivots. This will usually require irrigation every 3 or 4 days.

Two other factors affect the depth of water to apply each irrigation: the condition of the soil surface where the pivot must travel and the evaporation rate of the applied water. If large irrigations are applied the soil surface becomes much wetter, and in some conditions, the traction of the pivot suffers. Any water that runs off often accumulates in the tracks left by the pivot wheels. The water then either flows downhill in the track or ponds in the track and surrounding area. If the pivot still has to pass through the low spot for that irrigation, or if the water remains at the time of the next irrigation, the wheel tracks from the pivot can become very deep and the pivot may have difficulty moving through these areas. Applying smaller depths of water each irrigation can mitigate some of these factors leading to more dependable operation.

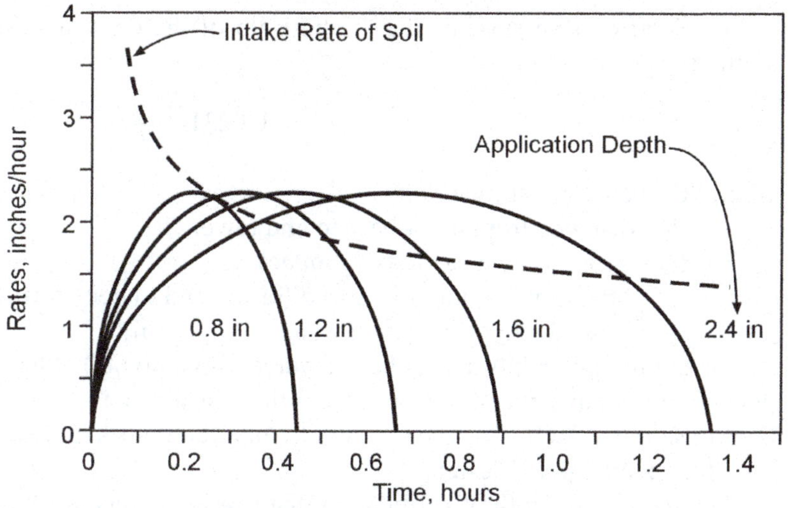

Figure 13.8. Effect of application depth on the potential for runoff (adapted from Martin et al., 2007).

The loss of water due to evaporation can be important for high-frequency irrigation. The amount of water that evaporates while the water droplets are in the air is much lower than many producers estimate. The maximum loss of evaporation while the drops are in the air is less than 5% for even the most severe wind and drying conditions. The major loss of water to evaporation comes after the water has reached the crop and soil. Research has shown that water on the canopy and bare soil will evaporate very quickly. In windy, arid conditions, such as the Great Plains of the U.S., corn canopies dry in about 1 hour after irrigation in the middle of the day. The water that evaporates from the canopy uses some energy that would have caused transpiration had the crop not been irrigated. Thus, not all of the canopy evaporation is truly a loss. However, high-frequency irrigations that wet the crop or soil will lead to increase evaporation and somewhat lower efficiency. Estimates are that under very windy and arid conditions in the southern High Plains of the U.S. the efficiency of pivots equipped with impact sprinklers is about 85%. The efficiency increases to about 90% for devices that apply water just above the crop canopy and to as high as 95% for LEPA systems that apply water near the soil surface without wetting the canopy. For application efficiencies to be this high, water must not runoff the field.

In any case, very high-frequency irrigation with small depths can lead to reduced efficiency if canopy evaporation becomes excessive. There have been reports that high- frequency irrigation maintains a wet soil surface that leads to reduced infiltration rates and increased runoff.

There are several conflicting conditions regarding the proper depth of application for pivots. It is critical that managers develop a routine of observing the performance of the pivot on the steepest areas of the field near the outer half of the pivot. Managers should experiment to determine the maximum depth that can be applied without runoff problems occurring. This depth may decrease during the season so monitoring during the season is important. Managers could then adjust the depth of application down from the maximum depth if desired. Irrigation intervals shorter than 2 days are probably impractical. If the system has to operate at that or shorter frequencies, the sprinkler package may be inappropriate or special tillage may be needed to prevent runoff.

The depth of application on pivots is adjusted by controlling the average speed of the end tower. On electric-drive pivots a timer can be set between 0 and 100%. At 100% the distal end tower is supplied power continuously and the tower moves at a constant speed. This setting produces the smallest depth of application possible. To apply larger irrigations the timer setting is reduced. The timer setting represents the percent of a 1-minute interval that the end tower receives power. For example, a 50% timer setting provides power to the end tower for 30 seconds and it moves at a constant speed. The end tower remains stationary for 30 seconds. Operation of hydraulically powered systems is slightly different. The end tower on these machines moves constantly. The control setting regulates the delivery of oil to the end tower and controls the speed. The control setting represents the relative speed of the end tower.

For electric drive systems the relationship of the control setting and the depth of water applied is given below:

$$C_s = 0.0231\frac{R_l C_g}{v_m d_g} \tag{13.7}$$

where: C_s = control setting (%),
R_l = distance from pivot base to end tower (ft),
C_g = gross system capacity (gpm/ac),
v_m = maximum continuous speed for the end tower (ft/min), and
d_g = gross depth of irrigation water to apply (in).

For example, to apply 1.3 inches of water with a pivot that has a maximum speed of 8 feet per minute, a capacity of 7 gpm/ac and the last tower is 1,280 feet from the pivot base; a control setting of 20% would be required. Manufacturers supply a tabular solution of Equation 13.7 for specific pivot designs.

The maximum depth of application that can be applied with a center pivot depends upon the soil infiltration, surface storage available, and peak application intensity of the system. The Natural Resources Conservation Service (NRCS, formerly the Soil Conservation Service) has categorized soils into intake families. Examples are shown in Table 13.5. In general, a low intake family, such as 0.1, is characterized by its high clay content and low infiltration rate. A high intake family, such as 3.0, is characterized by its high sand content and high infiltration rate.

As stated earlier, the storage of water on the soil surface in depressions can help avoid

Table 13.5. NRCS soil intake families (adapted from https://efotg.sc.egov.usda.gov//references/public/NE/NE_irrig_Guide_index.pdf).

Intake Family	Surface Soil Texture and Subsoil Permeability	Representative Soil Series	
0.1	Clays, silty clays, clay loam, silty clay loams (with slowly & very slowly permeable soils)	Albaton c Luton sic Wabash sic Filmore sicl	Crete sicl Pawnee cl Wymore sicl
0.3	Silt loam, loam silty clay loam, loams (with slow or moderately slow permeability)	Butler sil Colo sicl Wood River sil Belfore sicl	Burchard cl Hastings sicl Moody sicl Sharpsburg sicl
0.5	Silt loam, loam (with moderately slow or moderate permeability)	Hall sil Holder sil Holdrege sil	Judson sil Keith l Richfield l
1.0	Fine sandy loam, sandy loam, silt loam, loam, very fine sandy loam (with moderately slow to moderately rapid permeability) Loam, silt loam, very fine sandy loam, clay loam, sandy clay loam (with moderate or moderately rapid permeability)	Hord fsl Keith fsl Mitchell fsl	Crofton sil Monona sil
1.5	Fine sandy loam, loam, very fine sandy loam, sandy loam, silt loam (with moderate or moderately rapid permeability)	Anselmo vfsl Bayard vfsl Cass vfsl Alda fsl Brocksburg fsl	O'Neill fsl Rosebud fsl
2.0	Loamy fine sand, loamy very fine sand, loamy sand (with moderately rapid permeability)	Alice lfs Anselmo lfs Libory lfs Ovina lfs	Hersh lfs Jayem lfs Sarben lfs Otero lvfs
3.0	Loamy fine sand, loamy sand, fine sand, fine sandy loam, loamy very fine sand (with rapid permeability)	Bankard ls Dunday lfs Inavale lfs	Thurman lfs Valent lfs Valentine lfs

Figure 13.9. Influence of field slope on depressional storage. Photograph courtesy of USDA-NRCS (adapted from Martin et al., 2017).

Table 13.6. Allowable soil surface storage (without artificial storage) values for various slopes (from Dillon et al., 1972).

Slope (%)	Allowable Soil Surface Storage (in)
0–1	0.5
1–3	0.3
3–5	0.1
5	0.0

runoff in cases where application intensity exceeds soil infiltration rate. Figure 13.9 illustrates the concept of the storage in depressions. The amount of storage that is available depends upon field slope. For "conventional" tillage practices, this storage can be estimated from Table 13.6.

Peak application intensity is an important factor when considering the potential for runoff of water (maybe we've lost sight by now—we want to avoid runoff). Peak application intensity can be calculated from Equation 13.4. The results of Equation 13.4 are shown in Figure 13.10. Obviously, wetted diameter, illustrated in Figure 13.11, has a major influence on peak intensities as does system capacity and the distance from the pivot point (Figure 13.12).

Figure 13.13 provides a "management

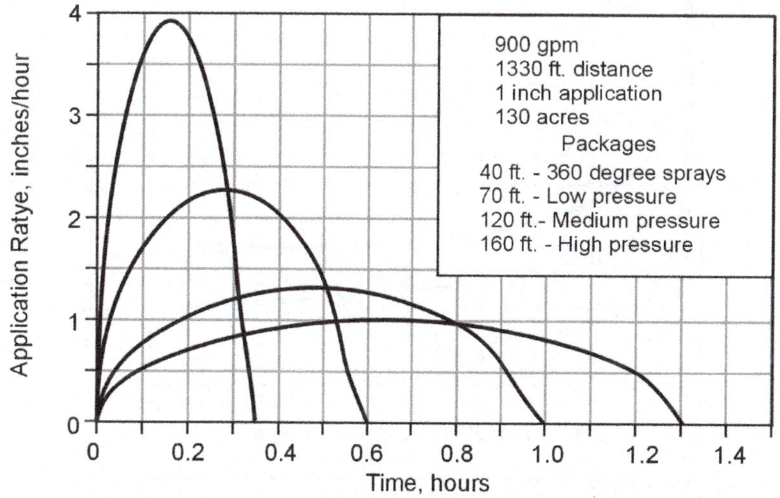

Figure 13.10. Effect of sprinkler packages on application rate.

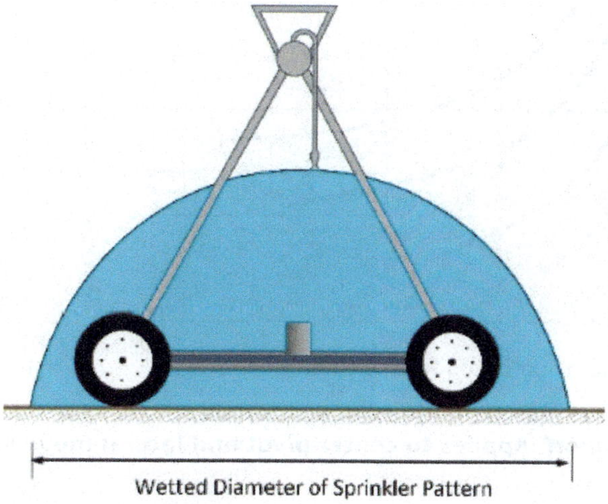

Figure 13.11. Illustration of wetted diameter (adapted from Martin et al., 2017).

guide" for avoiding runoff during water application (Gilley, 1984; Martin et al., 2007). The figure uses the important factors that we've just discussed to indicate how much water can be applied and yet avoid runoff. The use of Figure 13.13 is illustrated in the following Examples 13.1 and 13.2.

What if runoff is a problem? There are several design, management, and cultural practices that can be used if runoff is a problem. These practices are summarized in Table 13.7.

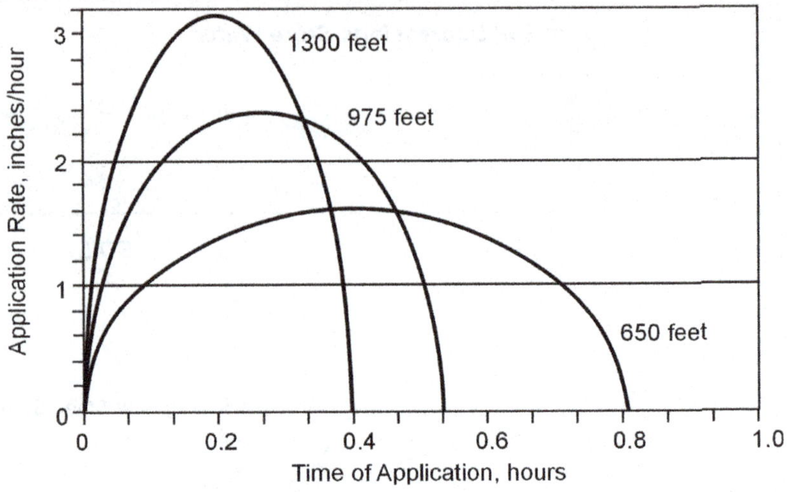

Figure 13.12. Effect of distance from pivot point on application intensity.

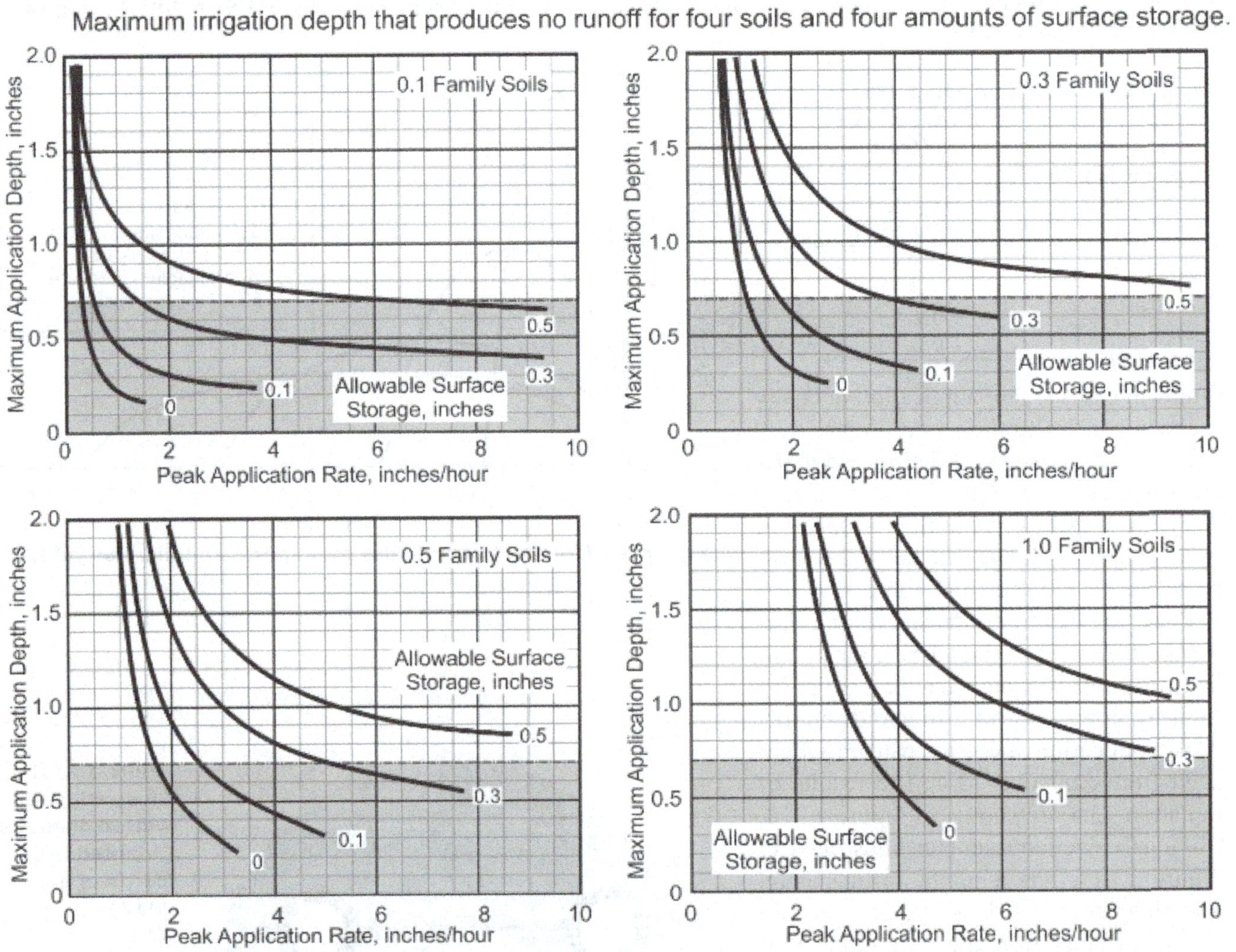

Figure 13.13. Maximum irrigation application depth (d_g) for different soils and peak application rates for zero potential runoff. Applies to center pivot and lateral move systems of any length.

Example 13.1

A center pivot operates with the following design features and field conditions:

Given: Q = 800 gpm

A = 130 ac

Sprinkler device = above canopy spray heads with 40 ft wetted diameter

System length = 1,300 ft

Field conditions: Soil—Holder silt loam, field slope 2%

Find: The maximum water application depth without runoff. Is this depth acceptable?

Solution:

$$A_p = \frac{0.0177 C_g R}{D_c} \quad \text{(Eq. 13.4)}$$

$$C_g = \text{gross system capacity} = \frac{800\ \text{gpm}}{130\ \text{ac}} = 6.15\ \text{gpm/ac}$$

Since the highest peak application rate ocurrs at the distal end R = 1,300 feet. Thus,

$$A_p = \frac{0.0177(6.15\ \text{gpm/ac})(1{,}300\ \text{ft})}{40\ \text{ft}} = 3.54\ \text{in/hr}$$

With a slope of 2%, the allowable surface storage is 0.3 inches (Table 13.6). The Holder silt loam is in the 0.5 intake family (Table 13.5).

Referring to Figure 13.13, we find that the maximum depth of application without runoff is 0.9 inches.

Is this acceptable? The 0.9 inches falls within the acceptable range of 0.70 to 1.25 inches per application, thus this system can be operated efficiently.

Example 13.2

Repeat Example 13.1 for a linear move system with the same capacity, 800 gpm, and lateral length, 1,300 feet.

Solution:

The peak application rate can be calculated with Equation 13.6:

$$A_p = \frac{122.5(800\ \text{gpm})}{(40\ \text{ft})(1300\ \text{ft})} = 1.88\ \text{in/hr}$$

From Figure 13.13 we find that we can apply 1.6 inches before runoff would occur, slightly higher than for the center pivot.

Table 13.7. Methods for reducing runoff under center pivot and lateral move sprinkler systems and their potential disavantages (bullet items).

1. Reduce system capacity
   - need to irrigate more hours per year
   - increases chances of soil water stress
2. Reduce application depth
   - requires more revolutions per year
   - increases frequency of leaf wetting
3. Change sprinkler package to increase wetted diameter
   - may require higher pressure
   - changes to pump and power unit may be needed
4. Increase surface storage
   - special interrow tillage practices may be needed
   - increased field operations
5. Increase soil surface cover with crop residues
   - may require significant change to farming operations

In Table 13.8 we present a method for estimating the required sprinkler wetted diameter to avoid runoff for various soil textures, surface storages, and desired application depths for the case of a 1300-ft center pivot lateral. The table is based on methods discussed in Martin et al. (2012), including the Green-Ampt approach for infiltration. The increase in soil surface storage using crop residues is also presented in Martin et al. (2017). One method for increasing wetted diameter is to use boom backs, illustrated in Figure 13.14. Boombacks have also been used to address problems with rutting in center pivot wheel tracks by keeping the wetting pattern from the sprinkler behend the wheels.

Table 13.8. Minimum allowable wetted diameter, ft. Applies to center pivots with a 1300-ft pivot lateral.[a][b]

Gross System Capacity, gpm/ac	Depth Applied, inch	Surface Storage, inch	Sand	Loamy Sand	Sandy Loam	Loam	Silt Loam	Sandy Clay Loam	Clay Loam	Silty Clay Loam	Sandy Clay	Silty Clay	Clay
4.0	0.8	0.1	< 10	10	18	39	46	110	> 150				
		0.3			< 10	19	22	53	88	88	> 150		
		0.5					< 10	19	32	30	63	73	120
	1.0	0.1	< 10	13	22	48	59	136	> 150				
		0.3		< 10	13	28	33	79	132	135	> 150		
		0.5			< 10	14	17	41	67	67	134	> 150	
	1.2	0.1	< 10	15	26	56	71	> 150					
		0.3		< 10	17	36	44	103	> 150				
		0.5		< 10	10	22	26	63	104	107	> 150		
5.0	0.8	0.1	< 10	13	22	48	58	137	> 150				
		0.3		< 10	11	23	27	67	110	109	> 150		
		0.5					< 10	24	40	38	79	91	> 150
	1.0	0.1	< 10	16	28	60	74	> 150					
		0.3		< 10	16	35	42	99	> 150				
		0.5			< 10	18	21	51	84	83	> 150		
	1.2	0.1	< 10	18	33	70	89	> 150					
		0.3	< 10	12	21	45	55	129	> 150				
		0.5		< 10	13	28	33	78	130	133	> 150		
6.0	0.8	0.1	< 10	16	27	58	70	> 150					
		0.3	< 10	a	13	28	33	80	132	131	> 150		
		0.5			< 10	10	11	29	48	45	95	109	> 150
	1.0	0.1	< 10	19	33	72	89	> 150					
		0.3	< 10	11	19	42	50	119	> 150				
		0.5			< 10	21	25	61	101	100	> 150		
	1.2	0.1	< 10	22	39	84	107	> 150					
		0.3	< 10	14	25	54	67	> 150					
		0.5		< 10	15	33	39	94	> 150				
8.0	0.8	0.1	< 10	21	35	77	93	> 150					
		0.3	< 10	10	17	37	44	107	> 150				
		0.5			< 10	14	15	39	64	60	126	146	> 150
	1.0	0.1	< 10	25	44	96	118	> 150					
		0.3	< 10	15	25	55	67	> 150					
		0.5		< 10	13	29	33	81	135	133	> 150		
	1.2	0.1	< 10	29	52	112	142	> 150					
		0.3	< 10	19	33	72	89	> 150					
		0.5	< 10	12	20	44	52	126	> 150				
10.0	0.8	0.1	< 10	26	44	96	116	> 150					
		0.3	< 10	13	21	47	54	133	> 150				
		0.5			< 10	17	19	49	79	75	> 150		
	1.0	0.1	10	32	55	120	148	> 150					
		0.3	< 10	19	32	70	83	> 150					
		0.5		< 10	16	36	41	102	> 150				
	1.2	0.1	12	36	65	140	> 150						
		0.3	< 10	24	42	90	111	> 150					
		0.5	< 10	15	25	55	66	> 150					

[a] Green shading denotes that the minimum wetted diameter is less than 10 feet.
[b] Gold shading denotes that the minimum wetted diameter is more than 150 feet.

13.6 Remote Monitoring of System Operation and Control

Figure 13.14. Boom backs used to increase wetted diameter for center pivot or lateral move irrigation systems (from Martin et al., 2017).

GPS and communication technology used by the center pivot and lateral (linear) move irrigation industry makes it relatively easy to remotely monitor system operation and control the water application using a computer, tablet, or mobile device. With little effort, irrigation managers can implement important water management decisions and detect system malfunctions or operational problems. For example, did it rain at the site and should the system be shut off? Does the soil moisture sensor suggest that the field is too wet or too dry? Are the system pressure and flow rate in agreement with what they should be? If the flow rate is high given the system pressure perhaps there is a broken sprinkler head. If both the flow rate and pressure are too low there could be a problem with either the well or pump. High pressure and low flow may indicate plugged nozzles in the system. The latter three cases will warrant a field inspection which could include the use of binoculars or unmanned aircraft (drone) to check for leaks or plugged nozzles.

13.7 Variable Rate Irrigation

Variable rate irrigation (VRI) technology allows for spatial management of soil water (Camp et al., 1998; Evans et al., 2013). Different depths of irrigation can be applied based on the irrigation requirement for each part of the field depending on variability in soils, topography, and ET. VRI technology is available for both center pivots and lateral moves, although it is more common for center pivots.

Sector control (speed control) is a lower cost option for VRI that simply varies the speed of the last tower on a center pivot. Increasing the speed results in a lower application depth. This method results in irrigation management zones shaped in sectors ("pie slices") (Figure 13.15).

In addition to varying the speed of the irrigation system, zone control VRI also controls the nozzle flow rate for individual sprinklers or groups of sprinklers along the pivot lateral (Figure 13.15). The nozzle flow rate is typically reduced from the design flow rate by using a valve to pulse the nozzle on and off. This provides a finer resolution of control compared to speed control, and irrigation management zones can be defined to follow the shape of irregular soil patterns. One drawback of zone-control VRI is the higher investment cost compared to speed control VRI. Zone control is most likely to be profitable in situations where VRI can be used to increase yield quantity and/or quality.

If the growing season starts at field capacity, soils with lower available water capacity with need to be irrigated before soils with a larger available water capacity. VRI can also be used to manage problems associated with topography, e.g. applying less water in a low spot that tends to be wet from accumulated runoff. Prescription maps for VRI are often based on soils maps and yield maps (Kranz et al., 2014). Irrigation scheduling for VRI is discussed in Chapter 6. VRI capability is being incorporated into cloud-based monitoring and control systems for center pivots. A thorough review of VRI, including advantages and disadvantages, is presented in O'Shaughnessy et al. (2019).

Mapping Symbol	Map Unit Name	Water Holding Capacity of Four-foot Root Zone, inches
HtB	Haxtun loamy sand, 0 to3%	6.2
HxB	Haxtun sandy loam, 0 to 3%	6.5
PeC	Platner-Eckley Assoc., 3 to 5%	5.9
Ra	Rago and Kuma loams	9.5

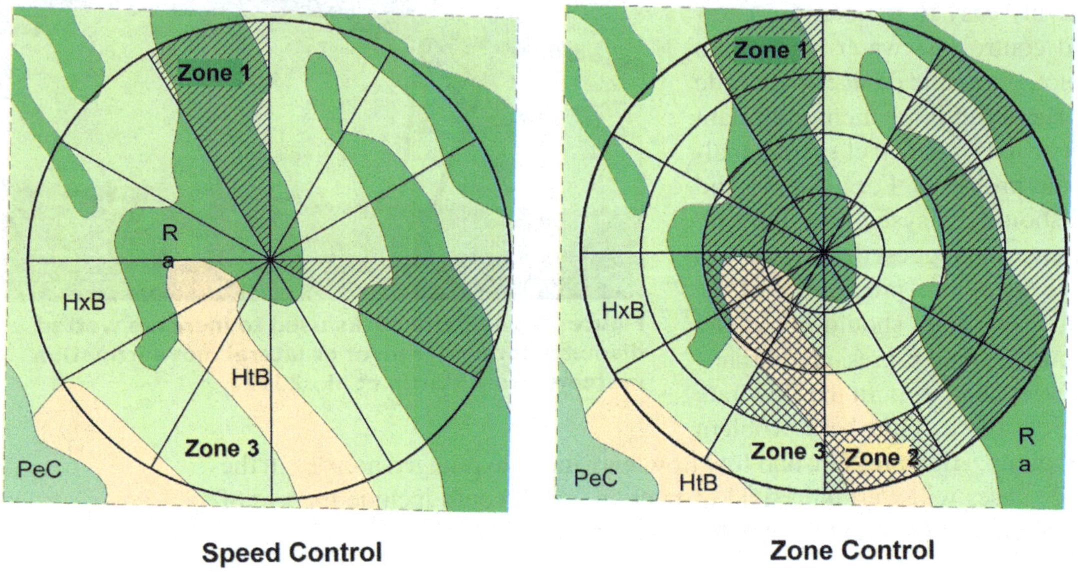

Figure 13.15. Example variable rate irrigation control scenarios: sector/speed control (left), and zone control (right).

13.8 Community Shared Center Pivot Systems

Since the cost of a center pivot increases proportionally with the length of the pivot, but the area irrigated increases proportionally with the square of the pivot length, the cost of the pivot per unit of land area ($/ha or $/ac) is lowest for large (longer) pivots. Therefore, pivots have been widely adopted on large fields (e.g., 60-70 ha). However, a large pivot can also be used to irrigate several small fields. This approach is being implemented on some smallholder farms in Sub-Saharan Africa (Figure 13.16; Chandra, 2020). Application depths can easily be changed for different sectors, either by changing the pivot speed for different angles in the pivot panel or using speed control VRI. Changing application depths enables the shared pivot to accommodate various crop types and planting dates in different sectors. It is ideal if fields

Figure 13.16. Shared center pivot irrigation systems for smallholder farms in Rwanda. (Photo courtesy of Ankit Chandra, DWFI.)

within a sector all have the same crop and planting date, resulting in similar irrigation needs. It is conceivable that zone control VRI could be used to provide unique irrigation management for many small fields of any shape and size, although zone control increases the complexity of the system and the level of management required. A shared center pivot required water users within the pivot to cooperate in a way similar to an irrigation district with canals for water deliver.

13.9 Summary

Center pivots are now used on more cropland than all of the other irrigation systems combined in the United States. Worldwide, center pivots and lateral move systems are gaining in popularity. Center pivots have grown in popularity because they can be very efficient (85 to 95% is possible), they can apply whatever depth of water is needed, and these systems have the capability to irrigate where surface and periodically-moved sprinkler systems are not feasible. Manufacturers have continued to improve pivots to operate at lower pressures and have improved sprinkler performance.

Equations to determine sprinkler discharge, area irrigated, and pressure distribution are presented. Minimization of surface runoff from pivot sprinklers is a major management concern discussed in detail. Application depths of 0.70 to 1.25 inches per irrigation resulting in irrigation every 3 to 4 days during peak periods of evapotranspiration are typical management decisions for center pivots.

Monitoring the operation and remote control of center pivot and lateral move systems is relatively easy using GPS and communication technology. Variable rate irrigation (VRI) technology permits different depths of water depending on variability in soils, topography, and evapotranspiration. VRI technology can be accomplished by speed control or varying the flow rate from the sprinklers. This technology is suitable for pivots shared to irrigate several small fields.

Questions

1. Describe the benefits and limitations of mechanized sprinkler irrigation systems.

2. A center pivot will be installed with 6 5/8-inch outside diameter pipe. The system will include 7 towers (178-ft spans), with a system length of 1,280 ft (including overhang). The flow rate of the entire system is 700 gpm. Determine the pressure loss due to friction (psi) in the pivot lateral without the end gun operating (Table 13.3).

3. For the center pivot in the previous question, determine the pressure required at the bottom of the pivot riser considering the nozzle pressure (10 psi), minor losses in regulator (assume 5 psi), pressure loss due to friction in pivot lateral, height of nozzle above ground (8 ft), and elevation change along the pivot lateral. The height of the pivot riser is 12 ft. The highest elevation in the field is at the north edge of the field, 14 feet higher than the elevation at the pivot point.

4. Calculate the nozzle flow rates and specify the nozzle sizes at the mid-point and at the outside end of the pivot lateral described in the previous questions. The sprinkler spacing (S) is 10 ft. Assume straight-bore nozzles.

5. For the irrigation system in the previous questions, determine the well pump total dynamic head, pump motor horsepower, and the number of pump stages required. The pump will be a Flowserve 12SKL (32.60) semi-open impeller (Figure 8.11). Choose the impeller diameter with the best efficiency. Assume that the well is located at the pivot point.

6. The predominate soil texture in the field is a Holdrege silt loam. What is the NRCS Intake

Family?

7. The diameter of coverage for the sprinklers is 25 ft. The typical slopes in the field are 1-2%. Using the nozzle flow rates calculated in Question 4, determine the peak application rate and the maximum application depth that can be applied (in) without causing runoff. Is this acceptable?

8. Find the pivot speed (timer setting) for the application depth determined in the previous question. The maximum tower speed is 6 ft/min. Use the system characteristics provided in Question 2.

References

Bittinger, M. W., & Longenbaugh, R. A. (1962). Theoretical distribution of water from a moving irrigation sprinkler. *Trans. ASAE, 5*(1), 26-30.

Camp, C. R., Sadler, E. J., Evans, D. E., Usrey, L. J., & Omary, M. (1998). Modified center pivot system for precision management of water and nutrients. *Appl. Eng. Agric., 14*(1), 23-31.

Chandra, A. (2020). Water-energy-food linkages in shared smallholder irrigation schemes. MS thesis. Department of Biological Systems Engineering, University of Nebraska-Lincoln.

Dillon Jr, R. C., Hiler, E. A., & Vittetoe, G. (1972). Center-pivot sprinkler design based on intake characteristics. *Trans. ASAE, 15*(5), 996-1001.

Evans, R. G., LaRue, J., Stone, K. C., & King, B. A. (2013). Adoption of site-specific variable rate sprinkler irrigation systems. *Irrig. Sci., 31*(4), 871-887.

Gilley, J. R. (1984). Suitability of reduced pressure center pivots. *J. Irrig. Drain. Eng., 110*(1), 22-34.

Heermann, D. F., & Hein, P. R. (1968). Performance characteristics of self-propelled center-pivot sprinkler irrigation system. *Trans. ASAE, 11*(1), 11-15.

Kranz, W. L., Irmak, Martin, D. L., Shaver, T. M., & van Donk, S. J. (2014). Variable rate application of irrigation water with center pivots. University of Nebraska-Lincoln Extension Circular EC2000.

Lamm, F. R., Porter, D. O., Bordovsky, J. P., Evett, S. R., O'Shaughnessy, S. A., Stone, K. C., Kranz, W. L., Rogers, D. H., & Colaizzi, P. D. (2019). Targeted, precision irrigation for moving platforms: Selected papers from a center pivot technology transfer effort. *Trans. ASABE, 62*(5), 1409-1415.

Martin, D. L., Kincaid, D. C., & Lyle, W. M. (2007). Design and operation of sprinkler systems. In G. J. Hoffman, R. G. Evans, M. E. Jensen, D. L. Martin, & R. L. Elliot (Eds.), *Design and operation of farm irrigation systems.* St. Joseph, MI: ASABE.

Martin, D., Kranz, W., Smith, T., Irmak, S., Burr, C., &Yoder, R. (2017). Center pivot irrigation handbook. University of Nebraska-Lincoln Extension Circular EC3017.

Martin, D. L., Kranz, W. L., Thompson, A. L., & Liang, H. (2012). Selecting sprinkler packages for center pivots. *Trans. ASABE 55*(2): 513-523.

O'Shaughnessy, S. A., Evett, S. R., Colaizzi, P. D., Andrade, M. A., Marek, T. H., Heeren, D. M., Lamm, F. R., & LaRue, J. L. (2019). Identifying advantages and disadvantages of variable rate irrigation: An updated review. *Appl. Eng. Agric., 35*(6), 837-852.

von Bernuth, R. D. (1983). Nozzling considerations for center pivots with end guns. *Trans. ASAE, 26*(2), 419-422.

Chapter 14 Microirrigation

14.1 Introduction

Microirrigation is a rapidly increasing method of irrigation, particularly for high value crops like vegetables and fruit and nut trees. A paramount question for a producer considering whether to invest in an expensive microirrigation system is whether or not the increase in crop production will be sufficient to pay for the system. A second concern is, Can the system be designed to filter the irrigation water to prevent the small emitters from clogging? Another important decision is which emitters, from the large array available, are appropriate for the intended purpose. Solving these issues will provide an excellent irrigation system for decades.

Microirrigation systems deliver water at low flow rates through various types of water applicators by a distribution system located on the soil surface, beneath the surface, or suspended above the ground. Water is applied as drops, tiny streams, or spray, through emitters, sprayers, or porous tubing and then flows through the soil by capillarity and gravity. Water pressure within the delivery lines is reduced by the design of the applicator to create a low discharge. Microirrigation is characterized by water being applied: (1) at low rates, (2) over prolonged periods of time, (3) at frequent intervals, (4) near or into the plant root zone, and (5) at relatively low pressure.

Most crops are adaptable to microirrigation. However, because the initial cost of these systems is high, microirrigation is best suited for high-valued crops, expensive land, or where environmental concerns are significant. Microirrigation systems are found on all soil types. These systems are particularly useful on very sandy and rocky soils that have a low water holding capacity or on salt-affected soils. Microirrigation is also excellent on steeply sloping land or where evaporation from the soil surface is a concern.

Potential advantages of microirrigation over other irrigation systems include: increased beneficial use of available water; enhanced plant growth, quality, and yield; reduced salinity hazards to plants; improved application of fertilizer and other chemicals; limited weed growth; decreased energy requirements; utilization of odd shaped areas; and improved cultural practices. Potential disadvantages include high initial costs, persistent maintenance requirements, restricted plant root development, and salt accumulation near plants (Bucks et al., 1982).

Microirrigation systems can be highly efficient and typically operate over prolonged periods of time; thus, low to moderate discharging water supplies can be utilized. This system offers maximum flexibility in chemigation. Frequent or nearly continuous application of plant nutrients, insecticides, fungicides, or other chemical amendments along with the irrigation water is feasible, and in most cases, beneficial for crop production. The low water discharge rates dictate a water applicator design with small openings; this can lead to clogging problems. Solutions to clogging include emitters that require less maintenance, adequate filtration of the water supply, and chemical treatment of the water.

14.2 History and Impact

Whereas surface irrigation by gravity began about 8,000 years ago, microirrigation is a relatively new concept. In 1913, E.B. House, at Colorado State University, experimented with subsurface trickle irrigation without raising the water table in the process, but concluded that it was too expensive. With the development of plastics during and following World War II, the idea of using plastic for irrigation pipe became plausible. The discovery of low-density polyethylene pipe in 1948 provided a suitable, economical material for drip irrigation. In the mid-1950s, an irrigation manufacturing firm in New York began supplying polyethylene tubing to water plants in greenhouses.

Publications on modern day trickle irrigation systems began to appear in Israel and the United States in the 1960s. Sterling Davis, working in California, installed the first field experiment with a subsurface trickle irrigation system in a lemon orchard in 1963. From 1968 into the 1970s, numerous inventors and companies developed drip irrigation emitters. By the mid-1970s more than 250 different water application devices had been produced.

In 1977, the plastic industry introduced linear, low-density polyethylene. This new plastic was less expensive, had improved strength, was resistant to stress cracking, and was flexible. With the addition of appropriate additives, such as carbon black, antioxidants, and stabilizers, clear plastic could be converted into a black, durable and economical pipe, well suited for microirrigation.

The first survey of microirrigation in the U.S. in 1974 (Irrigation Journal, 1974) reported about 70,000 acres of microirrigation. California had more than half of the national total (41,000 acres). Besides California, only Arizona, Florida, Hawaii, and Texas, had more than 1,000 acres of microirrigation at that time. By 1990 microirrigation had increased to 1.5 million acres. In the 2000 irrigation survey (Irrigation Journal, 2001), California still accounted for more than half of the 3.1 million acres of micro-irrigation in the U.S., but 40 states had at least 1,000 acres.

In 2018, nearly 56 million acres were irrigated in the U.S. (USDA, 2019). Of this total, microirrigation was used on nearly 6 million acres. Nearly 3 million acres were irrigated with surface drip; subsurface drip systems were installed on just over 1 million acres; and 1.5 million acres were irrigated by low-flow micro-sprinklers. Microirrigation was used on over 4 million acres in California. Fourteen other states had microirrigation on over 20,000 acres.

Although microirrigation accounts for less than 11% of the irrigated land in the United States, it is used frequently on high-value crops and for landscaping. The list of microirrigated crops include tree crops of avocado, citrus, stone fruit, nut, olive, coffee, and mango; and row crops like cotton, grape, melon, pineapple, sugar cane, tomato, and strawberry.

A global survey (FAO, 2021) indicates that in 2017, 20 countries had at least 1 million acres of microirrigation. The total area under microirrigation in these countries is about 25 million acres. Countries where the area of microirrigation exceeds that of sprinkler irrigation include Brazil, Chile, Cyprus, Egypt, Iran, Israel, Morocco, Palestine, Peru, Philippines, Poland, Portugal, Tunisia, and Uruguay. Microirrigation in Israel, for example, expanded from 25,000 acres in 1975 to about 400,000 acres in recent years. Estimates in 1991 indicated that microirrigation was used on 70% of the irrigated land in Israel (Stanhill, 1992).

14.3 System Types

Microirrigation includes trickle, drip, subsurface, bubbler, and spray irrigation systems (ASABE, 2019). Trickle and drip irrigation are often considered synonymous and we will refer to it as drip irrigation in this book. Here we differentiate these systems into four major types based on installation method, emitter design, or mode of operation. The four types are:

- Surface drip—water applied slowly through small emitter openings to the soil surface.

- Microspray—water sprayed over the soil surface at relatively low pressure (also called microsprinkler).
- Bubbler—a small stream of water applied to flood the soil surface in localized areas.
- Subsurface drip—water applied through emitter openings below the soil surface.

Examples of these four types of microirrigation are shown in Figure 14.1.

14.3.1 Surface Drip

A surface drip system consists of water applicators designed directly into or attached to lateral lines that are laid on the soil surface (Figure 14.1a). This method is also called trickle irrigation and is the most prevalent type. The maximum application rate for individual emitters is normally less than 3 gallons per hour (gal/hr). For porous tubing and other multiple outlet systems, the maximum application rate is generally less than 1 gallon per hour per foot of lateral. Advantages of surface drip irrigation over other microirrigation systems include ease of installation, inspection, repair, and cleaning of emitters. A major advantage is the ability to check soil surface wetting patterns and to measure individual emitter discharge rates. In some crops like grapes, which are trellised, the laterals are suspended above the soil surface by attachment to the trellis. This eliminates physical damage to the system from human traffic and cultural operations.

Figure 14.1. Examples of microirrigation systems: (a) surface drip (photo courtesy of Toro); (b) microspray (photo courtesy of Texas International Irrigation); (c) bubbler (photo courtesy of Anchor Commodities); and (d) subsurface drip (photo courtesy of Freddie Lamm, Kansas State University).

14.3.2 Microspray

Microspray systems apply water to the soil surface as a small spray, jet, fog, or mist (Figure 14.1b). Aerial distribution is a major component in the distribution of water by microspray systems compared to the other types where distribution of the water occurs primarily in the soil. Discharge of individual microspray applicators is typically less than 50 gal/hr. A comparison between the soil wetting pattern for surface drip and microspray systems is given in Figure 14.2. Microspray systems are used frequently to irrigate trees and other widely spaced agricultural crops. Microspray systems are vulnerable to drift and evaporation losses. Major advantages of microspray systems are minimal filtration needs and that maintenance requirements are small. Like drip, microspray systems are sometimes suspended above the soil surface.

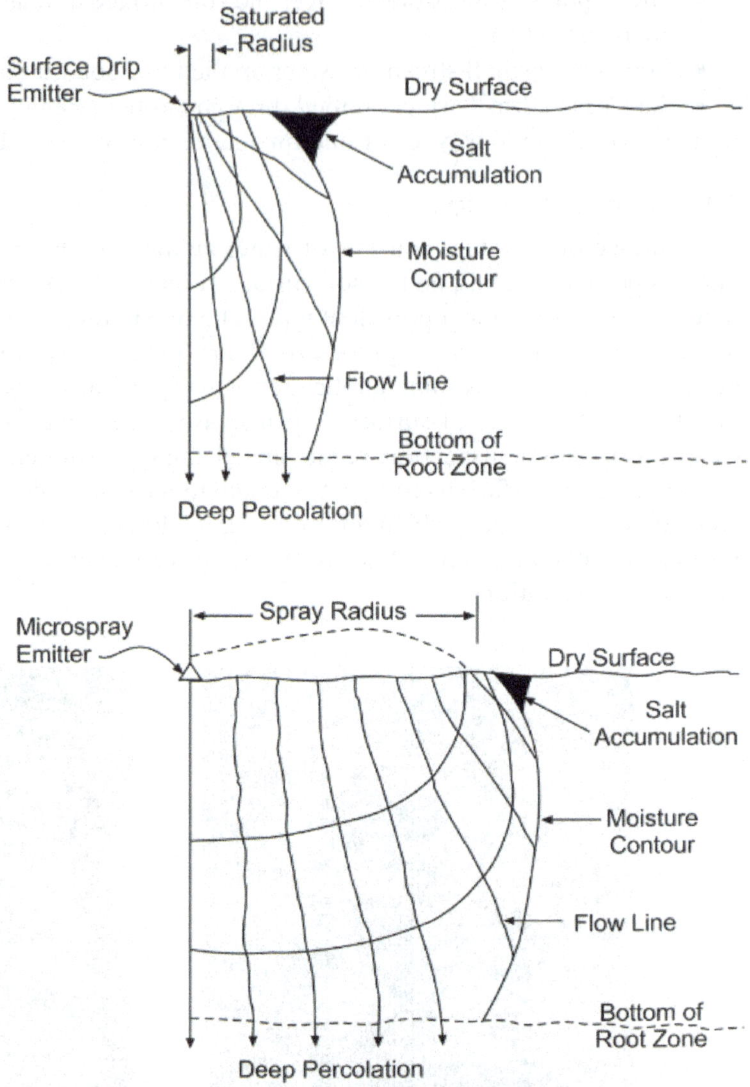

Figure 14.2. Comparison of one half of the soil wetting pattern between surface drip and microspray irrigation systems. Image adapted from Keller and Bliesner (1990) with permission from Blackburn Press.

14.3.3 Bubblers

With bubbler irrigation, water is applied to the soil surface in a small stream or fountain from a single discharge point (Figure 14.1c). Application rates are much greater than for drip systems but usually are less than 60 gal/hr. The discharge, applied at one point, normally exceeds the soil's infiltration rate and, therefore, a small basin is normally created by building an earthen dike to control the distribution of water on the soil surface. Advantages of the bubbler system include minimal filtration requirements, very little maintenance or repair, easy visual inspection, and extremely low energy requirements compared to other pressurized systems. However, large-size lateral lines are normally required to minimize pressure loss associated with high discharge rates. In well-designed systems only a few feet of pressure head are required to operate a bubbler system.

14.3.4 Subsurface Drip

A subsurface drip irrigation (SDI) system is basically a surface system that has been buried (Figure 14.1d). Comparison between the soil wetting pattern for surface and subsurface trickle systems is illustrated in Figure 14.3. This method of irrigation is not to be confused with subirrigation where an irrigation is accomplished by raising the water table to the bottom of the crop root zone. Subsurface systems are buried at a depth of a few inches to 18 inches or more. Shallow systems are installed in planting beds that are maintained over time. Deep installations do not require the crops to be placed in the same planting beds. Advantages of SDI include a permanently installed system that can last a decade or more, little interference with cultivation and other cultural practices, and irrigation and harvesting can occur at the same time on crops like melon and tomato. Figure 14.4 shows a schematic of a subsurface drip system with laterals placed in every other row between planting beds and photographs of equipment used to install subsurface drip laterals and a manifold that delivers water from the mainline to the laterals.

14.4 System Components

The basic components of a microirrigation system are the pump station, control station, mainline, manifold, laterals, and water applicators. Schematics of microirrigation systems are given in Figure 14.5. Typically, filters, chemical injection equipment, and control and monitoring devices are all located in close proximity at a control station. Refer to Chapter 15 for a discussion on chemicals that may be stored and injected into the irrigation system. Automatic controllers that activate

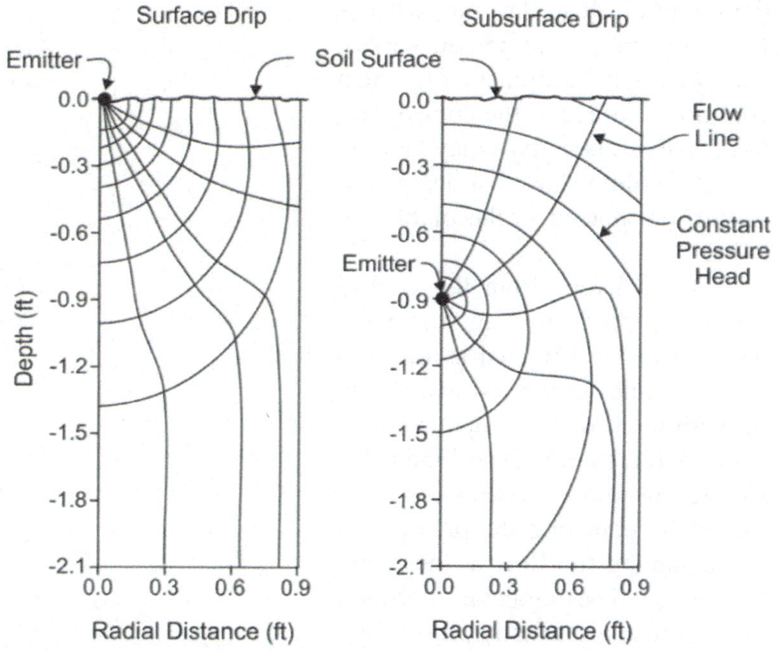

Figure 14.3. Typical wetting patterns for surface and subsurface drip systems. Image adapted from Coelho and Or (1997) with permission from John Wiley and Sons.

Figure 14.4. Installation of subsurface drip irrigation laterals: (a) schematic of a typical subsurface drip installation; (b) equipment for installing three subsurface drip laterals; and (c) laterals connected to the manifold. (Photos a and b courtesy of Suat Irmak, Nebraska Extension; photo c courtesy of Steve Melvin, Nebraska Extension.)

the pump or clean the filters, backflow prevention equipment, valves, and injection equipment, are also common features at the control station. Activating signals may be time or water volume based or dependent on sensors placed in the plant root zone.

Water delivery from the control station proceeds through the main pipeline to the irrigated area. From the mainline, water is distributed through manifolds to laterals where water application occurs. Water discharge, in most systems, is controlled by adjusting the pressure or by regulating the flow at the manifold inlets. The regulators used for this purpose are usually preset for a given pressure or flow rate and are not adjustable.

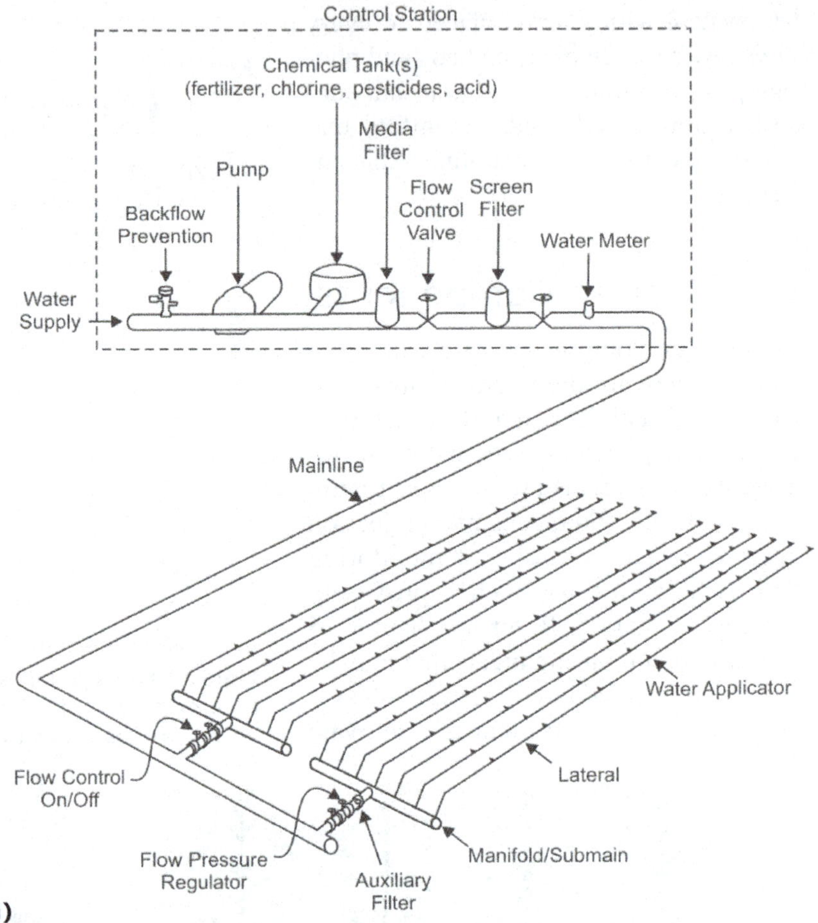

(a)

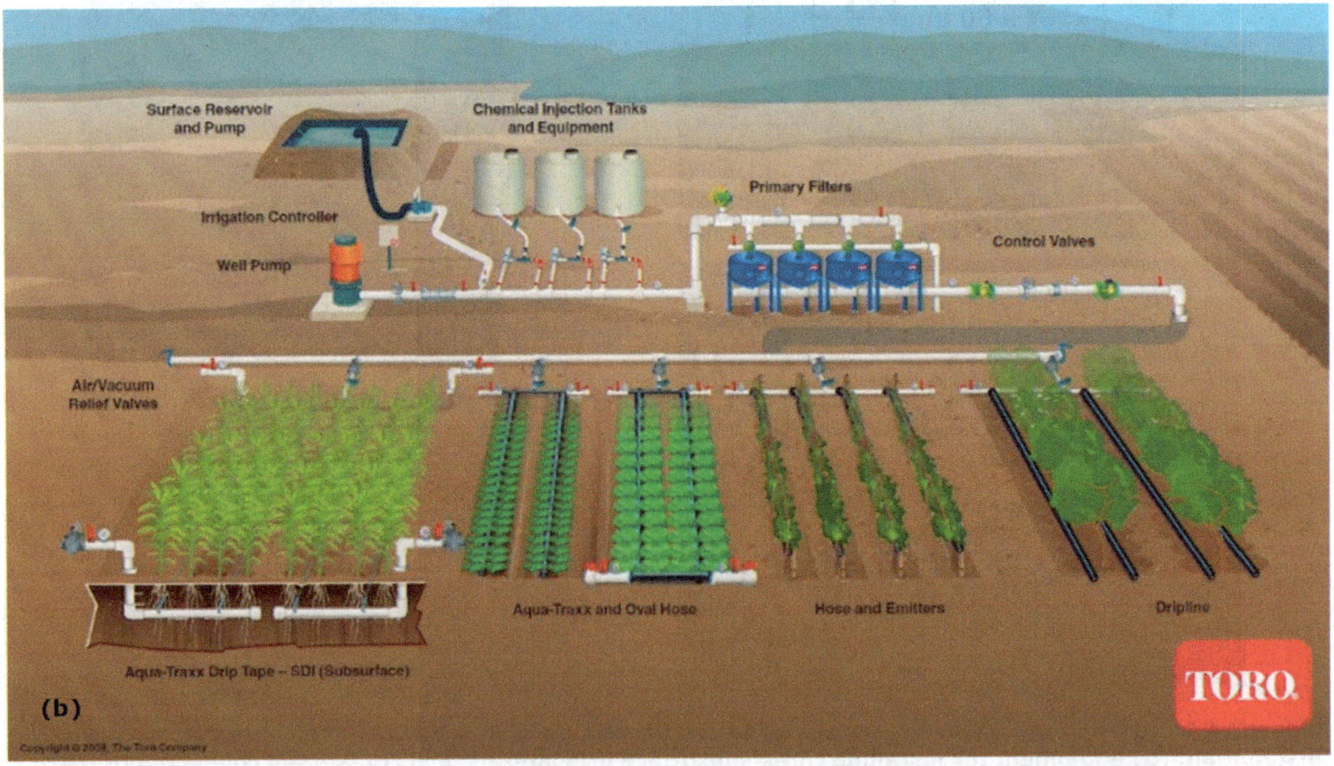

(b)

Figure 14.5. Basic components of (a) a control station and (b) an entire microirrigation system. (Image courtesy of Toro.)

14.4.1 Control Station

A sketch of a typical control station for microirrigation is given in Figure 14.6. A control station may include a pumping unit and controls, a backflow prevention device, water meters, filtration units, chemigation equipment, flow and pressure control devices, and irrigation controls. Depending upon the water source, a pump (no. 2 in Figure 14.6) may or may not be required. For some low-pressure surface drip irrigation applications, water from a small tank, barrel, or bucket can be delivered by gravity (Figure 14.7) and a pump is not needed. When the water source is an open body of water or a well, a pump may be at the control station or at some distance. When the water source is adjacent to the control station, the pump and its controls will typically be part of the control station (no. 2 in Figure 14.6). If the pressure is excessive, a pressure regulator or flow control valve will be required (no. 10 in Figure 14.6).

When an irrigation system is connected to a potential source of drinking water or when chemicals are to be injected into the irrigation water, a back flow prevention unit (no. 4 in Figure 14.6) is required. This device prevents any water and chemicals from flowing back into the water supply (see Chapter 15 for more detail).

Common at the control station is equipment for injecting chemicals into the mainline. Chemicals typically injected are for fertilization, water treatment, or pest control. Chemicals are injected into the irrigation water before the primary filter.

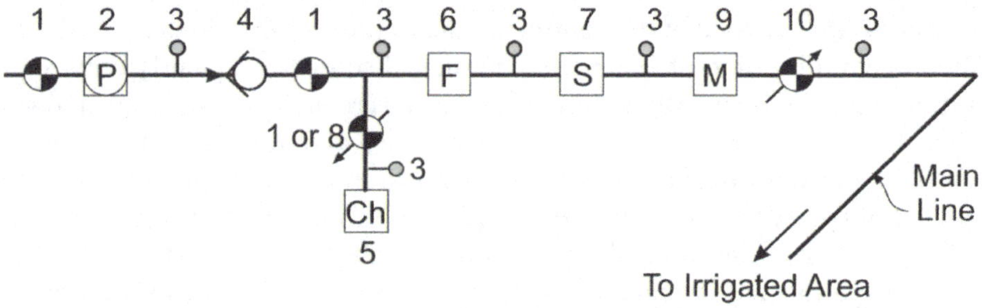

1 - Manual control valve
2 - Pump (not needed if water supply is pressurized)
3 - Pressure gauge
4 - Back flow prevention unit
5 - Chemical injection equipment
6 - Media filter
7 - Strainer or screen filter
8 - Automatic regulating valve
9 - Water meter
10 - Manual or automatic pressure regulating valve

Figure 14.6. Example of an arrangement at the control station for a microirrigation system.

Figure 14.7. Examples of a gravity water supply system for surface drip irrigation on smallholder farms. (Photos courtesy of iDE, International Development Enterprises.)

Filters are installed at the control station to prevent the passage of unwanted particles into the system. Filtration of water from municipal sources frequently only require precautionary filtration that is frequently accomplished by a screen filter. Water taken directly from wells or open bodies of water generally require primary and secondary filtration. The primary filter should be located after the pump and chemical injection equipment. Secondary filters are installed downstream from the primary filter to remove particles which may pass through the primary filter during normal or cleaning operations. The most popular filtration system for microirrigation is a primary media filter (no. 6 Figure 14.6) followed by a secondary screen filter (no. 7 in Figure 14.6). See Figure 14.10 for typical filtration systems.

The last major item at the control station is a water measuring meter (no. 9 in Figure 14.6). Types of water meters used for irrigation systems are described in Chapter 3.

Items in Figure 14.6 labeled as no. 3 indicate possible locations for pressure gauges. For well-designed control stations pressure gauges are placed at nearly all of these locations. Measures of pressure are particularly important on each side of the filters. A significant decrease in pressure on the discharge side of a filter compared to before the filter indicates that the filter is becoming clogged and cleaning is required. Of course, the pressure gauge reading as water leaves the control station will provide assurance that the pressure is appropriate for proper operation of the microirrigation system.

Also important at the control station are strategically located valves. The location of valves depends upon the complexity of the control station and the desire of the irrigator to shut down the system to clean, repair, or replace components.

14.4.2 Mainline and Manifolds

The mainlines and manifolds for example microirrigation systems are illustrated in Figure 14.5. Actual systems may be different and far more complex than the illustration. The mainline carries water from the control station to manifolds which distribute the water to each lateral. Normally, there are no fixtures along the mainline other than elbows or tees. If the system is large and there are a number of manifolds, flow control valves and shutoff valves are located at the head end of each manifold. These fixtures assure the correct flow enters each manifold and accommodates manual or automatic control of irrigation water. The diameter of mainlines and manifolds are normally large, in the range of 1 to 6 inches. To determine the friction loss in plastic pipe, both polyvinyl chloride (PVC) and polyethylene (PE), refer to Chapter 8. The same procedures can be used to determine the appropriate pipe diameter for mainlines and manifolds for microirrigation as described in Chapter 8.

Example 14.1

A mainline is required to convey 200 gpm, a distance of 500 ft from the control station to the microirrigation manifolds. What diameter of PVC pipe would you recommend if the pipeline velocity must be less than 5 ft/s?

Given: Q = 200 gpm
Pipe length = 500 ft
Pipe type = PVC
Velocity in pipe less than 5 ft/s

Find: Smallest pipe diameter recommended

Solution:

From Table 8.5, a 4-inch diameter pipe has a large enough cross sectional area to keep the velocity less than 5 ft/s.

Example 14.2

Determine the smallest diameter polyethylene pipe to be used for a manifold if the flow rate is to be 70 gpm. What will the friction loss be for the manifold if the length is 200 ft?

Given: Q = 70 gpm
Manifold length = 200 ft

Find: Smallest recommended pipe diameter
Friction loss for the manifold

Solution:

From Table 8.5, a 2-inch diameter pipe with a flow of 70 gpm will require a water velocity in excess of 5 fps and is, therefore, not recommended. A 2.5-inch diameter pipe satisfies the velocity requirement of 5 fps and the friction loss would be 0.96 psi/100 ft or a total of 1.92 psi of pressure loss (Table 8.2a). This would probably be acceptable for friction loss in for a manifold.

14.4.3 Laterals

Table 14.1. Friction loss for small diameter PE pipe based upon the Darcy-Weisbach equation for pipe with an *e* (absolute roughness) = 0.0005 in.

Nominal Size (in)	0.5	0.75	1.0	1.5
Inside Pipe Diameter (in)	0.622	0.824	1.049	1.61
Flow Rate, *Q* (gpm)	Friction Loss (psi/100 ft)			
0.5	0.13	0.03		
1.0	0.56	0.15		
1.5	1.13	0.30	0.09	
2.0	1.86	0.49	0.15	
2.5	2.76	0.72	0.23	0.03
3.0	3.81	0.99	0.31	0.04
4.0	6.38	1.64	0.51	0.07
5.0	9.55	2.43	0.76	0.10
6.0	13.3	3.37	1.05	0.14
7.0		4.45	1.38	0.18
8.0		5.67	1.76	0.22
9.0		7.02	2.17	0.28
10			2.62	0.33
15			5.47	0.68
20			9.26	1.14
25				1.71

Typically, emitters are spaced systematically along laterals in microirrigation systems. For row crops where plants are spaced uniformly in short intervals, emitters are spaced uniformly one to a few feet apart. Many laterals are currently manufactured with the emitters within the lateral itself as for single- or dual-chamber emitters (Section 14.4.4). For widely-spaced crops, like trees, emitters may be closely spaced near the tree with no emitters positioned between tree canopies. As the trees grow, additional emitters may be added. These emitter patterns apply for both surface and subsurface systems. Microspray and bubbler type emitters are also common for widely-spaced crops.

Regardless of the emitter type, the flow within the lateral decreases from the beginning of the lateral to zero at the lateral terminus. If the laterals are fairly long, it may be advantageous to decrease the size of the lateral as the flow decreases along the lateral. In most microirrigation systems, however, the laterals are relatively short so only one small-diameter lateral is used. As in Chapter 8, where friction loss and pipe size were determined for various types of irrigation pipe, the same procedure can be used for microirrigation laterals. In Chapter 8, the Hazen-William equation was used to calculate friction loss in pipes. For small-diameter, smooth-walled pipe used in microirrigation (e.g., laterals), the Hazen-Williams equation with a C value of 150 underestimates the friction loss (Keller and Bliesner, 1990). They recommend the Darcy-Weisbach equation for microirrigation laterals as was used in Table 14.1.

The flow rate within the lateral decreases as the flow moves past water applicators; thus, the friction loss changes. When the lateral has uniformly spaced and uniformly discharging outlets, the friction loss can be estimated by:

$$P_L = F\,L\,P_f \tag{14.1}$$

where: P_L = pressure loss due to friction for laterals with uniformly spaced and uniformly discharging outlets,

F = multiple outlet reduction factor (Table 8.3),

L = lateral length, and

P_f = pressure loss per unit length of a conveyance pipe without outlets.

For a pipe with no outlets, $F = 1.0$. There is a slight difference between values of F depending on the distance down the lateral from the manifold to the first outlet. If the spacing between the outlets is s, then the outlet factor is higher when the first outlet is a distance s from the manifold compared to a distance of one-half s for about the first 20 outlets on a lateral. Typical values of F are given in Table 8.3.

There are also minor pressure losses in laterals with emitters caused by flow constrictions for in-line emitters and by barbs for emitters inserted in the tubing. Keller and Bliesner (1990) present a method for estimating losses caused by in-line emitters and emitters with barbed insertions. Their method adds to the effective length of the pipe.

Example 14.3

What is the smallest recommended pipe diameter for a polyethylene lateral that is 200 ft long and has an emitter outlet spacing of 2 ft? Each emitter discharges 2 gallons per hour.

Given: L = 200 ft s = 2 ft

Emitter discharge = 2 gal/h

Assume medium length insertion barbs

Polyethylene pipe

Find: Smallest pipe diameter recommended

Solution:

Number of outlets, n = 200 ft/2 ft = 100 outlets

$F = 0.35$ (Table 8.3)

$Q = n$(2 gal/hr)

Q = 100(2 gal/hr) = 3.3 gpm

For d = 0.5 in: P_f = 4.7 psi/100 ft (Interpolated from Table 14.1)

Extra lateral length due to inserted barbs = 30 feet

For d = 0.75 in: P_f = 1.2 psi/100 ft (Interpolated from Table 14.1)

Extra lateral length due to inserted barbs = 20 feet

$P_L = F\,L\,P_f$ (Eq. 14.1)

For d = 0.5 in: P_L = 0.35(230 ft)(4.7 psi/100 ft)

P_L = 3.78 psi

For d = 0.75 in: P_L = 0.35(220 ft)(1.2 psi/100 ft)

P_L = 0.92 psi

If the design pressure in the lateral is 15 psi and the lateral is level, a maximum of 3 psi pressure loss would be acceptable if the criteria is that the allowable pressure variation be less that 20% of the average pressure. Thus the 0.75-inch tubing would be necessary.

Example 14.4

If microsprayers with a discharge rate of 0.5 gpm at a spacing of 8 ft were substituted for the emitters in Example 14.3, what would be the minimum recommended pipe diameter?

Given: L = 200 ft

Microspray discharge = 0.5 gpm

Assume medium length insertion barbs

Polyethylene pipe

Find: Smallest recommended pipe diameter

Solution:

n = 200 ft/8 ft = 25 microsprayers

$F = 0.365$ (Table 8.3)

$Q = n$(0.5 gal/hr) = 25(0.5 gal/hr) = 12.5 gmp

For d = 1.0 in: P_f = 4.0 psi/100 ft (Interpolated fromTable 14.1)

For d = 1.5 in: P_f = 0.51 psi/100 ft (Interpolated from Table 14.1)

Extra length due to inserted barbs (both tubing sizes) = 5 feet

$P_L = F\,L\,P_f$ (Eq. 14.1)

For d = 1.0 in: P_L = 0.365(205 ft)(4.0 psi/100 ft)

P_L = 2.99 psi

For d = 1.5 in: P_L = 0.365(205 ft)(0.51 psi/100 ft)

P_L = 0.38 psi

Using the same criteria as Example 14.3, the pressure loss of 2.99 psi in the 1-inch lateral is acceptable but only by a small amount.

14.4.4 Water Applicators

For microirrigation, adequate pressure must be maintained in the pipelines to overcome friction losses and elevation differences to distribute water throughout the field. Once the point of application is reached, the difference in pressure inside the lateral and the atmosphere must be dissipated by a water applicator device.

There are three common types of applicators: emitters, line-source tubing (drip tape and porous tubing), and sprayers. Many different emitter designs have been devised and manufactured with the requirements that the emitters be inexpensive and reliable. Emitters are designed to dissipate pressure while discharging small uniform flows of water at a constant rate. They are often classified according to the mechanism used to dissipate pressure. Long-path emitters have a long capillary-size tube or channel to dissipate pressure. Orifice emitters rely on an individual opening or a series of openings. Vortex emitters dissipate pressure by creating a whirling or circular motion that tends to form a cavity or vacuum in the center of the swirling.

Many emitters are designed with additional features. Some are designed to provide a flushing flow of water to clean the discharge opening every time the system is turned on. Continuous-flushing emitters permit the passage of relatively large particles while operating. Another special feature is pressure-compensating emitters which discharge water at a constant rate over a wide range of pressure. Some emitters have multiple outlets and supply water through small diameter auxiliary tubing at various points. Examples of various types of emitters are illustrated in Figure 14.8.

There are three types of line-source tubing, all of which are normally less than 1-inch in diameter. The wall thickness of tubing is available from 0.004 to 0.025 inches. The thin wall tubing is frequently discarded after each crop. The most common wall thickness is 0.008 to 0.010 inches. Recommended operating pressures depend on wall thickness. Several manufacturers recommend a maximum continuous operating pressure of 15 pounds per square inch (psi) for tube walls that are 0.015 inches thick, and 8 to 12 psi for thinner walls. Porous wall tubing is constructed of porous material with pores of capillary size that ooze water when under pressure. Single chamber tubing has orifices punched through the hose wall or emitters fabricated or inserted at fixed intervals along the hose. Double chamber tubing has both a main and an auxiliary passage. Widely spaced inner orifices are punched through the wall common to both passages. Typically, 3 to 6 exit orifices are punched at short intervals in the outer wall of the auxiliary passage for each inner orifice. Sketches of line-source tubing are provided in Figure 14.8. Early developments included double-chamber tubing and porous-wall tubing (illustrated in Figure 14.9).

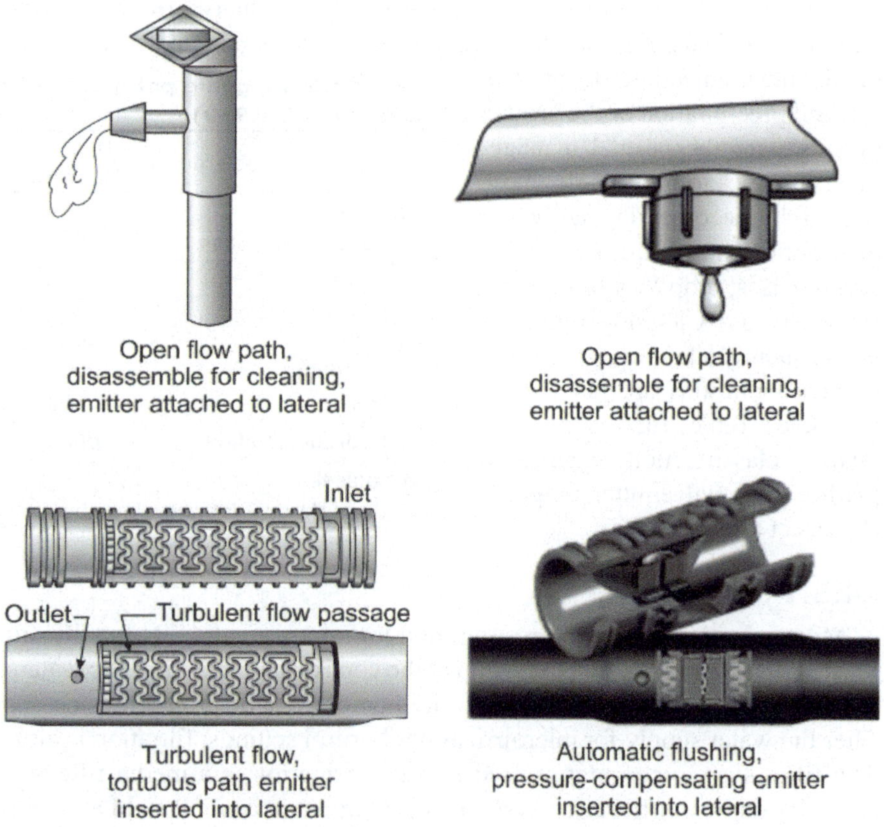

Figure 14.8. Example of emitters with various designs and features for line-source microirrigation laterals (images courtesy of Toro).

Sprayers are designed to discharge a small spray of water to cover an area of 10 to 100 ft^2. Aerosol emitters, foggers, spitters, misters, microsprayers, or miniature sprayers are examples of spray devices. Ideally, sprayers apply a relatively uniform depth of water throughout its wetted area and should have a low water trajectory angle.

The construction and materials used for water applicators are very important because they are exposed to sunlight, chemicals in and applied with the irrigation water, extremes in environmental conditions, and physical abuse. Emitter performance is a dominant factor in the uniformity of water applications and the life expectancy of the system.

Figure 14.9. Porous-wall tubing, an example of line-source tubing for microirrigation systems.

14.5 Preventing Clogs

Clogging of emitters is one of the major concerns for microirrigation. Obviously, the smaller the orifice or the longer the capillary section, the more prone the emitter is to clogging. Emitters can be clogged by particles, bacterial slimes, algae, water-borne organisms, or precipitation of various chemicals. Filtering to prevent mineral and organic particles from entering the system or chemical injections to prevent mineral precipitation or the growth of slime are the normal management schemes to prevent clogging. If emitters become clogged, water distribution is not uniform and in severe cases, crop loss from water stress occurs. Clogging problems are frequently site specific and economical solutions are not always available. Table 14.2 is a water quality classification scheme to predict potential emitter plugging (Bucks et al., 1979).

Table 14.2. Plugging potential of irrigation water for microirrigation (Bucks et al., 1979).

Potential Problem	Unit of Measure	Minor	Moderate	Severe
Physical				
Suspended solids	ppm	< 50	50–100	> 100
Chemical				
pH	-	< 7	7–8	> 8
Salts	ppm	< 500	500–2000	> 2000
Manganese	ppm	< 0.1	0.1–1.5	> 1.5
Iron	ppm	< 0.1	0.1–1.5	> 1.5
Hydrogen sulfide	ppm	< 0.5	0.5–2.0	> 2.0
Biological				
Bacteria populations	number/ml	< 10,000	10,000–50,000	> 50,000

14.5.1 Filtration

Most irrigation water requires filtration for microirrigation. Normally, filtration equipment is located just downstream of the pump at the control station. Domestic water, particularly municipal supplies, are already filtered so the homeowner or proprietor does not normally have to filter the water supply for microirrigation. In rural settings, filtration is almost always required. The filter system commonly used in microirrigation is a media filter followed by a screen filter. Examples of filtration systems for a large and a small field are shown in Figure 14.10. If the irrigation water has a heavy sand load, the water should pass through a sand separator or a settling basin before passing through media and screen filters.

Suspended particles that might plug a system can be either inorganic or organic. Algae, bacteria, diatoms, larvae, fish, snails, seeds and other plant parts are the major organic solids while sand and soil particles are the primary inorganic solids. Because a consistently clean water supply is vital, filtration and chemical treatment must be furnished for the worst possible conditions. In a few cases, chemical coagulants are required to control silt, clay, or suspended colloids. Chlorine may be required sometimes to control algae and other organic materials.

Well water is usually low in organic materials, but it can contain sand. Therefore, a screen filter is frequently adequate. Irrigation water may be saline or be chemically unstable thereby producing chemical precipitates. In some cases, water supplies contain chemical constituents that provide nutrients for bacterial growth. For these waters, chemical treatment is required.

The size of particle that can be tolerated by a water applicator should be indicated by the manufacturer because it depends on applicator construction. Typically, the recommendation is to remove all particles larger than one-tenth the diameter of the orifice or flow passage of the emitter. This is necessary because particles may become grouped and bridge the passageway. Many manufacturers recommend removing particles larger than 0.003 to 0.006 inches in diameter.

(a)

(b)

Figure 14.10. Typical control station and filtration system of media filters for (a) a subsurface drip system for a large field in Nebraska (photo courtesy of Laszlo Hayde, IHE Delft Institute for Water Education), and (b) a microirrigation system for a small field in India (image from Indiamart, https://www.indiamart.com/proddetail/drip-sprinkler-irrigation-system-20348028048.html).

In addition to the main filtration system, small screen filters should be installed at the inlet to each lateral or manifold as a precaution against plugging. These auxiliary screens prevent debris from entering the system when the main filters are cleaned or if breaks or openings occur in the distribution system.

Fine particles settle out when flow is slow or stops. The clogging that results may not be rapid, but it is inevitable. As a safeguard, either manual or automatic flushing devices should be installed at the end of each lateral. These protective devices are particularly important to clean the system after installation and repair.

Settling basins, ponds, or reservoirs can remove large quantities of sand and silt. They should be long and narrow with water discharged into the basin at one end and removed from the opposite end to provide settling time for the suspended materials. If water remains in the basin for at least 15 minutes, most inorganic particles larger than about 0.003 inches will settle out.

About 98% of the sand particles intercepted by a screen with 0.003-inch openings can be removed by a vortex separator. Centrifugal force is the principal employed by a vortex separator to remove high-density particles from the water. Organic materials, however, cannot be removed by this method because they have low density.

Media filters are used frequently in microirrigation systems. The filter consists of fine gravel and sand of selected sizes placed in graded layers inside a cylindrical tank (Figure 14.10). These filters are very effective in filtering inorganic and organic materials because they can be trapped throughout the depth of the media bed. Long, narrow particles, such as algae and diatoms, are more likely to be caught in the multilayered media bed than on the surface of a screen.

A drop in pressure of 2 to 3 psi occurs from the inlet to the outlet of a clean media filter. As the pores of the media become plugged with contaminants, the pressure drop increases. It is normally recommended that the media filter be flushed to remove the accumulated contaminants when the pressure drop reaches 10 psi. If the water is relatively clean and flushing is not needed frequently, manual flushing may be suitable. Where frequent cleaning is required, automatic flushing can be actuated by a timer or by sensing the pressure differential across the media filter.

Where suitable, screen mesh filters provide a simple and efficient means for filtering. Hole size and the total amount of open area in the screen determine a screen filter's efficiency and operational limits. Screen filters are used to remove fine sand or small amounts of algae. They are commonly used where the water is expected to be clean, i.e., pumped groundwater, municipal supplies, and following other filter systems.

Screen filters differ by their configuration for cleaning. The need for cleaning, as with media filters, is determined by the rate at which the filter clogs. This rate of plugging is normally monitored by the drop in water pressure across the filter. It is customary to clean screen filters whenever the pressure difference between the inlet and outlet to the filter is between 3 and 5 psi. Manual cleaning by opening the filter, removing the screen, and washing it is satisfactory when cleaning is not required frequently. If frequent cleaning is required, an automatic flushing system is normally installed. Back flushing, blow down, and gravity flow are examples of configurations for automatic cleaning. The flow of water is reversed in a backflushing filter to remove the collected materials. A high velocity jet of water is run over the screen to sweep away collected particles without opening the filter for blow down filters. A gravity flow filter functions by discharging the water supply onto and through a large screen before pumping it into the irrigation network. Some gravity flow filters have jets under the screen to lift particles and move them off the screen.

The screening material can be constructed of stainless steel, nylon, polyester, or other noncorrosive materials. A stainless steel screen offers strength. Nylon mesh in some blow down filters flutters during flushing which aids to dislodge collected particles.

The flow rate through a screen filter should not exceed 200 gallons per minute (gpm) per square foot of screen open area. The wire or plastic mesh itself obstructs much of the open area. For example, a screen constructed of stainless steel with 0.003-inch openings has 58% open area. An equivalent nylon mesh with the same size openings has only 24% open area. Thus, it is important to consider the actual open area of a screen when sizing a filter.

The total area of screen (A_s) needed for a screen filter can be calculated from:

$$A_s = \frac{Q / Q_m}{O_a} \tag{14.2}$$

where: Q = flow rate through the filter,
Q_m = minimum flow rate permissible per unit area, and
O_a = fraction of open area within the screen.

Example 14.5

A golf course manager has access to a municipal water supply and needs only a screen filter to protect a microirrigation system. If the system is designed for a maximum of 300 gpm and a stainless steel screen with 0.003-inch openings is to be used, what area of screen will be required?

Given: Q = 300 gal/min

Screen opening = 0.003 in

Find: Total filter screen area (A_s)

Solution:

Maximum permissible flow rate (Q_m) is 200 gal/min per ft² of screen open area

The fraction of open area of a stainless stell screen (O_a) with 0.003 in openings is 58% of the total screen area (A_s) is:

$$A_s = \frac{Q / Q_m}{O_a} = \frac{\left(300 \frac{\text{gal}}{\text{min}} / 200 \frac{\text{gal}}{\text{min ft}^2}\right)}{0.58}$$

$$A_s = 2.6 \text{ ft}^2$$

14.5.2 Precipitation of Dissolved Solids

Dissolved solids are a problem when they precipitate as a solid mineral or serve as a source of nutrients for algae and bacterial slime. Clogging and eventual plugging of water applicators by precipitates and organic deposits are problems that cannot be solved by filtration. Precipitates form inside pipes or emitters as a result of changing pH or temperature, but a major cause of mineral deposits is evaporation of water at the outlet of the water applicators during nonirrigation periods.

Calcium and iron precipitates are a common problem with many well waters. An analysis of the water can indicate if the bicarbonate (the typical source of calcium) or iron concentration is high enough to cause precipitation. A bicarbonate concentration greater than 2 meq/L (120 ppm) coupled with a pH greater than 7.5 is likely to produce calcium deposits. Injecting inexpensive acid to lower the pH to between 5.5 and 7.0 effectively prevents calcium precipitation. Acid treatments at the end of each irrigation or on a periodic basis is frequently practiced to reduce costs. Typically, acid is injected at roughly 0.02 to 0.2% of the system capacity. If more acid than 0.2% is required to lower pH where bicarbonate concentrations are high, it is generally more practical and less expensive to aerate the water and hold it in a reservoir until it reaches chemical equilibrium and the precipitate settles out rather than adding acid.

As little as 0.3 ppm of iron present in the soluble ferrous form in the irrigation water can cause precipitation in a microirrigation system. In the presence of oxygen, the iron oxidizes to the insoluble ferric form which causes a reddish-brown precipitate. If iron is a potential hazard, it should be precipitated deliberately and filtered out before the water enters the microirrigation system. A chemical feeder can be set to provide a measured volume of a chlorine solution to oxidize iron and other organic compounds present. A residual chlorine concentration of 1.0 ppm is normally provided to avoid precipitation. Sodium hyperchlorite is preferred over calcium hyperchlorite as a source of chlorine because of the potential for calcium precipitation. Where iron concentrations are as high as 10 ppm, aeration by a mechanical aerator and sufficient settling time in a reservoir is another practical method of controlling iron.

14.5.3 Organic Materials

Algae and slime created by bacteria can cause severe clogging. Algae is common in almost all surface waters. Small pieces of algae can pass through filters and grow inside a microirrigation system. Since sunlight is required for algae to grow, light must be prevented from entering the system.

Slime is a general term for long filament microorganisms produced primarily by bacteria. The slime acts as a "glue" for suspended particles to combine into larger particles that plug emitters. The more common microorganisms that result in slime problems are airborne, thus, systems using open water supplies are susceptible.

Both algae and slime can be controlled by chlorination. Maintaining a residual chlorine concentration of 1.0 ppm, measured at the far end of the system, usually prevents problems from organic materials. An alternative practice is to inject sufficient chlorine to bring the concentration in the irrigation system to between 10 and 20 ppm during the last 20 minutes of the irrigation cycle.

14.5.4 Flushing and Maintenance

Flushing is an important part of starting up the system after installation and for maintaining performance. After installation or repairs, the system should be flushed to remove any foreign materials. To ensure adequate flushing, valves placed at the ends of all pipelines should be opened momentarily. Flushing provisions, which can be very inexpensive, should also be placed on every lateral.

In addition to flushing periodically, adequate maintenance requires that filters be cleaned routinely to ensure that water applications are uniform and appropriate to meet crop water requirements. In addition to main filters, all secondary filters at inlets to manifolds and laterals must also be cleaned routinely. A water velocity of at least 1 foot per second is recommended to flush fine particles.

In addition to all these precautions, systematic field checks are required to detect malfunctioning emitters. Emitter discharge may be altered by blockage or wear of the emitter parts. Discharge from emitters should be checked periodically to maintain uniform applications.

14.6 Uniformity

The primary objective of a microirrigation system is to supply the prescribed amounts of water and chemicals to each plant at frequent intervals and in small volumes. For maximum uniformity the variation in discharge from the various water applicators must be acceptably low. Uniformity of manufacturing is critical when selecting emitters for a microirrigation system. Uniformity is also crucial in evaluating the performance of the system following installation and periodically during the life of the system.

14.6.1 Emitter Discharge

Flow variation among emitters is caused by differences in hydraulic, manufacturing, and field conditions. The discharge (q_e) in gal/hr for most emitters can be described by:

$$q_e = Kh^x \tag{14.3}$$

where h is the pressure head in feet at the emitter. The emitter discharge coefficient, K, contains the effects of the coefficient of discharge, emitter geometry, and the acceleration of gravity. The value of x, the emitter discharge exponent, characterizes the type and flow regime of the emitter. Orifice-type emitters are fully turbulent and have an emitter discharge exponent of 0.5. With long path emitters, $x = 0.5$ for those with fully turbulent flow and 1.0 for laminar flow. An x value of less than 0.5 indicates an emitter that compensates for changes in pressure.

To determine K and x for an emitter, the discharge must be measured at two different operating heads (h_1 and h_2). The x may be determined analytically from:

$$x = \frac{\log(q_{e1} / q_{e2})}{\log(h_1 / h_2)} \tag{14.4}$$

The value of x calculated from Equation 14.4 is used to calculate K from Equation 14.3.

It is impossible to manufacture any two items exactly alike. Very small variations in emitter passage size, shape, and surface finish can result in variations in discharge. The amount of variation also depends on emitter design, construction materials, and precision during manufacturing.

The coefficient of manufacturing variation for an emitter, v, is a measure of anticipated variations in the discharge for a sample of new emitters. The value of v should be available from the manufacturer. If not available, it can be determined from the discharge data of a sample set of at least 50 emitters operating at a constant reference pressure by:

$$v = \frac{\left[\left(q_1^2 + q_2^2 + \ldots + q_n^2 - n q_a^2\right)/(n-1)\right]^{\frac{1}{2}}}{q_a} \qquad (14.5)$$

where: n = number of emitters being tested,
q = discharge rate of an emitter, and
q_a = average emitter discharge rate.

For an emitter having a v of 0.06 and a q_a of 1 gal/hr, 95% of the emitters will have a discharge rate between 0.88 and 1.12 gal/hr. As a general guide, manufacturing variability can be classified in accordance with Table 14.3. A lower standard is given for line-source tubing because it is difficult to keep both the variation and price low. Line-source tubing is normally used for row crops which are relatively insensitive to moderate variations in discharge among closely spaced outlets.

Example 14.6

Determine the discharge exponent and the discharge coefficient for a vortex emitter.

Given: From laboratory measurements:
q_e is 0.75 gal/hr when h = 15 ft
q_e is 1.0 gal/hr when h = 30 ft

Find: Discharge exponent (x) and the discharge coefficient (K)

Solution:

$$x = \frac{\log(q_{e1}/q_{e2})}{\log(h_1/h_2)} \qquad \text{(Eq. 14.4)}$$

$$x = \frac{\log(0.75/1)}{\log(15/30)} = \frac{\log(0.75)}{\log(0.50)} = \frac{-0.125}{-0.301}$$

$$x = 0.42$$

$$q_e = Kh^x \qquad \text{(Eq. 14.3)}$$

Solving for K yields $K = \frac{q_e}{h^x}$

$$K = \frac{1}{30^{0.42}} = \frac{1}{4.2}$$

Table 14.3. Classification for manufacturing variation, *v*, of emitters (Solomon, 1979).

Classification Rating	Drip & Spray Emitters	Line-Source Tubing
Excellent	$v < 0.05$	$v < 0.1$
Average	$0.05 < v < 0.07$	$0.1 < v < 0.2$
Marginal	$0.07 < v < 0.11$	-
Poor	$0.11 < v < 0.15$	$0.2 < v < 0.3$
Unacceptable	$0.15 < v$	$0.3 < v$

14.6.2 Discharge Versus Pressure

The relationship between pressure head (h) and discharge (q_e) is an important characteristic of emitters (Equation 14.3). Figure 14.11 shows this relationship for various types of emitters. The discharge exponent, x, measures the flatness of the relationship between pressure and discharge. It shows clearly the desirability of an emitter that has a low value of x. Emitters that compensate for changes in pressure have the lowest values of x. Compensating emitters have some physical part that responds to pressure to keep discharge constant. Although having the advantage of compensating for changes in pressure, these emitters are prone to material fatigue and temperature change.

On hilly terrain the design of a highly uniform system is constrained by the sensitivity of the flow from emitters because of pressure differences in the laterals from changes in elevation. Pressure compensating emitters and pressure regulated flow in

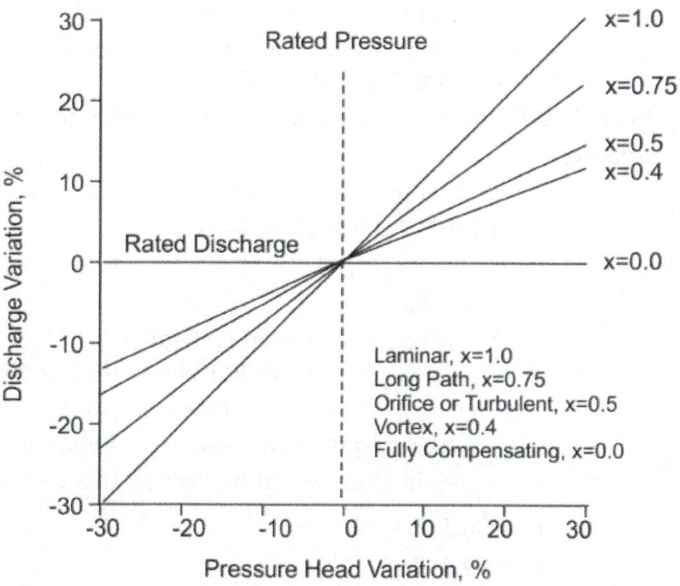

Figure 14.11. Discharge variation resulting from pressure changes for emitters having different discharge exponents. (Image courtesy of Keller and Bliesner, 1990, with permission from Blackburn Press.)

short laterals provide potential solutions. Even on level fields, the lateral length must be kept reasonably short to avoid excessive differences in pressure along the lateral.

For laminar flow emitters with a value of *x* near 1.0 the percent variation in pressure head results in about the same variation in discharge. Thus, variations in pressure throughout the system with laminar flow emitters should be held within about ±5% of the desired pressure to maintain water applications within acceptable limits.

For turbulent flow emitters, the change in discharge varies with the square root of pressure head; *x* is near 0.5. Consequently, to double the flow, the pressure must be increased four times. Thus, the pressure head for systems using turbulent flow emitters can vary up to ± 10% of the

Table 14.4. Characteristics of various types of emitters (Keller and Bliesner, 1990).

Emitter Type	Discharge Exponent, x	Coefficient of Manufacturing Variation, v	Flushing Ability
Orifice			
Vortex/orifice	0.42	0.07	None
Multiple flexible orifices	0.7	0.05	Continuous
Ball & slotted seat	0.50	0.27	Automatic
Compensating ball & slotted seat	0.25	0.09	Automatic
Capped orifice sprayers	0.56	0.05	None
Long-Path			
Small tube	0.70	0.05	None
Spiral path	0.75	0.06	Manual
Compensating	0.40	0.05	None
Compensating	0.20	0.06	Automatic
Tortuous	0.65	0.02	None
Short-Path			
Groove & flap	0.33	0.02	Automatic
Slot & disc	0.11	0.10	Automatic
Line-Source			
Porous pipe	1.0	0.40	None
Twin chamber	0.61	0.17	None

Example 14.7

The vortex emitter described in Example 14.6 is being considered for a microirrigation system to be installed in a field 1,000 feet long and sloping 1% up from south to north. The header for lateral lines is placed along the south edge of the field and the operating pressure in the header line is 15 psi. Assuming no pressure loss in the header or laterals due to friction loss, what is the difference in discharge for emitters near the header and at the north edge of the field?

Given: $K = 0.24 \qquad x = 0.42$

Pressure at south edge of field = 15 psi

Pressure loss at north edge of field due to elevation difference = 1,000 ft × 1% slope =1,000 ft × 0.01 = 10 ft

Find: Difference in discharge between south and north end of field

Solution:

$h_s = h$ at south end of field

$h_s = 15$ psi (2.3 ft/psi) = 34.5 ft

$q_{es} = Kh^x = 0.24(34.5)^{0.42} = 1.06$ gal/hr

$h_n = h_p + h_e + h_f + h_s$

where h_p = pressure head,

h_e = change in head due to elevation, and

h_f = loss in head due to friction loss

h_e is negative because the elevation is increasing in this example

h_f is assumed to be zero in this example

$h_n = 34.5$ ft − 10 ft − 0

$h_n = 24.5$ ft

$q_{en} = Kh^x = 0.24(24.5)^{0.42} = 0.92$ gal/hr

$\Delta q = q_{es} - q_{en} = 1.06$ gal/hr − 0.92 gal/hr

$\Delta q = 0.14$ gal/hr

desired pressure without unacceptable variations in water applications.

Flow compensating emitters provide some degree of flow regulation as pressure changes. When x is between 0.2 and 0.35, some regulation is possible and there is still some flexibility for adjusting the discharge rate. Compensating emitters are valuable on hilly sites where it is impractical to design for uniform pressure along the laterals.

Example 14.8

If a compensating, long path emitter without flushing ability is used in Example 14.5 rather than vortex emitters, what will the difference in discharge be between the north and south edges of the field?

Solution:

$\Delta q = 0$, since the emitter chosen is pressure compensating.

Characteristics of various types of emitters are given in Table 14.4. Refer to Figures 14.8 and 14.9 for examples of the types of emitters described in Table 14.4. The exponent, x, for Equation 14.3 is given in Table 14.4 along with typical values for the manufacturers' coefficient of variation. The remaining column in Table 14.4 indicates the flushing potential built into each type of emitter.

14.6.3 Emission Uniformity

Emission uniformity can be treated like distribution uniformity, DU, in Chapter 5 and is a measure of the uniformity of emissions from all the water applicators within the entire microirrigation system. For field tests,

$$DU = \frac{q_{LQ}}{q_a} 100\% \qquad (14.6)$$

where DU is the emission uniformity from a field test, %; q_{LQ} is the average discharge for the lowest one-fourth of the field measured emitter discharges (gal/hr); and q_a is the average discharge of all the emitters checked in the field (gal/hr).

The efficiency of an irrigation system is the relation between gross irrigation amounts and the net addition of water to the crop root zone. Distribution uniformity and the various sources of water loss that occur during the operation of the system are the two components of microirrigation efficiency. To estimate the distribution uniformity for a proposed design, the following formula was developed (Karmeli and Keller, 1975):

$$DU = 100\left(1.0 - 1.27\frac{v}{\sqrt{n}}\right)\frac{q_m}{q_a} \qquad (14.7)$$

where DU is design emission uniformity in %, v is the coefficient of manufacturing variation (Table 14.3 for typical values) and n is the number of emitters. The ratio q_m/q_a expresses the relationship between the minimum (q_m) and the average (q_a) discharges resulting from pressure variations within the system. The factor $(1.0 - 1.27\, v/\sqrt{n})$ adjusts for the additional nonuniformity caused by anticipated manufacturing variations between individual emitters.

Example 14.9

Determine the distribution uniformity for a microirrigation system designed for an average pressure of 20 psi. The system is being designed for spiral, long path emitters with automatic flushing. The field is flat and the friction head loss is a maximum of 10 ft. The system is being designed for 2 emitters per plant and the average design emission rate is 1.0 gallon per hour.

Given: $h_a = 20 \text{ psi} \times 2.31 \text{ ft/psi} = 46.2 \text{ ft}$

$x = 0.75$ and $v = 0.06$ (Table 14.4)

$q_a = 1.0$ gal/hr

Δh_f = change in pressure due to friction loss = 10 ft

h_e = change in pressure due to elevation = 0

$n = 2$

Find: The distribution uniformity, DU, for this design

Solution:

$$q = Kh^x \qquad \text{(Eq. 14.3)}$$

$$DU = 100\left(1.0 - 1.27\frac{v}{\sqrt{n}}\right)\frac{q_m}{q_a} \qquad \text{(Eq. 14.7)}$$

$$K = \frac{q}{h^x} = \frac{1}{46.2^{0.75}} = \frac{1}{17.72} = 0.056$$

$$q_m = Kh_m^{\ x}$$

$$h_m = h_p + h_e - h_f = 46 + 0 - 10$$

$$h_m = 36 \text{ ft}$$

$$q_m = 0.056(36)^{0.75}$$

$$q_m = 0.82 \text{ gal/hr}$$

$$DU = 100\left(1.0 - 1.27\frac{0.06}{\sqrt{2}}\right)\frac{0.82}{1.00}$$

$$= 100(1.0 - 0.5)(0.82)$$

14.7 Management

The success of any irrigation system, particularly microirrigation, depends on management. Irrigating by small quantities frequently is quite different from sprinkler and surface irrigation methods where larger, less frequent applications are normal. With microirrigation, precise information on crop water requirements is required to determine the appropriate irrigation amount. Feedback information on soil water or plant water status is frequently used to schedule irrigations for microirrigation systems.

14.7.1 Wetted Area

A major difference among irrigation systems for agronomic crops is the portion of the soil surface wetted. Most microirrigation systems wet only a portion of the cross-sectional area of the soil profile, as depicted in Figure 14.2. The percent of the surface area wetted, P_w, by microirrigation systems compared to the entire cropped area, depends on the volume and rate of discharge at each application point, the spacing of water applicators, and soil type. No best- or minimum-wetted area percentage has been found, but systems having high P_w values provide more stored water, which is a valuable advantage in the event of system failure. A reasonable design objective for widely-spaced crops such as vines, bushes, and trees is to wet between one-third and two-thirds of the soil surface dedicated to each plant. In regions that receive considerable supplemental rainfall, values of P_w less than one-third are acceptable for fine-textured soils. Maintaining P_w below two-thirds for widely-spaced crops maintains dry strips for cultural practices. In closely-spaced row crops with the laterals in every or every other crop row, Pw approaches full coverage.

Spray emitters wet a larger surface area than drip emitters. They are often used on coarse-textured soils where wetting a large surface area would require a large number of drip emitters. Figure 14.2 shows a comparison of wetting profiles for drip and spray emitters.

14.7.2 Salinity

Microirrigation has potential advantages where the soil or irrigation water is saline. The principal advantage is that with microirrigation the water content of the root zone is maintained high and nearly constant. As a result, the salt concentration of the soil solution is low and steady, and thereby not creating as much salt stress as a system where the soil dries between irrigations with congruent significant increases in salt concentration.

A potential disadvantage is the uneven salt distribution in the soil profile. Refer to Figure 14.2 to see comparable salt profiles for various irrigation methods. This uneven distribution can cause problems if the irrigation system fails and the crop roots begin extracting water from areas of high salt concentration, or if salts that are shallow in the profile are flushed into the root zone by rainfall. When used on annual crops, moving the plant row spacing geometry can cause problems if the salts have not been leached.

14.7.3 Water Requirements

The plant canopies of young or widely-spaced crops shade only a portion of the soil surface area and intercept only a portion of the incoming solar radiation. Conventional estimates of water requirements of young crops assume a portion of the applied water will be lost to non-beneficial consumptive use. This loss is through evaporation from the wetted soil surface or through transpiration from undesirable vegetation. Most microirrigation systems reduce evaporation losses to a minimum, so transpiration by the crop accounts for practically all of the water consumed.

Assuming that evaporation during application is minimal and no upward flow from groundwater into the root zone, the gross irrigation requirement can be expressed as:

$$d_g = ET + d_p + d_r - P_e - \Delta S \tag{14.8}$$

where d_g = gross irrigation requirement,
ET = evapotranspiration,
d_p = deep percolation,
d_r = runoff
P_e = effective precipitation, and
ΔS = the change in soil water storage.

All terms are normally expressed in units of depth. The volume equivalent for each term in Equation 14.8 is the product of the irrigated area and each term. Thus, if d_g was 2 inches and the irrigated area was 5 acres, the volume of water needed would be 10 acre-inch or 271,540 gallons.

One of the objectives of microirrigation is to maintain soil water content constant. If this is achieved, ΔS in Equation 14.8 is zero. For well managed microirrigation systems, runoff should be zero. If salinity is not a hazard, then the irrigation requirement does not need to include water for drainage (deep percolation).

Irrigation scheduling involves two primary decisions: when to irrigate (timing) and how much to apply (amount). Microirrigation inherently implies frequent irrigations. Depending on the system and the sophistication of the controls, irrigation frequency can be from once in several days to multiple times every day. Many commercial systems operate daily or every other day. The operational time for a microirrigation system should not exceed 20 hours per day. In case of repair or maintenance requirements, time is required to catch up. This is particularly critical during periods of peak crop water use.

One of the primary design considerations in microirrigation is determining how many emitters are required to meet the irrigation requirement. The number required can be determined by:

$$n = \frac{d_g A}{q_e \times t} \tag{14.9}$$

where: n = number of emitters,
d_g = applied (gross) depth of irrigation required,
A = irrigated area,
q_e = emitter discharge, and
t = application time.

For surface and subsurface drip systems the spacing between emitters can be specified to the manufacturer. Typical drip emitter spacings are from one to several feet. The spacing between emitters will depend upon the spacing between laterals, the irrigation requirement, and water availability. For example, if you wish to drip irrigate a field of tomatoes and the rows are 3 feet apart, you might place a lateral along each crop row or midway between adjacent rows. If you placed a lateral in every row and the irrigation requirement was 0.27 inches per day, you wanted to irrigate 1 hour per day, and you chose an emitter with a discharge of 2 gal/hr, then Equation 14.9 could be used. To determine the area to be irrigated by each emitter solve Equation 14.9 for A as illustrated in Example 14.10.

Example 14.10

Determine the area each emitter would irrigate in a trickle irrigated tomato field with one drip lateral serving every row and one lateral between two rows.

Given: d_g = 0.27 in
t = 24 hr
q_e = 2 gal/hr

Find: Area irrigated by each emitter with one drip lateral in each row

Solution:

Rearranging Equation 14.9 with n = 1:

$$A = \frac{q_e \times t}{n d_g} = \frac{2\text{gal/hr}/7.48\text{gal/ft}^3 \times 1\text{ hr}}{1 \times 0.27\text{in}/12\text{in/ft}} = 12\text{ ft}^2$$

Thus for 1 lateral per row each emitter will irrigate 12 ft². With 1 lateral per two rows that answer would be 6 ft² per emitter.

14.8 Summary

Microirrigation delivers low rates of irrigation through a wide variety of available water applicators. Application can be by surface drip, subsurface drip, microspray, or bubbler. Each type is illustrated and described. The components of a typical microirrigation system includes a control station, a mainline, manifolds, and the lateral lines that supply water to the water applicators. The necessity and selection of equipment at the control station are presented along with the procedures to select the appropriate diameter of the various sections of the pipeline. A major concern with microirrigation is the potential for clogging the emitters. The types of filters normally recommended are described and the types of materials that lead to clogging are discussed. The chapter concludes with procedures to determine the proper number of water applications required for various crop conditions.

Questions

1. What environmental and/or economic factors give microirrigation an advantage when selecting an irrigation system?
2. List the four major types of microirrigation systems and discuss scenarios where each type might be used.
3. What is the present land area under microirrigation in the U.S.? How does this compare with the total irrigated area in the U.S.? Do you think microirrigation will increase in the future? Why?
4. List and describe the major components of a microirrigation system.
5. Describe three important potential advantages of microirrigation and situations where these advantages are more likely to occur.
6. Describe three important disadvantages of microirrigation and situations where these advantages are more likely to occur.
7. From a water sample of your choice, evaluate the potential for the water to clog a microirrigation system.
8. You have installed a microirrigation system and during the first two days of irrigation, you note that the pressure gauge before the sand filter remains at 35 psi, but the one at the exit to the filter has dropped from 32 psi to 18 psi. Should you invest in an automatic flushing system for the sand filter?
9. If 500 gal of water are required to fill a microirrigation system, is it cheaper to inject chlorine continuously at 1.0 ppm during an irrigation lasting 4 hours or inserting 10 ppm for the last 20 minutes of the irrigation cycle? Assume the system applies 1,000 gal/hr.
10. Describe why microirrigation lends itself to the best control for irrigation.
11. An emitter from Europe has a discharge rate of 6 L per hour at a head of 10 m. In the laboratory you measure a discharge of 1 gal/hr at a pressure head of 15 ft. What is the discharge rate of this emitter at a head of 20 ft? Is this a pressure compensating emitter?
12. In Example 14.7, if the friction loss in the lateral is 0.01 psi/ft, what will the difference in emitter discharge be between the north and south ends of the field?
13. In Example 14.8 if the header pipe is placed at the north end of the field and the operating pressure remains 15 psi in the header and friction loss is zero, what will be the maximum difference in emitter discharge along the lateral? Which emitter along the lateral will have the highest discharge rate?

References

ASABE Standards (2019). ASAE EP405.1 APR1988 (R2019): Design and installation of microirrigation systems. St. Joseph, MI: ASABE.

Bucks, D. A., Nakayama, F. S., & Gilbert, R. G. (1979). Trickle irrigation water quality and preventive maintenance. *Agric. Water Manag, 2*, 149-196.

Bucks, D. A., Nakayama, F. S., & Warrick, A. W. (1982). Principles, practices, and potentialities of trickle (drip) irrigation. In D. Hillel (Ed.), *Advances in irrigation* (Vol. 1, pp. 219-298). New York, NY: Academic Press.

Coelho, E. F., & Or, D. (1997). Applicability of analytical solutions for flow from point sources to drip irrigation management. *Soil Sci. Soc. Am. J.,* 61, 1331-1341.

FAO. (2021). AQUASTAT database. Rome, Italy: Food and Agricultural Organization of the United Nations.

Irrigation Journal. (1974). 1974 Irrigation Survey. *Irrig. J., 24*(6), 15-22.

Irrigation Journal. (2001). 2000 Annual irrigation survey. *Irrig. J., 51*(1), 12-30, 40-41.

Karmeli, D., & Keller, J. 1975. *Trickle irrigation design.* Glendora, CA: Rain Bird Sprinkler Manufacturing Corp.

Keller, J., & Bliesner, R. D. (1990). *Sprinkle and trickle irrigation.* New York, NY: Van Nostrand Reinhold.

Solomon, K. (1979). Manufacturing variation of trickle emitters. *Trans. ASAE, 22*(5), 1034-1038, 1043.

Stanhill, G. (1992). Water use efficiency in agriculture. Irrigation in Israel: Past achievements, present challenges and future possibilities. In J. Shalhevet, L. Changming, & X. Yuenian (Eds.), *Proc. Binational China-Israel Workshop.*

USDA. (2019). 2018 Irrigation and water management survey. Vol. 3. Special Studies, Part 1. AC-17-SS-1. Washington, DC: USDA.

Chapter 15
Chemigation

15.1 Introduction

Chemigation is the application of chemicals with irrigation systems by injecting chemical solution into the irrigation water stream. The advantages and disadvantages of chemigation are important factors to be considered. All components of the chemigation system must be made of non-corrosive materials. The prevention of any chemicals from entering the water supply is crucial and the practice of chemigation is regulated by federal, state, and local agencies. An important component of a chemigation system is the injection device to assure accurate chemical application. As precision agriculture increases in popularity, the injection devices will become more sophisticated to account for spatial differences of chemical needs within a field.

In chemigation the term "chemicals" usually refers to fertilizers and pesticides with pesticides being inclusive of herbicides, insecticides, fungicides, nematicides, rodenticides, etc. According to data from USDA (2019), of the more than 25 million acres of irrigated field corn, vegetables, cotton, and orchards irrigated in the U.S., fertigation (application of fertilizers) was practiced on 32% of the irrigated area and pesticides were applied with chemigation on over 10% of the area. In some locations the term chemigation also includes the application of chemicals that are necessary for irrigation system maintenance, such as chlorination and acid treatment of microirrigation systems for preventing plugging of emitters by algae, slimes, and chemical precipitates (Chapter 14).

15.1.1 Advantages of Chemigation

The advantages of using chemigation can include better uniformity of chemical application, more timely application of chemicals, effective chemical incorporation, reduction in the number of field operations and the associated soil compaction and crop damage, improved efficacy of pesticides, and reduced environmental contamination (Threadgill et al.,1990 and van der Gulik et al., 2007). Timely application of soluble fertilizers using chemigation can reduce leaching losses on sandy soils, especially during years with greater than normal precipitation (Watts and Martin, 1981). Chemigation allows for the application of nitrogen at times that better match the time of crop uptake especially in taller crops such as field corn. Also, under the tall crop conditions, chemigation is often a viable alternative to aerial application of pesticides. During pest outbreaks timeliness of pesticide application can be critical, and rather than waiting for a commercial aerial sprayer irrigators can take advantage of using their irrigation system for the application. Even when irrigators could use their own field equipment to apply fertilizers or pesticides, using the irrigation system helps minimize field operations.

Precision agriculture depends on variable application of chemicals within a field to better match spatial differences in crop or pest control needs. Variable rate irrigation, discussed in Chapters 6 and 13, makes chemigation a viable application method for precision agriculture. Lo et al. (2019) documented how variable-rate chemigation is compatible with the needs in precision agriculture on a field scale.

15.1.2 Disadvantages of Chemigation

Water resource contamination, application onto non-target areas, increased risk of human exposure to chemicals, and limitations of the chemical products to be applied are some of the potential disadvantages of chemigation. Often with chemigation the chemical is mixed with the irrigation water through injection into the water stream. This poses an environmental risk when the irrigation system shuts off creating the potential of the chemical in the irrigation pipelines and chemical supply tank to backflow into the water source. In Section 15.3 we discuss methods to reduce the risk of this contamination.

In irrigated areas the public and field workers often become accustomed to working with and around irrigation systems and may regard the irrigation water as fresh water that they might consider safe for drinking or other uses and not know that hazardous chemicals might be mixed in the water. Also, with chemigation there is the potential risk of water contamination due to drift, runoff, or application of chemical onto non-target areas. An example of the latter case is a sprinkler system that applies water onto a stream or irrigation ditch that traverses the field.

As discussed by Threadgill et al. (1990) the chemical compatibility to chemigation must be considered. For example, by federal law in the U.S., pesticide labels must specifically state that it is legal to apply the chemical with an irrigation system. Also, there are fertilizers or pesticides that may not be good choices for chemigation if there is potential for precipitation of solids in the water when mixed with the chemical.

15.2 Chemical Injection System

15.2.1 Chemical Injection Pumps

To inject chemical solutions into a pressurized irrigation water stream requires the following equipment components: An injection pump, a chemical supply tank, injection tubing and associated valves, and calibration devices (Figure 15.1). There are three main types of pumps that are commonly used for chemical injection: piston, diaphragm, and venturi injectors (Figures 15.2 and 15.3). Piston and diaphragm pumps are classified as positive displacement pumps where pump discharge is not greatly influenced by the level of pressure in the irrigation

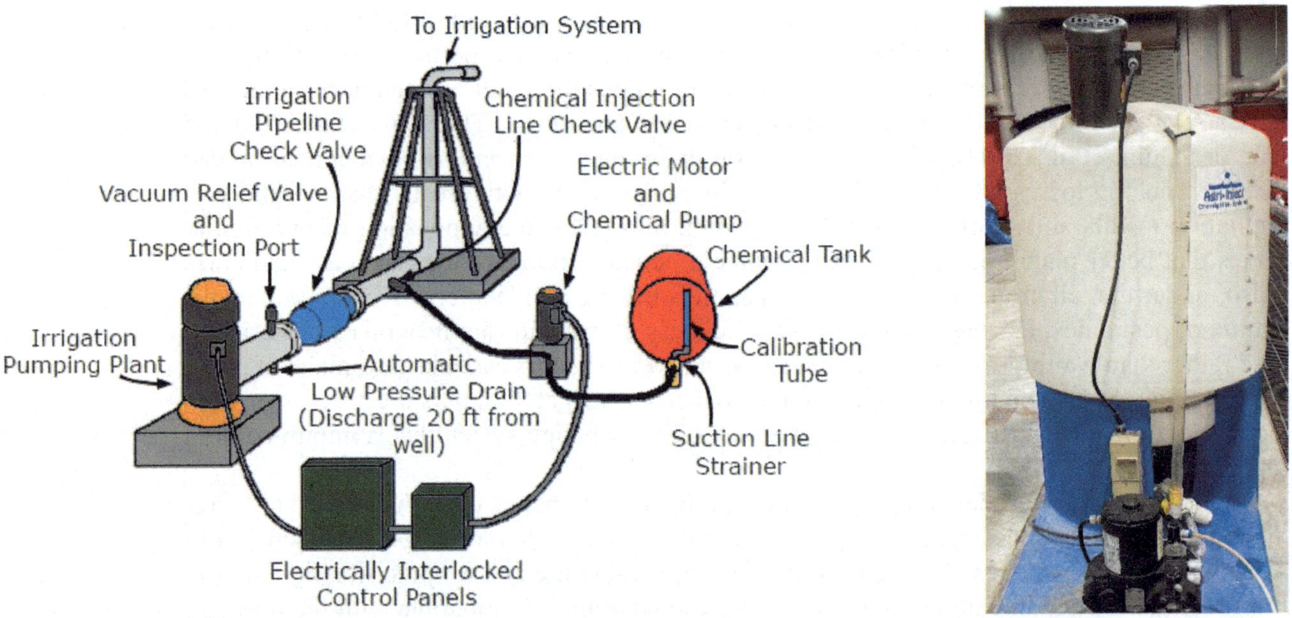

Figure 15.1. Chemigation system (drawing on left modified from Eisenhauer and Hay, 1989).

(a)
(b)

Figure 15.2. Positive displacement injection pumps: (a) piston, (b) diaphragm.

(a)
(b)
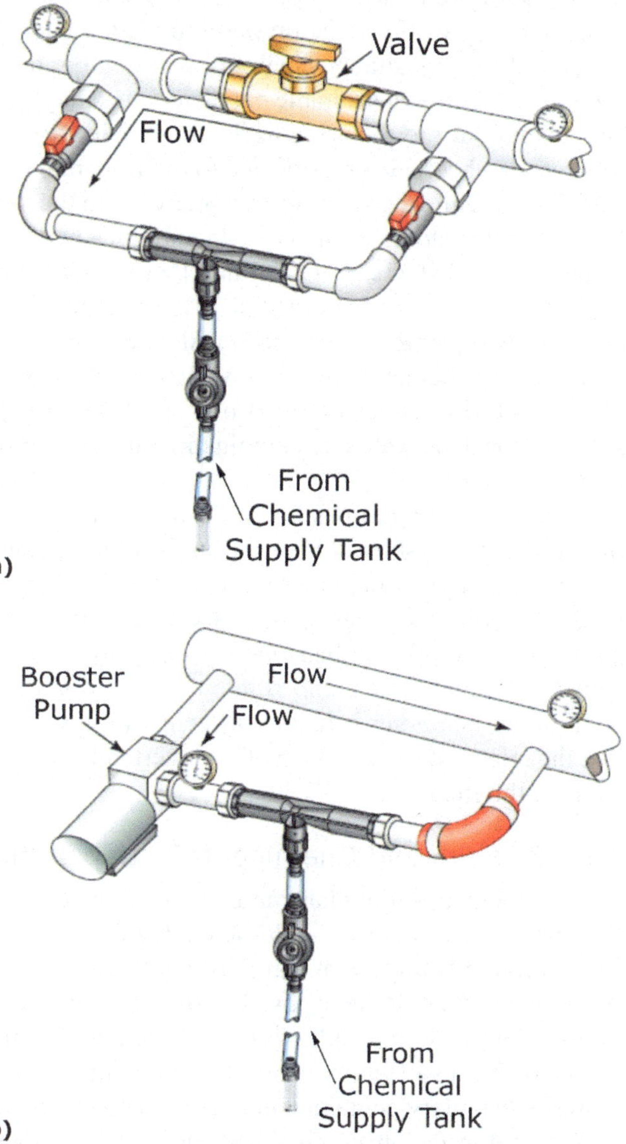

Figure 15.3. Venturi injectors (a) with valve in line to create pressure differential and (b) with booster pump to create pressure differential. Images courtesy of Mazzei Injector Company.

pipeline. The lower pressure in the throat of venturi injectors leads to the chemical solution being inducted into the water stream. The required pressure differential across the venturi makes the injection rate sensitive to the irrigation system pressure. In-line venturi injectors can be used for smaller irrigation pipelines while for larger systems, a by-pass line equipped with pressure reducing valves on the irrigation pipeline or an auxiliary booster pump in the by-pass line are necessary to create the required pressure differential for the venturi to function properly (Figure 15.3). Tests conducted by Kranz et al. (1996) found that regardless of the type of injection device, on-site calibration is necessary under the inlet and outlet pressure conditions at the site. Thus, the chemigation application system should be equipped with a calibration device, usually a calibration tube plumbed at the tank outlet.

For smallholder farmers who use gravity water supply systems such as those shown in Figure 14.7 an injection pump is not necessary. In this case the chemical can be mixed with the irrigation water in the supply reservoir and distributed with the irrigation system. Essential characteristics of the injection pump include material compatibility with the chemical being injected, flow adjustment capability within the range of its maximum capacity, and metering accuracy. Additional desirable characteristics may include flow proportional pump controllers and adaptability to precision agriculture. With flow proportional control the chemical injection rate can be made proportional to the flow rate in the irrigation pipeline. This is especially useful for chemical application accuracy when the land area irrigated per unit time varies, e.g, center pivot systems equipped with end guns and swing-boom corner water systems. Eisenhauer and Bockstadter (1990) found that without flow proportional injection rates, chemical application rate errors can exceed 20% with these scenarios. This error can be reduced to 4% or less when using flow proportional injection. Flow proportional injection also allows for variable rate chemical application when used with variable rate irrigation systems for either sector control or zone control (Lo et al., 2018). Injection pumping systems are available for sector control variable rate chemigation and do not require simultaneous variable rate irrigation.

There are many options for powering the chemical injection pump including belt connection to the power shaft of internal combustion engines (Figure 15.2a), electric motors (Figure 15.2b), and oil hydraulic motors. It is desirable that these power sources are connected to the irrigation power sources in such a way that if the irrigation system shuts off, the injection pump will shut off simultaneously, a one-way interlock. This prevents concentrated chemical from continuing to be pumped into the irrigation pipeline. Also, it is desirable, especially for continuously moving irrigation systems, to have the irrigation system shut off in the event that the injection system shuts off inadvertently. This two-way interlock will prevent untreated areas in the field.

15.2.2 Tanks and Chemical Injection Tubing

The chemical supply tank and chemical injection tubing, associated fittings, and all backflow and safety devices should be made of non-corrosive and chemically resistant materials. In addition, it is important that plastic tanks be made of sunlight resistant materials since it is common for them to be exposed to sunlight for long periods of time. Tank failure and tubing failure can result in a spill of concentrated chemicals resulting in expensive chemical losses and significant soil contamination near the injection site. Depending on the chemical and tank volume secondary containment may be required by regulations.

A common feature of chemical supply tanks is an agitator for mixing purposes. To avoid the accumulation of precipitates in the irrigation system the compatibility of the chemical with the irrigation water should always be evaluated before injecting the chemical. A simple "jar test" can be conducted. In a clear glass jar mix the chemical with the irrigation water to a concentration slightly higher than the planned concentration to be applied. After allowing the jar to sit undisturbed for 24 hours examine the contents for cloudiness, scums, and sediments, indicators of potential chemical precipitation problems.

15.3 Backflow Prevention and Other Safety Devices

As with all chemical applications in agriculture, there is always a concern about the potential for environmental contamination and for worker safety. With chemigation a primary concern is chemical contamination of the irrigation water source due to backflow of the water-chemical mixture in the irrigation pipeline and/or the flow of concentrated chemical from the supply tank to the water source. Another important matter is soil contamination with concentrated chemical in the injection area. Flow of chemical to the water source is not an issue as long as the irrigation system is operating since the flow direction is away from the water source. When the irrigation water flow stops there is potential for backflow.

In Figure 15.4 several soil and water source contamination possibilities that occur with chemigation are illustrated. In the first scenario (Figure 15.4a) the injection system could shut off unexpectedly while the irrigation pump continues to operate, causing water backflow through the chemical injection system and an overflow of the supply tank. This can lead to soil contamination near the injection site with subsequent potential for leaching to the groundwater, overland runoff of chemical, or flow of chemical to groundwater via the gravel pack of a well. Another possible occurrence is the flow from both the irrigation water supply and the injection system stopping resulting in backflow of the water-chemical mixture to the water source (Figure 15.4b). The most environmentally hazardous scenarios are when concentrated chemical is allowed to flow directly to the irrigation water source. This can occur by gravity-driven flow from the supply tank to the water source (Figure 15.4c) when both the irrigation water flow stops and the injection pump stops. Probably the worst case scenerio occurs when

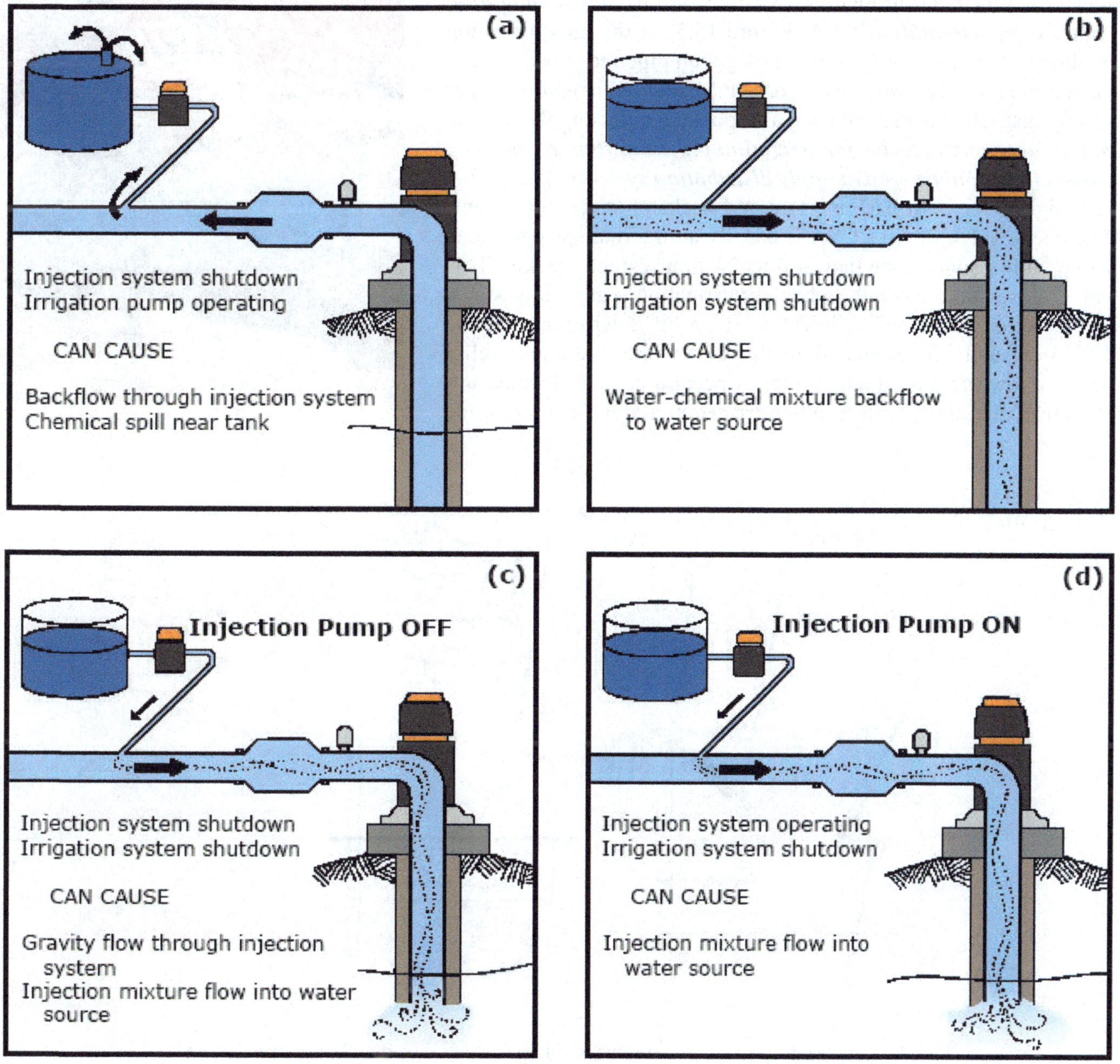

Figure 15.4. Soil and water source contamination scenarios with chemigation systems (modified from Eisenhauer and Hay, 1989).

the irrigation water flow stops but the injection pump continues to operate (Figure 15.4d). Eisenhauer et al. (1988) found that there was over 400 times as much pesticide in a full supply tank than was present in the water-chemical mixture in a 130-acre center pivot irrigation system lateral. This not only illustrated the environmental value but the monetary incentive of retaining the concentrated chemical in the tank.

The risk of contaminating the water source and soil near the injection site can be minimized by using the proper backflow prevention and chemigation safety devices (Eisenhauer and Hay, 1989, Kranz et al., 2015, and Threadgill et al., 1990). These devices will be discussed in the following sections.

15.3.1 Irrigation Pipeline Backflow Prevention Devices

Backflow prevention in the irrigation pipeline reduces the risk of direct chemical contamination of the water source caused by the scenarios illustrated in Figures 15.4 b, c, and d. The ***chemigation check valve assembly*** (CCVA, Figure 15.5) is the most common method of backflow protection on irrigation pipelines that are connected to privately owned wells or single function irrigation water supply districts that are not used as a potable water supply. ***This is not an acceptable device for irrigation pipelines that are directly connected to public water supply distribution systems***. The CCVA is designed for both backpressure and backsiphonage conditions. The check valve in the CCVA is usually an internal spring-loaded valve with a swing gate that is fitted with a resilient-gasket. This valve closes automatically when the irrigation water flow stops. The location of chemical injection must be downstream of the CCVA. Usually incorporated in the CCVA are a vacuum relief valve, a low pressure drain and an inspection port, all located upstream of the check valve. When irrigation water flow stops the

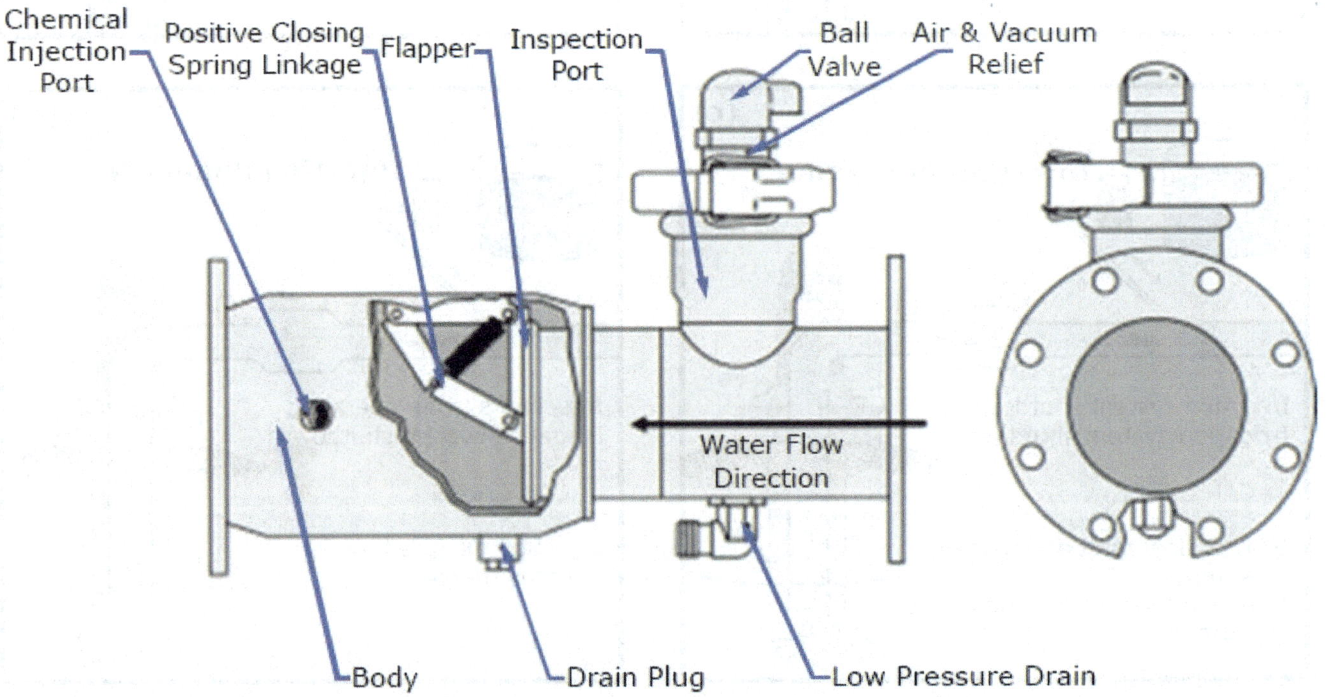

Figure 15.5. Chemigation check valve assembly. (Bottom image courtesy of Kranz et al., 2016.)

check valve closes preventing backflow of the water-chemical mixture in the irrigation pipeline (Figure 15.6a). The vacuum relief valve allows air into the system preventing the creation of a vacuum that could lead to siphoning. In the event that the check valve fails, the low pressure drain can discharge small leakage rates away from the water source (Figure 15.6b). The inspection port (Figure 15.7) can be used by the irrigator or regulatory agency personnel to visually check for leakage from the check valve.

An ***air gap*** is an acceptable method and alternative to the CCVA. An air gap is created by discharging the irrigation water supply into a tank, reservoir, or farm irrigation ditch in a manner such that there is a free atmospheric vertical separation between the discharge from the water supply pipeline and the water surface in the reservoir (Figure 15.8). The recommended air gap vertical distance is two times the inside diameter of the supply pipeline with a minimum distance of 1 in. (AWWA, 2015). The chemical is then either mixed in the water in the reservoir such as would be possible in the smallholder system illustrated in Figure 14.7 or injected into the water downstream of the reservoir or into the farm irrigation ditch. Depending on the reservoir elevation an irrigation pump may be necessary downstream of the reservoir.

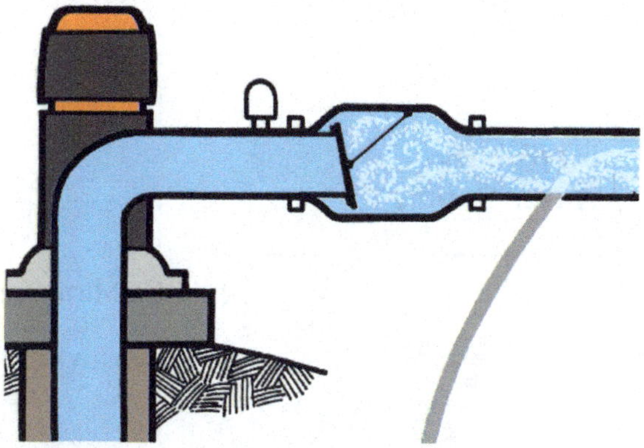
(a)

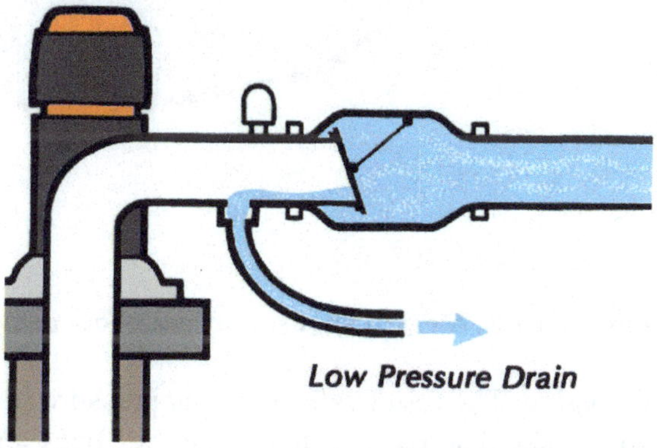

(b)

Figure 15.6. (a) Backflow prevention operations of CCVA check valve and (b) low pressure drain valve. (Images courtesy of DeLynn Hay, Nebraska Extension.)

Recommendations for backflow prevention when irrigation systems are connected to a public water supply system are presented by AWWA (2015). In general irrigation connections are considered a high hazard by AWWA (2015) which recommends four acceptable methods for backflow prevention: an air gap (discussed above), reduced-pressure zone backflow prevention assembly, a pressure vacuum breaker assembly, or an atmospheric vacuum breaker assembly.

The ***reduced-pressure zone (RPZ) backflow prevention*** assembly (Figure 15.9a) consists of two independently-acting internally-loaded check valves in series with one another and a differential pressure relief valve located in the chamber between the check valves and lower in elevation than the upstream check valve. The RPZ device is capable of preventing backflow in the event of either a backpressure or backsiphonage conditions. The loading or opening pressure of the upstream check valve (minimum 3 psi) creates a differential pressure across

Figure 15.7. Inspection for check valve leakage in CCVA.

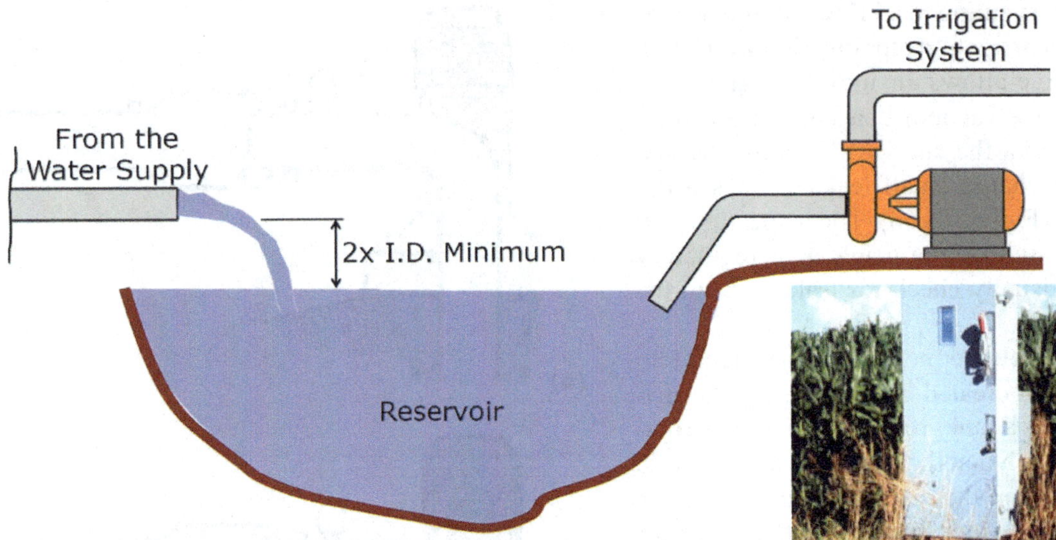

Figure 15.8. Air gap separation backflow prevention.

the valve. The relief valve opening pressure (minimum 2 psi) is less than the differential pressure created across the first valve. In a backsiphonage condition the lower pressure upstream of the first check valve will cause the relief valve to open to the atmosphere and prevent backflow to the water source even in the event of failure of the second (downstream) check valve. The relief valve will also open in a backpressure condition if the second check valve fails and allows leakage of pressure into the middle chamber.

The ***pressure vacuum breaker (PVB) assembly*** is the third alternative for preventing backflow in a high hazard environment (Figure 15.9b). It is only applicable in a backsiphonage condition. The assembly includes an internally loaded check valve and an internally loaded air-inlet vacuum relief valve that opens to the atmosphere. The PVB must be positioned so that the elevation of the discharge pipe is a minimum of 12 inches higher than the elevation of the highest irrigation outlet. Like the PVB the ***atmospheric vacuum breaker (AVB)*** can also be used under high hazard backsiphonage conditions, but it has more limited application in irrigation systems because shutoff valves downstream of the AVB are not allowed. Under normal irrigation flow conditions, the poppet in the AVB seals on the air-inlet seat. Under a back-siphonage condition the poppet opens and drops to seal on the check valve seat. The

(a) **(b)**

Figure 15.9. Backflow prevention assemblies, (a) reduced-pressure zone and (b) pressure vacuum breaker.

poppet is not spring loaded so if a closed downstream valve causes the poppet to be seated at the air inlet for long periods of time there is a risk for it to stick to the seat and not open when backsiphonage occurs.

Public suppliers of potable water and local plumbing codes almost invariably have their own requirements and specifications for backflow prevention which are usually based on recommendations by the American Water Works Association and the Foundation for Cross-Connection Control and Hydraulic Research at the University of Southern California.

15.3.2 Chemical Injection Pipeline Safety Devices

As discussed above it is imperative to stop the flow of concentrated chemical from the supply tank when the irrigation water flow stops. The one-way interlock between the irrigation water supply and the injection device, as discussed in Section 15.2.1, stops the injection pump when the water flow stops preventing the situation illustrated in Figure 15.4d. But the interlock will not stop the flow caused by gravity illustrated in Figure 15.4c. A chemical injection line check valve (Figure 15.10) will help prevent this flow. The check valve is internally loaded, usually by spring, so that it has an opening pressure of 10 psi or greater. At 10 psi the valve would block flow until 23.1 feet of water head is exceeded. The chemical injection line check valve will also stop the backflow through the injection system in the event that the injection system stops but the irrigation water continues to flow (Figure 15.4a). The chemical injection line check value is usually an integral part of the injection port with the discharge end near the center of the irrigation pipeline as would be the case for the valve shown in Figure 15.10a.

Whenever possible it is helpful to place the point of chemical injection at an elevation higher than the maximum liquid level in the supply tank Figure (15.11b). This will provide more protection against flow caused by gravity. If this is not possible another technique for additional protection is to create a vertical pipe loop with a vacuum relief valve at the apex. The apex must be at an elevation higher (minimum 12 in) than the maximum elevation of the liquid level in the supply tank (Figure 15.11a). The vacuum relief valve will break the siphon and stop the flow from the tank.

Another valve that is useful on the injection tubing is a bleed valve (Figure 15.10a). Upon system shutdown, pressure is usually locked into the tubing between the injection pump and the chemigation line check valve. The bleed valve can relieve this pressure before the tubing is disconnected preventing the operator from being sprayed with concentrated chemical. The bleed valve can also be helpful for removing air from the injection line when priming the system.

For further safety, a normally-closed solenoid valve on the inlet side of the injection pump can be electronically interlocked with the injection pump power supply to provide a positive shut off on the chemical injection line if the injection pump stops. This valve is sometimes included in regulatory requirements. Another device that is sometimes required by regulation is a flow sensor po-

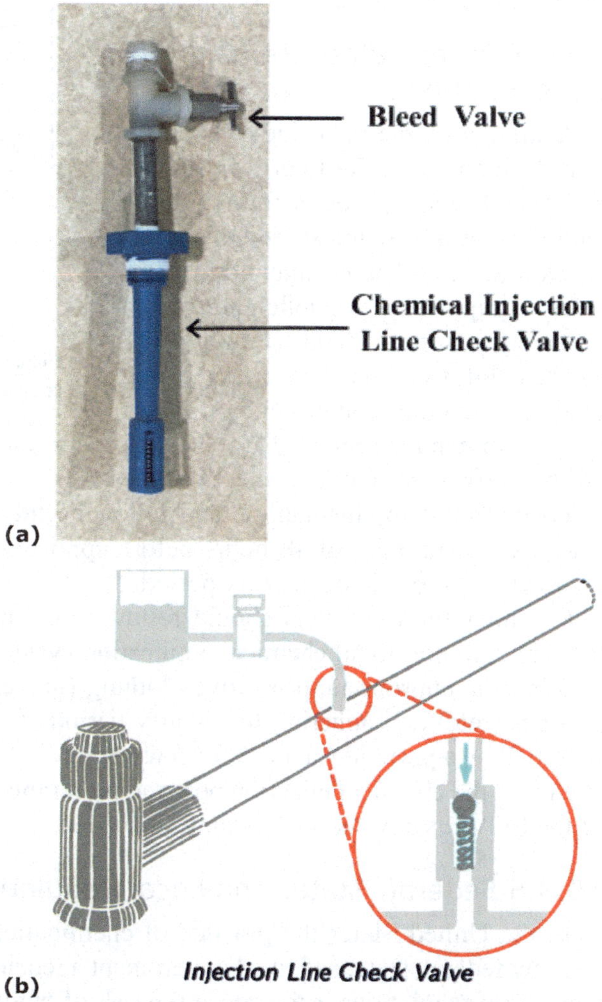

Figure 15.10. Chemical injection line check valve. (Image b courtesy of DeLynn Hay, Nebraska Extension.)

sitioned in the injection line just upstream of the injection port. The flow sensor safeguards against continued operation if there is a rupture of the injection line, injection pump failure, loss of prime, or the injection port is plugged.

15.3.3 Irrigation Pipeline Low Pressure Switch

A low pressure switch on the irrigation pipeline will shut the irrigation system and injection system off if the system pressure drops below a critical point. One potential advantage of chemigation is the uniformity of chemical application which is dependent on the irrigation application uniformity. If the pressure is too low the irrigation uniformity is compromised making it important to stop the application.

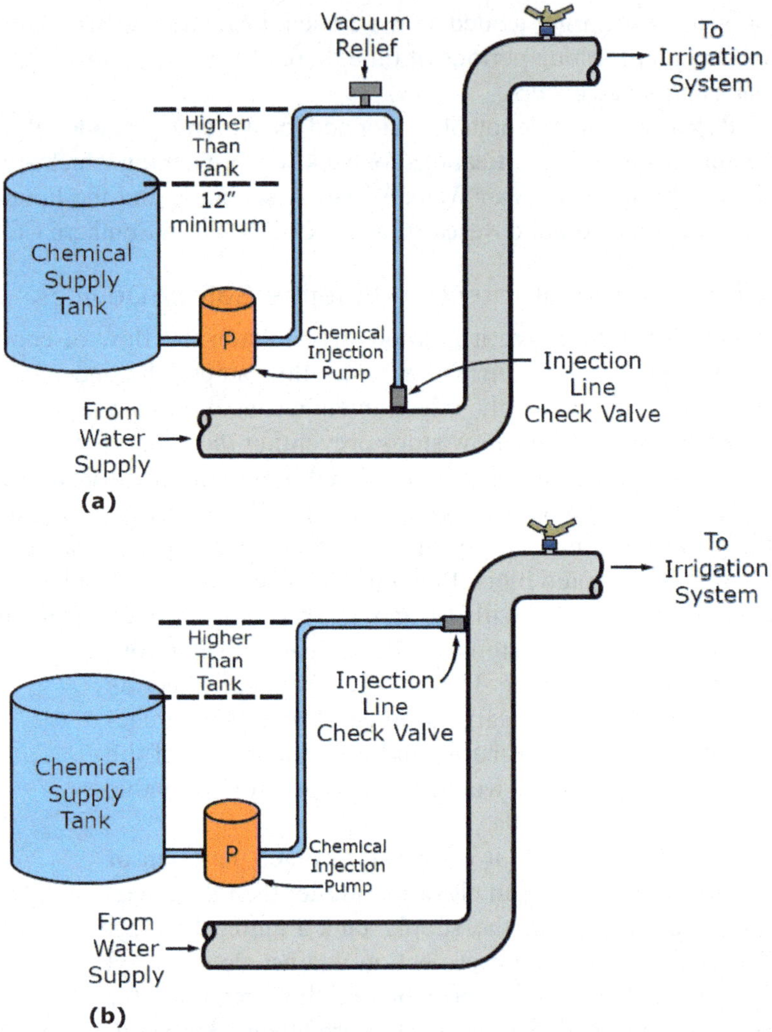

Figure 15.11. Chemical injection line options for providing additional protection from chemical flow due to gravity from the chemical supply tank into the irrigation pipeline. (Modified from Eisenhauer and Hay, 1989.)

15.3.4 Other Safety Items and Considerations

A strainer on the inlet side of the injection device is essential to prevent foreign materials from clogging or fouling the injection pump, chemical injection line check valve, or other injection system safety equipment. For public and applicator safety posting of fields (Figure 15.12) can be helpful and may be required by regulations. To avoid complacency it is usually recommended/regulated that the signs not be permanent but have specific times of posting prior to chemigation and following the event, for example a maximum of 48 hours before application and 48 hours after the pesticide re-entry period.

Chemigation applicators should follow safety procedures that are common to all chemical application systems such as wearing the appropriate protective clothing (gloves, goggle, rain gear, etc) and adhering to re-entry periods that may be specified on a pesticide label. A fresh water faucet located upstream of the chemical injection port and preferably upstream of the CCVA is advised for washing as needed.

Figure 15.12. Field posted for chemigation.

15.3.5 Federal, State, and Local Regulations

In the United States the practice of chemigation is regulated by federal, state, and local government agencies. The intent of the regulations is to reduce the risk of environmental contamination and to protect worker and public safety. Essentially all of the regulations contain some if not all of the backflow prevention and safety equipment discussed above and it is common that the regulations will reference the Engineering Practice, ASAE EP 409.1 Safety Devices for Chemigation (ASABE, 2018).

At the federal level the application of pesticides is regulated by the U.S. Environmental Protection Agency (EPA) through the Federal Insecticide, Fungicide, and Rodenticide Act (FIFRA). FIFRA requires that pesticides be applied according to the product label. Fertigation is not regulated by FIFRA. The Label Improvement Program, U.S. EPA Pesticide Registration (PR) Notice 87-1 and an updated list of alternatives provides pesticide manufacturers with generic statements that they can incorporate on the label of their product. For example the label will state whether or not the product can be applied using chemigation and if so what specific safety requirements must be followed. Are there field posting requirements? Is chemigation limited to specific types of irrigation systems, such as sprinklers? What are the specific backflow and safety equipment requirements? For products labeled for chemigation it is common to see the requirement of a CCVA, a chemical injection line check valve, a normally-closed solenoid valve on the inlet side of the injection pump, a one-way system interlock, irrigation system low pressure shutoff switch, and field posting. An example of an acceptable alternative device is to substitute the normally-closed solenoid valve with a chemical injection line check that has an opening pressure of 10 psi or higher. If the label allows for application of a pesticide using water from a public water supply it is likely to state that either an approved air gap or RPZ are required and that specific buffer distances from public places must be followed.

Many states in the U.S. have chemigation regulations and usually these apply to both fertilizers and pesticides. The state requirements can be more extensive than federal requirements but not less. An example is that some states require two CCVA assemblies placed in series if pesticides are to be applied. States usually provide lists of approved CCVA's that are commercially available. Permitting of chemigation sites is often required and the equipment at permitted sites may be regularly inspected for working performance. Chemigation applicator training and competency testing is sometimes required by states. Other items in some state regulations include accident reporting, secondary containment of the supply tank, and specific pre and post time requirements of posting fields.

Local government subdivisions may also have their own more chemigation regulations. For public water supply systems such as municipalities, the local plumbing code usually has specific regulations for connected irrigation systems.

15.4 Management of Chemigation Systems

Proper management of chemigation systems is necessary to apply chemicals uniformly in the correct amount and to ensure personal and public safety. If insecticides are being applied it is important to read and follow the product label for proper use and application and personal safety recommendations. Avoiding chemigation when wind speeds are high will help with uniform sprinkler applications and avoid drift onto non-target areas such as roads and areas with high public use. Two key management requirements are calibration of the injection system and flushing the injection and irrigation systems following chemigation events.

15.4.1 Injection Rates and Calibration of Injection Devices

The injection device must be calibrated so that the correct amount of chemical is applied per unit of land area. Kranz and Eisenhauer (1996) tested chemical injection pumps and found that even with positive displacement packed-piston and diaphragm pumps there is some sensitivity of pump discharge to irrigation pipeline pressure. Thus all types of injection devices should be calibrated in the field with the irrigation system operating at normal pressure.

For stationary irrigation systems e.g. surface irrigation systems, set-type sprinkler systems and microirrgation systems, the rate of chemical injection is calculated by:

$$q_i = \frac{G_p A_s}{t_i} \tag{15.1}$$

where: q_i = injection rate of solution (gal/h),
G_p = amount of chemical solution to apply (gal/ac),
A_s = area of the irrigation set or zone (ac), and
t_i = total time of injection during the irrigation set (h).

With set-type systems the time period for injection does not have to be equal to the set-time or zone run time; it can be equal to or less than the set-time. For certain chemicals such as insecticides or fungicides it may be advantageous to inject near the end of the set-time so more chemical remains on the plant leaves following the event. In other cases it may be advantageous to apply the chemicals at the beginning in the event and then flush the irrigation system at the end of the event. Some controllers can be programmed so that the injection pump is shut off near the end of the event for flushing purposes. In Example 15.1 we illustrate the use of Equation 15.1 for a subsurface drip irrigation system.

For continuously moving irrigation systems the injection rate is dependent upon the rate of land being irrigated per unit time as follows:

$$q_i = G_p R_i \tag{15.2}$$

where R_i = rate of land area irrigated per unit of time (ac/h).

For travelers and lateral move systems R_i is the area of the irrigated field divided by the total time of irrigation. For travelers the area and time can be for an individual set. R_i varies with the lateral pipeline position with center pivots that are equipped with end guns or swing-boom corner watering systems which leads to error in chemical application when using constant injection rates and the angular speed remains constant. Eisenhauer and Bockstadter (1990) found that for a typical $^1/_4$-section center pivot equipped with an end gun for irrigating a portion of each of the four corners, the average absolute injection rate error was 7.5% when R_i was calculated based on the total irrigated area and the total time to make a revolution and the injection rate and pivot speed were constant. The chemical would be under applied in the corners when the end gun is operating and over applied whenever the end gun is off. As discussed in Section15.2.1, using a flow proportional injection system is one way of reducing this chemical application rate error. Reducing the speed of the center pivot when the end gun is operating using an auxiliary timer or sector control variable rate center pivot is another way to reduce this error. Equation 15.2 is applied in Example 15.4.

Calibration of an injection pump is usually done with a calibration tube, a clear plastic tube with volume gradations marked on the outside. The tube is plumbed at the outlet of the chemical supply tank and using valves it can be isolated from the tank so that the liquid flowing into the injection device is only from the calibration tube. The calibration process is illustrated in Example 15.1.

Example 15.1

It is desired to inject a fertilizer solution into an irrigation system at a rate of 17.5 gal/h. The injection pump has a maximum capacity of 20 gal/h. The pump has a percentage dial that can be adjusted for the desired pumping rate. At what percentage would you set the dial? After setting the pump it is calibrated at the appropriate pressure. It requires 2.5 minutes (2 min:30 sec) to pump 3,000 mL from the injection tube. What is the injection rate?

Find: Pump dial setting and injection rate

Solution:

$$\frac{17.5\ \text{gal/h}}{20\ \text{gal/h}} \times 100\% = 88\%$$

$$\frac{3{,}000\ \text{mL}}{2.5\ \text{min}} = 1{,}200\ \text{mL/min}$$

$$1\ \text{gal/h} = 63.1\ \text{mL/min}$$

$$\frac{1{,}200\ \text{mL/min}}{63.1\dfrac{\text{mL/min}}{\text{gal/h}}} = 19.0\ \text{gal/h}$$

Thus the pump dial setting is initially set at 88%. Since 17.5 gal/h is desired and the injection rate is 19.0 gal/h the pump setting should be adjusted downward and followed by another test. Repeat these steps until the appropriate injection rate is reached.

15.4.2 Flushing the Injection and Irrigation System

At the end of a chemigation event it is important that the chemical injection pump and the injection line tubing and associated valves and the irrigation pipeline be flushed free of chemicals using fresh water. Flushing the injection system reduces the chance that chemical precipitates will foul the equipment components during future chemigation events. The irrigation pipeline system including the laterals should be flushed to prevent unexpected exposure of field workers to chemicals remaining in the line and to properly distribute the chemical in the field. The time required to flush a mainline is equal to the pipeline length divided by the mean velocity according to the following equation:

$$T_m = \frac{0.0408\,(ID)^2 \times L_m}{Q} \qquad (15.3)$$

where: T_m = flushing time of a mainline (min),
ID = inside diameter of the pipe (in),
L_m = the length of the mainline (ft), and
Q = irrigation system flow rate (gpm).

Equation 15.3 can also be used for manifolds and submains by solving it for each segment of the manifold.

Flushing time of a lateral is different than the mainline. DeTar (1983) provides the following equation for calculation of flushing times in laterals that have outlets with equal discharge:

$$T_L = \frac{0.0408\,(ID)^2 \times S}{q}\left[0.577215 + \ln\left(\frac{L_l}{S}\right)\right] \qquad (15.4)$$

where: T_L= flushing time in a lateral (min),
q = outlet discharge (gpm),
L_l = lateral length (ft), and
S = outlet spacing (ft).

The ratio L_l/S is the number of outlets, N, on the lateral and can be substituted into Equation 15.4 accordingly. Equation 15.4 is valid for laterals that have 10 outlets or more. Equation 15.4 can also be used for manifolds that have more than 10 laterals attached but the flow rate of each lateral must be the same. Figure 15.13 was developed using Equation 15.4.

Example 15.2

A 78-acre field is irrigated with a subsurface drip irrigation system. The flow rate of the system is 500 gpm. From the water source the water is conveyed 975 feet in an 8-inch mainline (ID = 7.762 in). The water then flows into a 4-inch manifold (ID 4.280 in). The manifold is 325 feet long (one side of the inlet tee) and has 65 laterals that are 5 ft apart. The flow rate in each lateral is 3.81 gpm.

Find: Total amount of time to flush the chemical out of the mainline and manifold.

Solution:

The mainline flush time can be calculated with Eq. 15.3:

$$T_m = \frac{0.0408(ID)^2 \times L_m}{Q} = \frac{0.0408(7.762)^2 \times 925}{500} = 5\text{ min}$$

The manifold flush time can be calculated with Eq. 15.4:

$$T_L = \frac{0.0408(ID)^2 \times S}{q}\left[0.577215 + \ln\left(\frac{L_l}{S}\right)\right]$$

$$T_L = \frac{0.0408(4.280)^2 \times 5}{3.81}\left[0.577215 + \ln\left(\frac{325}{5}\right)\right] = 5\text{ min}$$

So, the total time required to flush the mainline and manifold is 10 min.

This is also the time it will take to move chemical from the injection device to the last lateral on the manifold assuming the system has already been primed with water.

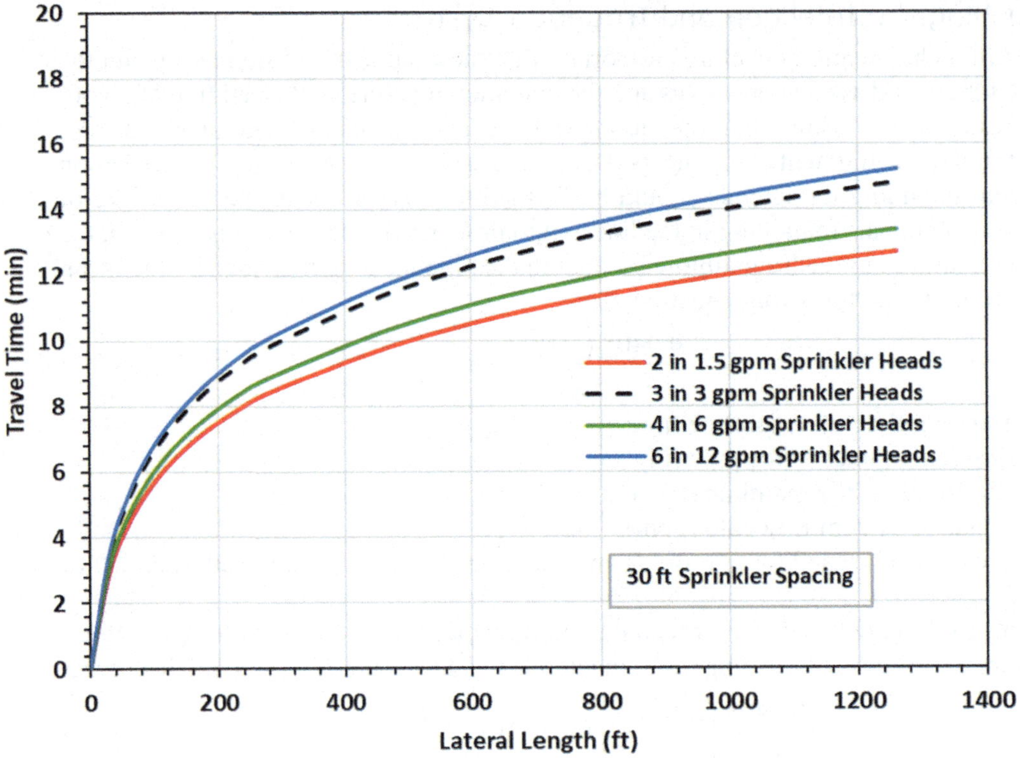

(a)

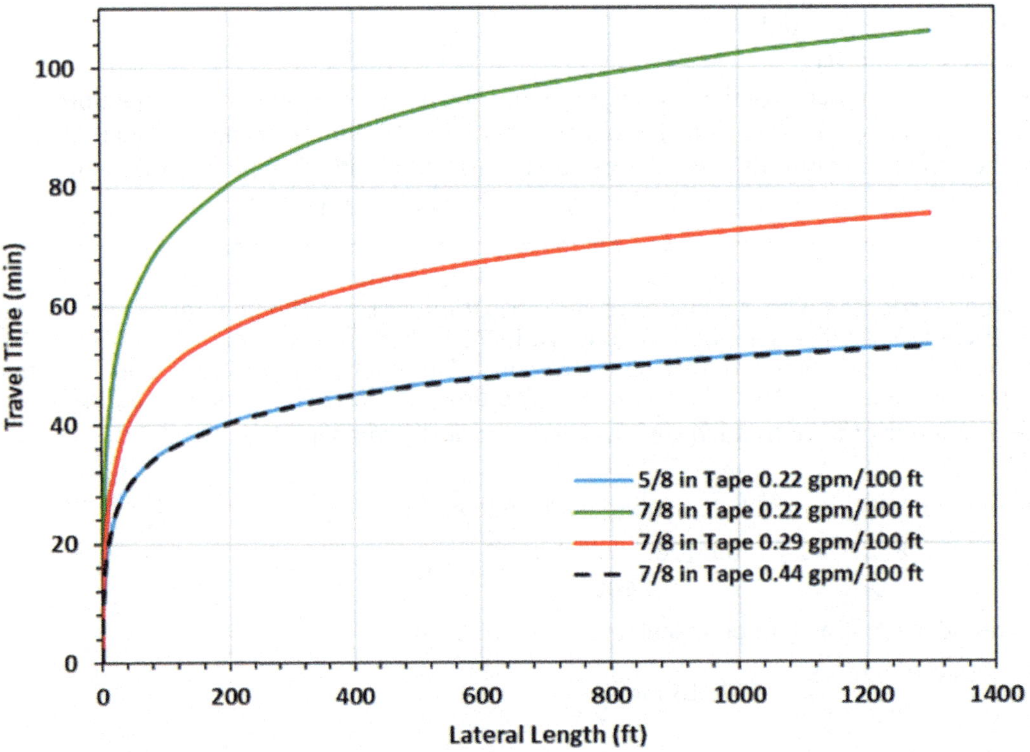

(b)

Figure 15.13. Travel or flush times in irrigation laterals, (a) aluminum sprinkler laterals, (b) drip irrigation tubing.

In Example 15.3 the three hours of flushing time is more than adequate since only 85 minutes total is necessary. The main point is that the time allowed for flushing must be equal to or greater that the total flushing time that was calculated. The travel time of chemical from the injection device to the most distant emitter equals the total flushing time. So the flushing time is important for the proper amount of chemical to be distributed in each zone. In Example 15.3 each zone will receive 4 hours of injected chemical but each emitter will only have discharged the correct amount of chemical when flushing is complete. It is important that the

Example 15.3

For the same field in Example 15.2, a farmer plans to apply 15 lbs/ac of nitrogen using fertigation with injection of 28% UAN nitrogen solution into a subsurface drip irrigation system.

Given:

Irrigation system characteristics:	
Field width = 2600 ft	Lateral length = 1300 ft
Field length = 1300 ft	Emitters per lateral = 866
Field area = 78 ac	Nominal lateral diameter = 7/8 in
Q = 500 gpm	ID = 0.859 in
Lateral spacing = 5 ft	Discharge per lateral = 3.18 gpm
Emitter spacing = 18 in	Laterals per zone = 130
Discharge per emitter = 0.264 gal/h	Number of irrigation zones in field = 4
Irrigation Management:	**Fertigation Plan:**
Irrigation interval = 2 d	Nitrogen per acre = 15 lb/ac
Set time per zone = 9 h	28% UAN (3 lb N/gal)
Zone area = 19.5 ac	

Find: Time required to flush laterals (through the emitters, not the downstream flushing manifold)
Injection rate in gal/h

Solution:

Application of the DeTar Equation, Eq 15.4 (applicable only when flushing through the emitters)

S = 1.5 ft

q = 0.264 gal/h = 0.0044 gpm

$$T_L = \frac{0.0408(ID)^2 \times S}{q}\left[0.577215 + \ln\left(\frac{L_l}{S}\right)\right]$$

$$T_L = \frac{0.0408(0.859)^2 \times 1.5}{0.0044}\left[0.577215 + \ln\left(\frac{1300}{1.5}\right)\right] = 75 \text{ min}$$

The same value can be obtained from Figure 15.13b.Thus, it will take 75 minutes to flush the laterals and it will take a total of 85 minutes to flush the mainline and manifold (Example 15.2) plus the laterals.

Assuming that an automatic controller and valves are used to change the water between zones the following approach is one alternative for managing this system.

Since the entire irrigation duration will take 36 hours one approach is to fertigate for the first 24 hours (four 6-hour sets) then terminate fertigation and flush each zone with fresh water for 3 hours giving a total of 12 additional irrigation hours. The injection rate would then be:

At 3 lbs N/gal, 15 lb/ac requires 5 gal/ac or solution to be injected.

Using Eq. 15.1:

$$q_i = \frac{G_p A_s}{t_i} = \frac{5 \text{gal/ac} \times 78 \text{ac}}{24 \text{h}} = 16.2 \text{gal/h}$$

Thus the injection rate needed is 16 gal/h.

irrigation system be primed with water prior to injecting chemicals so that the priming time is not part of the injection time. Example 15.3 illustrates that there is flexibility in managing chemigation with set-type systems. The solution given in Example 15.3 is only one of many acceptable management approaches.

For systems like center pivots, the discharge on the lateral varies by distance along the lateral as discussed in Chapter 13. Buttermore and Eisenhauer (1989) developed the following flushing time equation for the center pivot lateral problem:

$$T = \frac{0.0102(ID)^2\,(2L_l + S)}{Q}\ln\left(1 + 4\frac{L_l}{S}\right) \qquad (15.5)$$

The symbols L_l, S, and Q have the same meanings as in Equations 15.3 and 15.4. Equation 15.5 only applies to the case when an end gun is not operating and the diameter of the lateral is constant along its length. Buttermore and Eisenhauer (1989) presented equations that apply for the end gun condition and for the case where laterals have multiple diameters along their length. In general operating an end gun will result in shorter flush times than when not operating. The application of Equations 15.2 and 15.5 are illustrated in Example 15.4.

In Example 15.4 R_i was based on the total area irrigated in the field and the total time it takes to irrigate the field and the calculated injection rate was based on this average irrigation rate. In this example if an end gun operated in the four corners the R_i would be about 5.45 ac/h when the end gun is off and 6.05 ac/h when it is on. Thus with constant injection rate and constant angular speed of the pivot chemical application rate errors would be at least 5 to 6 % with the chemical application being too high when the end gun is off and too low when it is on.

Example 15.4

A farmer plans to apply 1.5 pints per acre of an insecticide through a center pivot system.

Given: Field area = 126 ac
Q = 860 gpm
Pivot length = 1280 ft
Sprinkler spacing = 20 ft
Nominal lateral diameter = 6 in ID = 6.395 in
Time for 1 revolution = 22 h
Injection occurs at the pivot point (no mainline)

Find: Injection rate in gal/h
Time required to flush the center pivot lateral following the application

Solution:

The injection rate can be calculated using Equation 15.2. We will assume that R_i is equal to the field area divided by the time to irrigate the field, i.e.,

$$R_i = \frac{126\text{ ac}}{22\text{ h}} = 5.72\text{ ac/h}$$

$$q_i = G_p R_i = 1.5\text{ pts/ac} \times 5.72\text{ ac/h} = 8.6\text{ pts/h} = 1.07\text{ gal/h}$$

So the injection device should be calibrated to a flow of 1.1 gal/h.

Apply Equation 15.5 for determining flush time:

$$T = \frac{0.0102(ID)^2\,(2L + S)}{Q}\ln\left(1 + 4\frac{L}{S}\right)$$

$$T = \frac{0.0102(6.395)^2(2 \times 1280 + 20)}{860}\ln\left[1 + 4\left(\frac{1280}{20}\right)\right] = 6.9\text{ min}$$

The time to flush the system, assuming the end gun is not operating while flushing is about 7 min.

Assuming that chemical injection started when the center pivot began moving the center pivot should continue to irrigate for 7 more minutes after injection has stopped so that the chemical is flushed and the beginning of the revolution obtains the correct amount of chemical.

15.5 Summary

Chemigation is the practice of applying chemicals with irrigation systems. The chemicals, which include fertilizers, pesticides, and system maintenance chemicals, are mixed with the irrigation water for application. The chemicals are often injected into the irrigation water stream with a pump, but in small-scale systems chemicals can be mixed with the water in the water supply tank. Chemigation offers many potential advantages to irrigators including more timely application of chemicals and reduced field operations. However the risk of contaminating the water supply is a concern because of the potential of backflow of chemical when the irrigation system shuts off. Backflow prevention equipment and safety devices are necessary to reduce the risk of contamination and the requirements for this equipment is often regulated by federal, state, and local agencies. Calibration of the injection system is essential for accurate chemical application. An important management practice is to adequately flush the injection system and the entire irrigation system following chemical application.

Questions

1. Define chemigation.
2. Locate the label of a herbicide and a insecticide that are approved for chemigation. Are the backflow prevention equipment requirements the same on each label? Can the products be legally applied in irrigation systems that are connected to a public water supply system?
3. Explain what it means to have a flow proportional injection system and under what conditions or situations would it be useful to use a flow proportional injection system.
4. What does the term positive displacement pump mean?
5. List three potential advantages and three possible disadvantages of the practice of chemigation.
6. List the components of a chemigation check valve assembly and explain their function.
7. What are the two functions of the chemical injection line check valve?
8. An injection rate of 1.5 gal/h is required to apply the desired amount of a pesticide by chemigation. If the injector pump has a maximum capacity of 2 gal/h, what is the estimated correct percent setting of the device? At this setting the liquid level in a calibration tube is timed to check the flow rate of the injector under field operating conditions. 500 ml of chemical is pumped in 4 min and 54 sec (4 min:54 sec). Determine the actual injection rate in gal/h. Does the injection device setting need to be increased or decreased to match the desired 1.5 gal/h injection rate?
9. A 122-ac center pivot will be used to apply 28% UAN nitrogen solution (3 lb of N per gal). The planned application is 25 lb/ac of nitrogen. The timer on the pivot is set at 35% which will result in a making a revolution in 60 hours. Determine the required injection rate in gal/h.
10. A herbicide is applied using chemigation with a stationary sprinkler system. The system has 3-inch diameter aluminum irrigation laterals which are 1000 feet long with 3-gpm sprinklers spaced at 30-foot intervals. How long will it take to flush the laterals with fresh water after chemical injection has stopped?

References

ASABE Standards. (2018). ASAE EP409.1 MAR 1989 (R2018): Safety devices for chemigation. St. Joseph, MI: ASABE.

AWWA. (2015). *Backflow prevention and cross-connection control, recommended practices.* Manual of water supply practices M14 (4th ed.). American Water Works Association.

Buttermore, G. W., & Eisenhauer, D. E. (1989). Calculation of chemical flushing times in center pivot irrigation systems. *Trans. ASAE, 32*(4), 1193-1196.

DeTar, W. R. (1983). Travel time for chemicals in an irrigation system. *Trans. ASAE, 26*(2), 495-496.

Eisenhauer, D. E., & Bockstadter, T. L. (1990). Injection pump flow considerations for center pivots with corner watering systems. *Trans. ASAE, 33*(1), 162-166.

Eisenhauer, D. E., & Hay, D. R. (1989). Anti-pollution protection when applying chemicals with irrigation systems. EC 89-730-B. Nebraska Cooperative Extension.

Eisenhauer, D. E., Munir, H. M., & Gilley, J. R. (1988). Chemigation back flow prevention assemblies. *Proc. Planning Now for Irrigation and Drainage. ASCE Irrigation and Drainage Division Conf.*, (pp. 69-78).

Kranz, W. L., Eisenhauer, D. E., & Parkhurst, A. M. (1996). Calibration accuracy of chemical injection devices. *Appl. Eng. Agric., 12*(2), 189-196.

Kranz, W., Burr, C., Hay, J., Schild, J., & Yonts, D. (2016). *Using chemigation safely and effectively—Training manual*. University of Nebraska Extension and the Nebraska Department of Environmental Quality.

Lo, T. H., Rudnick, D. R., & Shaver, T. M. (2019). Variable-rate chemigation via center pivots. *J. Irrig. Drain. Eng., 145*(7), 04019012. https://doi.org/10.1061/(ASCE)IR.1943-4774.0001394.

Threadgill, E. D., Eisenhauer, D. E., Young, J. R., & Bar-Yosef, B. (1990). Chemigation. In G. J. Hoffman, T. A. Howell, & K. H. Solomon (Eds.), *Management farm irrigation systems.* St. Joseph, MI: ASAE.

USDA. (2019). 2018 Irrigation and water management survey. Vol. 3. Special Studies, Part 1. AC-17-SS-1. Washington, DC: USDA.

Van der Gulik, T. W., Evans, R. G., & Eisenhauer, D. E. (2007). Chemigation. In G. J. Hoffman, R. G. Evans, M. E. Jensen, D. L. Martin, & R. L. Elliott (Eds.), *Design and operation farm irrigation systems* (2nd ed.). St. Joseph, MI: ASABE.

Watts, D. G., & Martin, D. L. (1981). Effects of water and nitrogen management on nitrate leaching loss from sands. *Trans. ASAE, 24*(4), 911-916.

Glossary

Excerpted with permission from ASABE Standards. ASAE S526.4 SEP2015(R2019). Soil and water terminology. American Society of Agricultural and Biological Engineers: St. Joseph, Mich. https://elibrary.asabe.org/toc.asp.

allowable depletion: That part of soil water stored in the plant root zone managed for use by plants, usually expressed as equivalent depth of water in mm (acre-inches per acre, or inches).

application efficiency (E_a): The ratio of the average depth of irrigation water infiltrated and stored in the root zone to the average depth of irrigation water applied, expressed as a percentage. Also referred to as AE.

application efficiency of low quarter: The ratio of the average depth of irrigation water infiltrated and stored in the root zone as determined from the lowest 25% of the area, to the average depth of irrigation water applied, expressed as a percent.

application rate: Rate that water is applied to a given area. Usually expressed in units of depth per time.

aquifer: A geologic formation that holds and yields useable amounts of water. Aquifers can be classified as confined or unconfined.

available water capacity, AWC: The portion of soil water that can be readily absorbed by plant roots of most crops, expressed in mm water per mm soil (inches per inch, inches per foot, or total inches) for a specific soil depth. It is the amount of water stored in the soil between field capacity (FC) and permanent wilting point (WP). It is typically adjusted for salinity (electrical conductivity) and rock fragment content. Also called available water holding capacity (AWHC), or available soil water.

backflow prevention device: Safety device which prevents the flow of water from the water distribution system back to the water source.

backpressure: Increase of pressure in the downstream piping system above the supply pressure at the point of consideration which would cause, or tend to cause, a reversal of the normal direction of flow.

backsiphonage: Reversal of flow (backflow) due to a reduction in system pressure which causes a negative or subatmospheric pressure to exist at a site in the water system.

basin irrigation: Irrigation by flooding areas of level land surrounded by dikes. Used interchangeably with level border irrigation, but usually refers to smaller areas.

border irrigation: Irrigation by flooding strips of land, rectangular in shape and cross leveled, bordered by dikes. Water is applied at a rate sufficient to move it down the strip in a uniform sheet. Border strips having no downfield slope are referred to as level border systems. Border systems constructed on terraced lands are commonly referred to as benched borders.

centrifugal pump: Pump consisting of rotating vanes (impeller) enclosed in a housing and used to impart energy to a fluid through centrifugal force.

Christiansen's uniformity coefficient: A measure of the uniformity of irrigation water application. The average depth of irrigation water infiltrated minus the average absolute deviation from this depth, all divided by the average depth infiltrated.

consumptive use: The total amount of water taken up by vegetation for transpiration or building of plant tissue, plus the evaporation of soil moisture, snow, and intercepted precipitation associated with vegetal growth.

conveyance efficiency (E_c): The ratio of the water delivered to the total water supplied to an open channel or pipeline at the upstream end, expressed as a percentage.

crop coefficient (K_c): The ratio of the actual crop evapotranspiration to reference evapotranspiration.

cumulative intake: The depth of water infiltrated into the soil from the time of initial water application to the specified elapsed time.

deep percolation: Water that moves downward through the soil profile below the root zone and cannot be used by plants.

discharge: Volume flux or flow rate of water, measured in units of volume per unit time. Discharge can be used to describe water flow in pipes, streams or groundwater.

distribution uniformity of low quarter: The ratio of the average of the lowest one-fourth of measurements of irrigation water infiltrated to the average depth of irrigation water infiltrated, expressed as a percent.

doctrine of appropriation: Water law developed in the arid Western states, where water supplies are limited and often inadequate (also known as the Appropriation Doctrine). This doctrine is essentially a rule of capture (first in time of use is first in right), where application of the water to a beneficial use is the basis and measure of the right.

drawdown: Lowering of the water surface, water table, or piezometric surface resulting from the withdrawal of water from a well or drain or principal spillway; also the elevation of the static water level in a well minus the elevation of the pumping water level (at the well) at a given discharge rate.

drip irrigation: A method of microirrigation wherein water is applied to the soil surface as drops or small streams through emitters. Discharge rates are generally less than 8 L/h (2 gal/h) for single-outlet emitters and 12 L/h (3 gal/h) per meter for line-source emitters.

effective precipitation: That portion of total precipitation which becomes available for plant growth.

electrical conductivity (EC): A measure of the ability of the water to transfer an electrical charge. Used as an indicator for the estimation of salt concentration, measured in dS/m (mmhos/cm), at 25 °C (77 °F).

emission uniformity: An index of the uniformity of emitter discharge rates throughout a microirrigation system. Takes account of both variations in emitters and variations in the pressure under which they operate.

emitter: A small microirrigation dispensing device designed to dissipate pressure and discharge a small uniform flow or trickle of water at a constant discharge, which does not vary significantly because of minor differences in pressure head. Also called a "dripper" or "trickler."

evapotranspiration: The combination of water transpired from vegetation and evaporated from the soil, water and plant surfaces.

exchangeable sodium percentage (ESP): The fraction of the cation exchange capacity of a soil occupied by sodium ions determined as: exchangeable sodium (meq/100 gram soil) divided by CEC (meq/100 gram soil) times 100. It is unreliable in soil containing soluble sodium silicate minerals or large amounts of sodium chloride.

field capacity: Amount of water remaining in a soil when the downward water flow due to gravity becomes negligible. An estimate of field capacity ranges between soil water contents at matric potentials of –10 to –33 kPA (–0.1 to –0.33 bar).

flow meter: An instrument used to measure the volume and/or rate of flow of water in a conduit or channel.

flume: A specially calibrated structure for measuring open channel flows.

friction head: Energy required to overcome friction caused by fluid movement relative to the boundaries of a conduit or containing medium.

furrow irrigation: Method of surface irrigation where the water is supplied to small ditches or furrows for guiding across the field.

gross irrigation: Total water applied to a given area that may or may not equal total irrigation water requirement.

groundwater: Water occurring in the zone of saturation in an aquifer or soil.

head: The energy in the liquid system expressed as the equivalent height of a water column above a given datum.

hydraulic conductivity: The ability of a porous medium to transmit a specific fluid under a unit hydraulic gradient; a function of both the characteristics of the medium and the properties of the fluid being transmitted. Usually a laboratory measurement corrected to a standard temperature and expressed in units of length/time. Although the term hydraulic conductivity is sometimes used interchangeably with the term permeability, the user should be aware of differences.

infiltration: The downward entry of water through the soil surface into the soil.

intake family: A grouping of intake characteristics into families based on field infiltrometer tests on many soils, developed by the SCS (now the Natural Resources Conservation Service). Used to analyze and design border and furrow irrigation systems.

interlock injection device: Safety equipment used to ensure that a chemical injection pump will stop if the irrigation pumping plant stops to prevent the entire chemical mixture from emptying from the supply tank into the irrigation pipeline. An injection device may include check valves to prevent water from flowing back through the injection pump and overflowing.

irrigation efficiency: The ratio of the average depth of irrigation water that is beneficially used to the average depth of irrigation water applied, expressed as a percent. Beneficial uses include satisfying the soil water deficit and any leaching requirement to remove salts from the root zone.

lateral: Secondary or side channel, ditch, or conduit. Also called "branch drain" or "spur." Also, water delivery pipeline that supplies irrigation water from the main line to sprinklers or emitters.

leaching: Removal of soluble material from soil or other permeable material by the passage of water through it.

manifold: Pipeline that supplies water to the laterals.

manufacturer's coefficient of variation: A measure of the variability of discharge of a random sample of a given make, model, and size of microirrigation emitter, as produced by the manufacturer and before any field operation or aging has taken place; equal to the ratio of the standard deviation of the discharge of the emitters to the mean discharge of the emitters.

net irrigation: The actual amount of applied irrigation water stored in the soil for plant use or moved through the soil for leaching salts. Also includes water applied for crop quality and temperature modification; i.e., frost control, cooling plant foliage and fruit. Application losses, such as evaporation, runoff and deep percolation are not included. Generally measured in mm (inches) of water depth applied.

opportunity time: The time that water inundates the soil surface with opportunity to infiltrate.

Penman-Monteith Method: A method used to estimate reference crop evapotranspiration (ET_o) using current climatic data including air temperature, relative humidity, wind speed, and solar radiation. Also referred to as the FAO Penman-Monteith Method.

permanent wilting point: Soil water content below which plants cannot readily obtain water and permanently wilt. Sometimes called "permanent wilting percentage" or WP. Often estimated as the water content corresponding to a matric potential of –1.5 MPa (–15 bar).

pump efficiency: Ratio of the water power produced by the pump, to the power delivered to the pump by the power unit.

reasonable-use rule: A concept of water law in which a landowner is given the right to the reasonable use of water for domestic or similar needs.

reference evapotranspiration: The evapotranspiration from a dense, well-watered hypothetical crop with specified height, albedo, and surface resistance. The ASCE Standardized Reference Evapotranspiration Equation recognizes two reference surfaces: short crop reference ET (ETo), which represents a clipped, cool-season grass (this is equivalent to FAO 56 reference ET), and a tall crop reference (ETr), which represents alfalfa. Reference ET is a measure of the atmospheric demand for water and is used in conjunction with a crop coefficient to estimate crop ET.

return flow: That portion of the water diverted from a stream which finds its way back to the stream channel, either as surface or subsurface flow.

riparian doctrine: This doctrine is in effect in most eastern States, some mid-western and southern States, and the State of California (which also uses the appropriation doctrine). In almost all jurisdictions, the doctrine has been modified to fit local conditions. It applies to all bodies of water including streams, lakes, ponds, and marshes, and grants to all riparian owners the right to make reasonable use of the water so long as the water use does not interfere with the reasonable use of water by other riparian users.

saline soil: Nonsodic soil containing soluble salts in such quantities that they interfere with the growth of most crops. The electrical conductivity of the saturation extract is greater than 4 dS/m (0.01 mho/in.), and the exchangeable-sodium-percentage is less than 15.

saturation: To fill all (100%) voids between soil particles with water.

sodic soil: A nonsaline soil containing sufficient exchangeable sodium to adversely affect crop production and soil structure. The exchangeable sodium- percentage is greater than 15 and the electrical conductivity of the saturation extract is less than 4 dS/m (0.01 mho/in.).

soil-water deficit: Amount of water required to raise the soil-water content of the crop root zone to field capacity. It is measured in mm (inches) of water. Also called soil-water depletion.

soil-water tension: A measure of the tenacity with which water is retained in the soil. It is the force per unit area that must be exerted to remove water from the soil and is usually measured in bars, or atmospheres. It is a measure of the effort required by plant roots to extract water from the soil.

solar radiation (R_s): Radiation from the sun that passes through the atmosphere and reaches the combined crop and soil surface. The energy is generally in a waveband width of 0.1 to 5 microns. Net R_s is incoming minus reflected radiation from a surface.

specific capacity: Well discharge divided by the water level drawdown after a specified pumping duration.

sprinkler head: A device for distributing water under pressure.

sprinkler irrigation systems:

center pivot: An automated irrigation system consisting of a sprinkler line rotating about a pivot point and supported by a number of self-propelled towers. The water is supplied at the pivot point and flows outward through the line supplying the individual outlets.

lateral move: An automated irrigation machine consisting of a sprinkler line supported by a number of self-propelled towers. The entire unit moves in a generally straight path and irrigates a basically rectangular area. Sometimes called a "linear move."

permanent: Underground piping with risers and sprinklers.

portable (hand move): Sprinkler system which is moved by uncoupling and picking up the pipes manually, requiring no special tools.

side-roll sprinkler: The supply pipe is usually mounted on wheels with the pipe as the axle and where the system is moved across the field by rotating the pipeline by engine power.

solid set: System which covers the complete field with pipes and sprinklers in such a manner that all the field can be irrigated without moving any of the system.

towed sprinkler: System where lateral lines are mounted on wheels, sleds, or skids, and are pulled or towed in a direction approximately parallel to the lateral.

subsurface drip irrigation: Application of water below the soil surface through emitters, with discharge rates generally in the same range as drip irrigation. This method of water application is different from and not to be confused with subirrigation where the root zone is irrigated by water table control.

water table: The upper surface of a saturated zone below the soil surface where the water is at atmospheric pressure.

weir: Any of a group of flow measuring devices for open-channel flow. Weirs can be either sharp-crested or broad-crested. Flow opening may be rectangular, triangular, trapezoidal (Cipolletti) or specially shaped, e.g., to make the discharge linear with flow depth (sutro weir).

well casing: Pipe installed within a borehole to prevent collapse of sidewall material, to receive and protect pump and pump column, and to allow water flow from the aquifer to pump intake.

well screen: That part of the well casing which has openings through which water enters.

Index

Made in United States
Orlando, FL
04 November 2026

90206771R00203